Environmental Science

Revised Edition

MARTIN SCHACHTER

Former Assistant Principal, Science
New York City Board of Education

AMSCO SCHOOL PUBLICATIONS, INC.

315 Hudson Street
New York, N.Y. 10013

The publisher wishes to acknowledge the helpful contributions of the following reviewers in the preparation of this book:

Nancy Affeltranger
Environmental and Physical Sciences Teacher
Tioga High School, Pineville, LA

Martin Allen, Ph.D.
Assistant Principal, Science
John Bowne High School, Queens, NY

Ed Braddy
Science Teacher
J. W. Mitchell High School, New Port Richey, FL

Emil Fogarino
Environmental Science Instructor
George Washington High School, San Francisco, CA

Howard Gottehrer
Biology, Earth and Environmental Sciences Teacher
Martin Van Buren High School, Queens Village, NY

Dorothy L. Gregg
Science Department Team Leader
Blanche Ely High School, Pompano Beach, FL

Shirley L. Griffin, Ed.D.
Environmental Science Teacher
Oakmont Regional, Ashburnham, MA

Joan O. Hall
Chemistry and Environmental Science Teacher
Oak Ridge Military Academy, Oak Ridge, NC

Michael Lopatka
Biology, Earth/Space, Environmental and Marine Sciences Teacher
Edgewater High School, Orlando, FL

Louis P. Pataki, Ph.D.
Master Teacher of Physical Sciences
New York University, New York, NY

Robert Zottoli, Ph.D.
Professor of Biology
Fitchburg State College, Fitchburg, MA

Editor: Margaret Pearce
Design: Betty Binns/Betty Binns Design and Howard Leiderman/One Dot, Inc.
Composition: Brad Walrod/High Text Graphics, Inc.
Cover photograph: Mt. Ranier, Washington. © Craig Tuttle, Corbis Stock Market

Please visit our Web site at: ***www.amscopub.com***

When ordering this book, please specify:
either **R 782 H** *or* ENVIRONMENTAL SCIENCE, REVISED EDITION

ISBN 0-87720-192-7

Printed in the United States of America.

1 2 3 4 5 6 10 09 08 07 06 05

To the Student

Have you ever dreamed of blasting off into space, and looking back at the beautiful blue color of Earth's oceans? Did you ever dream of being one of the first astronauts to visit Mars? And did you ever dream of sailing over the ocean to a tropical paradise where there are exotic animals and plants?

Well, in some important ways you *are* on a voyage in space, although you have not visited Mars, yet. Earth is a space voyager that moves in a year-long circuit around the sun. Earth carries a precious cargo of people and all the other kinds of life that call Earth home, along with all the things that make life possible. And, in some ways, modern technology can allow you to visit a tropical island, the depths of the ocean, or the tops of tall mountains to see the life that lives there.

In the past, people would not have been able even to dream the kinds of dreams mentioned above. We live in interesting times and we face some enormous problems. The large human population strains Earth's available resources. The wastes produced by our modern industrial society build up in ever-increasing amounts and affect the quality of life. The actions of people have, in some cases, poisoned our drinking water, the air we breathe, and the soil in which we grow our crops. We are still learning about the complex relationships that exist among the many forms of life on Earth.

Today, we are learning ways to protect and preserve the life that exists on our home planet. Our actions in this country may profoundly affect life in countries far away. In the future, we will know more, and the work of caring for Earth may be easier. The knowledge you acquire in this course can be put to good use. But always keep in mind that we are the caretakers of this planet, not its owners.

The first unit of this textbook introduces you to the work of environmental scientists and to planet Earth: its place in the universe, its structure, and its history. The second unit introduces Earth's living things. The cycles of nature are covered in the third unit. The fourth unit explains how organisms are adapted to their environment. Ecosystems and biomes are discussed in the fifth unit. In the sixth unit, you will learn about the changes that have occurred in

Earth's human population. The seventh and eighth units focus on Earth's resources and energy needs. In the ninth unit, you will learn about strategies for preserving Earth for future generations.

At the end of each chapter, you will find a Chapter Review section that includes a vocabulary list along with fill-in, multiple-choice, short-answer and essay questions, and in some chapters a Research Project. The questions help you review the material you have learned. Research Projects help you connect your science concepts with topics that you may have studied in other subject areas. You will find that doing a Research Project can be a challenging and creative endeavor, often one that will help you think like a scientist. You will ask questions, make observations, develop explanations, and draw conclusions. We hope you will enjoy doing these projects. Now, let us begin our exploration of the world we share with all other forms of life.

CONTENTS

UNIT II
THE LIVING PLANET

UNIT III
NATURE'S CYCLES

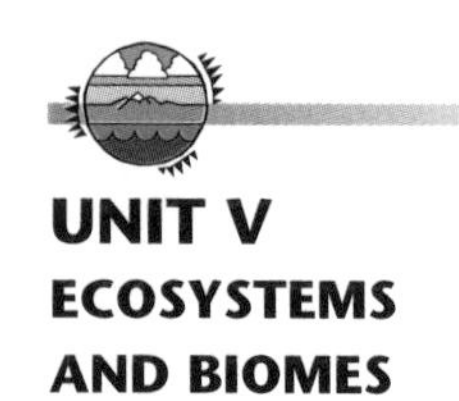

UNIT V
ECOSYSTEMS AND BIOMES

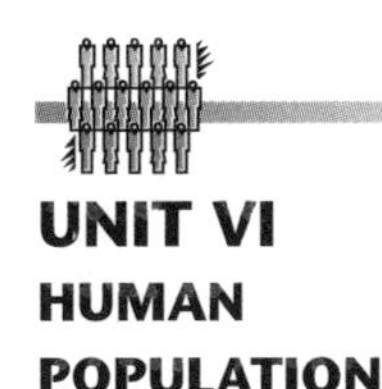

UNIT IX PRESERVING EARTH FOR THE FUTURE

UNIT ONE

EARTH IN SPACE

On November 20, 1998, the Russian-built control module *Zarya* was launched. *Zarya* was the first module of the International Space Station (ISS). On December 4 of that year, the Space Shuttle *Endeavour* carried the second component, *Unity,* into orbit where the two modules were joined. The first crew to live on the ISS left Earth on October 31, 2000. However, before the crew was sent, other missions supplied the space station with air, food, and water.

All creatures on Earth need air, food, a place to live, and water to survive. We rarely worry about these necessities because we believe that Earth will supply them. Like the space station, Earth's resources are limited. To ensure that there is enough for all, we must care for our planet. That is what environmental science is about.

CHAPTER 1
The Environmental Scientist

When you have completed this chapter, you should be able to:

Define ecology.

Apply the scientific method to solve a problem.

Calculate the population size of a species from sampling data.

When scientists study an area, they include all the organisms present in that place. They also study all the nonliving things that the living things come in contact with. Scientists who study the environment must take the living and nonliving factors and their interrelationships into account to reach accurate conclusions about the natural world.

1.1 THE WEB OF LIFE

Every living thing, or **organism**, gets energy and life-support materials from its surroundings. As shown in Figure 1-1, on the following page, life support may include, for example, light, heat, oxygen, carbon dioxide, nutrients, and water. To get these materials, living things interact with, depend upon, and thus affect other living things around them. In addition, living things affect the nonliving portions of their surroundings. An organism's surroundings are called its **environment**.

The interactions of living things and nonliving things form an interconnected "web of life" called an **ecosystem**. Within ecosystems energy is processed and nutrients are recycled. Earth's overlapping mosaic of ecosystems forms the **biosphere**, the thin layer of life that surrounds Earth. **Ecology** is the study of the relationships between the living and nonliving things. **Environmental science** is the study of human interactions with the environment.

Ecologists

Ecologists are scientists who study the interactions that occur within the biosphere. To gain information about the behavior of plants and animals, ecologists study individual organisms. In natural settings, individual organisms rarely live by themselves. They usually live with other members of their **species**, a group of similar organisms that in their natural environment can interbreed to produce fertile offspring. As organisms go about their daily activities, they interact with members of their species and other species. These interactions form ecological relationships among the species. **Intraspecies interactions** occur between members of the same species. **Interspecies interactions** occur between members of different species. Ecologists use the data they gather from the study of these interactions to understand better how the biosphere works.

People have an inborn curiosity about the world around them. From birth, infants crawl about, using sight, hearing, smell, taste, and touch to investigate and learn about their world. This practice of learning continues throughout life.

Figure 1-1 In an ecosystem, the living and nonliving parts interact.

Primitive humans were the first ecologists. These people were curious about the plants and animals that shared their world. Some plants and animals were sources of food, medicine, clothing, and shelter. Other animals and plants used humans as sources of food. Early humans had to observe nature to insure their survival. This curiosity persists.

Charles Darwin

The English scientist Charles Darwin served as the naturalist aboard the exploration vessel H.M.S. *Beagle*. Darwin traveled for several years recording detailed observations of the animal and plant life of South America and the South Pacific. In his diary, he recorded data on Patagonian fossils and their relative positions in rock layers, the distribution of giant Galápagos tortoises, and the wide variety of life among the rain forest animals. He also made detailed drawings and descriptions of the plants and animals he collected in his explorations.

Ground finch eats seeds

Catches insects with beak

Tool-using finch digs insects from under bark with thorn

Figure 1-2 Darwin recognized that Galápagos finches were similar to the seed-eating finches of Ecuador. However, some Galápagos finches ate insects, and others ate fruit.

In 1835, the *Beagle* landed in the Galápagos Islands, where Darwin observed the birds. These birds resembled the seed-eating finches found in Ecuador, some 970 kilometers east. Even though some Galápagos finches ate seeds, there were also fruit eaters and insect eaters. One specialized insect eater was able to use a cactus spine to remove insects from under the bark of trees. (See Figure 1-2.) He found big finches and small finches, ground-dwelling finches and some that lived in shrubs. These birds filled many of the environmental roles occupied elsewhere by a wide variety of other organisms. For example, the finch that used the cactus spine occupied the role filled by woodpeckers in other ecosystems.

Darwin's observations of the birds of the Galápagos Islands stimulated his curiosity and raised many questions. How did the islands' plants and animals originate in an open ocean? Why were the finches similar in many ways yet different in some particular characteristic? Were all the finches in some way related to a mainland species?

After many years of study, Darwin found evidence that the distribution of plants and animals was related to the distribution of ancestral organisms. He also found that differences among individ-

RESEARCH

Amateur Ecologists

Ecologists are at the forefront of a battle to protect the natural world. The interactions that occur in ecosystems are like the pieces of a giant jigsaw puzzle, which must be put together before the puzzle can be understood. The role of the ecologist is to study environmental changes as they occur and to predict the effects that interactions will create in the future. This is important to society since the changes that are made to the environment can impact on the health and welfare of the human population.

The chief tools of the ecologist are a pen and notebook or a notebook computer. The basic mode of study is that of making careful, detailed, and accurate observations of the world around them. Many ecologists first gained a love and appreciation of the natural world when they were young by studying the life found in ponds, streams, woods, fields, and their own backyard.

The growth of cities has created an artificial environment containing a new array of plants and animals that can be studied by an amateur ecologist. Suburban lawns and gardens are highly engineered ecosystems that have been sculpted by human hands. These lawns are unnaturally carpeted with rye, red fescue, Bermuda, or zoysia grasses, or blends of grasses. In reality, a thick, healthy lawn prevents the growth of other plants. This manufactured setting is maintained through the use of chemical fertilizers and pesticides. Trees, shrubs, and flowering plants that are not native to the area often surround the lawn. Yet lawns and gardens are unique ecosystems which support a wealth of animal life. Insects, small reptiles, mammals, and a wide variety of birds will make a garden their home. An amateur ecologist can learn a great deal by studying the interactions in a lawn.

To study the interactions in a garden, the amateur ecologist needs but a few tools. These include the ever-present pen and notebook, as well as collecting bottles or plastic bags, a magnifying lens, a net, clippers, and tweezers. A pocketknife can be substituted for the clippers. Using binoculars, the amateur ecologist can study small birds and animals from a distance, so as not to interfere with their daily activities. For the more advanced ecologist, a camera with a telephoto lens is a must. This enables the observer to keep a pictorial record of the flora and fauna he or she encounters. Finally, field guides serve as sources for identification of the wildlife that has been seen.

uals within a species can cause changes within that species. Darwin theorized that a group of finches from South America had been carried accidentally to the Galápagos Islands by a hurricane or some other natural occurrence. Because there was no competition from other organisms, the descendants of these birds were able to spread out into a wide variety of environments. Variations in individuals led to the development of different bill shapes and sizes. These adaptations allowed the birds to tap into the different food sources available on the island. By using different food sources, the birds no longer competed for the same food, and more birds could survive on the island. Over many generations, these differences in bill shape and size resulted in one species of bird evolving into different species. These species interbred only with their own kind, maintaining the adaptations in future generations.

In later years, Darwin was able to piece together all the observations he made during his years on the *Beagle*. In his book *On the Origin of Species,* Darwin theorized that the many species in the world evolved through a gradual accumulation of adaptations brought about by natural selection. In the process of natural selection, only those animals, plants, and other creatures that are best adapted to their environment live to reproduce and pass their adaptations to their offspring.

1.1 Section Review

1. Describe the different life-support materials organisms get from their environment.
2. Why could ecosystems be called "webs of life"?
3. List some examples of intraspecies and interspecies interactions.

1.2 THE SCIENTIFIC METHOD

Scientists use a variety of problem-solving procedures to study the biosphere. Scientists want to learn the what, when, where, and why of nature. We ask the same questions regarding our everyday lives and use many of the same problem-solving methods that scientists use. The scientific method, which comprises many different procedures, is not just for scientists.

As ecologists gather data, the facts are fit together like the pieces of a giant jigsaw puzzle. Some areas are complete, but there are still many pieces missing. Every observation that adds a new piece to the puzzle brings us closer to understanding the world we live in.

An **observation** is information gathered directly by the senses, measurements, or experiments. Observations that describe something are **qualitative data**. These observations tell what but not how much. They may describe an ecosystem or the organisms that live in it, for example that a flower is red. Very often, instruments such as binoculars, telescopes, microscopes, cameras, night-vision goggles, and video equipment are used to help make qualitative observations. This information is obtained without taking measurements, and this type of data usually requires support by several observers.

By contrast, **quantitative data** are obtained through direct measurements. They answer questions such as how many, how large, what temperature, or how fast. If collected carefully and verified by repetition to eliminate possible human errors made while measuring, quantitative data usually are definite and not open to interpretation. Metersticks, scales, balances, stopwatches, and thermometers are some of the instruments used to make quantitative measurements.

Ecologists use **fieldwork**, the study of living organisms in their natural setting, to gather information. The information may be used to determine population numbers, diet preferences, size differences, territorial boundaries, and behavior patterns. (See Figure 1-3.) Fieldwork also includes the study of the physical aspects of the environment that the species inhabits, such as soil, rocks, air, and water. Laboratory studies may follow, focusing on questions not answered by the data collected in the field.

Sampling Methods

One aspect of fieldwork is sampling a population (all the organisms of one species that live in a particular area). **Sampling** uses a representative portion of a population to determine a single characteristic of the entire population. Sampling can also be used to estimate the number of individual organisms of one species that live in an

Figure 1-3 Some interspecies and intraspecies interactions in a meadow ecosystem are shown here.

area. By knowing how many of each species are in an ecosystem, scientists can begin to understand how organisms interact.

In a small area, trees and shrubs can be counted easily because they cannot run away or hide. Over a large area, counting each tree takes too long. When a direct count is not possible, ecologists often use indirect sampling methods. They count samples from a small portion of the area, and then these numbers are extrapolated, or projected beyond the known values, to arrive at a description of the whole area. Suppose there are 10 oak trees in a 1-hectare portion of a 200-hectare forest. (A hectare is 10,000 square meters, about 2.5 acres.) You can set up a proportion:

$$\frac{10 \text{ oaks}}{1 \text{ hectare}} = \frac{x}{200 \text{ hectares}}$$

$$\frac{10 \text{ oaks}}{1 \text{ hectare}} \times 200 \text{ hectares} = x$$

$$2000 \text{ oaks} = x$$

Therefore, there are approximately 2000 oak trees in the whole forest. The result of indirect sampling is accurate only if the area sampled is representative of the whole area. For example, if the area chosen was the place with the fewest oaks, the scientist would not get an accurate representation of the forest. How could this sampling technique be improved?

Large animals can usually be counted directly by sight. But counting populations of small animals such as field mice, insects, or earthworms is a much more difficult task. A population of small animals is trapped, marked, and then released. The animals are marked with spots of paint or, for larger animals, tags can be attached to their ears. A second trapping at some later time supplies information used to predict the size of the total population.

For example, in a given area 25 field mice were trapped, marked, and then released. Several days later in a second trapping, 20 field mice were collected. Of these 20 mice, 5 had been marked in the first trapping. If you assume that both samples were random, then the first trapping must have contained $\frac{5}{20}$ of the total population of field mice. This means that the total population of field mice in the area is about 5×20 or 100. (See Figure 1-4.)

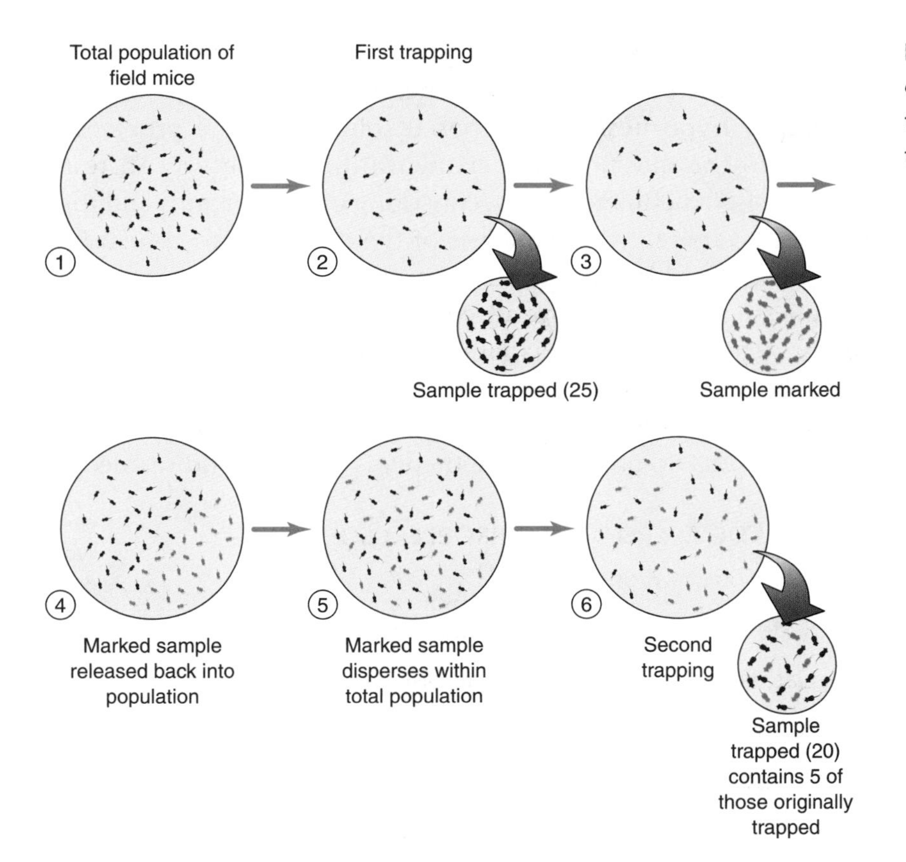

Figure 1-4 How could you estimate the size of a population of field mice?

$$\text{First trapping} = \frac{5}{20} \times \text{total population}$$

$$25 = \frac{5}{20} \times \text{total population}$$

$$25 \times \frac{20}{5} = \text{total population}$$

$$100 = \text{approximate total population}$$

How Scientists Work

Scientists seek information about how the world around them works. They gather data by conducting research, observing, and asking questions. The **scientific method** is a systematic approach to answering questions or solving problems.

First scientists clearly state the problem to be solved or the question to be answered. Next they gather information they can use to develop a **hypothesis**, or proposed solution to the problem or answer to the question. **Experimentation** is one way to test a hypothesis. Experimental data, consisting of observations and measurements, are recorded. Analysis of the data leads to conclusions that are a plausible solution to the problem or a possible answer to the question. From a proven hypothesis, scientists develop a **theory**, which is a logical explanation based on the data gathered and verified by many other scientists. After a theory has been tested many times with the same results, it is generally accepted as true. If the test results do not agree, the theory is changed or discarded.

SYSTEMATIC APPROACH TO PROBLEM SOLVING

1. Make observations.
2. State the problem or question.
3. Gather available information about the problem.
4. Form a hypothesis you can test.
5. Devise and conduct research to test the hypothesis.
6. Carefully collect qualitative and quantitative data.
7. Analyze the data.
8. Based on the analysis of the data, was your hypothesis supported or not supported?
9. State a conclusion or a solution to the original problem.
10. Repeat the research to verify the conclusion.

1.2 Section Review

1. Why do ecologists often find it necessary to use indirect sampling methods?
2. Explain why scientists repeat research many times before their findings are accepted.
3. Describe how you could apply the scientific method to the purchase of a gift for a friend.

LABORATORY INVESTIGATION 1

Using the Scientific Method

PROBLEM: *A student noticed that although she used the same brand of soap at home and at her grandparents' house next door, the soap lathered better at her grandparents' home. She knows that her grandparents have a water softener. She devised the following activity to try to understand the difference.*

SKILL: *Applying the scientific method.*

MATERIALS: *liquid soap (not detergent), 7 test tubes, calcium bicarbonate, heat source, distilled water, tap water, test tube rack, test tube holder, 7 stoppers, marking pencil, water softener*

PROCEDURE

1. Propose a hypothesis to explain the student's observation.
2. Place 50 mL of tap water in a test tube. Mark the test tube A.
3. Place 50 mL of distilled water in a test tube. Mark the test tube B.
4. Place 50 mL of distilled water and 1g of calcium bicarbonate into a test tube. Mark this test tube C.
5. Place 2 drops of liquid soap in each test tube, stopper the tubes, and shake the tubes vigorously for 15 seconds. Record your results.
6. Place 50 mL of distilled water and 1 g of calcium bicarbonate into a test tube. Mark this test tube D.
7. Heat this test tube to boiling, and then place it back into the test tube rack. Record the result. (CAUTION: Wear an apron and eye protection. Point the mouth of the test tube away from yourself or any others near you.)
8. Repeat steps 1 through 4. Mark these test tubes E, F, and G.
9. Add 1 drop of a water softener to each test tube. Stopper each test tube and shake the tubes vigorously for 15 seconds. Record your results.

OBSERVATIONS AND ANALYSES

1. Base your answers to the following questions on the results of steps 1 through 4 for test tubes A, B, and C.
   a. Did soap suds form in all 3 test tubes? ________
   b. Which test tube contained the most suds? ________
   c. Which test tube contained the least suds? ________
2. What do you observe on the inside of the tube D?

   ____

   a. What caused this?

   ____

   b. Complete the chemical equation for the reaction that occurred.

$$Ca(HCO_3)_2 \xrightarrow[\text{HEAT}]{} CaCO_3 + H_2O + \_\_\_\_\_\_$$

3. Did soap suds form in test tubes E, F, and G?

   ____

4. How did the amount of suds that formed in test tubes E, F, and G compare with the amount of suds that formed in test tubes A, B, and C?

   ____

5. State the problem in the form of a question that can be answered by the activity.

   ____

6. Based on the data, was your hypothesis supported?

   ____

7. State a conclusion or solution to the original problem.

   ____

Chapter 1 Review

Answer these questions on a separate sheet of paper.

Vocabulary

The following list contains all of the boldfaced terms in this chapter.

biosphere, ecologists, ecology, ecosystem, environment, environmental science, experimentation, fieldwork, hypothesis, interspecies interactions, intraspecies interactions, observation, organism, qualitative data, quantitative data, sampling, scientific method, species, theory

Fill In

Use one of the vocabulary terms listed above to complete each sentence.

1. _______ occur between members of the same species
2. A group of similar organisms that can interbreed to produce fertile offspring is a(an) _______.
3. The method used to study living things in their natural surroundings is _______.
4. Data gained by direct use of the senses are a(an) _______.
5. A logical explanation based on experimentation is called a(an) _______.

Multiple Choice

Choose the response that best completes the sentence or answers the question.

6. Environmental life-support materials include all the following except *a.* light. *b.* heat. *c.* the moon. *d.* water.
7. Which of the human senses is least safe to use in making scientific observations? *a.* taste *b.* sight *c.* touch *d.* hearing
8. Which of the following is not a step in the Systematic Approach to Problem Solving? *a.* stating the problem *b.* finding a solution in a book *c.* devising a research design *d.* retesting the results
9. The observations made by Charles Darwin during his voyage on the *Beagle* did not include *a.* Patagonian fossils. *b.* Hawaiian honeycreepers. *c.* Galápagos tortoises. *d.* South American rain forest animals.
10. In which of the following cases is direct sampling not possible? *a.* determining the number of lions in an area of South Africa *b.* counting the population of prairie dogs in a colony *c.* determining the size of a bison herd *d.* finding the number of oak trees on Green Mountain

Short Answer (Constructed Response)

Use the information you learned in this chapter to respond to the following items.

11. What are biomes?
12. Describe two intraspecies interactions that may occur in a pack of wolves.
13. What types of relationships occur between the organisms within an ecosystem?
14. Why are dogs considered to belong to a single species, even though they may be very different in appearance?
15. What are some of the tools of an ecologist?
16. Explain why the technique of sampling is used in the study of species populations.
17. Describe some observational techniques used by ecologists.
18. How does a scientific hypothesis differ from a scientific theory?
19. Explain what Charles Darwin meant by "natural selection."
20. Describe some of the differences Charles Darwin saw in Galápagos finches.

Essay (Extended Response)

Use the information you learned in this chapter to respond to the following items.

21. How did Charles Darwin explain the differences he found in the Galápagos finches?
22. A sampling of a 4-kilometer-square section of a 200-kilometer-square forest finds 10 red squirrels. Approximately how many red squirrels live in the whole forest?
23. Within a 10-hectare farm, 50 field mice were trapped, marked with dye, and released. In a later trapping, 40 field mice were collected, including 8 mice that had been marked with dye. What is the approximate population of field mice on the farm?
24. Observe an animal in its natural environment, in a zoological park, around your home or school, in an aquarium or terrarium, or a pet at home. Make a list of the interspecies and intraspecies interactions that you observe.

CHAPTER 2
Earth's Place in Space

When you have completed this chapter, you should be able to:

Describe the Doppler effect.

Distinguish between stars and galaxies, latitude and longitude.

Explain the cause of tides and seasons.

People have always been fascinated by the night sky. Early in human history, people made up supernatural explanations for the lights they observed in the sky. As our knowledge increased, scientists proposed theories to explain what they saw. Astronomy is the study of the position, size, and composition of objects in space. As with all sciences, astronomy is based on observations and interpretation of data. In addition to their unaided eyes, astronomers use many instruments, including ground-based telescopes and the Hubble Space Telescope, which is shown above.

2.1 THE UNIVERSE

To a meteorologist, the sky is limited to the region where weather occurs. But to the astronomer, the sky has no limit: it continues ever outward, seemingly without any boundaries. This view of the sky defines the universe. The universe consists of everything that ever existed or that still is physically present. Most of the universe is a near-vacuum, an endless sea of empty space. Floating in this sea are stars and clusters of stars, called galaxies.

The distances between stars and galaxies are so great our conventional measures of distance are of little use. Astronomical distances are measured in units called **light-years**, the distance light travels in one year. Light travels at a velocity of 300,000 kilometers per second. One light-year equals approximately 9.5 trillion kilometers. Our sun's closest stellar neighbor, Proxima Centauri, is 4.3 light-years away (more than 40 trillion kilometers). The Andromeda galaxy, the closest galaxy to our Milky Way galaxy, is more than 2 million light-years away.

Stars and Galaxies

The first stars and galaxies formed from clouds of hydrogen and helium. Stars are spheres of gaseous matter that glow due to the energy they produce. They are born when gravitational compression of a gas cloud heats its core to temperatures high enough to start hydrogen fusion. As the nuclear reaction changes some matter to energy, it also creates heavier elements. As a result of these nuclear processes, stars change in composition over time, eventually leading to the star's death. The heavier elements, such as oxygen and iron, found in stars now are the remnants of dead stars.

A **galaxy** is a star cluster, a family of between a million and a trillion stars. Galaxies are surrounded by cosmic dust and gases held together by mutual attraction. Galaxies tend to be found in groups, called galactic clusters. The Milky Way and the Andromeda galaxy, together with 27 other smaller galaxies, make up a cluster called the

Figure 2-1 Galaxy NGC 4414, about 60 million light-years away, is a spiral galaxy as is the Milky Way.

Local Group. Some astronomers theorize that gravitational attraction from the larger galaxies pulled the cluster together. Another theory is that these galaxies formed at the same time within the same gas cloud and have been traveling together ever since.

The hazy band of star-studded light that cuts across the night sky is our disk-shaped Milky Way galaxy. It is a cluster of three to five billion stars that includes our sun. The Milky Way appears as a band of light because we are observing it edge on from within the disk. When compared with the other stars in the Milky Way, the sun is of moderate size and brightness. It appears so large and bright to us because it is closer to Earth than are the other stars. The sun and its planets occupy a position in one of the spiral arms of the galaxy, about two-thirds of the way from the edge to the center. Figure 2-1 is a photograph of the Andromeda galaxy.

Electromagnetic Waves

Light is part of a family of waves called electromagnetic waves. Electromagnetic waves are produced by the interactions of various forms of energy with atomic particles. Unlike light, most members of this family of waves are invisible to human eyes and can be detected only with special devices. The family of electromagnetic waves is called the electromagnetic spectrum. (See Figure 2-2 on page 22.) In a vacuum, all electromagnetic waves travel at the speed of light.

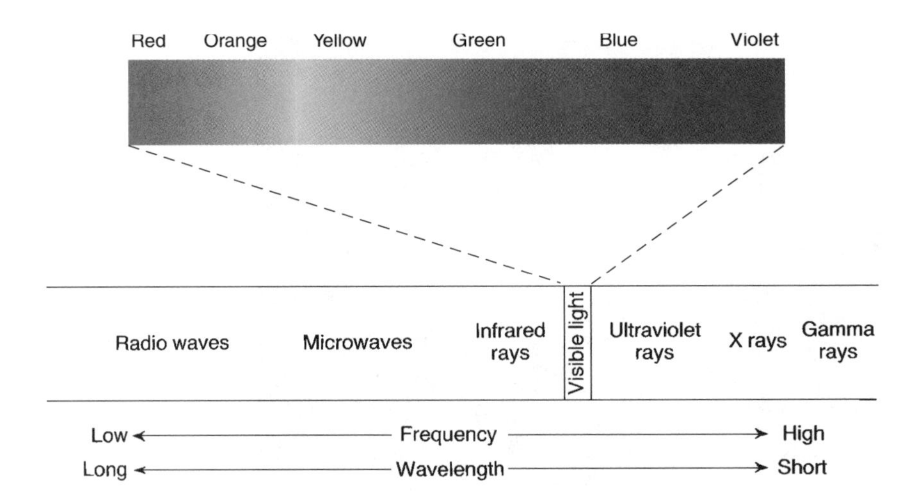

Figure 2-2 Visible light makes up only a small part of the electromagnetic spectrum.

However, they all have different frequencies and wavelengths. Gamma rays have the smallest wavelength and the highest frequency. The gamma rays are followed by X rays and ultraviolet light. Visible light is next. In the visible range, violet light has the highest frequency and the shortest wavelength, while red light has the lowest frequency and longest wavelength. In order of decreasing energy and increasing wavelength, the rest of the spectrum is made up of infrared and radio waves.

Besides visible light, stars emit a wide range of electromagnetic waves. A study of the electromagnetic waves emitted by a star enables astronomers to determine many of the star's properties, such as the star's temperature, relative brightness, chemical composition, size, and distance from Earth.

Doppler Effect

The Doppler effect occurs with light and sound. You may be more familiar with its effect on sound. You have probably noticed that the sound of a car's horn seems to change as the car first approaches and then passes you. As the car approaches you, the pitch of the horn appears higher. When the car passes you, the pitch of the horn seems to get lower. The pitch of a sound is related to its frequency: The higher the frequency, the higher the pitch; the lower the frequency, the lower the pitch. The frequency of the sound produced by the horn does not change. The change in frequency that you hear is caused by the motion of the car. As the car moves, the sound waves produced by the horn pile up in front of the car and stretch out behind it. This causes you to receive a higher frequency as the car approaches and a lower frequency as the car goes away. This apparent change in the frequency of a sound emitted by an object due to the motion of the object and/or the observer is called the Doppler effect.

Electromagnetic waves also show the Doppler effect when the source of the light and/or the observer is in motion. The color of light depends on its frequency. Light at the blue end of the spectrum has a higher frequency than light at the red end of the spectrum.

The spectrum of radiation emitted by stars and galaxies indicates whether they are moving toward or away from the solar system. The light emitted by a star or galaxy that is moving toward us is shifted toward the blue end of the spectrum. When a star is moving away, its light is shifted toward the red end of the spectrum.

Since most galaxies are too far away for us to be able to record the spectrum emitted by a single star, we measure the spectrum of a whole galaxy to determine its Doppler shift. The spectrum of most of the galaxies that have been observed indicates a red shift—they appear to be moving away from us. Further analysis of data indicates that the farther galaxies are from us, the faster they seem to be moving away.

This phenomenon can be demonstrated by blowing up a polka-dot balloon. As the balloon is inflated, the spots move farther apart.

Likewise, at any point in the universe, all the other galaxies would appear to be receding. This suggests that if we trace the movement of the galaxies back in time, in the distant past they were together at a single point. The big bang theory suggests that an explosive event must have sent the galaxies flying apart. Over time, they reached their present position. The big bang theory is the most widely accepted of several theories that try to explain the formation of the universe.

Data indicating the velocity at which galaxies are receding from one another allow astronomers to estimate the age of the universe. In February 2003, astronomers announced that, based on new satellite data, the universe is 13.7 billion years old.

2.1 Section Review

1. How long would it take to send a radio message to and receive an answer from a civilization on a planet that circles the nearest star to the sun?
2. Explain why astronomers think that the universe is expanding.
3. Traveling at the rate of 100 million kilometers per year, how long would it take to reach the star Proxima Centauri?

2.2 THE SUN'S FAMILY

Astronomers calculate that the solar system is about 6 billion years old. One hypothesis is that the solar system began as a cloud of dust and gases that was created by exploding stars. Gravitational forces caused the cloud to collapse into a spinning disk. At the center of the disk, heat and pressure from the accumulating matter started up the nuclear reactions that created the sun. Farther out along the disk, gases and solids slowly collected into clumps that formed the planets, their satellites, comets, and asteroids.

Planets

The solar system contains nine planets and their moons. Planets are large bodies that do not radiate light, but they do reflect it. They travel around the sun in elliptical paths called **orbits.** An ellipse is a flattened circle. The sun, Earth's star, is the largest body in the solar system. The sun's diameter is approximately 1,390,000 kilometers. Compare this with the diameter of the largest planet, Jupiter, 142,800 kilometers, and Earth, 12,750 kilometers. There is evidence that planets orbit some other stars in the Milky Way.

In the region closest to the sun, its intense heat drove off most of the gaseous matter. This left behind particles of dust and molecules of some of the heavier elements. Collisions between particles caused the material to form clumps over time. This eventually led to the formation of the four Earthlike, or terrestrial, planets. These rocky,

Figure 2-3 Jupiter is the largest planet in our solar system. The Great Red Spot, visible in this photograph, is larger than Earth.

inner planets are Mercury, Venus, Earth, and Mars. In the region farther away from the sun's heat, gases condensed to form the Jupiter-like, or Jovian, planets. These are the giant, outer planets: Jupiter, Saturn, Uranus, and Neptune. (See Figure 2-3.)

Pluto is the farthest planet from the sun. According to available data, it is composed of a small, rocky core that is covered by a layer of frozen liquid and gas. Pluto travels in a very elliptical orbit that crosses that of Neptune. Some astronomers say Pluto is not a true planet, but is a misplaced satellite or asteroid that was captured by the sun after the solar system had formed. Other theories make Pluto a wanderer from the Kuiper belt, a reservoir of comets and other frozen bodies beyond the orbit of Neptune. In 2002, several large bodies were found in the Kuiper belt. One of these bodies, Quaoar, is more than half the diameter of Pluto. It is possible that other planet-sized objects exist further out in this region

Interplanetary Debris

Other members of the sun's family include comets, meteoroids, and asteroids. Comets are thought to be composed of frozen gases mixed with interplanetary dust. These huge snowballs travel in highly elliptical orbits. Comets spend most of their lives far from the sun in the Kuiper belt. Periodically a comet's orbit takes it near the sun where some of its frozen matter warms and evaporates. The gases glow and surround the mass of the comet with a halo. As the comet nears the sun, some of the gases stream away from the comet to form a tail.

Asteroids are clumps of rocky material. They are usually found in the asteroid belt, a region between Mars and Jupiter. Astronomers believe that the asteroids in the asteroid belt may be material that was kept from forming into a planet by Jupiter's strong gravitational effects.

Meteoroids are solid, rocklike objects that orbit around the sun and sometimes strike Earth. When meteoroids enter Earth's atmosphere they are called meteors. We see meteors, or shooting stars, as they burn due to friction with the atmosphere, producing streaks of light. When meteors strike Earth's surface, they are called meteorites. Scientists theorize that the impact of a very large meteorite 65 million years ago led to the extinction of the dinosaurs.

Life in the Solar System

Earth's biosphere is the result of a combination of factors found nowhere else in the solar system. Earth's carbon-based life-forms require oxygen, nitrogen, and water, as well as a narrow range of temperatures. Life as we know it has very specific requirements.

The terrestrial planets all have solid crusts, but only Earth has liquid water on its surface. The atmosphere of each planet is different, but only Earth's atmosphere contains life-supporting oxygen. Because of their proximity to the sun, Mercury and Venus are too hot for life as we know it to exist. Mars is thought to be too far from the sun to receive enough heat energy to support life. Mars has little water in its atmosphere, but supplies may exist as permafrost. Studies of the data from Pathfinder and Odyssey missions may eventually give us clues about whether there is or was life on Mars.

The Jovian planets are believed to be composed mainly of the gases hydrogen and helium surrounding a small rocky core. Methane, ammonia, and water ice are also present. Surrounding each planet is a system of rings and a variety of large and small moons, or satellites. Some of these satellites may be capable of supporting some form of life. On October 15, 1997, the *Cassini* spacecraft was launched to study the planet Saturn and its moon Titan. *Cassini* reached Saturn in June 2004. Since previous planetary probes had hinted at the existence of water on Titan, this international venture contains instruments to detect the presence of water, determine whether the surface is solid or liquid, and search for complex organic molecules.

2.2 Section Review

1. Explain the differences between the terrestrial and Jovian planets.
2. What is a comet?
3. Explain why astronomers think Pluto is not a true planet.

2.3 EARTH'S FAMILY

Satellites are smaller objects that orbit larger objects such as planets. The number of natural satellites, or moons, in orbit around the planets of the solar system varies from none for Mercury and Venus to, at last count, 61 for Jupiter. Earth's only natural satellite is the moon. The moon has a nearly circular orbit. Like Earth, the moon creates no light of its own. We see the moon "shining" in the night sky because it reflects light from the sun. As the moon revolves around Earth, the lighted portion of the moon facing Earth changes. The changing pattern of light is known as the phases of the moon. (See Figure 2-4 on page 28.) The moon's period of rotation, turning on its axis, is the same as its period of revolution around Earth. As a result, the same side of the moon is always turned toward Earth.

The phases of the moon have a cycle of 29.5 days. The divisions of the year known as **months** are loosely based on the time it takes the moon to complete its cycle of phases.

Tides

The moon's gravity pulls on Earth, creating a bulge on the surface facing the moon. This bulge follows the moon around Earth. The bulge in the solid portion of Earth is slight. However, the bulging of the oceans is large because they are liquid. The bulging of the ocean is called **high tide**. The oceans bulge in two places, the side facing the moon and the side opposite. The tidal bulge occurs on the side opposite the moon because there is less force pulling the water toward Earth, therefore, the water rises. As shown in Figure 2-5, **low tides** occur midway between the two high-tide bulges.

Figure 2-4 The moon takes the same amount of time, 27.3 days, to revolve once around Earth as it takes to rotate once on its axis. In this diagram, "Face" means the side of the moon that faces Earth. The "×" represents a feature on the moon's surface. Notice how the "×" makes one rotation as the moon makes one revolution.

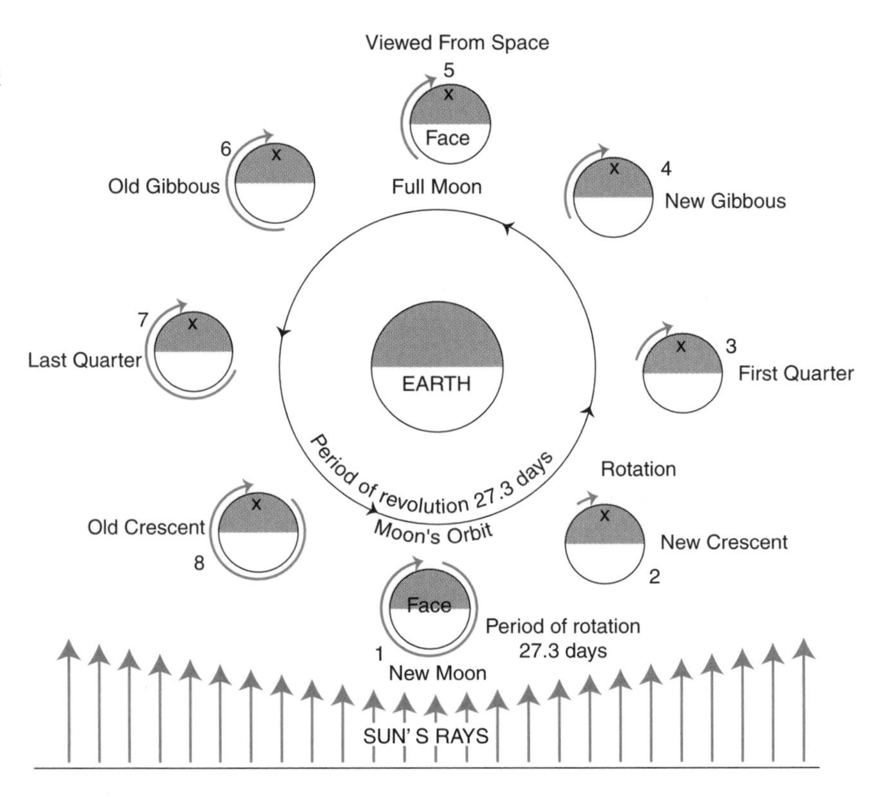

Because Earth rotates on its axis every 24 hours, the moon's gravity affects different parts of Earth at different times. There are two high tides and two low tides during every 24-hour, 50-minute period. The period between successive high or low tides is 12 hours, 25 minutes.

The sun also affects tides, as illustrated in Figure 2-6. When the sun and moon are in line, during the new moon and full moon phases, their gravitational forces combine to produce very high tides called **spring tides.** Similarly, when the sun and moon are at a right

Figure 2-5 Tides are mainly the result of the gravitational attraction of the moon on Earth's oceans.

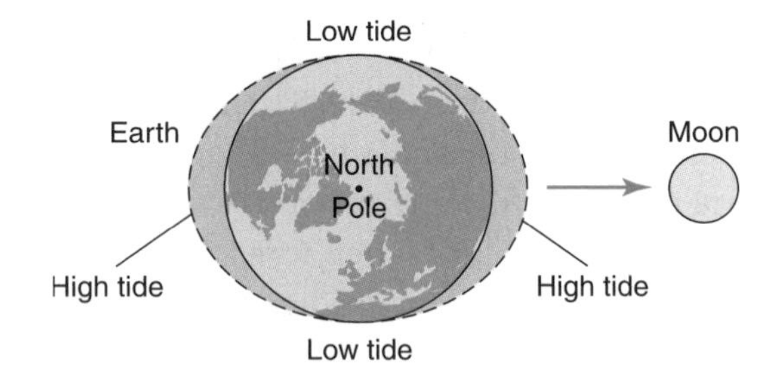

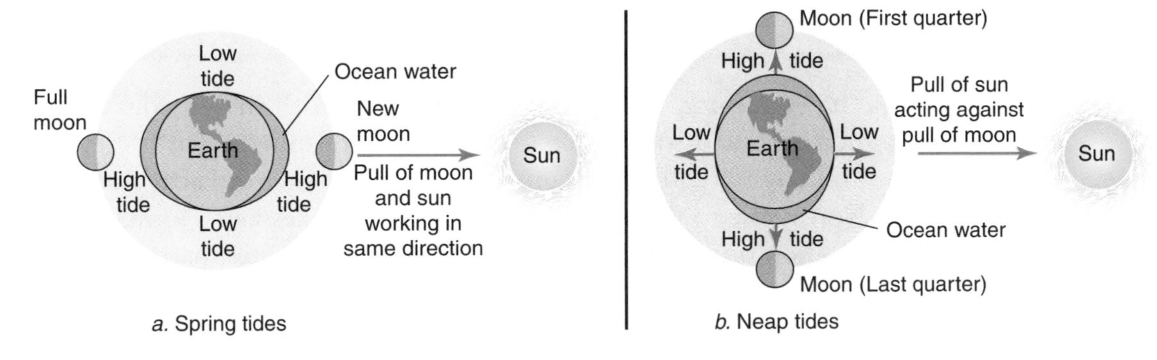

Figure 2-6 The sun also affects tides.

angle to Earth, during first- and third-quarter phases, the result is lower than normal tides called **neap tides.**

Many organisms regulate their life cycle to tidal changes in their environments. Marine organisms that live along the seashore find food or take shelter in step with tidal changes. Grunion are fish that spawn during the higher tides at the full or new moon. Above the normal high-tide line, eggs are laid and fertilized in the sand. At the next full or new moon, the young hatch and enter the water. Many other animals link their reproductive cycle to the phases of the moon.

2.3 Section Review

1. What causes high tide?
2. There are usually four tides each day. Explain why there are only three tides on some days.
3. Observe the moon for one week. Each night, draw the moon.

2.4 THE YEAR

Earth, like all planets in the solar system, revolves around the sun in an elliptical orbit. It takes Earth, the third planet from the sun, 365.25 days, one **year**, to complete one revolution around the sun. Earth rotates on its axis once every 24 hours, as it moves around the sun. The two factors of rotation on its axis and revolution around the sun create Earth's day and night and its seasons. (See Figure 2-7 on page 30.)

Day and Night

The **equator** is an imaginary line around Earth that divides it into northern and southern halves, called hemispheres. Earth's **axis** is an imaginary line that runs through the center of Earth from the North Pole to the South Pole. The time it takes Earth to make one rotation on its axis is called a **day**. As Earth rotates, part of it faces the sun, and is bathed in sunlight, while the rest of Earth is in darkness. Rotation causes daylight and night to alternate every 24 hours. As seen from above the North Pole, Earth rotates counterclockwise, from west to east. The sun therefore appears to rise, or come up, in the east and to set in the west.

The length of daylight and darkness changes during the year. This is because Earth's axis is tilted at an angle of 23.5 degrees. If the axis were straight up and down, all parts of Earth would experi-

Figure 2-7 The time it takes Earth to rotate once on its axis determines the length of the day. What determines the length of the year?

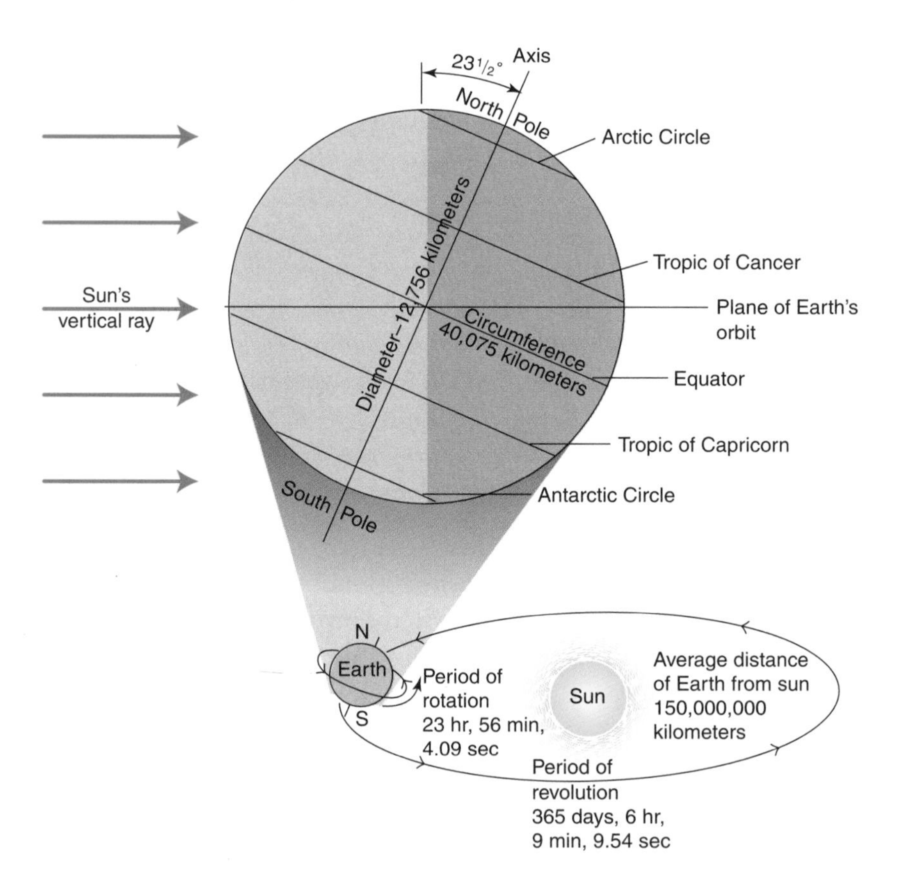

ence 12 hours of daylight and 12 hours of darkness every day. When one pole is tilted toward the sun, the other pole is inclined away from it. As a result, the number of hours of daylight varies throughout each hemisphere. The hemisphere that is inclined toward the sun has long periods of daylight and short nights. The opposite hemisphere has short periods of daylight and long nights.

Due to the tilt of Earth's axis, most of its surface is subject to variations in the amount of solar energy received. Long periods of daylight lead to increased warming of Earth's surface. The daily and yearly variations in solar energy received are responsible for changing weather patterns, or **seasons.** These effects also define Earth's climate zones, creating different habitats.

The Four Seasons

As Earth revolves around the sun, its axis is tilted toward the sun for part of the year and away from the sun for another part. This is illustrated in Figure 2-8 on page 32. When the Northern Hemisphere is tilted toward the sun, that part of Earth has summer. At the same time, the Southern Hemisphere, which is tilted away from the sun, experiences winter. Since Earth follows an elliptical path around the sun, its distance from the sun varies during the year. It is interesting to note that summer in the Northern Hemisphere occurs when Earth is farthest from the sun.

The hemisphere tilted toward the sun receives the more direct rays of the sun. Direct rays distribute their energy over a smaller area than indirect rays, which intensifies their warming effect. Thus the combination of longer days and more direct sunlight warms this hemisphere more than the other and causes its summer.

The first day of summer in the Northern Hemisphere occurs on or about June 21. This day has the longest period of daylight. It is called the **summer solstice**, because at noon on this day, the sun is at its highest point in the sky. The Northern Hemisphere also is at its maximum tilt toward the sun on this day. The sun is directly over the Tropic of Cancer at 23.5°N latitude. On the summer solstice, every point on Earth from the Arctic Circle at 66.5°N to the North Pole experiences 24 hours of daylight.

The Southern Hemisphere experiences its longest period of

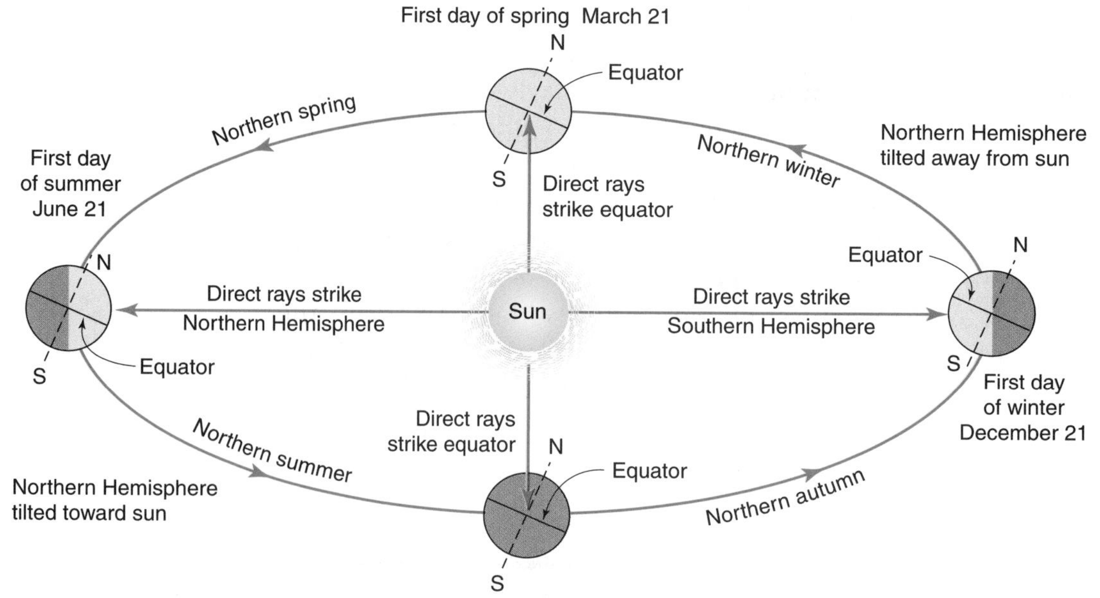

Figure 2-8 The tilt of Earth's axis causes seasonal changes to occur.

daylight on the first day of its summer, on or about December 21. This is the shortest period of daylight in the Northern Hemisphere and is called the **winter solstice**. The sun is now straight over the Tropic of Capricorn at 23.5°S latitude. Every point north of the Arctic Circle has 24 hours of darkness. Every point within the Antarctic Circle, 66.5°S latitude, has 24 hours of daylight.

There are two days midway between the solstices when neither hemisphere is tilted toward the sun. Since the periods of daylight and night are of equal length, they are called equinoxes. The **spring equinox** is on or about March 21, and the **autumnal equinox** is on or about September 23. The sun is directly overhead at the equator on these dates.

2.4 Section Review

1. Explain why the period of daylight varies over Earth's surface.
2. Describe how the tilt of Earth's axis causes the seasons.
3. Compare your lifestyle with that of someone who lives in the opposite hemisphere. Compare climate, weather, type of housing, occupations.

LABORATORY INVESTIGATION 2

The Distribution of Radiant Energy

PROBLEM: *Is the distribution of radiant energy affected by the angle at which it strikes a surface?*

SKILLS: *Measuring, manipulating*

MATERIALS: *Thumbtack, graph paper, ruler, string, glue, cardboard with square cutout, cardboard tube, flashlight, pencil*

PROCEDURE

1. Use a thumbtack to mark the center of a sheet of graph paper.
2. Attach one end of a 20-centimeter-long string to the thumbtack. Glue the other end of the string to a square cardboard cutout. Attach the cardboard tube to the flashlight. Place the cutout over the flashlight beam.
3. Shine the flashlight directly (at a 90° angle) on the paper, 20 centimeters above the point you marked.
4. Trace the outline of the beam as it shines on the paper.
5. Count the number of graph squares that fall within the outline of the beam.
6. Shine the flashlight at an angle of 45° to the mark on the center of the paper, 20 centimeters above the paper.
7. Trace the outline of the beam on the paper.
8. Count the number of graph squares that fall within the outline of the beam.

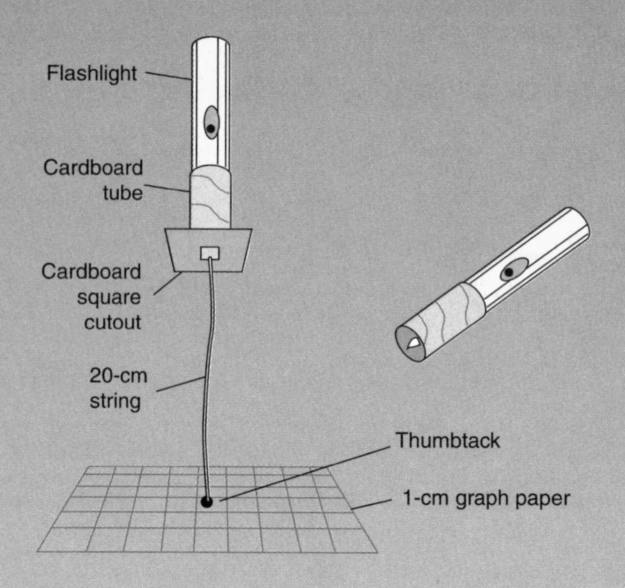

OBSERVATIONS AND ANALYSES

1. Did the direct or indirect beam of the flashlight illuminate more squares?
2. In which instance was the light energy spread over a wider area?
3. Why does direct radiation have a more concentrated warming effect than indirect radiation?

Chapter 2 Review

Answer these questions on a separate sheet of paper.

Vocabulary

The following list contains all the boldfaced terms in this chapter.

autumnal equinox, axis, day, equator, galaxy, high tide, light-years, low tides, months, neap tides, orbits, seasons, spring equinox, spring tides, summer solstice, winter solstice, year

Fill In

Use one of the vocabulary terms listed above to complete each of the following sentences.

1. The distance between stars is measured in a unit called a(an) ________.
2. The tilt of Earth's axis causes changing weather patterns called ________.
3. The imaginary line that divides Earth into the Northern and Southern Hemispheres is the ________.
4. Because of Earth's tilt on its axis, the sun appears directly over the equator during the ________ and ________.
5. The period of time that it takes Earth to make one revolution around the sun is a(an) ________.

Multiple Choice

Choose the response that best completes the sentence or answers the question.

6. Most of the universe is composed of *a.* stars. *b.* galaxies. *c.* empty space. *d.* electromagnetic energy.
7. The age of the universe is estimated to be about *a.* 6 million years old. *b.* 6 billion years old. *c.* 12 to 15 million years old. *d.* 12 to 15 billion years old.
8. How many galaxies make up the Local Group? *a.* 2 *b.* 15 *c.* 29 *d.* 100
9. Stars glow due to heat produced through *a.* nuclear reactions. *b.* chemical reactions. *c.* gravity. *d.* the big bang.
10. What are the rocky space objects that strike Earth's surface called? *a.* asteroids *b.* meteoroids *c.* meteors *d.* meteorites
11. The phases of the moon undergo one complete cycle every *a.* 24 hours. *b.* 24 days. *c.* 29.5 days. *d.* 31 days.

12. There are two high tides and two low tides every *a.* 12 hours and 25 minutes. *b.* 21 hours. *c.* 24 hours and 50 minutes. *d.* 27.3 days.
13. The period of Earth's revolution around the sun is closest to *a.* 24 hours. *b.* 29.5 days. *c.* 365 days. *d.* 3 years.
14. What is the tilt of Earth's axis? *a.* 0° *b.* 23.5° *c.* 45° *d.* 90°
15. What is the latitude of the equator? *a.* 0° *b.* 23.5° *c.* 45° *d.* 90°

Short Answer (Constructed Response)

Use the information you learned in this chapter to respond to the following items.

16. Is the Milky Way galaxy the only galaxy? Explain.
17. What is the closest star to the sun?
18. Describe the shape of the orbits the planets follow around the sun.
19. Explain the reason for Earth's changing seasons.
20. Explain the basis for our division of time into months and years.
21. What is the cause of our cycle of tidal changes?
22. What is indicated by the red shift observed in the light from other galaxies?
23. What relationship exists between the spring and autumnal equinoxes and the position of the sun in the sky?
24. What do the summer and winter solstices represent?
25. At what time of year is Earth closest to the sun?

Essay (Extended Response)

Use the information you learned in the chapter to respond to the following items.

26. How long does it take light from the moon to reach Earth (384,000 kilometers)?
27. Explain why the red shift in the light from distant galaxies indicates that the universe is expanding.
28. How does the big bang theory describe the formation of the universe?

CHAPTER 3
Earth's Rocky Crust

When you have completed this chapter, you should be able to:

Identify the divisions of Earth.

Explain the theory of continental drift.

Define earthquake and Richter scale.

If you could travel millions of years back in time, you would encounter a planet that would not look familiar at all. Earth's face is constantly changing. However, most of these changes occur too slowly to be witnessed during a person's lifetime. Some natural forces cause mountains to rise, others wear them down.

Through the use of modern instruments, geologists can detect small changes. Over millions of years these many small changes translate into huge differences. The modifications caused by earthquakes, floods, and storms seem catastrophic to us. However, they are insignificant when compared with what has occurred to Earth during its geologic history.

3.1 EARTH'S DIVISIONS

Scientists divide the outer portion of Earth into three distinct layers: the lithosphere (solid), the hydrosphere (liquid), and the atmosphere (gas). The **lithosphere** is the dense, solid outer layer of Earth. It is composed of rock and soil, which surround the fluid inner layers. Oxygen and silicon are the two most common elements that make up the compounds of the lithosphere. Aluminum, iron, calcium, sodium, potassium, and magnesium are other important elements in crustal minerals.

Scientists theorize that as Earth formed, the denser elements, such as molten iron and nickel, sank into the core. This forced the less dense rocks to the surface, where they cooled and formed a single large continent. Lava flows from erupting volcanoes constantly brought new material to the surface. Some of this material contained gases that eventually created an atmosphere that surrounded the planet. As Earth cooled, water vapor in the atmosphere condensed to form the oceans. This is one possible explanation; no one knows for sure. Today's continents are very different from those that first formed on Earth's surface. In the billions of years since the formation of the first continents, natural processes have been at work altering them.

Almost three-quarters (71%) of the crust is covered by liquid water in the form of oceans, seas, lakes, rivers, and streams. The surface water, along with water that exists below ground, make up the layer called the **hydrosphere.**

A blanket of gases, called the **atmosphere** or air, covers the lithosphere and hydrosphere. The atmosphere is composed mostly of nitrogen (78%) and oxygen (21%). Argon, carbon dioxide, water vapor, ice crystals, dust particles, and small amounts of several other gases make up the remaining 1 percent of the atmosphere. The lithosphere, hydrosphere, and atmosphere are shown in Figure 3-1.

Earth's Rocks and Minerals

Earth's rocky **crust**, the layer that covers the mantle, is composed of more than 2000 different compounds, called minerals. Most min-

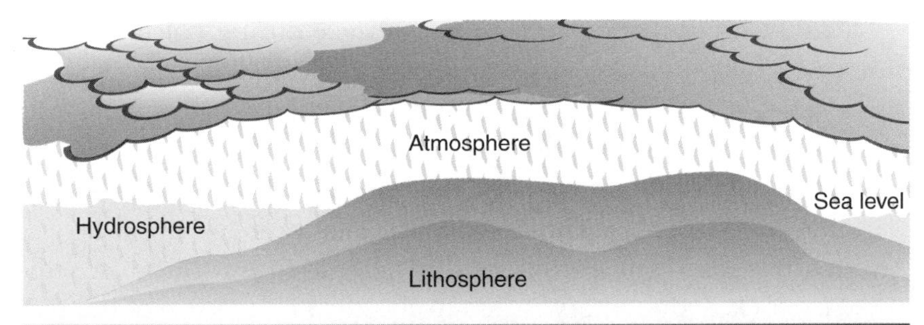

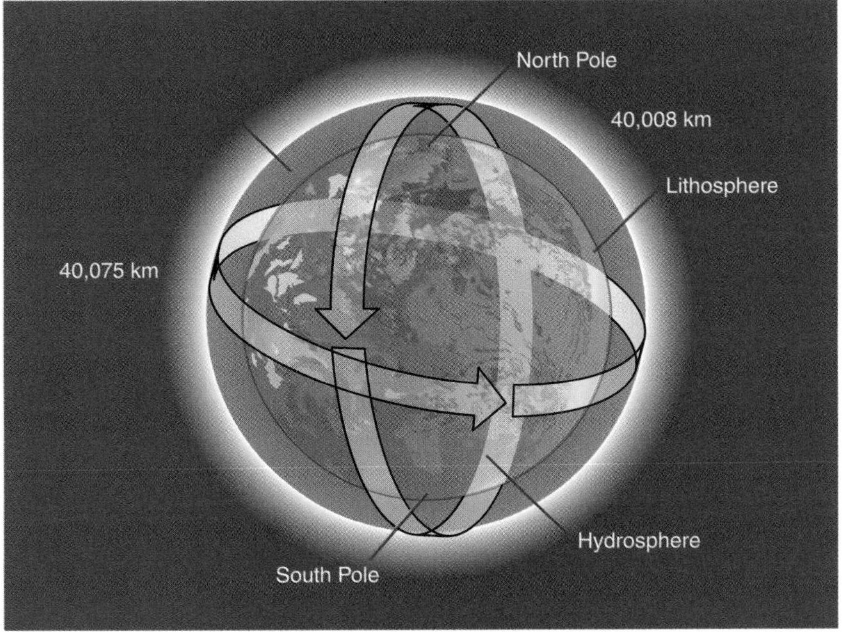

Figure 3-1 The lithosphere, hydrosphere, and atmosphere are the three main layers of Earth.

erals are chemical compounds, though some are elements, such as gold, copper, and carbon. Some of the more common minerals are feldspar, mica, quartz, and calcite. Minerals occur in rocks as crystals, which can be identified by their individual chemical properties. Crystals form as molten material cools or as liquids that contain dissolved salts evaporate. Minerals can be changed into different minerals by heat, pressure, or chemical processes. Rocks are mixtures of minerals. The rocks that make up Earth's crust form in three ways.

Igneous rocks form by the cooling and solidification of **magma**, the hot, molten rock from Earth's interior. Igneous rocks contain large interlocking crystals.

Magma that flows out onto Earth's surface during volcanic eruptions is called **lava**. Because they cool rapidly, volcanic igneous rocks contain very small, fine crystals. Volcanic igneous rocks can also form from fine dust and ash.

Sedimentary rocks form by the hardening and cementing of rock particles or fragments. These rocks are usually made up of layers. The fossil remains of ancient organisms are often found in sedimentary rock. Sedimentary rocks are classified by the size of the sediment particles of which they are composed. The largest particles are pebbles and gravels. When these sediments are cemented they form conglomerate.

Metamorphic rocks form as heat and pressure from geologic processes deep within Earth change the composition of existing sedimentary and igneous rocks.

3.1 Section Review

1. What are the most common elements in Earth's crust?
2. Explain how geologic processes change minerals into other minerals.
3. As a class field trip or an individual effort, collect several varieties of rocks from the school grounds or surrounding area. Determine whether each rock is sedimentary, igneous, or metamorphic. Give one scientific reason for your classification.

3.2 DRIFTING CONTINENTS

Alfred Wegener, a German scientist, proposed a revolutionary theory in 1912: that all the continents had once been joined together as a single supercontinent, which he called Pangaea. Over millions of years, Pangaea split into pieces. The pieces drifted apart and eventually assumed their present-day positions. Evidence Wegener used to support his theory included the interlocking shapes of the western coast of Africa and the eastern coast of South America. He also showed that rock formations and plant and animal fossils from these coasts had similar origins. He called this theory **continental drift.** Continental drift is illustrated in Figure 3-2. However,

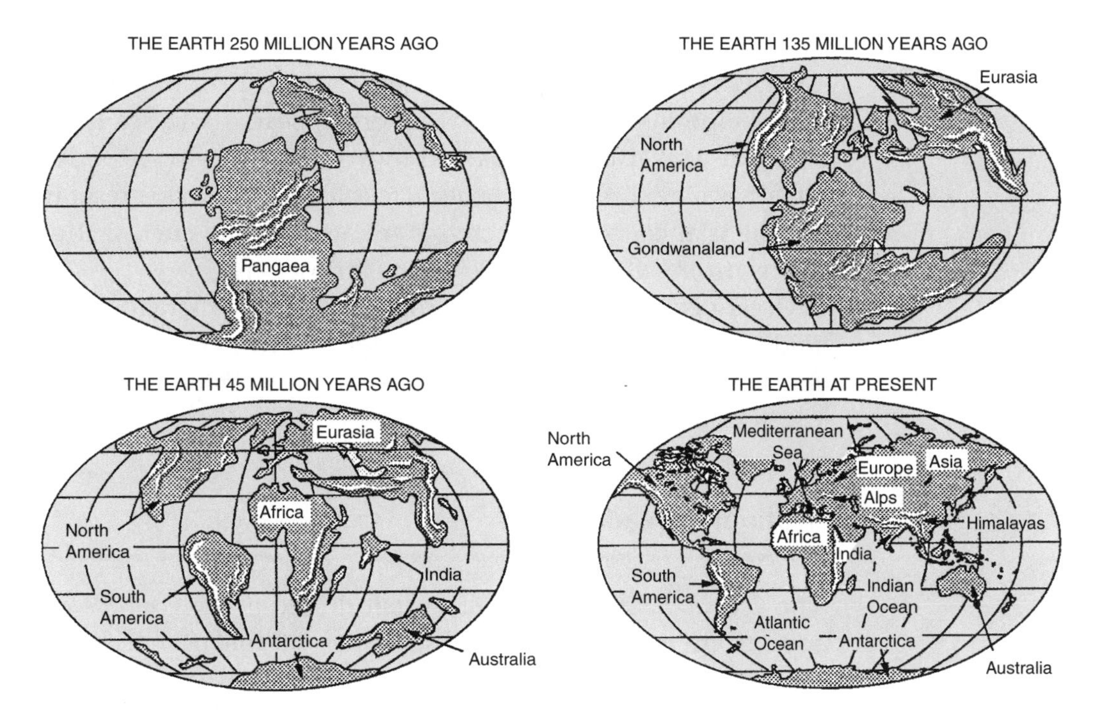

Figure 3-2 The breakup of Pangaea and continental drift over time have moved the continents into their present positions.

Wegener could not explain how the continents could move through Earth's solid crust.

Scientists did not accept Wegener's theory, and for many years he was ridiculed. Then in the early 1960s, evidence was found that supported his idea that the continents were indeed drifting across Earth's surface. Underwater exploration revealed a network of ridges and rifts, or cracks on the seafloor, that extend along the middle of the Atlantic Ocean. Molten rock, rising from within Earth, was found to have oozed out through the rifts and cooled to form new seafloor. Scientists surmised that as the seafloor widened, the continents on either side were pushed apart.

Mid-Ocean Ridges

During the 1950s, the United States government sponsored oceanographic research projects to unlock the mysteries of the ocean depths. Scientists discovered a large system of underwater volcanic mountains, which they named the **mid-ocean ridge**. This system extends through all the oceans and forms the largest mountain chain on Earth. (See Figure 3-3.) Throughout much of this system there is a central rift valley, a deep crack that runs along the crest of the mountains. The rift valley seldom appears above sea level, but where it does it is accompanied by volcanic activity. A portion of the rift valley surfaces and cuts across Iceland, and it is responsible for the island's volcanic activity and geothermal energy, the heat energy from within the rocks of the lithosphere.

Oceanographic research also discovered faults that cut across the mid-ocean ridges. **Faults** are cracks in Earth's crust along which movement has occurred. The faults across the mid-ocean ridge are called transform faults.

Seafloor Spreading

The *Glomar Challenger* is a ship that drilled many cores of crustal rock from the seafloor on both sides of the mid-ocean ridge. Radioactive dating techniques used to determine the age of the rocks in the cores indicated that rock samples near the ridge were younger than rock samples farther away. In fact, rocks were found to be progressively older the farther away they were from the mid-ocean ridge.

Scientists theorized that new ocean floor was forming at the mid-ocean ridges. As lava erupts from a rift valley, it is cooled rapidly by

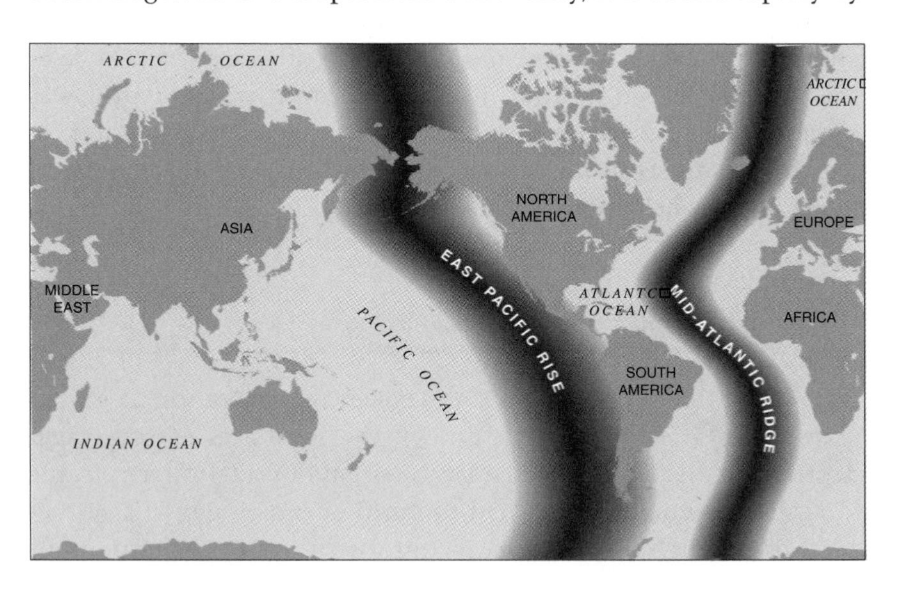

Figure 3-3 The farther you travel east or west of the Mid-Atlantic Ridge, the older the rocks become.

the seawater and hardens to form igneous rock. This produces new seafloor on both sides of the rift valley and pushes the older seafloor away from the rift. You might think that Earth must be growing larger. However, further exploration revealed that older rock is pushed down into Earth's crust along deep ocean trenches, long V-shaped valleys located around the edges of the Pacific Ocean. In the trenches, rock melts and rejoins the mantle. Intense volcanic activity takes place in these trenches. This area is called the "Ring of Fire," because of the many volcanoes and earthquakes that occur there.

Plate Tectonics

Scientists now think that the seafloor and continents rest on huge slabs of rock called plates. Earth's crust is composed of a dozen large plates and several smaller ones. Look at Figure 3-4. **Plates** are rigid slabs of crustal rock that float on Earth's mantle. The lighter continental plates carry Earth's landmasses; the more dense oceanic plates are composed mainly of seafloor. The plates are in constant motion and interact with one another at their boundaries. **Plate tec-**

Figure 3-4 This illustration shows Earth's major crustal plates. The arrows indicate the direction of plate motion. The triangles represent subduction zones.

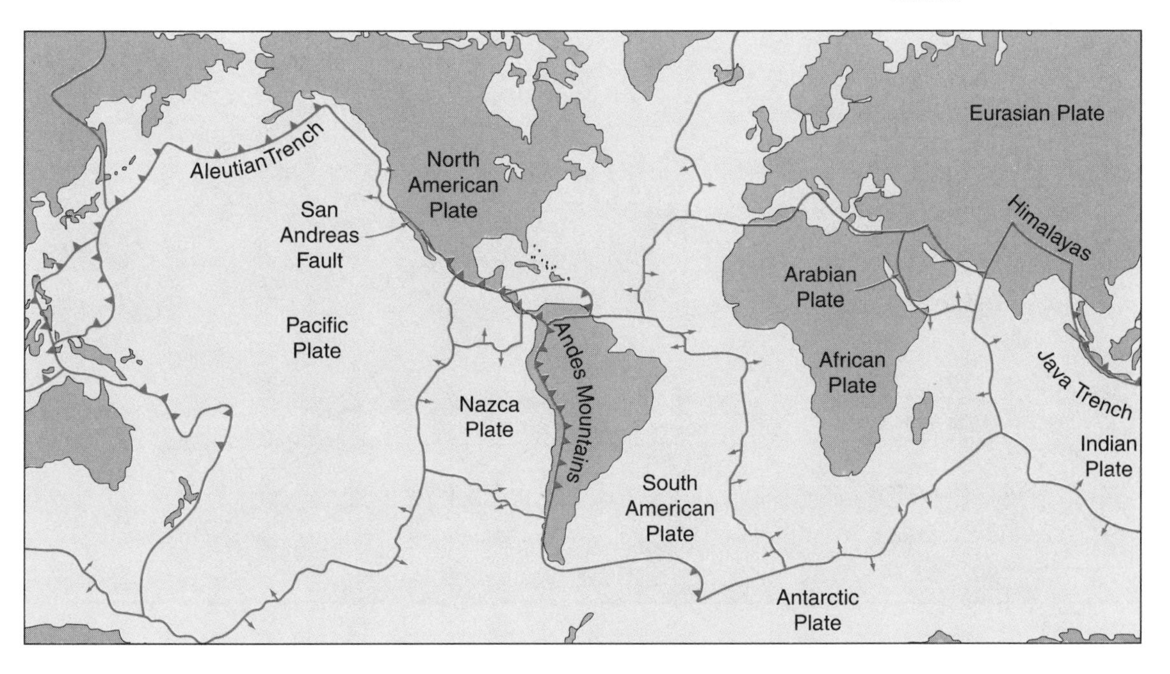

tonics describes the interactions of the moving plates that make up Earth's crust.

The process of plate tectonics is always at work, slowly and steadily changing the shape of the continents. As North and South America continue to move westward, the Atlantic Ocean will widen as the Pacific Ocean narrows. Activity along the San Andreas Fault will split California and form a new elongated island off the west coast of North America. Africa will move northward into Europe, which will cause the Mediterranean Sea to narrow and dry up. The collision of Africa with Europe will create a chain of mountains that will extend from northwest Africa to Turkey. Australia will move north and eventually cross the equator. Its climate will change from subtropical to tropical. Of course, this activity will happen slowly, over many millions of years.

Subduction and Plate Collisions

Subduction is the process by which the seafloor plunges through a trench into Earth's interior. (See Figure 3-5.) The trenches where crustal rock descends into the mantle are called subduction zones. Here, heavier oceanic plates bend downward and dive beneath the

Figure 3-5 In subduction zones, oceanic plates melt. Some of the molten material may rise to the surface and form volcanoes.

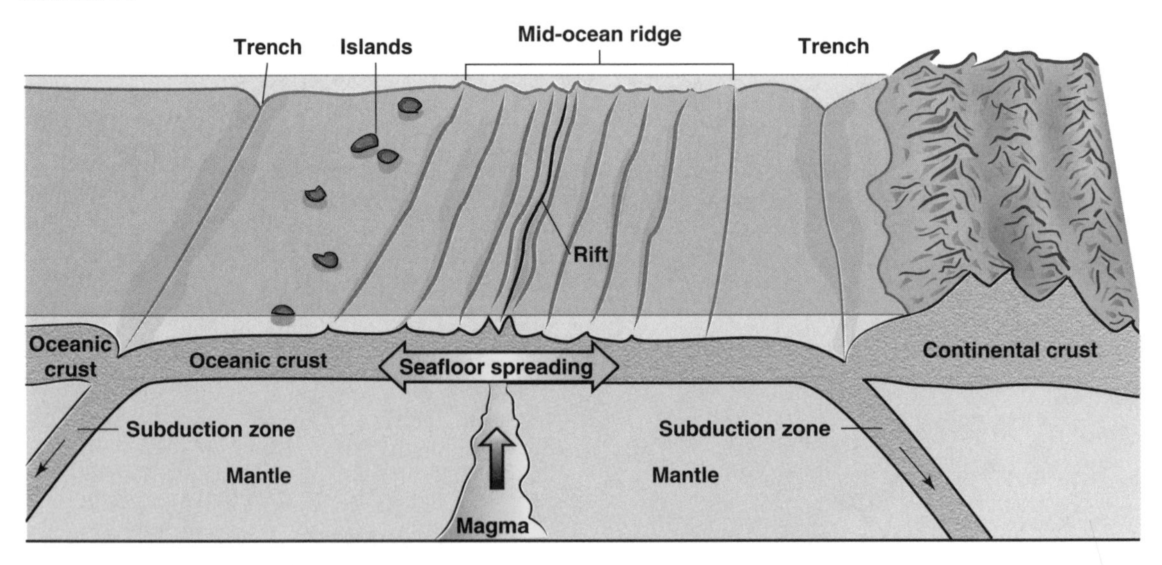

lighter continental plates. The oceanic plates melt as they plunge deep into subduction zones, and some of this molten material may rise to the surface to form volcanoes. Mount Rainier, Mount Adams, and Mount Saint Helens, in the state of Washington, are volcanoes that formed as a result of subduction.

Two continental plates may collide with neither undergoing subduction. As landmasses push slowly against each other, their edges crumple and are forced upward. The Himalaya Mountains were formed as the Indo-Australian plate collided with the Eurasian plate. These mountains are still rising as the plates continue to push against one another. The Himalayas are among the youngest mountains on Earth.

Transform boundaries are areas where plates move past each other in opposite directions. As the plates slowly grind against each other, they cause many earthquakes. The San Andreas Fault in California is an example of a transform fault.

3.2 Section Review

1. Describe the evidence Alfred Wegener used to support his theory of continental drift.
2. How do scientists explain the continued rise of the Himalaya Mountains?
3. Prepare a report about the use of geothermal energy in Iceland.

3.3 MOUNTAINS OF FIRE AND A FRACTURED CRUST

A **volcano** is an opening in Earth's crust through which molten magma reaches the surface. At the surface, magma is called lava. Some volcanoes erupt explosively; these volcanoes blast ash, dust, and gases into the atmosphere. Mount Vesuvius in Italy and Mount St. Helens in Washington are well-known explosive volcanoes. Because these volcanoes are made of alternating layers of lava and ash, they are called composite volcanoes. Krakatoa is a composite volcano in the Indonesian island chain. The most violent volcanic eruption in modern times took place there on August 27, 1883. More than half its island disappeared in that eruption, and the sound was heard as far away as the eastern coast of Africa. This kind of violent eruption is a typical characteristic of composite volcanoes.

A cinder cone volcano is formed when ash and cinders expelled from the volcano's opening mound up. Paricutín in western Mexico is a cinder cone volcano. It began to form in 1943 when a crack in the ground opened up. In 1953 its eruptions ended. At that time, Paricutín stood 410 meters tall.

Shield volcanoes ooze magma out of cracks, or vents, and from the crater on top. The volcano increases in height as lava flows. When lava reaches the sea, it forms new land. A shield volcano has gently sloping sides made of cooled lava, which may be colonized by plants and animals. Mauna Loa, Kilauea, and the rest of the volcanoes in the Hawaiian Islands are shield volcanoes.

Mount Rainier is a dormant volcano in northwest Washington State. It is the tallest peak in the Cascades, a chain of mountains that parallel the Pacific coast. Even though dormant, it poses a threat to surrounding cities and towns. Mount Rainier's summit holds the largest crest of glacial ice within the continental United States. Landslides triggered by melting glaciers can race down the mountainside sweeping away everything in their path. Past floods of glacial meltwater have destroyed roads, bridges, and facilities in the surrounding area. Because of the rapidly growing population in the area around the mountain, it is considered the most dangerous volcano in the United States.

Since the written history of the area goes back less than 200 years, it is difficult to predict future activity. An eruption in the 1840s is the only one on record. But, sedimentary evidence suggests that it has erupted at least 11 times over the past 10,000 years. Eruptions melt glacial ice, which in turn mixes with ash and other debris to form a lahar. This is a flow that looks and flows very much like wet concrete. Lahars can flow very fast and reach areas far from the volcano. These lahars pose the greatest threat to populated areas. More than 60 lahars have occurred around Mount Rainier in the past 10,000 years. Geologists hope that monitoring the activity within Mount Rainier will supply warning.

Hot Spots

Stationary hot spots in the mantle are often responsible for hot springs and island chains. For example, Yellowstone National Park in Wyoming sits above a huge magma chamber. As water seeps down through the crust it is heated by the molten magma. The water is changed into steam and percolates up to the surface to form geysers and hot springs.

The Hawaiian Islands were formed conveyor-belt style as the Pacific plate moved over a hot spot in the mantle. This is illustrated in Figure 3-6. The youngest island in this chain of volcanoes, Hawaii,

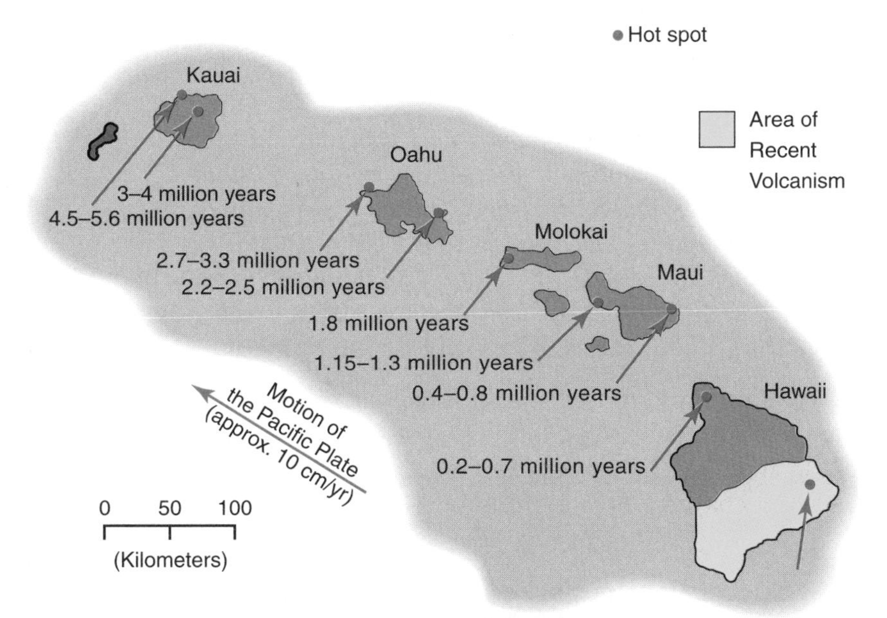

Figure 3-6 (A) Progression of age of the Hawaiian Islands. More islands may be formed by the hot spot as the plate continues to move.

Figure 3-6 (B) The Hawaiian Islands as seen from space.

DISCOVERIES

Mount Saint Helens, Washington

Mount Saint Helens made headlines in the fall of 2004 when it produced a small eruption. However, its eruption in 1980 sent hundreds of cubic meters of ash into the atmosphere. It killed 70 people, blocked the Toutle River with ash, blew down millions of trees, destroyed more than a thousand homes, killed a million and a half birds and other animals, and the ash cloud brought daytime darkness to surrounding cities. Within three days the cloud of ash had drifted as far east as Maine. In several weeks, it had encircled Earth. The ash lingered in the atmosphere for more than two years, producing beautiful sunsets. However, the ash also caused a reduction in the amount of sunlight that reached Earth's surface, which led to worldwide cooling.

The area around the volcano was devastated. Not a sign of life could be found. But within a year of the eruption, plants and animals had begun to colonize the barren landscape. Pine seedlings, fireweed, insects, and birds were among the first organisms to return. Spirit Lake, 8 kilometers away from the blast, was choked with mud and debris. The high temperatures and reduction in available oxygen destroyed aquatic life. However, by 1993, the lake's waters had cleared and it was again supporting an ecosystem of varied plant and animal life.

Ecologists have used the area as a laboratory to test biological succession, the process by which areas renew themselves. In the past scientists thought that succession occurred in an orderly predictable fashion. But research at Mount Saint Helens showed that chance was most important in rebuilding the ecosystem.

According to theories of succession, mosses and lichens should have been the first to arrive, followed by wildflowers and herbs, then deciduous trees would take hold, and finally a coniferous forest would develop. But instead, pockets of survivors that made it through the blast were the first to colonize the damaged environment. These organisms are now known as "biological legacies." They include gophers and ants that survived the blast because they were underground; small trees and shrubs that were buried by the snow and did not experience the force of the blast; and roots, bulbs, and wildflowers that were buried close enough to the surface for them to germinate. From these few organisms much of the new life has sprung. Ecologists now recognize this as a crucial part of the recovery of an ecosystem.

sits directly over the hot spot. As you move to the northwest, the islands in the chain get progressively older. Most of the native plants and animals living there are found nowhere else on Earth.

Earthquakes

Crustal movements along faults cause **earthquakes**. This movement results in the ground shaking as rock breaks or shifts, releasing energy. An earthquake's intensity is a measure of the amount of damage it causes. The magnitude of an earthquake is defined as the amount of energy it releases. Earthquakes cause widespread destruction by causing land movements, mudslides, and **tsunamis**, huge ocean waves or series of waves usually associated with an earthquake. Tsunamis are sometimes called tidal waves, although they are not related to tides.

The Richter scale measures the shaking caused by an earthquake as recorded by a seismograph. The scale identifies the magnitude of an earthquake in terms of its energy level. On the Richter scale, an increase in one unit indicates a tenfold increase in shaking. A magnitude 5 earthquake is 10 times stronger than a magnitude 4 event; a magnitude 5 earthquake causes 100 times the shaking of a magnitude 3 quake. The smallest quakes that can be felt measure between 2 and 3 on this scale. Major quakes usually measure 6 or higher. The largest quakes have measured between 8 and 9.

On December 26, 2004 a magnitude 9.0 earthquake occurred off the coast of the Indonesian island of Sumatra, causing a tsunami that affected coastal areas of Sri Lanka, India, Thailand, Indonesia, Maldives, Malaysia, and even Somalia, on the eastern coast of Africa. Walls of water 12 meters high (40 feet) came ashore, washing away everything in their path. The death toll from this tsunami was estimated to be 150,000 people.

3.3 Section Review

1. Describe the formation of the Hawaiian Islands.
2. How do shield volcanoes and cinder cone volcanoes differ?
3. Explain why there are many geysers and hot springs around the area of Yellowstone National Park.

LABORATORY INVESTIGATION 3

Demonstrating Continental Drift

PROBLEM: *How can you demonstrate continental drift?*

SKILLS: *Manipulating, drawing*

MATERIALS: *Pencil, tracing paper, scissors, outline of continents*

PROCEDURE

1. Use the continental outlines provided by your teacher.
2. Cut out the continents along the dotted lines.
3. Place the paper continents on your desk in their approximate present-day position.
4. The arrows on each continent indicate the direction of movement due to continental drift. Move each continent in the direction opposite that indicated by the arrow. This will trace their paths back to a starting point where all were joined as one landmass.
5. Trace the outline of this landmass on a sheet of paper.

OBSERVATIONS AND ANALYSES

1. What was the position of the single landmass that contained our present-day continents?
2. Which continents seem to fit together best?
3. Was India always a part of Eurasia? What was the effect of India colliding with Eurasia? Why is India referred to as a subcontinent?
4. Draw a diagram that shows the possible future positions of the continents.

Chapter 3 Review

Answer these questions on a separate sheet of paper.

Vocabulary

The following list contains all the boldfaced terms in this chapter.

atmosphere, continental drift, crust, earthquakes, faults, hydrosphere, lava, lithosphere, magma, mid-ocean ridge, plates, plate tectonics, subduction, tsunamis, volcano

Fill In

Use one of the vocabulary terms listed above to complete each of the following sentences.

1. The solid outer layer of Earth is the ______.
2. A(an) ______ is an opening in Earth's crust through which molten magma reaches the surface.
3. Earth's continents rest on huge slabs of rock called ______.
4. Magma that reaches Earth's surface is called ______.
5. Movements of Earth's crust along faults often cause ______.

Multiple Choice

Choose the response that best completes the sentence or answers the question.

6. Rocks that have been changed by heat and pressure are called *a.* igneous. *b.* sedimentary. *c.* metamorphic. *d.* all of the above.
7. Which statement is not true about the mid-ocean ridges? *a.* They form the largest mountain chain on Earth. *b.* There is a central rift valley. *c.* Part of the rift valley can be seen in Iceland. *d.* The seafloor shrinks along the ridges.
8. Which of the following is not a volcano formed as an oceanic plate melts in a subduction zone beneath the state of Washington? *a.* Mount Saint Helens *b.* Mount Vesuvius *c.* Mount Rainier *d.* Mount Adams
9. The collision of the Eurasian plate with the Indo-Australian plate produced the *a.* Alps. *b.* Andes. *c.* Himalayas. *d.* Rocky Mountains.
10. The lithosphere is composed mainly of compounds that contain *a.* silicon and oxygen. *b.* iron, magnesium, and nickel. *c.* iron, magnesium, and silicon. *d.* iron, nickel, and oxygen.
11. Which of the following is not a common mineral found in Earth's crust? *a.* mica *b.* meteorite *c.* quartz *d.* calcite

12. Which gas makes up 78 percent of the atmosphere?
    a. oxygen *b.* carbon dioxide *c.* nitrogen *d.* water vapor
13. Which rocks form by the cooling and solidification of magma? *a.* igneous *b.* sedimentary *c.* metamorphic *d.* all of the above
14. According to Figure 3-6 (A) on page 47, the next island in the Hawaiian Island chain will form near *a.* Kauai. *b.* Oahu. *c.* Maui. *d.* Hawaii.
15. An example of a shield volcano is *a.* Mount Vesuvius. *b.* Mount Saint Helens. *c.* Kilauea. *d.* Mount Etna.

Short Answer (Constructed Response)

Use the information you learned in this chapter to respond to the following items.

16. Explain the difference between Earth's crust and mantle.
17. How do igneous rocks form?
18. Describe how magma reaches Earth's surface.
19. What are some of the minerals that make up Earth's crust?
20. What are earthquakes?
21. How is magma different from lava?
22. What is the cause of tsunamis?
23. What is meant by plate tectonics?
24. What happens when moving crustal plates collide?
25. How do sedimentary rocks form?

Essay (Extended Response)

Use the information you learned in the chapter to respond to the following items.

26. Describe how Earth's atmosphere and hydrosphere formed.
27. Describe the process that causes seafloor spreading.
28. What is meant by "Ring of Fire"?
29. Explain how geological changes that create new environments trigger evolutionary changes within the biosphere.

CHAPTER 4
Earth's Air and Water

When you have completed this chapter, you should be able to:

Describe the atmosphere.

Compare weather and climate.

Describe El Niño.

Relate hydrogen bonding to the properties of water.

Of all the planets in our solar system, only Earth has liquid water and oxygen gas that support life. Oxygen is in the atmosphere, the blanket of gases that surrounds Earth. Liquid water forms the rivers, lakes, and oceans that make up the hydrosphere. In fact, so much of Earth's surface is covered by water that as seen from space Earth looks blue. The atmosphere cannot be seen. However, puffy white clouds in the atmosphere floating over the blue waters cause Earth to look like a big blue marble.

4.1 THE ATMOSPHERE

We live at the bottom of an invisible ocean of air, the atmosphere, which is a mixture of gases. As illustrated in Figure 4-1, the most abundant atmospheric gases by volume are nitrogen (78%), oxygen (21%), argon (0.93%), carbon dioxide (0.03%), and water vapor (amount varies). Earth's atmosphere is divided into layers based on temperature structure. The layer closest to Earth's surface is called the troposphere, and it contains about 75 percent of the total mass of the atmosphere. This is the layer in which weather occurs. The lower troposphere is warmed by heat radiating from Earth's surface. This radiating heat forms convection systems that transfer heat to the upper troposphere.

The stratosphere is the layer above the troposphere. The air temperature here is relatively stable. In the lower stratosphere, strong eastward winds, called the jet stream, blow horizontally around Earth. **Ozone**, a form of oxygen, is most abundant within this region. The upper boundary of the stratosphere is called the stratopause.

The mesosphere extends above the stratosphere. In this zone, temperature steadily decreases with increasing altitude. The upper mesosphere is the coldest region in the atmosphere, with a temperature below –100°C. Due to friction with the atmosphere, meteoroids usually burn up in this region of the atmosphere.

Beginning in the mesosphere and extending into the thermo-

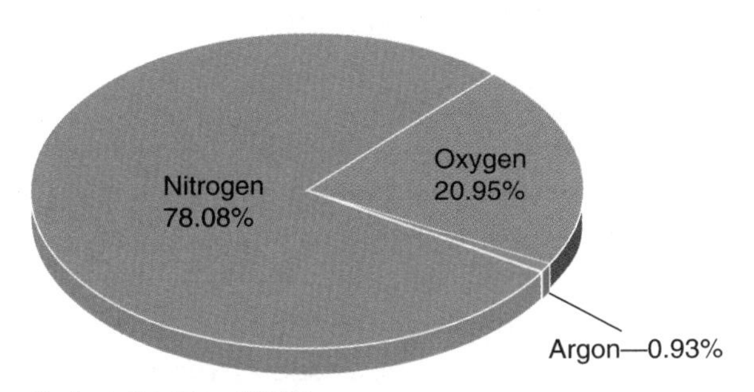

Figure 4-1 The chart shows the composition of dry air.

sphere is a region of ionized gases called the ionosphere. Here, atmospheric gases absorb electromagnetic radiation from the sun, lose electrons, and become electrically charged ions. The aurora borealis (northern lights) and aurora australis (southern lights), natural-light shows that occur when ionized gases are stimulated to emit visible light, occur in this region. (See Figure 4-2.)

The thermosphere is above the mesosphere. In this layer, the temperature increases with altitude as the widely scattered oxygen and nitrogen atoms absorb solar energy. The exosphere is the last layer of the atmosphere. It extends into outer space. Figure 4-3 on page 56 shows a cross section of the atmosphere.

Oxygen

Earth's early atmosphere of volcanic gases contained little oxygen. Scientists theorize that oxygen entered the atmosphere when organisms evolved the ability to carry out photosynthesis. In **photosynthesis,** organisms use chlorophyll and energy from the sun to convert water and carbon dioxide into food. As a waste product of photosynthesis, oxygen gas is released into the atmosphere. Oxygen is, therefore, a direct product of the chemical activity of the

Figure 4-2 **This photograph, taken from the Space Shuttle *Endeavour,* shows the aurora and the constellation Orion.**

Figure 4-3 Earth's atmosphere is layered.

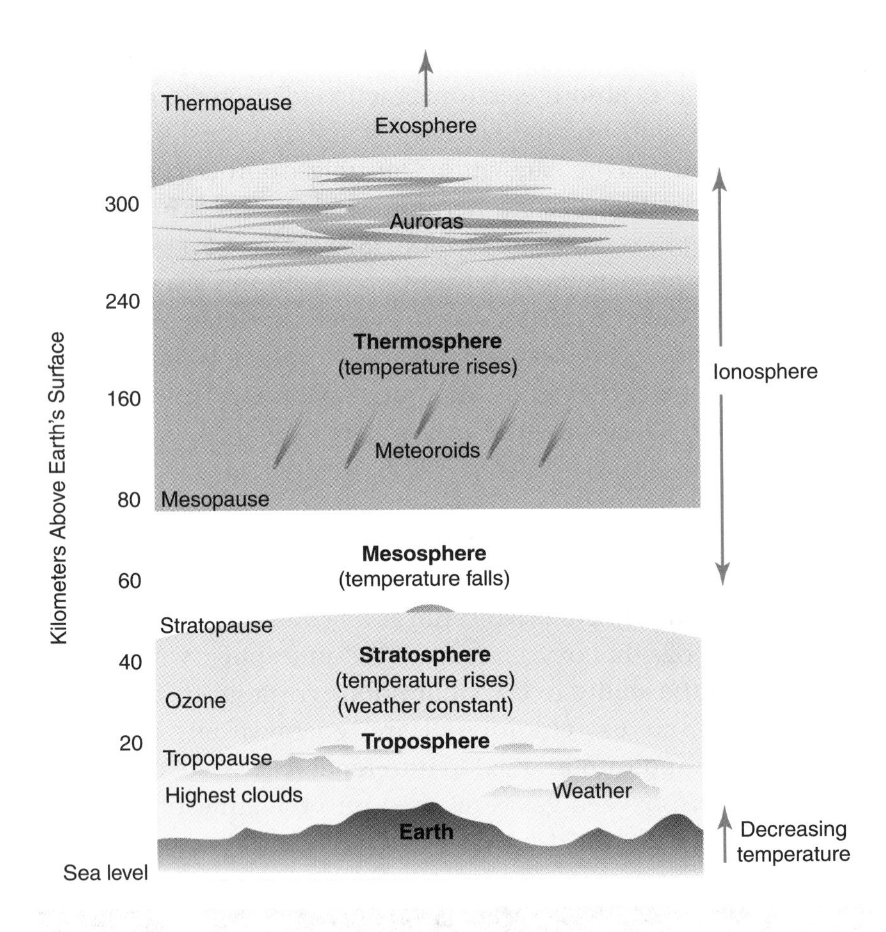

biosphere. It took billions of years for the oxygen concentration of the atmosphere to reach its present level.

Ozone also was not present in Earth's early atmosphere. It is, like the rest of the oxygen in the atmosphere, a product of Earth's biosphere. Ozone forms a layer within the stratosphere that absorbs ultraviolet radiation from the sun, which can be deadly to many living things. It is also a cause of skin cancer in humans. By absorbing much of the sun's ultraviolet radiation, ozone partially protects the biosphere.

Atmospheric Water

The amount of water in the air varies from place to place and from day to day. In the atmosphere, water can exist as a gas, a liquid, and a solid. Water molecules that evaporate from surface water enter the atmosphere as water vapor. As water vapor rises in the atmosphere and cools, it may condense. Condensation forms liquid water, which is seen as droplets of rain, mist, or fog. Droplets of water may crystallize and form bits of solid water, such as snow, ice, hail, or sleet. Some clouds are composed of water droplets while others are composed of ice crystals. Water that falls from the atmosphere to Earth's surface is called **precipitation**.

4.1 Section Review

1. How do the northern lights and southern lights form?
2. How do scientists theorize that oxygen formed in Earth's atmosphere?
3. Explain the relationship between ozone and the biosphere.

4.2 WEATHER

Many of our daily activities are affected by the weather. **Weather** is the short-term (hours or days) condition of the troposphere over a given area. Weather includes atmospheric factors such as moisture, temperature, air pressure, and wind speed and direction. Weather patterns distribute the sun's energy over Earth's surface.

Atmospheric moisture can be present as humidity, precipitation, and cloud cover. **Humidity** is a measure of the moisture content of the air. As radiation from the sun heats Earth's surface, water absorbs energy and evaporates into the atmosphere. The heated atmosphere forms convection systems that create huge air masses. Generally, air masses move in an easterly direction over North America. Moisture that enters the atmosphere above the oceans is transported across the continents. As warm air rises and cools, clouds form and the water vapor eventually returns to Earth as water, snow, or ice. Wind systems help to distribute precipitation over Earth's surface.

At Earth's surface, changes in air temperature tend to occur in two cycles: the daily weather pattern and the seasonal pattern. In general, the atmospheric temperature is lowest just after sunrise and warmest at midafternoon.

Cloud cover describes the type and quantity of clouds above Earth's surface. During the daytime, cloud cover tends to reduce temperatures by reflecting sunlight back into space. At night, cloud cover acts as a blanket, reflecting heat back to Earth and keeping it warm. Latitude, the distance from the equator, also has an effect on seasonal temperatures throughout the course of the year.

Air pressure is the weight of the atmosphere pressing on Earth's surface. With increasing altitude above sea level, air pressure decreases. As you descend below sea level, air pressure increases. At any given point, air pressure is the same in all directions. The average air pressure at Earth's surface is 1 atmosphere. The device used to measure air pressure is the barometer.

Temperature affects air pressure. Warm air is less dense than cold air. When a warm air mass replaces a cold air mass, there is a decrease in air pressure. Similarly, the replacement of a warm air mass by a cold air mass causes an increase in air pressure.

Humidity also affects air pressure. A molecule of water is lighter than a molecule of nitrogen or oxygen. Atmospheric moisture replaces the nitrogen and oxygen molecules in the air. The more water vapor air contains, the lighter it is. Humid air has a lower air pressure than dry air. Therefore, a rise in air pressure indicates good weather is coming. When the air pressure falls, get out the umbrella because there will be precipitation.

Winds are caused by uneven heating of the atmosphere. Wind blows from areas of high pressure to areas of low pressure. The greater the difference in pressure, the stronger the winds. Earth's rotation deflects the path of winds and forms circular patterns called the Coriolis effect. Winds are turned to the right of their path in the Northern Hemisphere and turned to the left in the Southern Hemisphere. In the Northern Hemisphere, the Coriolis effect causes winds to flow clockwise around high-pressure areas and counterclockwise around low-pressure areas. Winds are illustrated in Figure 4-4.

Severe Weather

Storms are atmospheric disturbances that often cause changes to the biosphere. Prolonged or heavy rainfall can erode topsoil, flood low-

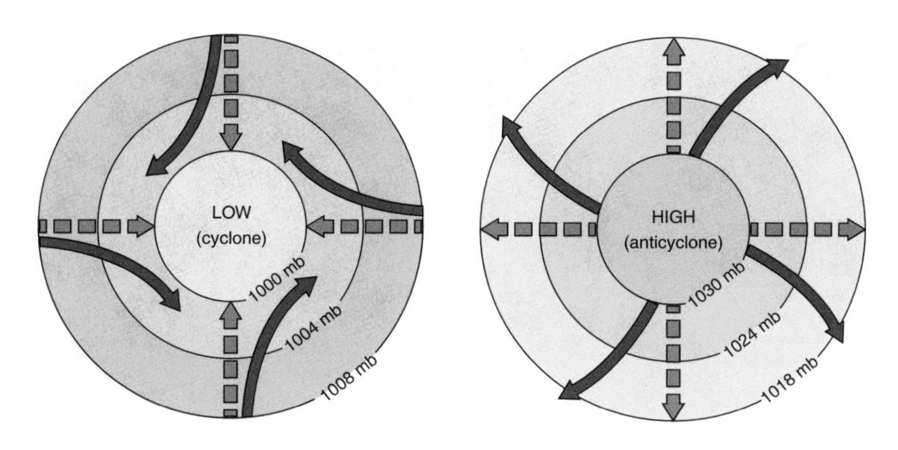

Figure 4-4 Winds flow out of high-pressure areas and into low-pressure areas. Blue arrows indicate how the Coriolis effect affects the path of the wind. Red arrows indicate the unaffected path of the wind.

lying land, raise river levels, and destroy crops and vegetation. The damage done to the environment, as well as to the economy, often takes many years to repair.

Thunderstorms are the most common severe storms. They usually occur in warm, moist air masses during spring and summer. The heavy rains do not last long but often cause severe damage from high winds, lightning, and hail. Hailstones, which can be as large as golf balls, can flatten crops and damage homes. High winds often blow down trees, while lightning may start forest fires.

Tornadoes, also called twisters and cyclones, are the most destructive of all storms. They are usually associated with thunderstorms and occur most frequently throughout the Great Plains of North America and the southeast United States. As shown in Figure 4-5 on page 60, a tornado is a localized, rapidly spinning funnel-shaped cloud. Wind velocities within the tornado are usually about 500 kilometers per hour, enough to devastate anything in their path. Atmospheric pressure at the center of the tornado is very much lower than the surrounding air pressure. The damage from a tornado is usually confined to a narrow strip several hundred meters wide and frequently less than 3 kilometers long. Waterspouts are similar to tornadoes but are smaller and occur over water.

Hurricanes, also known as typhoons, are large, destructive tropical storms. Hurricanes have wind speeds in excess of 125 kilometers per hour. Hurricanes cover very large areas. Most hurricanes form over warm ocean waters in late summer. Warm air spirals toward the center of the hurricane, where it then rises, creating the eye, a low-pressure area with little or no wind. As winds spiral inward, they

Figure 4-5 The tornado pictured here occurred in Alfalfa, Oklahoma.

increase in velocity. The fastest winds are those around the eye. Figure 4-6 shows a hurricane.

Water is often the most destructive aspect of a hurricane, and this makes hurricanes especially destructive to the ecology of islands and low-lying shore areas. Waves can wash over small islands. On Sep-

Figure 4-6 This picture of a hurricane was taken from space. The diameter of a hurricane averages about 600 kilometers.

tember 8, 1900, a storm surge from a hurricane hit Galveston, Texas, drowning more than 6000 people. This was the greatest natural disaster in the history of the United States.

Blizzards are severe winter storms that have heavy snowfalls, high winds, and low temperatures. For a snowstorm to be called a blizzard, temperatures must be below –7°C and winds greater than 56 kilometers per hour.

4.2 Section Review

1. What are the factors in the atmosphere that determine weather?
2. How does cloud cover affect temperature?
3. Chart the weather in your area over two weeks. Record the temperature, precipitation type, humidity, cloud cover, and wind speed and direction.

4.3 CLIMATE

Ecosystems are regulated by climate. Climate and weather should not be confused with each another. Weather is the condition of the atmosphere on a short-term or immediate basis. **Climate** is the long-range temperature and precipitation patterns over a particular area. These factors are often affected by the number of days and hours of sunlight (which is determined by latitude), the direction and prevalence of winds, and the nearness to large bodies of water and ocean currents. (See Figure 4-7 on page 62.)

The diversity of climates over Earth's surface, interacting with the topography of each area, produces a wide variety of environments. **Topography** refers to physical features such as mountain ranges, valleys, plains, and plateaus.

The equatorial region, located between 30° North and 30° South of the equator, is characterized by a warm, moist climate. Near the equator, there is little variation in the amount of solar energy received daily. The region experiences daily periods of rainfall. The farther one moves from the equator, the more the climate begins to

Figure 4-7 **When rays from the sun strike Earth on a slant, they are less intense than direct rays.**

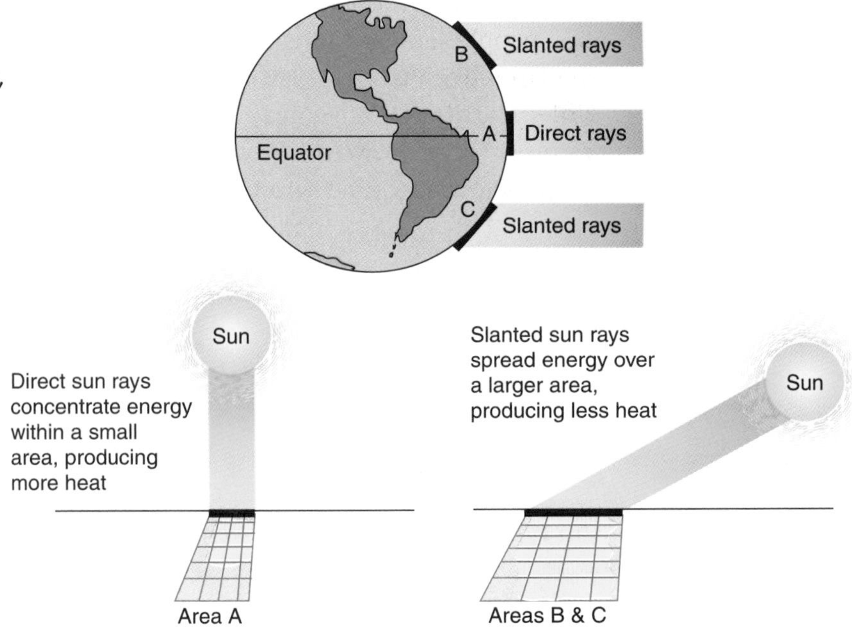

exhibit seasonal variations. Rainfall usually occurs only during the warmest parts of the year and the cooler seasons are usually dry.

The temperate regions are located between 30° and 60° North and 30° and 60° South of the equator. Temperature and rainfall may vary greatly with the seasons, and there are many different types of climates, from warm and dry to cold and wet.

At about 60° North and South of the equator, the polar regions begin. They have a cold, dry climate and are considered cold deserts. The small amount of moisture present in the air usually precipitates as snow.

The distribution of continents and oceans influences climate zones. Oceans, which occupy about three-quarters of Earth's surface, undergo limited temperature change. Climates modified by the oceans therefore experience relatively small variations in temperature throughout the year. These are called **maritime climates**. Tampa, Florida, on the Gulf of Mexico and Seattle, Washington, on the Pacific coast are examples of cities with maritime climates.

The interiors of continents undergo significant variations in temperature throughout the year. These areas are called **continental climates** and include Minneapolis, Minnesota, and St. Louis, Missouri.

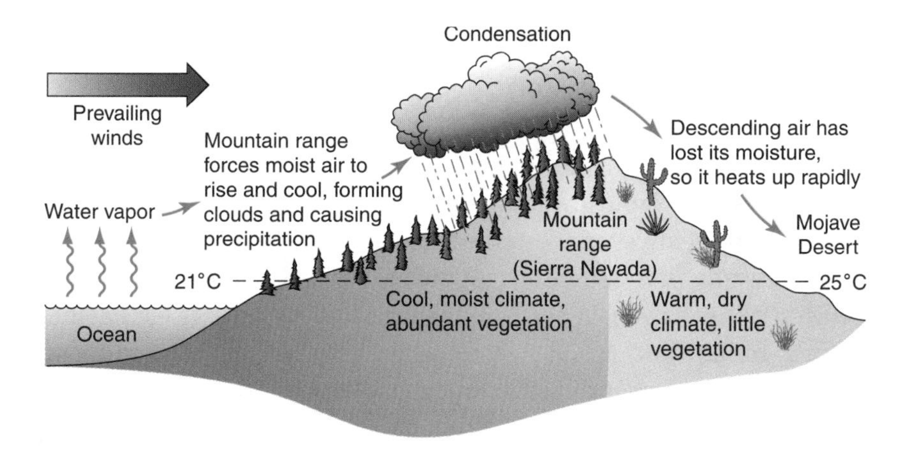

Figure 4-8 As moist air moves up the windward side of a mountain, it expands, cools, and drops its moisture. The air moving down the other side of the mountain is dry, forming a desert in the rain shadow of the mountain.

Warm, moist air from tropical regions or cold, dry air from polar regions may modify both maritime and continental climates. Buffalo, New York, has a continental, polar climate, while Miami, Florida, exhibits a maritime, tropical climate.

Mountain ranges often disrupt the movement of air masses. (See Figure 4-8.) The side of a mountain that faces the prevailing winds is the **windward** side. As moist air moves up the windward slopes of mountains, it expands and cools. This causes precipitation and may create lush forests on these slopes. The air moving down the leeward slopes tends to be warm and dry, often leading to desert conditions. These desert regions are said to be in the rain shadow of the mountain range. The western slopes of the Sierra Nevada range in California are heavily forested while the eastern slopes are arid.

Mountain climate differs from that of the surrounding region. As you ascend a mountain, the temperature decreases and the precipitation increases. It is as if you were moving from the equator to the poles. In fact, Mount Kilimanjaro, near the equator in Africa, is permanently capped with snow.

4.3 Section Review

1. Compare and contrast weather and climate.
2. Describe the climate of Earth's equatorial region.
3. Identify the climate of your area and the factors that modify that climate.

4.4 THE HYDROSPHERE

The hydrosphere is the liquid part of Earth; it includes oceans, seas, lakes, rivers, and groundwater. Of Earth's surface, 71 percent is covered by water. However, 97 percent of this water is salt water.

The saltwater part of the hydrosphere is divided into four vast interconnected oceans and 21 smaller seas. The four oceans are the Pacific, Atlantic, Indian, and Arctic. A **sea** is an ocean division that is partly or totally enclosed by land. Some of the major seas are the Caribbean, Mediterranean, Bering, Ross, North, Caspian, Black, and Arabian.

Freshwater is crucial for all life, but three-quarters of Earth's freshwater is locked in cold storage. It exists in the form of glaciers, ice caps, icebergs, and permafrost. The small amount of freshwater available for living things is present as standing water, flowing water, and groundwater. Standing water includes lakes and ponds. Flowing water includes rivers, streams, and brooks. Groundwater refers to the extensive supply of water stored in soil and in porous rock. Groundwater is usually obtained only through wells and natural springs.

The Ocean Floor

At the edge of the continents, where they touch the oceans, starts a vast undersea world. Mountains bigger than Everest, a chain of mountains that encircles the globe, underwater volcanoes, deep trenches, and submarine canyons are found on the ocean floor. The continental shelf is a broad, flat extension of the continents that gently slopes away from the continents toward the deeper oceans. Sediments that have been eroded from the continents and washed into the seas cover the continental shelf. The continental shelf teems with marine life. It is the most productive part of the oceans. Most commercial fishing takes place here.

Farther out from the shoreline, a steep slope called the continental slope begins. The continental slope is scarred by submarine canyons, formed when landslides flow down the continental slope and as river systems cut their way out to sea. The Hudson Canyon was formed as the Hudson River eroded the land on its way to the

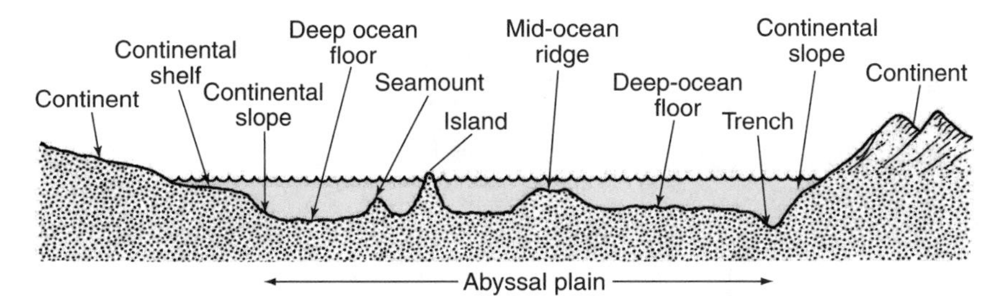

Figure 4-9 **The ocean floor has many mountains, valleys, and other topographic features.**

Atlantic Ocean. In some areas, the steepness of the slope decreases; this gentler slope is called the continental rise.

The deep ocean basins stretch away from the continental slopes to form the broad abyssal plain. (See Figure 4-9.) Like the continents, the abyssal plain has various topographic features. The mid-ocean ridge is an undersea mountain chain that extends around Earth. There are deep trenches parallel to some of the continental margins. Flat-topped seamounts, called guyots, and volcanoes and island chains rise up from the ocean floor. The tallest mountain on Earth is not Mount Everest. It is Mauna Loa, a volcano on the island of Hawaii. It rises more than 9.2 kilometers from the bottom of the Pacific Ocean.

The average depth of the ocean basins is about 4.5 kilometers. Sunlight cannot penetrate to this depth. Pressure is very high and temperature very low, yet many organisms are able to survive in this environment, which seems hostile to us.

The Oceans' Water

Waves are rhythmic motions of the water caused by winds. As winds move across the surface of a body of water, they give energy to the water. This energy causes the water to rise and fall and produces waves. Waves have a high point called a crest and a low point called a trough. When a wave reaches shallow water, its energy is released, and it rolls over and forms breakers. Along the coasts the force of waves crashing onto beaches and cliffs causes erosion.

Ocean waters are in constant motion. In addition to waves, they contain long, wide rivers of moving water called **currents**. Like rivers on land, ocean currents always keep to the same course. Ocean currents are driven by winds as well as the Coriolis effect, which makes the currents take curved paths. This causes currents to

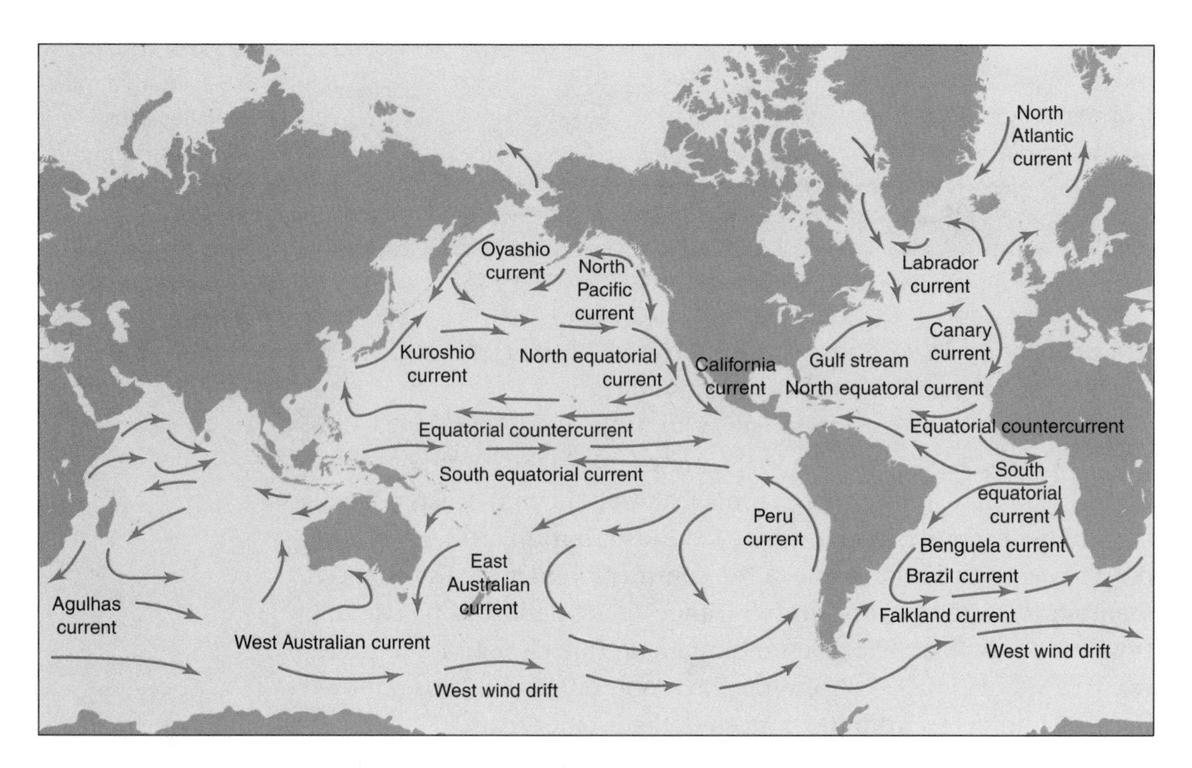

Figure 4-10 **Earth's major ocean surface currents are produced by winds and the Coriolis effect.**

rotate, or turn, clockwise in the Northern Hemisphere and counterclockwise in the Southern Hemisphere. Beside surface currents, many other currents flow at varying depths throughout the oceans. This creates a complex circulation pattern.

Ocean currents strongly affect the weather and climate of the continents. Warm Gulf Stream currents off the east coast of the United States cause moderate weather conditions in the British Isles. Figure 4-10 illustrates some of Earth's major currents.

Off the Pacific coasts of North and South America, winds create surface currents that cause warm surface water to flow away from the shore. These currents, in turn, cause an upwelling, or upward movement, of cold, deep ocean waters. The vertical mixing brings nutrients to the surface and leads to a rich marine ecosystem and productive fishing industry.

Every few years, the wind pattern off the Pacific coast of Chile changes. Warm surface waters move eastward along the equator and block the upwelling of deep coastal waters. This creates an event called El Niño, which causes worldwide changes in climate. (See Figure 4-11.) Because nutrients are no longer brought to the surface,

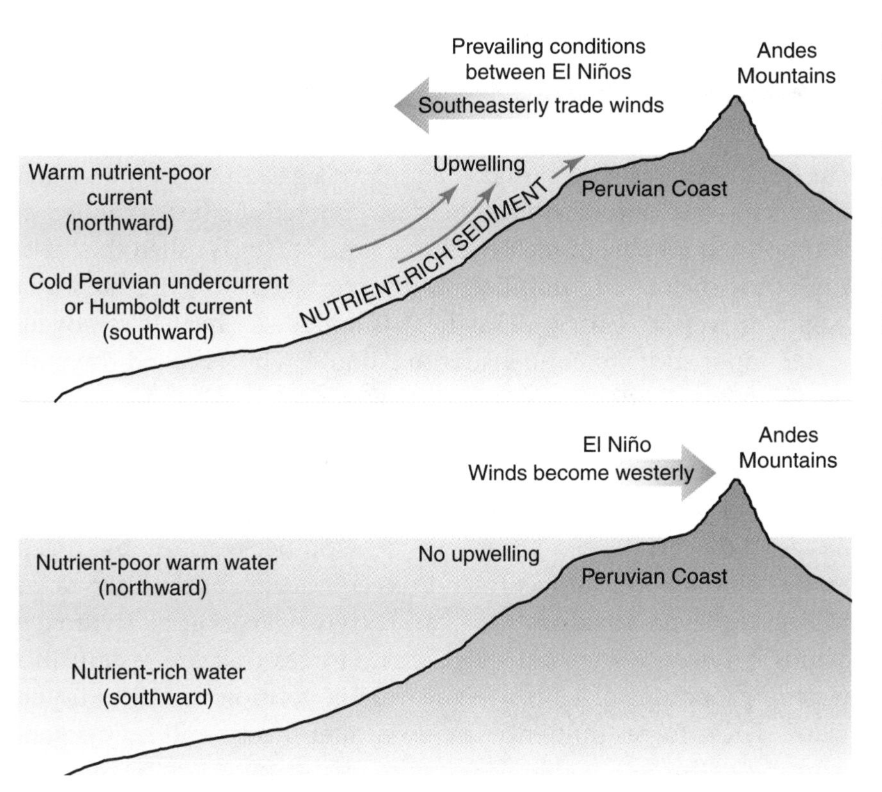

Figure 4-11 **The onset of El Niño blocks the upwelling of nutrient-rich waters along the Peruvian coast. This leads to a decline in plankton and fish populations.**

many organisms in the coastal waters die. Marine life-forms and the fishing industry suffer.

Marine ecosystems are found in the oceans and seas. Marine organisms are subjected to much greater pressures than organisms living on land. Water pressure is much greater than air pressure, and water pressure increases with depth. Because pressures vary greatly throughout the depths of the oceans, marine organisms are often restricted to specific depths. Deep-sea organisms cannot survive at the surface, and surface organisms usually cannot dive to great depths. Exceptions include the sperm whale, elephant seal, and chambered nautilus.

4.4 Section Review

1. What is the difference between standing water and groundwater?
2. What are the divisions of the ocean floor?
3. Examine a sample of seawater with a microscope. Draw what you observe. Identify the organisms using a reference sheet provided by your teacher or with the help of a biology textbook.

4.5 PROPERTIES OF WATER

On Earth, water can be a solid, a liquid, or a gas. At 0°C, liquid water begins to change to a solid as heat is removed. As heat is added at 0°C, solid ice begins to melt to form a liquid. This is called the melting point or freezing point of water (ice). At 100°C, liquid water begins to vaporize, or boil, as heat is added. As heat is removed, water vapor starts to condense into a liquid. This is called the boiling point or condensation point of water. At lower temperatures, molecules of water evaporate, or escape into the air as a gas. But the process of evaporation is relatively slow compared to vaporization.

Water molecules are made up of two hydrogen atoms and one oxygen atom. However, attractions develop between the hydrogen atom of one water molecule and oxygen atoms of another water molecule. These attractions are called **hydrogen bonds**. Hydrogen bonds between water molecules create forces that are responsible for the properties of viscosity and surface tension found in liquid water. These forces influence the way water moves and how organisms move through water.

Viscosity determines how a liquid will flow. It should not be confused with density. Viscous liquids flow slowly. Molasses and honey are highly viscous liquids while alcohol and water are less viscous. Viscosity affects the amount of energy organisms must expend to swim. Moving through water creates drag, which holds back an organism's forward motion. The more viscous a liquid, the greater is the drag.

Surface tension is caused by the hydrogen bonds of the surface water molecules. The forces on the surface molecules are unbalanced. The surface molecules are pulled down and together to create a very thin skin of tightly packed water molecules between the air and other water molecules. Surface tension gives a water droplet its round shape. Some insects, such as the water strider, whirligig beetle, and water boatman, can walk or swim on this thin film without breaking through the surface. But most insects cannot overcome the forces created by surface tension and drown if they fall below the surface of the water.

Hydrogen bonding also gives water a high heat capacity, the ability to absorb, store, and release heat energy. This property enables water to resist rapid temperature changes. During the daytime heat of summer, lake waters warm up more slowly than the air or land, which makes a summer swim cool and refreshing. When night comes, lakes cool more slowly than air or land, remaining pleasantly warm.

Water is called the universal solvent because of its ability to dissolve other substances. So many substances are able to dissolve in water that even freshwater contains many impurities. As water moves over and through the soil, minerals dissolve in it. (See Table 4-1.) The type and quantity of impurities found in water affect the types of living things that can exist there.

Oxygen, carbon dioxide, and other gases are absorbed from the atmosphere through the surface of the water. Cold water can hold more dissolved gases than warm water can. Dissolved oxygen and carbon dioxide are necessary for the existence of aquatic life.

Turbulence describes the degree of movement of water. The more turbulent a body of water is, the more surface area is exposed to the atmosphere, and thus the more oxygen and other gases that are absorbed. Fast-flowing rivers are rich in dissolved oxygen; the still waters of stagnant lakes and ponds contain little dissolved oxygen. The amount of oxygen affects the type and numbers of aquatic organisms that are present.

From the atmosphere and lithosphere, water also picks up solid particles that cannot be dissolved. These undissolved particles, called sediment, may sink to the bottom or remain suspended in the water. The more turbulent a body of water is, the more suspended particles of sediment it is likely to contain. As rivers slow down, they deposit these sediments, which form deltas, alluvial deposits, and floodplains.

TABLE 4-1 COMMON MINERALS OFTEN FOUND DISSOLVED IN WATER

Mineral	Common Name
Sodium chloride	Table salt
Magnesium sulfate	Epsom salt
Calcium carbonate	Limestone
Calcium sulfate	Gypsum
Sodium bicarbonate	Baking soda

The amount of suspended particles in the water affects how far sunlight penetrates through the water. Sunlight does not penetrate all the way to the bottom of a lake or sea. The average depth of light penetration is about 50 to 100 meters. When waters are murky with suspended sediments, sunlight cannot penetrate at all. Green plants and algae need sunlight to carry on photosynthesis; therefore they live at or near the water's surface. The lack of sunlight affects the living things of the whole ecosystem.

Seawater

On average, about 3.5 percent of the weight of seawater is in the form of dissolved salts. About 2.7 percent is sodium chloride, or table salt. Most of the remaining 0.8 percent consist of other salts. (See Table 4-2.)

Seawater does not freeze at 0°C. The freezing point of seawater depends on its **salinity**, the measure of the amount of salt present in water. The higher the percentage of salt, the lower the freezing point. Some bottom waters may be cold as –2°C. Most of the oceans' waters are colder than 7°C. Warmer water temperatures occur only in a narrow surface layer in the oceans and in shallow bays, lagoons, and estuaries.

The salt in seawater creates living conditions that are different from those in freshwater. Seawater becomes more dense as it cools, until it reaches its freezing point. These subfreezing temperatures can be hostile to life. Some fish that exist in this region produce an "antifreeze" in their blood to protect themselves from freezing.

Salinity affects the density of seawater and the types of life-forms

TABLE 4-2 SALTS FOUND IN SEAWATER

Salt	Percent of Total Salt Content
Sodium chloride	67.0
Magnesium chloride	14.6
Sodium sulfate	11.6
Calcium chloride	3.5
Potassium chloride	2.2
Other salts	1.1

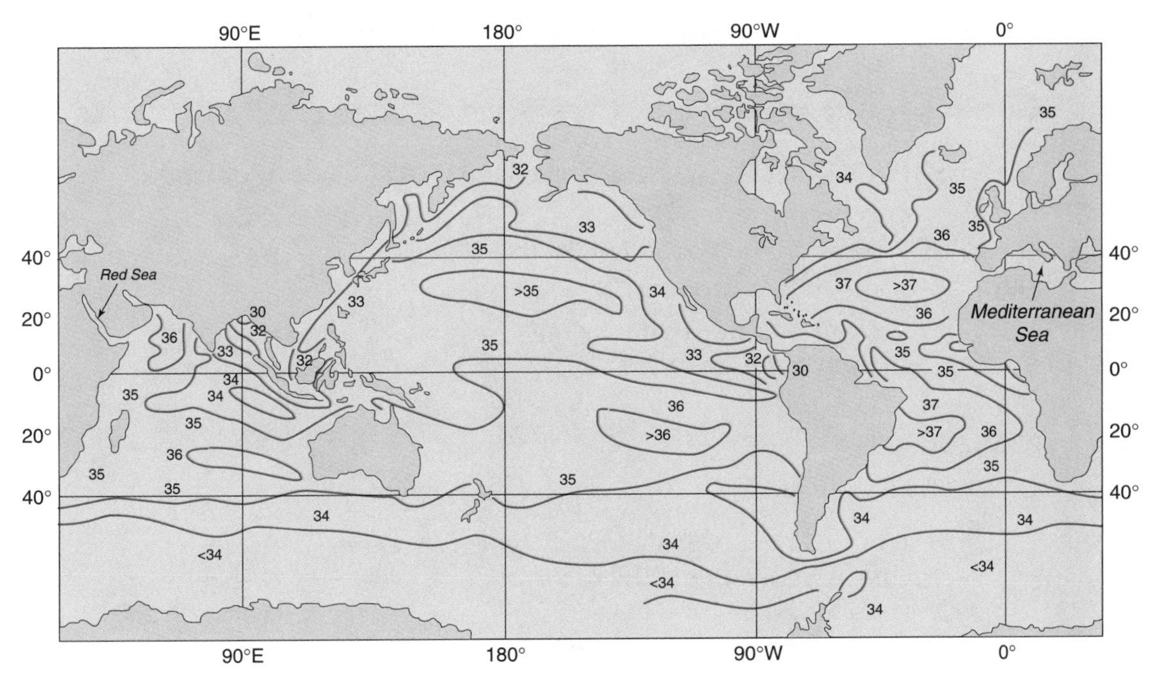

Figure 4-12 The numbers on this map give the salinity of ocean waters in parts per thousand.

that survive there. The higher the percentage of salt, the more dense the water. The density of the water affects its **buoyancy**, the upward force water exerts on floating objects. Salt water exerts greater buoyancy than freshwater does. Floating is easy in Utah's Great Salt Lake because the high salinity makes the buoyancy of the water high. In estuaries where fresh and salt waters meet, the freshwater, which is less dense, often floats on top of the denser seawater. Figure 4-12 shows the salinity of the oceans and some seas.

4.5 Section Review

1. How do hydrogen bonds affect the properties of water?
2. Explain why lake waters remain cooler than the air or land during the summer months.
3. Why is water called the universal solvent?

LABORATORY INVESTIGATION 4

Demonstrating Wave Action on a Shoreline

PROBLEM: *How does wave action affect a shoreline?*

SKILLS: *Manipulating, observing*

MATERIALS: *Dissecting tray or pan, water, clear plastic wrap (heavy), marker, sand, gravel, paper towels*

PROCEDURE

1. Half fill a dissecting tray or 30-centimeter baking pan with water.
2. Gently lift one end of the pan 2 to 3 centimeters and then slowly let it down so that a wave forms in the pan. Notice how the wave changes direction each time it reaches the end of the pan. Repeat this activity. Note how many times the wave moves back and forth across the pan.
3. Pour sand into one end of the pan to form an 8-centimeter-wide beach and shoreline.
4. Place a sheet of plastic wrap over the pan. Use the marker to trace the shoreline. Carefully remove the plastic wrap.
5. Create wave action by gently raising and lowering one end of the pan two or three times.
6. Place your shoreline tracing over the pan again. Note any change in the beach and shoreline. Label this tracing "sand beach." Follow your teacher's directions to empty the pan.
7. Repeat Steps 3 through 6 using pebbles or gravel instead of sand. Label this tracing "pebble beach."
8. Repeat steps 3 through 6 using a mixture of sand and pebbles or gravel. Label this tracing "mixed sand and gravel beach."

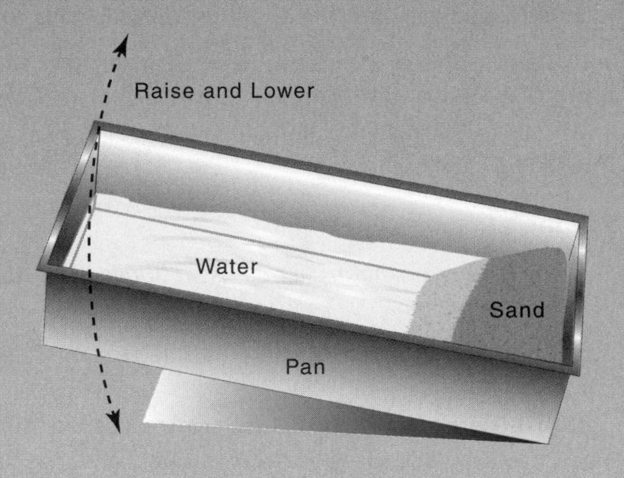

OBSERVATIONS AND ANALYSES

1. What effect did the wave action have on the sand beach?
2. How did wave action affect the gravel or pebble beach?
3. How did wave action affect the beach composed of sand and gravel?
4. Which were moved more by the wave action, the sand particles or the gravel particles? Which forms a more permanent beach? Why?
5. Using your tracings as a guide, draw a diagram of your demonstration. Label the beach and shoreline.

Chapter 4 Review

Answer these questions on a separate sheet of paper.

Vocabulary

The following list contains all the boldfaced terms in this chapter.

air pressure, buoyancy, climate, continental climates, currents, humidity, hydrogen bonds, maritime climates, ozone, photosynthesis, precipitation, salinity, sea, topography, waves, weather, windward

Fill In

Use one of the vocabulary terms listed above to complete each of the following sentences.

1. The short-term conditions in the troposphere are called ______.
2. A measure of the moisture content of the air is called ______.
3. Climates modified by the oceans are called ______.
4. ______ are rhythmic motions of the oceans caused by winds.
5. The upward force water exerts on floating objects is called ______.

Multiple Choice

Choose the response that best completes the sentence or answers the question.

6. The most abundant gas in the atmosphere is *a.* oxygen. *b.* nitrogen. *c.* carbon dioxide. *d.* ozone.
7. The average air pressure at Earth's surface is *a.* 14.7 atmospheres. *b.* 10 atmospheres. *c.* 5 atmospheres. *d.* 1 atmosphere.
8. Which of the following is not a major sea? *a.* Caribbean *b.* Mediterranean *c.* Bering *d.* Indian
9. The equatorial region experiences *a.* variation in solar radiation. *b.* little rainfall. *c.* a warm, moist climate. *d.* a warm dry climate.
10. When mountain ranges disrupt the movement of moist air masses, *a.* rain falls on the windward side of the mountain. *b.* rain falls on the leeward side. *c.* the leeward side is dry. *d.* *a* and *c* are correct.
11. In which atmospheric layer do the northern lights occur? *a.* troposphere *b.* stratosphere *c.* ionosphere *d.* exosphere

12. In which layer of the atmosphere does weather occur?
    a. troposphere *b.* stratosphere *c.* ionosphere
    d. exosphere
13. A destructive tropical storm that has winds in excess of 125 kilometers per hour is called a *a.* tornado. *b.* blizzard. *c.* hurricane. *d.* thunderstorm.
14. Surface tension in water is the result of *a.* buoyancy. *b.* hydrogen bonds. *c.* heat capacity. *d.* turbulence.
15. Precipitation is *a.* fog. *b.* sleet. *c.* dew. *d.* frost.

Short Answer (Constructed Response)

Use the information you learned in this chapter to respond to the following items.

16. How is the climate on the western slope of the Sierra Nevada Mountains different from that on the eastern slope?
17. How is the biosphere protected from the harmful effects of ultraviolet light?
18. Describe some of the various forms of precipitation.
19. How is the moisture content of the air measured?
20. What is the cause of air pressure?
21. Explain the differences between maritime and continental climates.
22. What type of air mass is associated with thunderstorms?
23. What are the causes of ocean currents?
24. What is the salinity of salt water?

Essay (Extended Response)

Use the information you learned in the chapter to respond to the following items.

25. Explain the effect of an increase in humidity on air pressure.
26. What factors affect climates?
27. Why is a large portion of Earth's freshwater not available for use in agriculture or as drinking water?
28. The Norse called the large island in the North Atlantic Ocean, Greenland. The island is now covered with ice and shows little sign of any greenery. Discuss how this may be related to changes in Earth's climate.

CHAPTER 5
Earth Through the Ages

When you have completed this chapter, you should be able to:

Distinguish between relative and absolute time.

Explain the significance of index fossils.

Identify the eras in geologic time.

Earth's crust is made up of layers that geologists can read like the pages of a book. These rock layers contain information that covers Earth's 4.5-billion-year history. The layers contain the remains of ancient organisms that once inhabited Earth's biosphere. The remains provide clues to Earth's past. Based on these clues, scientists can reconstruct Earth's history and trace the development of the biosphere.

5.1 WRITTEN IN STONE

Since the beginning of geologic time, processes have been at work creating vast changes in Earth's surface. Prolonged rains filled in the ocean basins. Continents formed, drifted apart, collided, and split apart to drift away again. The shifting landmasses created mountains, altered climates, formed deserts, and produced new environments. Living things responded to these changes with waves of extinctions followed by bursts of evolution. These turning points in Earth's history have been identified, categorized, and charted in sequential order to produce a geologic time scale.

Geologists are able to determine the age of many of the rocks that make up Earth's crust through the use of various scientific processes. The results are used to place the events of the past in sequential order. By establishing the relative ages of various rock strata, their age of formation and place in geologic time can be determined. The ages are usually confirmed using several different testing methods.

The time during which events occurred can be indicated in two ways. **Relative time** puts events in sequence but does not give their actual age. **Absolute time**, or measured time, identifies the actual date an event occurred and establishes its absolute age. Through continuous testing, the approximate dates of most of the divisions that make up geologic time have been established.

Superposition

One method of determining the relative ages of the strata, or layers, in sedimentary rocks is by their position. Sediments are deposited underwater in horizontal layers, and over time, temperature and pressure change sediments into rocks. Since the bottom layers are deposited first, they are the oldest layers; succeeding layers are progressively younger. This is the principle of **superposition**.

At times, geological processes cause exceptions to this rule. Folding can cause rock strata to be overturned, and thus older layers end up above younger layers. Faulting also may thrust older layers above younger ones.

Figure 5-1 To determine whether rock strata in different locations formed at the same time, geologists use the process called correlation.

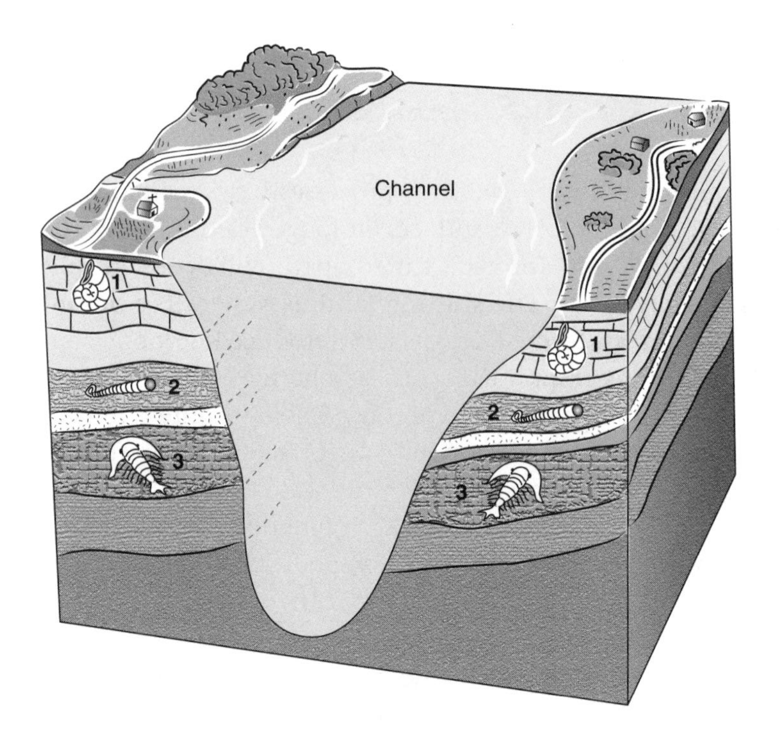

When lava flows onto Earth's surface and cools, it forms a feature called an igneous extrusion. The extrusion is younger than the rocks below it but is older than any rocks that may later form above it.

Magma may be injected into older rock strata, where it cools and solidifies. This rock feature is known is an igneous intrusion. Intrusions are younger than all the rock layers they pass through.

Geologists use a process called **correlation** to match the rock strata in different locations to determine if they formed at the same time. (See Figure 5-1.) Following an exposed rock layer or rock formation throughout its length is called correlation by continuity. The properties of rocks in one location are often compared with the properties of the rocks in other locations. Sequencing the types of strata found in two locations can also be used to determine correlation. Some volcanoes erupt explosively, covering a large area with a layer of ash. These ash layers allow geologists to make accurate correlations throughout the region covered by the ash fall. The ash fall is always younger than the rocks it covers.

Radioactive Dating

The most important process used to determine absolute time and absolute age is radioactive dating. **Radioactive dating** uses the rate of decay of naturally occurring, radioactive isotopes of elements to determine the absolute age of rocks and fossils. Isotopes are atoms of the same element that have different atomic masses. The nuclei of radioactive isotopes are unstable and break down to form lighter elements. The rate of disintegration of these isotopes remains constant through time. This rate is called the half-life.

The **half-life** is the period of time it takes for half the atoms in a radioactive sample to break down into simpler, stable atoms. Figure 5-2 illustrates the half-life of a radioactive isotope. Every radioactive isotope has a particular half-life. Potassium-40 has a half-life of 1.3 billion years, uranium-235 has a half-life of 713 million years, and carbon-14 has a half-life of 5700 years. The length of the half-life determines the type of material the isotope can be used to date. The age of a sample can be estimated by determining the ratio between the amount of radioactive isotope and the amount of decay product present in the sample.

The half-life of uranium-238 is 4.5 billion years. At the end of 4.5 billion years (one half-life), half of the atoms of uranium-238 in a sample have broken down to form lead-206, and half remain unchanged. At the end of another 4.5 billion years (two half-lives), half of the previously unchanged uranium-238 has broken down. Therefore, one-fourth of the original uranium-238 remains, and three-fourths has become lead-206. After another 4.5 billion years (3 half-lives), only one-eighth of the original uranium-238 remains,

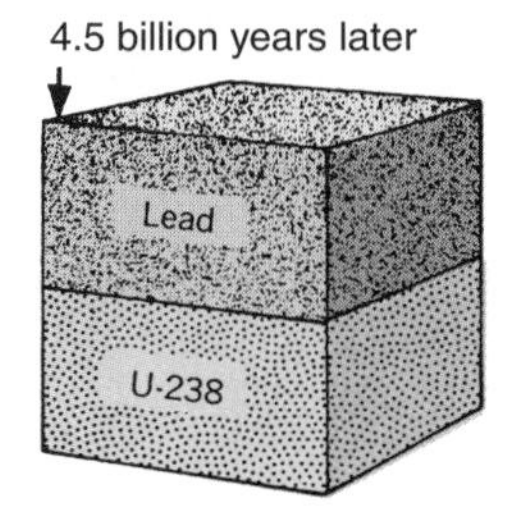

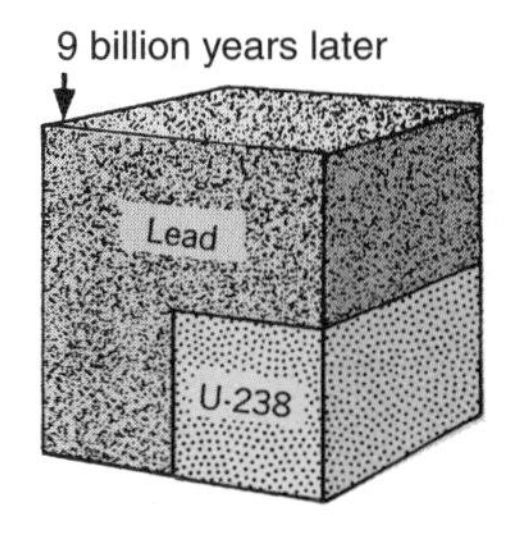

Figure 5-2 **One-half of the uranium-238 present in a sample decays to lead during its 4.5-billion-year half-life.**

TABLE 5-1 HALF-LIVES OF RADIOACTIVE ISOTOPES USED TO ESTABLISH THE AGE OF ROCKS AND FOSSILS

Element	Half-Life
Rubidium-87	50.00 billion years
Thorium-232	13.90 billion years
Uranium-238	4.51 billion years
Potassium-40	1.30 billion years
Uranium-235	713 million years
Carbon-14	5770 years

and seven-eighths has been changed to lead-206. The ratio between uranium-238 and lead-206 in a sample is a measure of its age.

Radioactive isotopes with a long half-life are used to date most rocks because rocks are millions or billions of years old. During these long periods, isotopes with a short half-life would have decayed to such an extent that the amount remaining would be too small to measure accurately.

Elements with a shorter half-life are used to date the remains of organic materials and relatively young rocks. Carbon-14 is an accurate tool for dating materials up to 50,000 years of age. This process is called **radiocarbon dating**. The half-lives of several common isotopes are shown in Table 5-1.

Index Fossils

Fossils are naturally preserved remains or impressions of living things. They are generally found in sedimentary rock. Since igneous and metamorphic rocks are formed under intense pressure and extremely high temperatures, any fossils present are usually deformed or destroyed. The hard parts of organisms, such as bones, teeth, and shells, are the most common structures that are preserved. The original chemical structure is usually replaced by dissolved minerals from groundwater. Imprints of whole organisms, as well as footprints and tracks, are often preserved as impressions in hardened mud. Fossils can often be used to determine the absolute age of the rocks in which they are found.

Index fossils are the remains of organisms that existed for a brief period of geologic time and had a wide geographic distribution. Since their age can be determined, their presence is used to establish the absolute age of the rock strata in which they are found.

Trilobites were crablike animals, whose remains are used as index fossils. They flourished during the middle Cambrian Period; then they became extinct. Sediments laid down during the mid-Cambrian often contain trilobite fossils. Radioactive dating of these rocks provides an absolute age of between 500 million and 600 million years. Any rocks that contain trilobite fossils can be inferred to be of this age. Graptolites are ancient marine organisms that lived in colonies between 350 million and 450 million years ago. Their remains are also used as index fossils.

5.1 Section Review

1. Why is radiocarbon dating not used to determine the age of most rock formations?
2. Describe some of the exceptions that may make relative dating by superposition confusing.
3. Develop a theory to explain why rocks that represent the first 600,000 years of geologic time have never been found.

5.2 GEOLOGIC TIME

Earth's history of more than 4.5 billion years has been divided into units of geologic time called eras, periods, and epochs. These are arbitrary divisions that are not equal in length. The largest divisions of geologic time are called **eras.** There are four eras. The boundaries between eras mark extreme changes that occurred in Earth's surface or climate. These changes were wide in scale and may have taken millions of years to occur.

Eras are divided into periods. **Periods** are blocks of geologic time characterized by the appearance, disappearance, or dominance of various life-forms in the fossil record. Periods are further divided into shorter intervals called epochs. The biospheric time scale traces the evolution of the biosphere through geologic time.

Early geologists observed that rock formations could be identified by the fossils they contained. They noted that certain formations were always located directly above or below other formations. From these observations, they were able to establish a relative time scale and a sequence of fossil groups from oldest to youngest. Some divisions of the geologic time scale are named for the location where the fossils characteristic of that time were first observed in the rocks. For example, fossils commonly found in rock formations from Pennsylvania were said to have lived in the Pennsylvanian Period.

The following discussion represents the current theories of how our planet changed and life developed. New data may cause scientists to modify these theories.

Precambrian Era

The **Precambrian Era** is the earliest division of geologic time. It lasted 4 billion years, which is 87 percent of geological time. Rocks formed during this era can still be found at the core of the continents. Early geologists didn't find any fossils in Precambrian rocks, so they concluded that this was the time before life began.

However, later scientists discovered evidence of fossil algae, called stromatolites, in Precambrian rocks that are more than 2.5 billion years old. Older, soft-bodied organisms, which left no fossils, must have existed before this. Therefore, the biosphere is thought to be about 3.8 billion years old and must have begun early in the Precambrian Era.

Scientists think that the Precambrian was a time of widespread volcanic activity. There is evidence that there were extensive plate movements and several different periods of mountain building. The oceans and seas formed and new life-forms, including bacteria, algae, jellyfish, corals, and clams, evolved in the waters. During this era, photosynthetic organisms enriched the atmosphere with oxygen, which set the stage for more complex life-forms that followed.

Paleozoic Era

The Precambrian Era was followed by the **Paleozoic Era**, which scientists calculate lasted for almost 350 million years. Scientists divide the Paleozoic Era into six geological periods: the Cambrian, Ordovician, Silurian, Devonian, Carboniferous, and Permian. In North America, the Carboniferous Period is often divided into the Mississippian Period and Pennsylvanian Period. Whereas Precambrian rocks contain few fossils, Paleozoic rocks have an abundance of fossil remains. Evidence indicates that during the Paleozoic Era many changes took place: the continents moved, climates changed, and many new life-forms appeared.

CAMBRIAN PERIOD

The early part of the Paleozoic Era is called the Cambrian Period. Scientists calculate that it lasted for 70 million years. Geologists theorize that there was little mountain building during this period. It is thought that the continents were close together, and many shallow seas covered the continents. These conditions allowed marine life to flourish. Fossils from this period are numerous and varied. The most common organisms were invertebrates, animals without backbones, including trilobites, sponges, worms, and brachiopods. (See Figure 5-3.) During this period there was an evolutionary explosion that created a wide variety of life-forms.

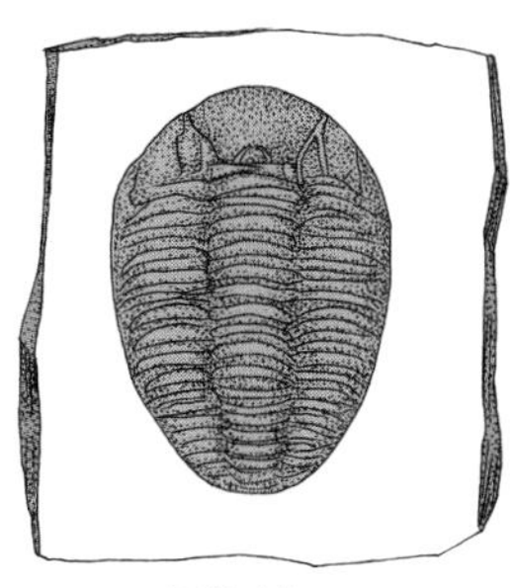

Figure 5-3 **Trilobites were abundant during the Cambrian Period.**

ORDOVICIAN PERIOD

The Ordovician Period lasted for 70 million years. Scientists think that a wide variety of marine organisms evolved during this time. The dominant organisms included graptolites, bryozoans, cephalopods (similar to the modern nautilus), gastropods (snails), and echinoderms (sea stars). Fossil evidence indicates that the first vertebrates appeared during this period. They were primitive, jawless fish, called ostracoderms. This was a period of extensive mountain building and volcanic activity. Geologists theorize that the northern landmasses were coming together, and a single landmass had formed in the south. North Africa lay over the South Pole.

SILURIAN PERIOD

The Silurian Period lasted for about 35 million years. The oldest coral reefs come from this time. Coral reefs indicate a warm climate, since they do not form in cool waters. Eurypterids (sea scorpions) were common marine invertebrates. (See Figure 5-4.) The most significant event in the late Silurian was the colonization of land, first

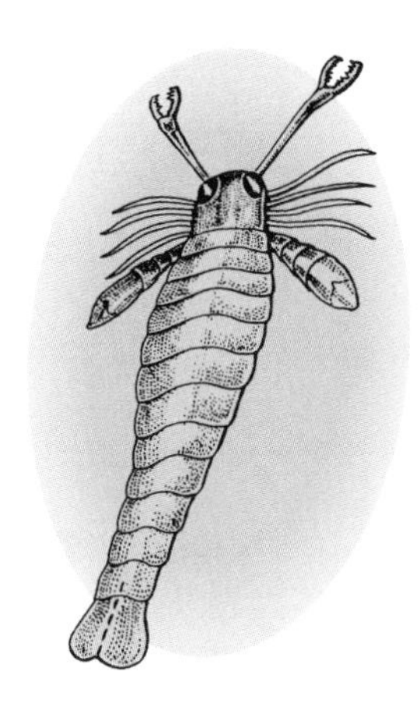

Figure 5-4 **Eurypterids, or sea scorpions, lived during the Silurian Period.**

by plants and then by the first air-breathing animals. Fossils of spiders, millipedes, and scorpions are found in Silurian rocks. Armored, jawed fish developed in the seas during the Silurian. Evidence indicates that the climate of what is now the northern United States became dry, and thick beds of rock salt and gypsum formed. The arrangement of the continents at this time is unknown.

DEVONIAN PERIOD

The seas of the Devonian Period were dominated by bony fish and sharks, which diversified and multiplied tremendously. The Devonian marked the emergence of lobed-finned fishes and lungfish. Scientists theorize that lungfish were the first air-breathing vertebrates. They were able to crawl out of the water onto land, so it is thought that they later gave rise to the first amphibians. This period is called the Age of Fishes. (See Figure 5-5.) It lasted for 50 million years.

The Devonian was a time of extensive reef building by corals and sponges. Giant ferns, horsetails, and primitive scale trees covered the swampy lands with the first forests. The first wingless insects, springtails and bristletails, appeared in these forests, along with spiderlike arachnids. Geologists theorize that the supercontinents Laurasia and Gondwanaland drew closer together. A collision between the North American plate and the African plate led to the uplift of the White Mountains in what is now Vermont.

CARBONIFEROUS PERIOD

The Carboniferous Period lasted for 65 million years. The climate of North America was warm and rainy, and the lowlands were covered with swampy areas. Amphibians, some as large as alligators, lurked in the shallows. The Carboniferous Period is called the Age of Amphibians.

Tree ferns, scale trees, and rushes filled the swamps with forests. As these plants died, they formed deposits of peat, which later changed

Figure 5-5 In the Devonian Period, lobe-finned fishes evolved.

into thick beds of coal. The extensive North American coal deposits originated during the Carboniferous. The first true land vertebrates, the reptiles, appeared later during this period. They were able to move into areas where amphibians could not live. Giant insects were common, including winged dragonflies, cockroaches, and centipedes. The Appalachian Mountains in North America and the Ural Mountains in Europe formed as these continents moved together.

PERMIAN PERIOD

The Permian Period, which lasted for 55 million years, marks the close of the Paleozoic Era. Corals thrived in the warm waters of the oceans and inland seas. A large-scale glaciation occurred in the Southern Hemisphere. Conifers appeared, and forests of cycads and ginkgoes were common. Reptiles took over from the amphibians to dominate the land. The *therapsids* combined the features of reptiles and mammals. (See Figure 5-6.) New, major insect groups emerged, such as beetles. Scientists theorize that as the continents formed a single giant supercontinent, Pangaea, the greatest mass extinction of all time occurred. Nearly half of the known animal groups became extinct at this time, as did seed ferns, scale trees, and primitive conifers.

Two important groups that survived this extinction were the cephalopods, such as squids and octopuses, and the reptiles. The reptiles that survived are of particular importance, because they were the ancestors of the dinosaurs and modern-day birds, snakes, crocodiles, turtles, lizards, and mammals.

Figure 5-6 The Lystrosaurus, a herbivore, had features of both reptiles and mammals.

Mesozoic Era

Scientists divide the **Mesozoic Era** into three periods—Triassic, Jurassic, and Cretaceous. This era lasted for 160 million years and ended 65 million years ago. Geologists theorize that during the early Mesozoic, Pangaea, which had formed at the end of the Paleozoic Era, broke into Laurasia in the north and Gondwanaland in the south. Laurasia eventually broke up into Eurasia and North America. Gondwanaland became Africa, Antarctica, Australia, India, and South America. Evidence indicates that the climate of the Mesozoic was mild. The poles were free of ice, and corals grew in what is now Europe. The surface temperatures of the oceans were 10°C to 20°C higher than they are today. Deserts occupied vast inland areas. The most common land vertebrates during the Mesozoic were the dinosaurs, a group of terrestrial reptiles. The Mesozoic Era is known as the Age of Reptiles.

Dinosaurs arose at the start of the Mesozoic, during the Triassic Period. The *Brachiosaurus* was one of the largest animals ever to walk the land. *Tyrannosaurus rex* was among the largest carnivores and was surely a fearsome animal. The dinosaurs shared the land with other reptiles, such as lizards and tortoises, while skin-winged pterosaurs dominated the skies. (See Figure 5-7.) The first mammals, small, shrewlike animals, appeared during the Jurassic Period.

Figure 5-7 Edmontosaurs browsed in a conifer forest during the Cretaceous Period of the Mesozoic Age.

Archaeopteryx and *Protoavis,* the oldest known birds, also made their appearance. Ammonites, a type of cephalopod, filled the seas. These were preyed on by marine reptiles, such as the ichthyosaurs.

Plant life during the early Mesozoic included giant tree ferns, rushes, and conifers such as pines, yews, and cypresses. The first flowering plants, or angiosperms, appeared during the Cretaceous and became the most abundant plants. Chief among the flowering plants were deciduous trees, which shed their leaves. They spread rapidly until they were the dominant plants in most of the forests. The early Cretaceous forests consisted of magnolia, willow, fig, and tulip trees; later came oak, maple, beech, and chestnut.

The Mesozoic Era closed with another great mass extinction. It is known as the K-T extinction, because it occurred at the boundary between the Cretaceous Period and the Tertiary Period. All of the dinosaurs vanished, along with the flying reptiles and marine reptiles. Many other species of plants and animals also disappeared.

The discovery of iridium in the thin layer of rocks that formed at this time has led to some interesting speculation. Iridium is not abundant in Earth's crust, but high concentrations are found in extraterrestrial objects such as meteorites. Scientists suggest that the iridium found in the boundary rocks came from the impact of a large meteorite or comet. The dust from the impact would have filled the atmosphere and blocked out sunlight for several years. Plants, which require sunlight, would have died first, then animals that fed on plants, and finally the meat-eating animals.

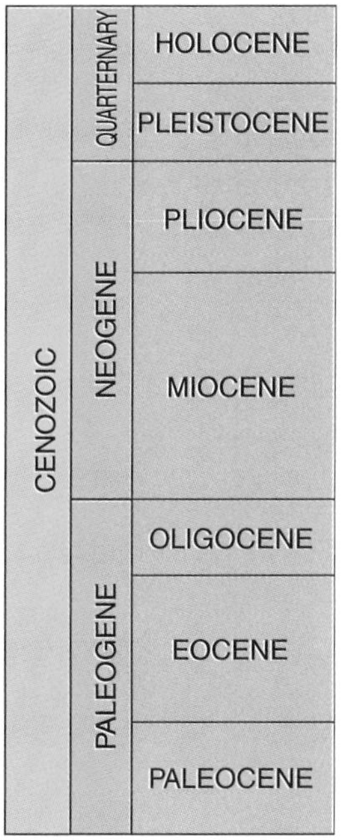

CENOZOIC	QUARTERNARY	HOLOCENE
		PLEISTOCENE
	NEOGENE	PLIOCENE
		MIOCENE
	PALEOGENE	OLIGOCENE
		EOCENE
		PALEOCENE

Figure 5-8 The International Committee on Stratigraphy, which has the responsibility for setting geologic names and dates, recently eliminated the Tertiary from the Geologic Time Scale.

Cenozoic Era

The **Cenozoic Era** is the most recent era, the one we live in. (See Figure 5-8.) It began 65 million years ago. Geologists theorize that plate movements, which began in the Mesozoic with the breakup of Pangaea, carried the continents to their present locations and set the limits of the present-day oceans. Many of today's mountain ranges were uplifted during this era. Scientists think that during the Cenozoic there was a general cooling and drying of Earth's climate that caused periods of worldwide glaciation called **ice ages**.

Some scientists divide the Cenozoic Era into two periods, the Tertiary Period and the Quaternary Period. Others divide the Cenozoic

Era into the Paleogene, Neogene, and Quaternary periods. The Paleogene Period is divided into the Paleocene, Eocene, and Oligocene Epochs. The Quaternary covers the past 1.8 million years and is divided into two epochs, the Pleistocene, or ice age, and the Holocene, or Recent. The Recent covers the last 11,000 years, since the end of the last ice age.

The Pleistocene Epoch started 1.8 million years ago. Four times during this epoch, glaciers covered large areas of the Northern Hemisphere. (See Figure 5-9.) Each time they formed an ice sheet that was between 2500 and 3000 meters thick. Each of the four ice ages ended as Earth's climate warmed and the glaciers that made up the ice sheet retreated northward. The last retreat ended about 11,000 years ago. The movement of the glaciers affected Earth's cli-

Figure 5-9 **This illustration shows the extent of glaciation during the last ice age.**

mate and geological features, leading to the evolution of new life-forms in the biosphere. There is evidence to suggest that Earth has been experiencing recurring ice ages for more than 600 million years.

The Cenozoic is called the Age of Mammals, since mammals are the dominant, large, land-dwelling life-forms. The tiny mammals that survived the K-T extinction evolved. Most of the mammals that first appeared in the Cenozoic are now extinct. (See Figure 5-10.)

Figure 5-10 Mammoths and saber-toothed cats lived in the Cenozoic Era.

However, some evolved to become the mammals we are familiar with today. Many of the early birds were large and flightless, resembling the modern ostrich and emu. As the climate cooled during this era, grasslands replaced forests. This led to the explosive evolution of grazing mammals such as horses, cattle, and antelope.

During the Cenozoic, our human ancestors appeared. Scientists think that, over time, these ancestors evolved into modern humans. Human evolution is linked in many ways to the changes in climate that occurred during the Cenozoic. As ecosystems changed, humans evolved to meet the new conditions they had to face. Today it is human activity that is responsible for many changes occurring in the biosphere.

5.2 Section Review

1. Describe how the climate of the Carboniferous Period contributed to the formation of the coal beds that are mined today.
2. Explain the relationship between iridium concentrations in the K-T boundary and theories regarding the causes of mass extinctions.
3. Construct and illustrate a time line that indicates how long each of the following groups of living things existed on Earth: algae, sponges, cephalopods, ferns, bony fish, amphibians, reptiles, insects, dinosaurs, mammals, birds, flowering plants, humans. Indicate the era, period, or epoch in which the organisms first appeared and, as applicable, when they became extinct.

LABORATORY INVESTIGATION 5

Radioactive Dating and Half-Life

PROBLEM: *How can half-life be represented?*

SKILLS: *Manipulating, observing*

MATERIALS: *100 pennies, shoe box, 100 washers*

PROCEDURE

1. First trial: Put 50 pennies, head side up, into the shoe box. Place the cover on the box. The pennies represent atoms, and the shoe box represents a sample of radioactive isotope.
2. Shake the box so that some of the pennies flip over. Open the shoe box. Consider those pennies that are tails side up to be atoms that have decayed. Count the "decayed" pennies and record the number. Remove these pennies from the shoe box and replace them with washers.
3. Repeat steps 1 and 2 until all the pennies have decayed.
4. Try the procedure again, starting with 100 pennies.

OBSERVATIONS AND ANALYSES

1. What percent of the pennies decayed after each trial?

$$\text{percent decayed} = \frac{\text{number decayed}}{\text{total in box for trial}} \times 100\%$$

2. Why did you replace the "decayed" pennies with washers?
3. Why does the time it takes to complete each trial represent the half-life of the radioactive isotope?
4. How many trials were needed for all the pennies to decay?
5. If the half-life is 5 minutes, what is the age of the shoe box when all the pennies have decayed?

Chapter 5 Review

Answer these questions on a separate sheet of paper.

Vocabulary

The following list contains all the boldfaced terms in this chapter.

absolute time, Cenozoic Era, correlation, eras, half-life, ice ages, index fossils, Mesozoic Era, Paleozoic Era, periods, Precambrian Era, radioactive dating, radiocarbon dating, relative time, superposition

Fill In

Use one of the vocabulary terms listed above to complete each of the following sentences.

1. Using radioactive isotopes to determine the age of rocks is called ________.
2. The ________ of an isotope is the time it takes for half the atoms in a sample to break down.
3. The only era in which there was a time during which no life existed was the ________.
4. Humans first appeared during the ________.
5. Periods of worldwide glaciation are called ________.

Multiple Choice

Choose the response that best completes the sentence or answers the question.

6. The most abundant plants during the Cretaceous Period were *a.* angiosperms. *b.* conifers. *c.* cycads. *d.* ferns.
7. In which period did vertebrates first appear? *a.* Cambrian *b.* Ordovician *c.* Silurian *d.* Devonian
8. There is evidence to indicate that Earth has been experiencing ice ages for about *a.* 600 thousand years. *b.* 6 million years. *c.* 60 million years. *d.* 600 million years.
9. Earth's age is estimated to be about *a.* 4.5 million years. *b.* 45 million years. *c.* 4.5 billion years. *d.* 45 billion years.
10. *Archaeopteryx* and *Protoavis* were early *a.* dinosaurs. *b.* mammals. *c.* fishes. *d.* birds.
11. The two periods of the Cenozoic Era are the *a.* Mississippian and Jurassic. *b.* Triassic and Jurassic. *c.* Tertiary and Quaternary. *d.* Permian and Devonian.

12. Which animal group benefited most by the change from trees to grasses that occurred during the Cenozoic? *a.* grazing mammals *b.* amphibians *c.* carnivorous dinosaurs *d.* flying reptiles
13. Scientists think the ancestors of the first amphibians were *a.* dinosaurs. *b.* flying reptiles. *c.* lungfish. *d.* scorpions.
14. Which era is known as the Age of Reptiles? *a.* Precambrian *b.* Paleozoic *c.* Mesozoic *d.* Cenozoic
15. Which era was divided into six geologic periods? *a.* Precambrian *b.* Paleozoic *c.* Mesozoic *d.* Cenozoic

Short Answer (Constructed Response)

Use the information you learned in this chapter to respond to the following items.

16. What are the differences between absolute and relative time?
17. How are radioactive isotopes used to determine the age of rocks?
18. Explain the principle of superposition.
19. How is correlation used to determine the age of rock strata?
20. What are index fossils?
21. Large deposits of coal formed during which geologic period?
22. When did dinosaurs first appear in the geologic record?
23. When did the last ice age end?
24. Which animal group disappeared following the K-T extinction?
25. What type of life dominated the Devonian Period?

Essay (Extended Response)

Use the information in the chapter to respond to these items.

26. Why do Precambrian rocks contain few fossils?
27. Explain how fossils can be used to determine past climates.
28. What became of Pangaea?
29. Visit a natural history museum and look at the dinosaur and other ancient vertebrate skeletons. Choose five organisms and show how they survived in their environment.

UNIT TWO

THE LIVING PLANET

Earth is the only planet in our solar system and possibly our galaxy that supports intelligent life. On Earth, living things are found in some unexpected places, for example, near the tops of mountains, in deep ocean waters, in deserts, in the cold regions around the North and South poles, and even deep under the ground. Of course, living things are also found in more familiar places such as lakes, meadows, farms, parks, and cities.

The living planet is home to a wide variety of creatures. There are one-celled protists, giant redwood trees, and elephants. We are most familiar with those creatures that live with us: house plants, pets, and other people. In this unit, you will learn about the web of life and what all living things need to survive.

CHAPTER 6
Life on Earth

When you have completed this chapter, you should be able to:

Define biotic environment and abiotic environment.

Compare plant cells and animal cells.

Distinguish between producers and consumers.

Because all living things are made of cells, they carry out similar life processes and have similar needs. All of Earth's varied creatures depend on one another. For example, flowers depend on insects for pollination. People depend on other animals and plants for clothing and food. However, we often forget that we depend on the rain forests in Brazil and the grasses of the prairies. We must be concerned about their survival, too.

6.1 THE BIOSPHERE

Scientists think that Earth is about 4.5 billion years old. To learn about the ancient Earth, scientists study rocks and fossils, the remains of ancient life. Fossil evidence from ancient rocks indicates that bacterialike organisms were present about 3 billion years ago. Fossil stromatolites, mats produced by cyanobacteria, show that these organisms first appeared about 2.5 billion years ago. **Cyanobacteria** are a type of blue-green bacteria that make their own food. Since cyanobacteria are relatively complex organisms, other, simpler organisms must have existed before them. Scientists estimate that life on Earth first appeared in the sea about 3.8 billion years ago. Over the course of time, Earth and the creatures that live on it have changed.

Now, living things can be found almost everywhere on Earth. Life can exist even under the most harsh conditions. In the extreme cold of Antarctica, cyanobacteria have been found between the sand grains of sedimentary rock. In addition, emperor penguins lay eggs and raise their young in the freezing cold of the Antarctic ice shelf. Fish, shellfish, tube worms, and other sea creatures inhabit the cold, black depths of the oceans, where pressures are great enough to crush a submarine. The hot, dry deserts are home to lizards and mice that never drink water. They get all the water they need from the foods they eat. Drilling operations have even found bacteria living within the pores in rocks buried many kilometers below Earth's surface.

The Environment and the Biosphere

Everything that surrounds a living thing is its environment. The environment is divided into two parts. The **biotic environment** is everything in an organism's surroundings that is or was alive, such as plants and animals. The **abiotic environment** is composed of everything in an organism's surroundings that is not alive, including soil, sunlight, precipitation, atmospheric gases, sources of water, dissolved mineral salts, and geologic formations.

The part of Earth where there are living organisms is called the biosphere. The biosphere can be thought of as a thin skin, like the peel of an apple, that covers Earth's surface. The biosphere is thickest at the oceans, where it extends from the deepest ocean floor up into the atmosphere. At the poles, where life exists on the surface of the ice and to a depth of only several centimeters below it, the biosphere is the thinnest. On the continents, life is found mainly in the space between the land's surface and the tops of the trees. Living things may also be found in the top several meters of soil and in underground caves. Recently, microorganisms have been found living in rock more than 2.8 kilometers below Earth's surface. In the atmosphere, most life remains close to Earth's surface, but some insects have been found at altitudes as high as 9 kilometers.

Earth's Changing Climate

Figure 6-1 **Scientists theorize that dinosaurs, including *Allosaurus*, became extinct when they could not survive in the changing climate.**

Earth's climate is constantly changing, and these changes affect the whole biosphere. Meteorological processes occur within the atmosphere and affect Earth's surface. Changes in the sun's energy output can change climate. Changes in climate can cause ice ages, lead to the formation of deserts, or cause a rise or fall in sea level. Factors within the atmosphere cause weathering and erosion of Earth's surface. Changes in climate can lead to the extinction of existing species of plants and animals or create unique environments for new species to fill. (See Figure 6-1.)

Ice ages are recurrent cycles of worldwide cooling, also known as glacial periods. At these times, glaciers cover a large portion of Earth's surface. During the last glacial period, 10 to 20 thousand years ago, North America was inhabited by many species of animals that no longer exist. Giant cave bears, elephantlike mammoths, huge ground sloths, fierce saber-toothed cats, and herds of giant bison roamed the cool plains of the continent. (Look at Figure 5-9 again.) Scientists think that the climate warmed following this glacial period. These animals could not adapt to the changing conditions and died. The warmer climate produced different environments, and evolutionary changes led to the biota present today. The **biota** are the plants and animals in a region. Another school of thought is that these large animals may have been hunted to extinction by humans.

INTERACTIONS

Gaia—The Living Earth

The British scientist James Lovelock and the American biologist Lynn Margulis have proposed that Earth behaves as if it is an individual, self-regulating superorganism. They named this superorganism Gaia, after the ancient Greek goddess of Earth. They define Gaia, "the Living Planet," as a complex being that includes the biosphere, atmosphere, oceans, and soil. This does not mean that the entire Earth is alive, but it can be compared to a tree that is 99% dead bark and wood and 1% living tissue. The interactions of its living and dead parts and their interactions with the outside world keep the tree alive. Earth's living portion is sustained by sunlight, geothermal energy, and chemical energy and by the interactions among the living things and their interactions with Earth.

The temperature and composition of the atmosphere suggest that it is a product of, and is maintained by, the biosphere—especially by Earth's microorganisms. Scientists theorize that photosynthetic microorganisms in the sea first released oxygen into Earth's primitive atmosphere and now supply 70% of it. Oxygen is crucial to all life since it is necessary for plant and animal respiration. Oxygen also maintains the protective ozone layer. Carbon dioxide gas can trap heat within the atmosphere, and so it affects global temperatures. Photosynthesis by marine microorganisms regulates the amount of carbon dioxide in the air and thus maintains its ratio to the other atmospheric gases.

People have significantly changed Earth. We have sharply increased the amount of carbon, sulfur, and nitrogen oxides in the atmosphere. Toxins have been released into the air and water. Deforestation has reduced Earth's green cover. Yet conditions within the atmosphere show that it is not easily disturbed. Earth tries to heal the wounds inflicted upon it. But one day the control system may become overloaded and unable to handle the harm done to it. Environmentalists are trying to teach us how to work with Gaia.

Evidence suggests that climate changes have been occurring throughout Earth's history and will continue in the future. At present, we are between glacial periods, experiencing a warm interglacial period. But scientists predict that the ice sheets could return in about 20 thousand years as the glacial cycle continues.

Because Earth has been changing since its formation, the biosphere has also experienced constant change. But no matter how it may change, the biosphere is a self-renewing, self-sustaining system. The organisms in the biosphere interact with one another and with the environment. These interactions furnish the organisms with the materials necessary for life and maintain Earth's supply of resources.

6.1 Section Review

1. Contrast the thickness of the biosphere at the following locations:
   a. the Antarctic ice cap
   b. the Brazilian rain forest
   c. the Atlantic Ocean
2. How did scientists arrive at an age for the biosphere?

6.2 LIFE'S BUILDING BLOCKS

Individual organisms can vary greatly in size—from a huge blue whale or giant redwood tree to a microscopic bacterium or protozoan. Regardless of their size, all living things interact with one another and with the abiotic environment. To understand these interactions better, you must study them from the inside as well as from the outside.

Cells

All living things are composed of a basic, internal unit called the **cell**. Cells are the building blocks of life, and they are involved in all the activities carried out by an organism. New cells can arise only

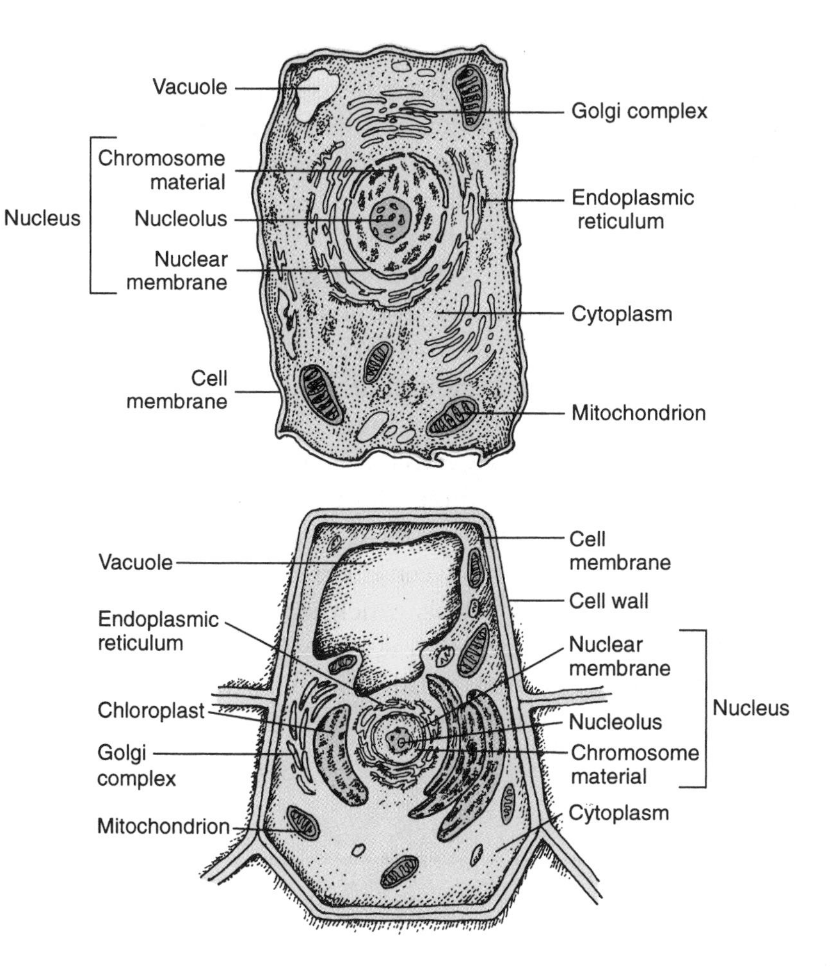

Figure 6-2 The typical features found in an animal cell (top) and a plant cell (bottom) are shown here.

from existing cells. Most cells reproduce by dividing in half, which produces two identical daughter cells.

Even though all living things are made up of cells, all cells are not the same. The cells found in plants differ from those found in animals. Plant cells are surrounded by a rigid cell wall that gives plants their structure. The cell wall is composed of cellulose. All cells have small structures called organelles. An organelle carries out a specific function within a cell. Most plant cells contain chloroplasts, organelles that aid in making food. Animal cells do not have a cell wall or chloroplasts. The centrosome, which aids in cellular reproduction, is an organelle that is in many animal cells but is not in plant cells. Figure 6-2 illustrates a plant cell and an animal cell.

All cells are filled with a semiliquid material called cytoplasm, a rich mixture of proteins and other materials used by the cell to carry out its functions. The organelles are suspended in the cytoplasm. The boundary between the cell and its environment is called the cell membrane. It is composed of proteins and fatty acids. The cell membrane acts as a gateway to regulate the materials that enter and leave the cell. Animal and plant cells have a cell membrane. The endoplasmic reticulum is the transport system within a cell. It is a network of channels within the cytoplasm that connects to the cell membrane.

Some other organelles are found in most plant and animal cells. The **nucleus** is the cell's control center. It sends out instructions in the form of chemical messages that keep the cell alive. The nucleus contains DNA, long, coiled molecules that carry inherited information in the form of genes. The nucleus is surrounded by the nuclear membrane, the boundary between the nucleus and the cytoplasm. The nuclear membrane controls which materials can enter or exit the nucleus. **Ribosomes**, which line the endoplasmic reticulum, are the sites of protein synthesis within the cell. Sausage-shaped **mitochondria** are the sites of energy production.

Life's Varied Forms

The building blocks of life combine in various ways to form the many kinds of living things you see on Earth. If you were to examine the living things in a drop of pond water with a microscope, you would see thousands of individual organisms of many different varieties. (See Figure 6-3.) There would be **unicellular**, or one-celled organisms, such as paramecia (paramecium is the singular form), amebas, or euglenas, and **multicellular**, or many-celled organisms, such as rotifers or daphnias. Some organisms, such as spirogyra, produce their own food. Others, such as the ameba, prey on other organisms. Each organism has its own particular way of life, yet each lives in a delicate balance with every other organism and with the abiotic factors in its environment.

In multicellular organisms, including people, cells differ based on the function they perform. Waferlike red blood cells carry oxygen throughout an animal's body. Muscle cells stretch and contract to aid an organism in movement. Nerve cells, called neurons, have

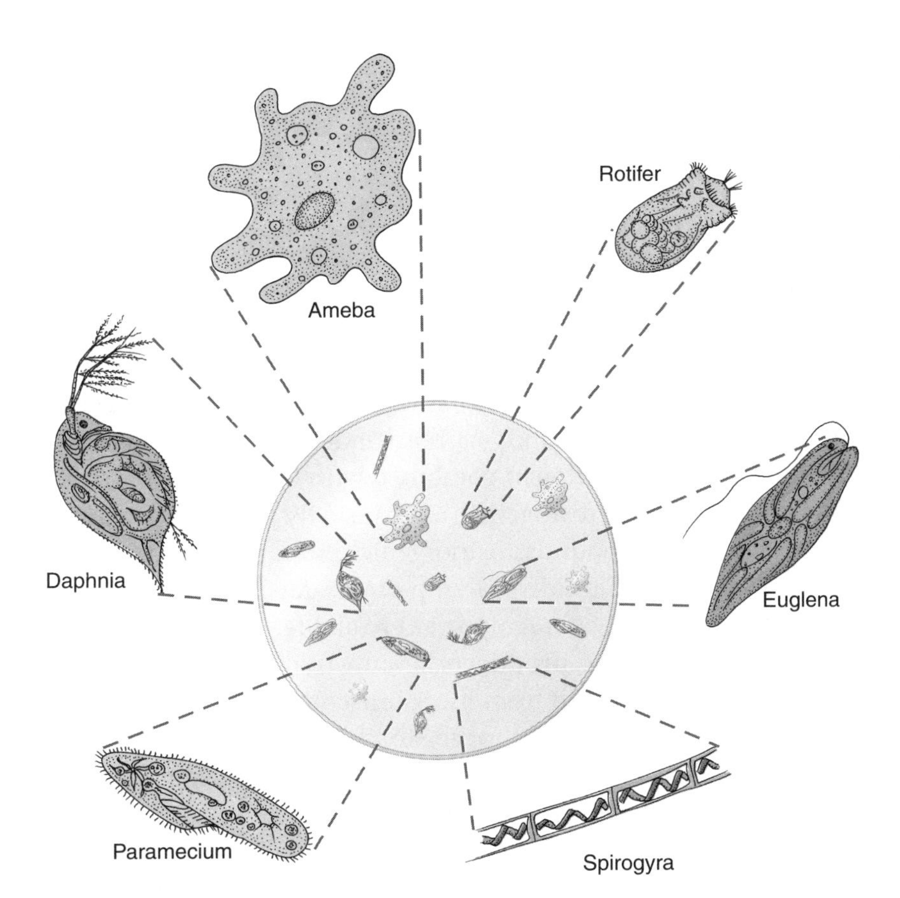

Figure 6-3 **You can observe a community in a drop of pond water if you use a microscope.**

long fibers and spreading branches that connect with other neurons and carry messages in the form of nerve impulses. Flat epithelial cells act as internal and external linings.

Figure 6-3 shows several different types of organisms. Each type represents a separate species that contains many similar individuals. A species consists of similar organisms that can interbreed and produce fertile offspring. The amebas, paramecia, cats, dogs, and humans are examples of species. Organisms within a species may differ in appearance, but all have similar genetic makeup and traits.

6.2 Section Review

1. List some of the organelles that are common to both plant and animal cells.
2. How are plant cells different from animal cells?

3. Read more about the ameba, paramecium, or euglena. From the one-celled creature's point of view, write a story about a day in its life.

6.3 LIFE'S VARIED ROLES

Through photosynthesis, green plants, including algae, use energy from the sun to change inorganic compounds (carbon dioxide and water) into organic compounds (carbohydrates). Organic compounds contain carbon and hydrogen combined with other elements, such as oxygen, sulfur, and nitrogen, and are usually produced by living things. All other compounds are inorganic compounds, and most do not contain carbon. Carbohydrates, proteins, fats, and oils are some organic compounds that are important to all life.

Green plants are called **producers** because they use photosynthesis to produce their own food from inorganic compounds. **Consumers** are those organisms that cannot make their own food. Organisms that use green plants as food are called **primary consumers**. Primary consumers are also known as **herbivores**. Organisms that feed on primary consumers are called **secondary consumers**. Secondary consumers are also known as **carnivores** because they eat other animals. On the North American prairie, the many species of grasses are producers. Grass is food for bison, which are primary consumers. Wolves, which feed on the bison, are secondary consumers. At each step, only about 10 percent of the energy that is produced is passed on. In this way, the energy from the sun, which was first stored by green plants, is passed through the ecosystem.

Almost all living things carry out cellular respiration, which occurs in the mitochondria. During **cellular respiration**, living things use oxygen from the atmosphere to release energy from food and in the process form carbon dioxide gas. In photosynthesis, green plants use the carbon dioxide and release oxygen to the atmosphere, where it is again available for respiration. The processes of respiration and photosynthesis maintain a fairly constant ratio of carbon dioxide to oxygen in the atmosphere.

To build new tissues, plants and animals must make proteins. To make proteins, plants use the carbohydrates they manufacture and inorganic nitrogen compounds that they get from the air, water, or soil. Other important inorganic compounds required by plants contain potassium, phosphorus, sulfur, and magnesium. Animals make proteins from the foods they eat. When plants and animals die, decay organisms called **decomposers** break down their tissues and recycle the organic and inorganic compounds back into the soil. These compounds can then be used by a new generation of plants. Animals that eat plants or other animals thus pass these materials along through the ecosystem.

Life on Mars?

A potato-shaped meteorite was picked up from the Antarctic ice during the 1984 annual search for meteorites. It was labeled ALH 84001, since it was the first meteorite discovered that year. Of the 24,000 meteorites discovered on Earth, 24, including ALH 84001, have been identified as possibly originating on the planet Mars. The clues to their Martian origin are based on their chemistry. They are similar to the Mars rocks evaluated by the Viking Landers in 1976. These rare meteorites have created a stir among space scientists.

Figure 6-4 **This meteorite may contain evidence of life on Mars.**

When the scientists at the NASA/Johnson Space Center examined ALH 84001 they found evidence that suggested the presence of microfossils buried within the rock. The meteorite contained pure crystals of magnetite imbedded in carbonate granules. (See Figure 6-4.) These appear to be similar to crystals produced by terrestrial bacteria. This hints at the possibility of life on Mars. Debate over the origin of these crystals continues. The implications of this discovery may answer some of our oldest questions and yet pose still others. The recent discovery of frozen water beneath Mars's surface adds to the possibility of life.

6.3 Section Review

1. Why are producers the key organisms within an ecosystem?
2. How do animals and plants maintain the oxygen–carbon dioxide ratio of the atmosphere?

LABORATORY INVESTIGATION 6

Making a Model of the Biosphere

PROBLEM: *To make a model of the biosphere*

SKILLS: *Measuring, approximating*

MATERIALS: *6 sheets of 28 × 43 centimeter paper, string 40 centimeters long, pencil, clear tape*

PROCEDURE

1. Tape together 6 sheets of 28 × 43 centimeter paper to form a sheet approximately 80 centimeters × 80 centimeters.

2. Tie one end of the string to a pencil. You will use the string and pencil as a compass to draw a cross section of Earth. Use a scale of 1 centimeter equals 100 kilometers.

3. The following dimensions will help you complete your drawing:

radius of inner core	=	1300 kilometers
thickness of outer core	=	2100 kilometers
thickness of mantle	=	2900 kilometers
thickness of crust	=	10 to 60 kilometers, average 35 kilometers
thickness of atmosphere	=	500 kilometers
thickness of biosphere	=	several centimeters to several kilometers, average 1 kilometer

OBSERVATIONS AND ANALYSES

1. How thick in centimeters does the biosphere appear in your cross section?

2. How does the thickness of the biosphere compare with the thickness of Earth's crust?

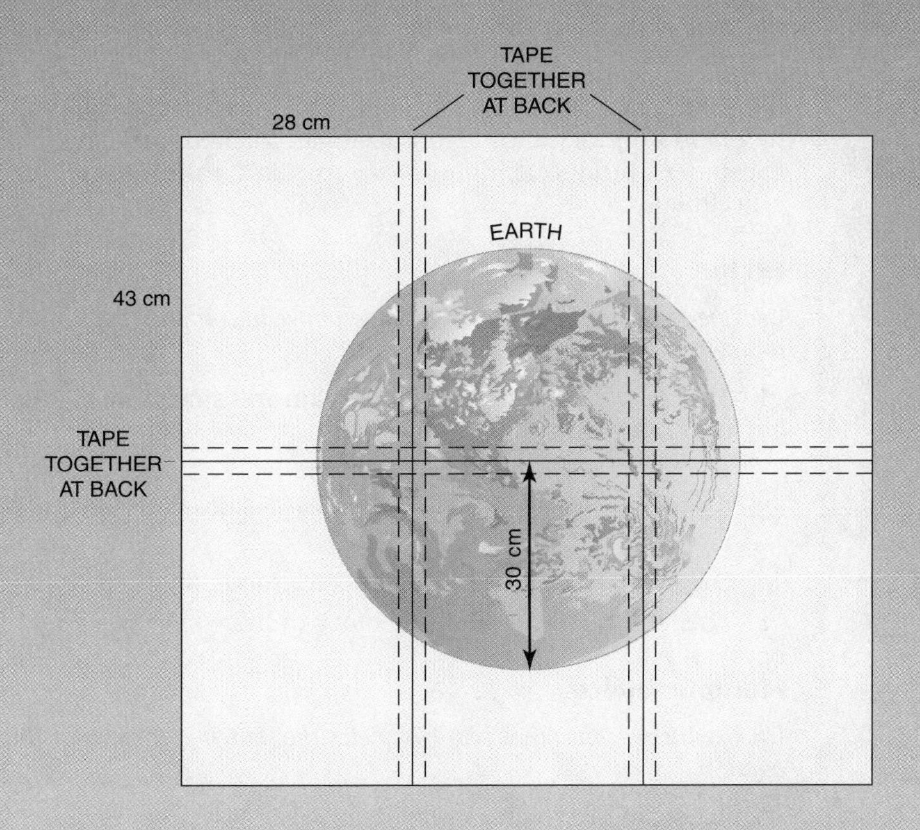

3. Why is the biosphere compared to the skin of an apple?
4. The tallest mountain on land rises 9.7 kilometers above sea level and the deepest trench descends 11 kilometers beneath the ocean's surface. How would these appear on your cross section?

Chapter 6 Review

Answer these questions on a separate sheet of paper.

Vocabulary

The following list contains all the boldfaced terms in this chapter.

abiotic environment, biota, biotic environment, carnivores, cell, cellular respiration, consumers, cyanobacteria, decomposers, herbivores, mitochondria, multicellular, nucleus, primary consumers, producers, ribosomes, secondary consumers, unicellular

Fill In

Use one of the vocabulary terms listed above to complete each of the following sentences.

1. The ______ is the part of an organism's surroundings that is or was alive.
2. The control center of the cell is the ______.
3. Green plants are called ______ because they make their own food.
4. Organisms that use green plants for food are called ______.
5. The ______ is the building block of life.

Multiple Choice

Choose the response that best completes the sentence or answers the question.

6. According to the rock record, about how many years ago did fossil stromatolites appear? *a.* 4.5 billion *b.* 3.8 billion *c.* 2.5 billion *d.* 1.3 billion
7. Plant cells have *a.* cell membranes. *b.* chloroplasts. *c.* cell walls. *d.* all of the above.
8. The biological process that releases carbon dioxide gas into the atmosphere is *a.* photosynthesis. *b.* recycling. *c.* cellular respiration. *d.* precipitation.
9. The biosphere is thinnest *a.* over the United States. *b.* at the poles. *c.* at the oceans. *d.* in caves.
10. Cells that are adapted to carry messages are *a.* red blood cells. *b.* muscle cells. *c.* nerve cells. *d.* epithelial cells.
11. Which of the following is not part of an organism's abiotic environment? *a.* water *b.* soil *c.* wind *d.* other organisms

12. The biota of a region includes *a.* rocks and minerals. *b.* plants and animals. *c.* lakes and rivers. *d.* only microscopic organisms.
13. The powerhouse of the cell is the *a.* mitochondria. *b.* endoplasmic reticulum. *c.* nucleus. *d.* ribosome.
14. The transport system of the cell is the *a.* mitochondria. *b.* endoplasmic reticulum. *c.* nucleus. *d.* ribosome.
15. What type of cells have a cell wall? *a.* plant *b.* animal *c.* blood *d.* lung

Short Answer (Constructed Response)

Use the information you learned in this chapter to respond to the following items.

16. How is the biotic environment different from the abiotic environment?
17. What are cyanobacteria?
18. What is the function of the cell's nucleus?
19. What are three examples of unicellular organisms?
20. Explain the function of ribosomes.
21. Explain the differences between producers and consumers.
22. What are herbivores?
23. What function do decomposers have in an ecosystem?
24. Explain the function of mitochondria within a cell.
25. Describe the process of cellular respiration.

Essay (Extended Response)

Use the information in the chapter to respond to these items.

26. Describe the extent of the biosphere.
27. How do changes in climate affect the biosphere?
28. What would happen to life on Earth if all the green plants died?
29. Describe some of the biotic and abiotic factors that are important in your personal environment.

CHAPTER 7
The Five Kingdoms

When you have completed this chapter, you should be able to:

Define kingdom, phylum, species.

Identify the five biological kingdoms.

There are so many different living things that even in an entire lifetime one person could not see them all. Many living things cannot be seen with the unaided eye; they can only be seen with a microscope. On the other hand, some, such as the giraffe shown here, are hard to miss. To organize the many living things on Earth, scientists developed a classification system based on structural similarities. This chapter presents a brief overview of the types of living things that inhabit the biosphere.

7.1 A PLACE FOR EVERYTHING

When you look at the huge number of organisms that share the biosphere, you notice the many differences among them. But when you study these organisms up close, you can see that they share many similarities in form and function. Scientists use these similarities as the basis for classification systems that organize living things into groups that share similar characteristics.

At first, classification systems were based solely on an organism's appearance or function. As our knowledge about living things grew, systems were developed that were based on structural similarities instead of appearance.

Carolus Linnaeus, an eighteenth-century Swedish naturalist, laid the groundwork for the classification system we use today. (See Figure 7-1.) He developed binomial nomenclature, which gave uniformity to the scientific names of living things. Because Linnaeus and all other early scientists wrote in Latin, binomial nomenclature consists of a two-part Latin name. For example, the wolf is *Canis lupus.* In every science book, whether it is American, Russian, or Japanese, scientific names are the same, even though the language in the rest of the book is different.

Figure 7-1 Carolus Linnaeus developed the system of binomial nomenclature.

Kingdoms

Based on the work of Linnaeus, **taxonomy** is the modern method of classification that identifies groups of organisms and gives each a scientific name. Organisms are classified by common characteristics and the evolutionary relationships they share. There are several different classification systems. Within most of them, organisms are first placed into one of the major categories, called **kingdoms**. The criteria for placement of an organism into a kingdom is based on the presence or absence of a nucleus within the cells, whether the organism is unicellular or multicellular, and whether it makes its own food or eats other organisms. The classification system used in this book has five kingdoms: Monera, Protista, Fungi, Plantae, and Animalia.

Each kingdom is divided into smaller groups called phyla

(phylum is the singular form). The **phylum** is the largest division within a kingdom. Within each phylum, organisms are separated into smaller and smaller categories: class, order, family, genus, and species. Within each smaller category, organisms show a greater similarity in structure and function. The smallest classification category, the species, includes only one type of organism.

Species

A species is a population of closely related organisms that have related physical and behavioral characteristics, a similar genetic composition, and can freely interbreed with each other (but not with members of other species) to produce fertile offspring.

Organisms from two different but closely related species will sometimes mate when placed together in an artificial setting such as a zoo. The offspring produced usually are infertile and cannot reproduce. For example, lions and tigers are often mated in zoos to produce offspring known as ligers or tiglons. Even when this type of union can produce fertile offspring, it is not likely to occur in nature because of differences in the species' habits, mating and nesting behavior, or habitat preferences.

For many years, taxonomists had a difficult time classifying the giant panda, a bamboo-eating animal in Asia. Are giant pandas related to bears? They do resemble bears. However, bears eat meat; giant pandas normally do not. Or are giant pandas related to the lesser panda and raccoons? Giant pandas share many behavioral traits with these animals. It was thought that a check of their DNA might help. DNA is a molecule that stores and transmits genetic information. DNA tests indicate that the giant panda is more closely related to bears than to raccoons, since bears and pandas share a more recent common ancestor.

Classification of Living Things: Cladistics

Cladistics is an approach to the study of the evolutionary history of living things. Organisms are classified based on shared characteristics, or homologous features. The strict use of this theory is in debate among taxonomists. Cladistics theorizes that common fea-

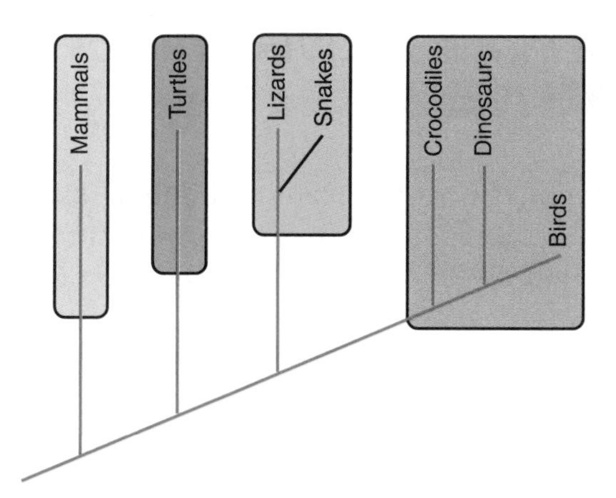

Figure 7-2 A cladogram shows how different groups diverged from a common ancestor.

tures are due to descent from the same ancestor. Organisms that share several characteristics are placed in groups called clades for taxonomic purposes. A clade, or taxon, is composed of an ancestor and all its descendants. They can be species or higher levels of classification.

Every taxon has two types of homologous features: primitive traits and derived traits. Primitive traits are those that existed in the common ancestor. Derived traits are those that represent evolutionary changes from the ancestral features. All organisms are a mixture of primitive and derived traits. Cladistics allows us to judge which species are most likely to share a common ancestor. This validates their placement in the same genus or higher taxon. In practice, many traits are used to examine the relationships between many taxa. Analysis of clades allows for the development of theories regarding the evolution of life on Earth.

As shown in Figure 7-2, a cladogram is a branching diagram that shows how different groups of organisms diverged from a common ancestor. Cladograms portray relative relationships between living things, rather than direct information about ancestors and descendants.

7.1 Section Review

1. Explain why dogs that vary greatly in appearance, for example Great Danes and poodles, still belong to the same species.
2. Make a collection of leaves. Develop a classification system for your collection.

7.2 THE ROLL CALL OF LIVING THINGS

You are probably most familiar with the plant and animal kingdoms. In addition, there are kingdoms that contain only single-celled organisms. This section will introduce you to members of all five kingdoms.

Monera

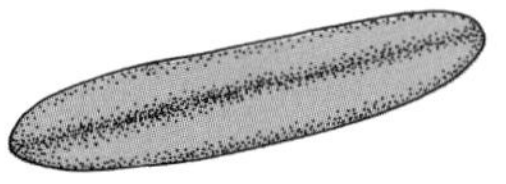

Bacilli
(rod shaped)

Cocci
(round)

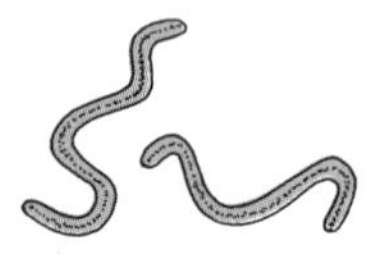

Spirilla
(spiral shaped)

Figure 7-3 Bacteria are monerans.

The Monera are single-celled **prokaryotes**, organisms whose DNA is distributed throughout the cytoplasm of their cells. These are the only organisms whose cells do not have a nucleus. Bacteria and cyanobacteria, which used to be known as blue-green algae, are the most common types of monerans. Bacteria have a cell wall; however, they lack a nucleus, mitochondria, and chloroplasts. Some bacteria are **aerobes**, which require oxygen for respiration. Others are **anaerobes**, which use other chemicals to release energy. Anaerobes are poisoned by oxygen. Figure 7-3 illustrates some monerans.

Bacteria are widespread and abundant microorganisms. There can be tens of billions of them in just a handful of soil. They inhabit deserts, hot springs, hydrothermal vents on the ocean floor, and spaces within the polar ice. Some bacteria live on and in the bodies of other organisms, which are called the **host** of the bacteria. There may be more bacteria living on the skin and within the digestive system of a person than there are cells in that person's body. In some cases, bacteria play an important role in the digestion of food for their host organism. They may supply the host with important nutrients. *Escherichia coli* is a species of bacteria that lives in the human intestine. This bacteria produces vitamin K, which is essential to blood clotting. Ruminants, such as cattle and deer, have huge populations of bacteria in their digestive system. The bacteria digest cellulose and make nutrients available to their host.

Bacteria are important in the formation and composition of soil. They also take part in the breakdown of organic materials to form nitrates. Photosynthetic bacteria are important producers in most ecosystems. Bacteria are used in the production of cheeses, yogurt, and antibiotics, in tanning leather, and in sewage treatment facilities.

The majority of bacteria use nutrients that were synthesized by other organisms. Some, such as the pneumoncoccus, are **pathogens**, which cause disease in their host organism.

Protista

Protista are single-celled **eukaryotes**, organisms whose DNA is located within a nucleus that is enclosed by a nuclear membrane. Some protists form multicellular colonies, a trait that may have led to the formation of multicellular organisms. Protozoa are the most common members of the kingdom Protista, which also includes slime molds, euglenoids, algae, diatoms, dinoflagellates, and sporozoa. The Protista differ from the Monera in that their DNA is enclosed within a nuclear membrane, their cytoplasm contains organelles that perform various metabolic functions, and many lack a cell wall. The earliest eukaryotes were single-celled organisms similar to modern protozoa. These are believed to have given rise to multicellular organisms, composed of many cells arranged to form tissues and organs. Figure 7-4 illustrates some protists.

It is thought that organelles in protists came about through accidental partnerships between free-living organisms and invading bacterial species. Mitochondria, organelles that aid in energy production, resemble bacteria in size and structure, and they operate somewhat independently of the cell that houses them. Mitochondria contain their own DNA and reproduce independently from nuclear DNA. Scientists theorize that primitive protozoa that preyed on aerobic bacteria ingested a species of bacteria that was able to resist digestion. Within the protected, nutrient-rich environment, the bacteria changed and became dependent on the host. In time, that species of bacteria became incapable of independent existence. Chloroplasts, which resemble photosynthetic bacteria, might have become incorporated into protists in the same way.

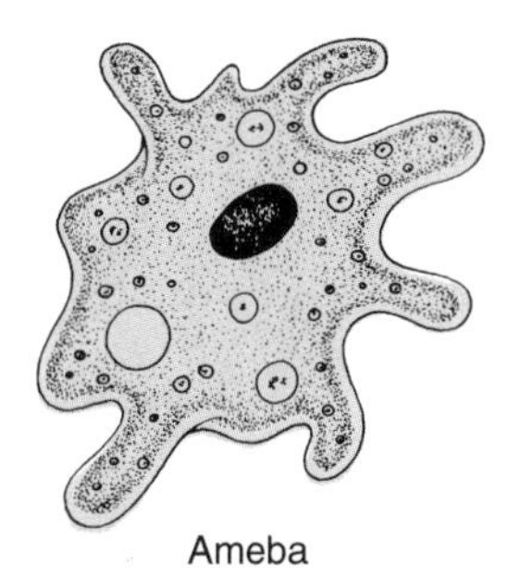
Ameba

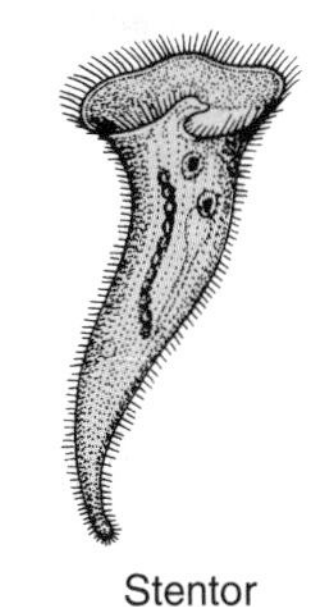
Stentor

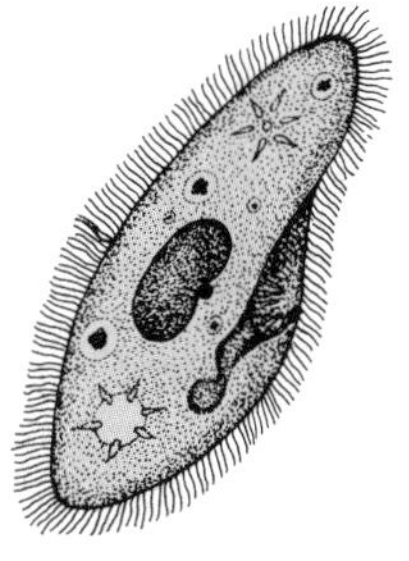
Paramecium

Figure 7-4 The ameba, stentor, and paramecium are organisms in the kingdom Protista.

Fungi

The cells of **fungi** are enclosed within a cell wall, contain a nucleus, but lack chloroplasts. Fungi get their nutrients from other organisms. Some fungi obtain nourishment from nonliving organic

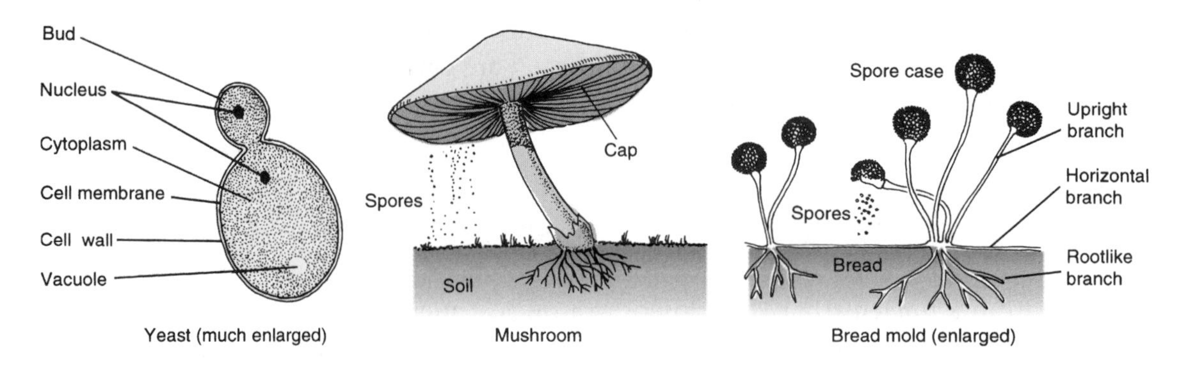

Figure 7-5 Yeast, mushrooms, and bread mold are in the kingdom Fungi.

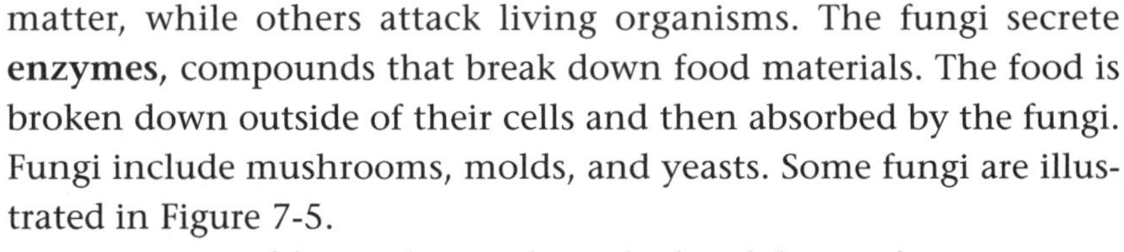

matter, while others attack living organisms. The fungi secrete **enzymes**, compounds that break down food materials. The food is broken down outside of their cells and then absorbed by the fungi. Fungi include mushrooms, molds, and yeasts. Some fungi are illustrated in Figure 7-5.

Many types of fungi play a role in the breakdown of organic matter, which recycles nutrients. Other fungi form relationships with plant roots that aid the plant in absorption of minerals from the soil. Fungi are responsible for human diseases, such as ringworm and athlete's foot, and plant diseases, such as Dutch elm disease, corn smut, potato blight, and wheat rust. Some yeasts are used in the production of bread, wine, and beer.

Figure 7-6 The kingdom Plantae is composed of photosynthetic organisms.

Plantae

Plantae, the plant kingdom, is composed of a wide variety of multicellular, photosynthetic organisms. Plants first evolved in aquatic habitats and then colonized the land as they developed specialized tissues that include roots to absorb water from the soil, stems for support, tissues for the transport of water and nutrients, protective coverings to prevent drying in the air, and reproductive structures. Plants are classified into two main groups, those with vascular tissue and those without vascular tissue. Some members of the plant kingdom are illustrated in Figure 7-6.

The multicellular algae are considered to be the most primitive members of the plant kingdom. They have no true leaves, stems, roots, or vascular tissue. Most of the multicellular algae are aquatic plants that carry on photosynthesis.

DISCOVERIES

The Bristlecone Pine—Ageless Wonder

Bristlecone pines are among the oldest living things known. Examination of annual rings and carbon-dating tests indicate that some of these trees are more than 4000 years old. The oldest tree tested was more than 4600 years old.

Bristlecone pines grow very slowly. They rarely reach a height of more than 10 meters. Bristlecones have dense clumps of short green needles. Their cones are about 7 centimeters long and very prickly, as the name suggests. The age of many trees can be obtained by counting the annual rings formed within the trunk. The woody cells formed in the spring are generally larger than the cells formed in summer. The contrast between the two is visible as a line that encircles the woody, inner part of the tree. A look at the cross section will reveal that the growth rings are densely packed. This type of growth pattern helps protect the tree from attacks by insects and fungi. The wood is so dense it is virtually impenetrable. When the tree finally dies, it is so resistant to decay that its woody skeleton will remain standing for many years.

Bristlecone pines grow at very high altitudes in the subalpine forests of the western United States. Pruned by the harsh wind and weather, they sprawl in gnarled, shrubby shapes, which make them appear dead. The inner growing layer, called the cambium, and the bark may be scoured away on the windward side of the tree. The transport of food and water is maintained by a thin strip of cambium left on the leeward side of the tree. Though bristlecones are considered to be the oldest living organisms on Earth, some shrubs and trees, such as aspens that spread from the same root system or that form colonies, may actually be older.

Bryophytes, which include liverworts and mosses, are plants that have developed simple leaves but still lack vascular tissue. Unlike algae, bryophytes and mosses are equipped to survive in land environments, although they need to keep moist.

Tracheophytes, which include ferns, cone-bearing plants, and flowering plants, have roots, stems, leaves, and a vascular system for transporting materials through the plant.

Animalia

Animalia, the animal kingdom, is composed of multicellular organisms. Except for sponges, the cells of these organisms are organized into tissues, and their tissues are further organized to form organs and organ systems. In general, the members of the animal kingdom eat other organisms to meet their nutritional needs. Most animals are motile, able to move from place to place on their own. They have organs to help them move.

The members of the animal kingdom that are most familiar include mammals, birds, reptiles, amphibians, and fish. But these creatures make up only a small portion of the animal kingdom. This group is called **vertebrates**, because they have a backbone. Of the approximately 1.5 million species and 30 phyla of animals known to inhabit the biosphere, only about 50 thousand species and 1 phylum are vertebrates. **Invertebrates**, animals without a backbone, are much more numerous. They include sponges, jellyfish, coral, flatworms, roundworms, mollusks, segmented worms, lobsters, insects, spiders, and sea stars. Figure 7-7 illustrates some members of the animal kingdom.

Figure 7-7 The frog, whale, and snake are vertebrates; the sea star is an invertebrate.

7.2 Section Review

1. Explain the differences between the Monera and the Protista.
2. Explain the theory used to account for the emergence of organelles in the Protista.
3. Make a collage with pictures or drawings of the members of the five biological kingdoms.

LABORATORY INVESTIGATION 7

Identifying Characteristics of Organisms

PROBLEM: *To describe a group of organisms so that a classmate will recognize them*

SKILLS: *Observing, describing*

MATERIALS: *Specimens or pictures of the members of each group of organisms listed below*

PROCEDURE

1. Select one of the following groups of organisms and describe the characteristics of each member that distinguish it from the others.
   a. English ivy, philodendron, arrowleaf
   b. Grasshopper, cricket, cockroach
   c. Frog, toad, newt
   d. Butterfly, moth, mayfly
   e. Oak, maple, chestnut leaves
   f. Snake, anole lizard, salamander
2. Give your descriptions, without identifying the organisms, to a classmate. Have him or her use the descriptions to identify the organisms they represent.

OBSERVATIONS AND ANALYSES

1. What were some of the types of characteristics you used to differentiate organisms in the group you chose?
2. Were your descriptions accurate enough to enable someone else to distinguish among the organisms within your group?

Chapter 7 Review

Answer these questions on a separate sheet of paper.

Vocabulary

The following list contains all the boldfaced terms in this chapter.

aerobes, anaerobes, enzymes, eukaryotes, fungi, host, invertebrates, kingdoms, pathogens, phylum, prokaryotes, taxonomy, vertebrates

Fill In

Use one of the vocabulary terms listed above to complete each of the following sentences.

1. The widest category within the modern classification system is the ______.
2. Multicellular animals that do not have backbones are ______.
3. ______ are organisms whose DNA is distributed through their cytoplasm.
4. The kingdom that includes organisms whose cells have a cell wall but no chloroplasts and who must get their nutrients from other organisms is ______.
5. Animals that have backbones are classified as ______.

Multiple Choice

Choose the response that best completes the sentence or answers the question.

6. The giant panda is most closely related to the *a.* lesser panda. *b.* raccoon. *c.* coati. *d.* black bear.
7. Organelles that are thought to have resulted from an accidental partnership between protists and free-living bacteria are *a.* fungi. *b.* ribosomes. *c.* Monera. *d.* mitochondria.
8. Fungi cause human diseases such as *a.* ringworm. *b.* tuberculosis. *c.* tapeworm. *d.* measles.
9. Which of the following may be a pathogen? *a.* bacteria *b.* mitochondria *c.* DNA *d.* chloroplasts
10. Animals with a backbone are called *a.* anaerobes. *b.* invertebrates. *c.* vertebrates. *d.* prokaryotes.
11. Which of the following is not an animal? *a.* bird *b.* sponge *c.* insect *d.* tracheophyte
12. Which of the following is an example of a plant? *a.* sponge *b.* coral *c.* moneran *d.* bryophyte

13. The smallest division in the modern taxonomy used in this book is *a.* kingdom. *b.* phylum. *c.* order. *d.* species.
14. In the animal kingdom, which of the following are most numerous? *a.* vertebrates *b.* invertebrates *c.* mammals *d.* birds
15. All members of the plant kingdom *a.* are multicellular and photosynthetic. *b.* have stems. *c.* have roots. *d.* have flowers.

Short Answer (Constructed Response)

Use the information you learned in this chapter to respond to the following items.

16. List the five kingdoms used in the classification of living things.
17. What factors is the modern method of taxonomy based on?
18. How is cladistics different from the five kingdom classification scheme?
19. What are protists?
20. How are vertebrates different from invertebrates?
21. Describe the differences between anaerobic bacteria and aerobic bacteria.
22. What characteristics do prokaryotes possess?
23. What is an enzyme?
24. What are some characteristics of eukaryotes?
25. What are some of the characteristics of fungi?

Essay (Extended Response)

Use the information you learned in the chapter to respond to the following items.

26. Why are Latin names used in our modern system of classification?
27. Describe the differences between aerobes and anaerobes.
28. What are some of the natural controls that keep one species from interbreeding with another species?

CHAPTER 8
What Living Things Need

When you have completed this chapter, you should be able to:

Describe the process of photosynthesis.

Distinguish between the light and dark reactions in photosynthesis; autotrophs and heterotrophs.

Contrast aerobic and anaerobic respiration.

Explain the predator-prey relationship.

All living things need food to survive. Green plants make their own food. Animals, on the other hand, must obtain their food. Cows and other herbivores eat green plants. Carnivores eat herbivores. Humans eat plants and animals. In this chapter, you will learn how plants make food for themselves and the animals.

8.1 FOOD PRODUCTION IN THE BIOSPHERE

Photosynthesis is the process by which green plants make carbohydrates. **Chlorophyll** is the compound that enables a plant to trap the energy in sunlight and use it to make carbohydrates. **Carbohydrates** are composed of the elements carbon, hydrogen, and oxygen. Sugars and starches are carbohydrates. Within the cells of green plants are organelles called **chloroplasts**, which contain the green pigment chlorophyll.

Within the chloroplast, energy from sunlight drives the chemical reactions that produce carbohydrates. As a result of these reactions, 6 molecules of carbon dioxide (CO_2) are combined chemically with 6 molecules of water (H_2O) to produce 1 molecule of glucose ($C_6H_{12}O_6$). As a waste product of this reaction, six molecules of oxygen (O_2) are released into the air:

$$6\,CO_2 + 6\,H_2O \xrightarrow[\text{SUNLIGHT}]{\text{CHLOROPHYLL}} C_6H_{12}O_6 + 6\,O_2$$

CARBON DIOXIDE + WATER → GLUCOSE + OXYGEN

Photosynthesis is illustrated in Figure 8-1 on page 124.

Visible light is a small part of the electromagnetic spectrum. The process of photosynthesis uses only a small portion of the solar energy present in visible light. Figure 2-2 on page 22 illustrates the electromagnetic spectrum. The wavelengths used in photosynthesis are in the ranges we see as red and blue light. Some of this light energy is used for **transpiration**, the process by which water evaporates through pores in the leaves of plants. These pores are called **stomata** (singular form, stomate). Thus only a small portion of the light energy that strikes the surface of a green leaf is used for photosynthesis and converted into carbohydrates. The yellow and green light is reflected, which makes leaves appear green.

Light and Dark Reactions

Photosynthesis takes place in two stages, called the light reaction and the dark reaction. During the light reaction, chlorophyll absorbs light

Figure 8-1 **Photosynthesis changes sunlight to chemical energy.**

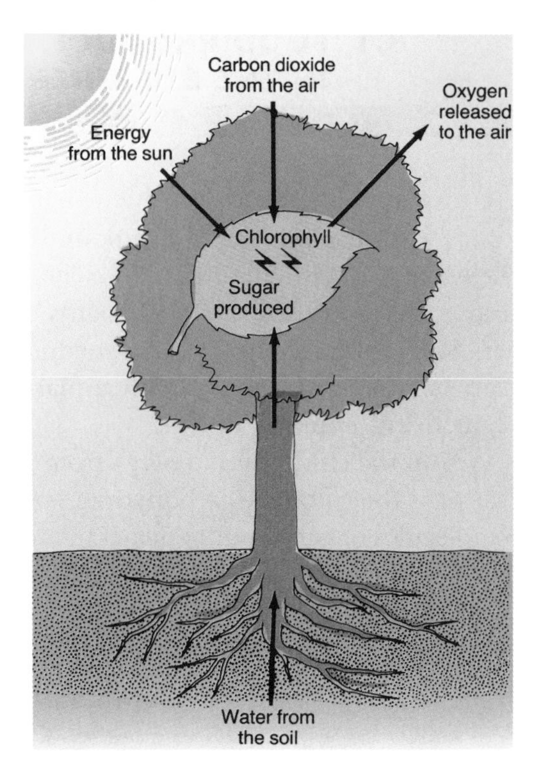

energy and uses it to split water molecules (H_2O) to form oxygen (O_2) and two hydrogen ions ($2H^+$). The oxygen (O_2) is released into the air. The dark reaction does not require light, but it can and does occur in the light. In the dark reaction, the hydrogen ions (H^+) combine with carbon dioxide (CO_2) to produce the simple sugar glucose ($C_6H_{12}O_6$).

Plants are able to make other nutrients by using glucose as the base. Complex sugars and starches are produced by joining molecules of glucose together. Fats are synthesized by breaking down glucose into smaller carbon-containing molecules and then rearranging them to form fatty acids and glycerol. When fatty acids are combined with nitrogen, they form amino acids. **Proteins** are composed of chains of amino acids.

8.1 Section Review

1. Describe the process of photosynthesis.
2. Compare the products made during the light reaction with those made during the dark reaction.
3. Explain how carbohydrates are the base for other nutrients.

8.2 USING ENERGY

The process by which cells release the chemical energy stored in foods is called cellular respiration. Cellular respiration involves splitting carbon and hydrogen atoms from glucose molecules and recombining them with oxygen to form carbon dioxide and water. Cellular respiration is the reverse of photosynthesis:

$$C_6H_{12}O_6 + 6\ O_2 \longrightarrow 6\ H_2O + 6\ CO_2 + \text{energy}$$

In photosynthesis, energy is captured, while in cellular respiration energy is released. Organic materials in food are broken down through cellular respiration to produce energy. In the process, oxygen is used and carbon dioxide released. The processes of photosynthesis and cellular respiration cycle carbon dioxide and oxygen through the biosphere. (See Figure 8-2.)

Anaerobic Respiration

In most situations, cellular respiration is an **aerobic** process, which means it requires oxygen. When an animal uses a great deal of energy, for example as it runs, it must breathe faster to increase its oxygen supply. When energy is used faster than oxygen can be supplied to the cells, an anaerobic condition results. **Anaerobic** refers to

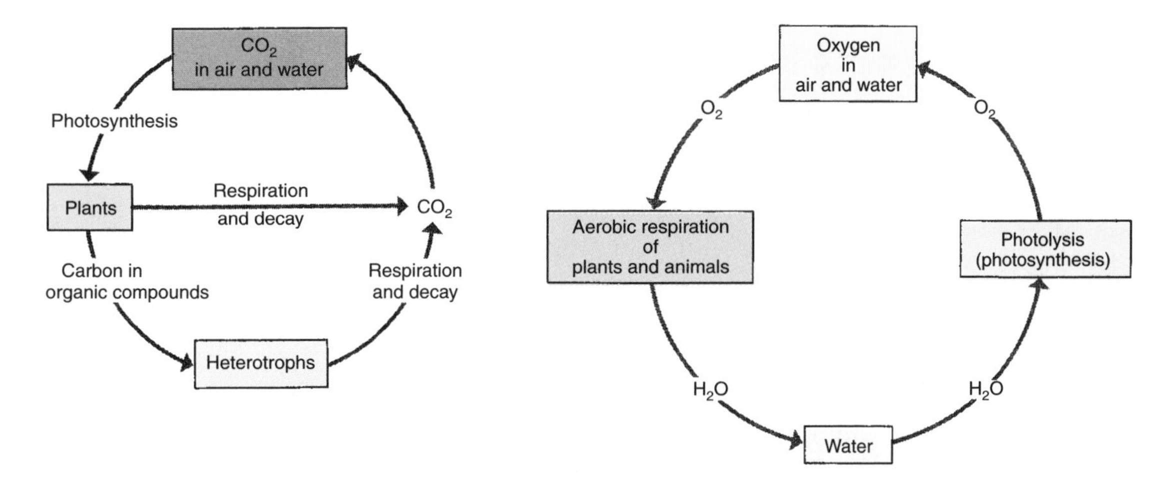

Figure 8-2 The carbon cycle (left) and the oxygen cycle (right) distribute carbon dioxide, oxygen, and water in the biosphere.

a process that does not require oxygen. Without enough oxygen, the breakdown of glucose is inefficient, and little usable energy is produced.

Under anaerobic conditions, cells use a process that does not require oxygen to break down glucose. During this process, lactic acid is produced, and as it builds up in the muscle cells, fatigue occurs. When the oxygen supply catches up with the demands of the cells, breathing returns to normal.

Fermentation is the anaerobic release of energy. In animal cells, fermentation produces lactic acid; in fungi, alcohol and carbon dioxide are the products:

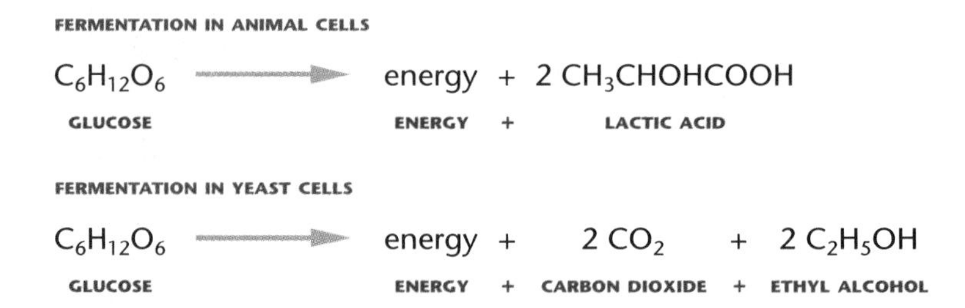

Wineries and bakeries make use of fermentation. Yeast is a unicellular fungus. Yeast cells ferment the sugar in grapes to produce the alcohol in wine. Bakers add yeast to dough because yeast produces carbon dioxide through fermentation. The carbon dioxide makes bread dough rise by producing tiny pockets of gas. The alcohol evaporates when the bread is baked.

Energy Flow Through the Biosphere

In the biosphere, energy is used over and over again, or **recycled**. As energy is recycled, some energy is lost to the ecosystem in the form of heat. Because energy is lost, every ecosystem must have a constant source of incoming energy. Cycles are important processes in maintaining ecosystems. Energy, oxygen, carbon dioxide, water, and nitrogen are recycled by living things within an ecosystem. (See Figure 8-3.)

Every ecosystem contains a nonliving environment and a community of living organisms. The living community is composed of populations of organisms of many different species. The popula-

Figure 8-3 The energy cycle.

tions within the community interact in various ways to sustain the ecosystem. Most **plankton** are microorganisms that live near the surface of the water. There are two types of plankton: in a pond, **phytoplankton**, such as algae, produce food by photosynthesis; **zooplankton** feed on the phytoplankton. Larger organisms, such as insect larvae, crustaceans, and small fish, feed on zooplankton.

Figure 8-4 Some interactions that occur in a pond ecosystem are shown here.

Larger fish, frogs, and birds feed on the insects and small fish. In turn, these creatures are preyed upon by raccoons, hawks, and snakes. (See Figure 8-4.) Where there is something to be eaten, there will be something to eat it.

Through photosynthesis, producers are able to convert only about 1 percent of the incoming solar energy into plant tissue. Cellular respiration in consumers is also an inefficient process. Only about 40 percent of the energy stored by producers during photosynthesis can be used by primary consumers. The rest is lost as heat during cellular respiration and thus is not available to cells. Secondary consumers are even more inefficient in their use of stored energy. The fewer steps there are in the food-getting process between producer and consumer, the less energy is lost to the ecosystem.

Some of the largest consumers, for example baleen whales, basking sharks, and manta rays, are the most efficient consumers. Their food chain is short. They feed directly on tiny phytoplankton and zooplankton. In our agricultural system, cows are fed corn. Only a small part of the solar radiation that reaches Earth is captured by the corn and stored as food energy. Cows are able to convert only a small portion of this energy into beef. Thus a great deal of energy has been lost before we ever get to use the beef as food. This is the reason why beef, chicken, and pork cost more than corn, grain, and beans. Farmers get less food energy production from their land in animal-based agriculture than they do from plant-based agriculture.

As the human population increases, meat and other animal products may become more and more scarce. In the future, society may have to rely on more efficient approaches to meet its energy needs. By increasing productivity and making greater use of green plants for food, more energy can be made available to the world's ever-increasing population.

Predator and Prey

Animals get the energy they need by eating plants, eating animals that eat plants, or eating animals that eat plant-eating animals. When one animal kills and eats another animal, the killer is called the **predator** and the animal that is killed is the **prey**. The interaction between the two is a predator-prey relationship. A wolf eating a deer, an osprey eating a trout, and a frog eating a fly are examples of predator-prey relationships. All animals are hunters or hunted, and some of them are both.

Predator-prey relationships often seem cruel and savage. But these relationships can be beneficial for the prey species as well as the predator. In evaluating these relationships, the terms *harmful, beneficial,* and *neutral* are used. A harmful predator-prey relationship causes a decline in a prey population. A beneficial relationship leads to an increase in the prey population. A neutral relationship causes no change in the prey population. Predator-prey relationships are beneficial or neutral most of the time, since harmful relationships lead to the disappearance of the prey species and eventually the disappearance of the predator species as well.

It is clear that the death of the prey supplies energy to the predator. The energy allows the predator to live and produce offspring to continue the species. A readily available source of food is a necessity in raising a healthy generation of new individuals. But how does the death of a prey animal benefit its own species?

In their search for food, predators usually weed out the weak and the sick members of the prey species. This process reduces the spread of disease in prey populations and insures that only the healthiest prey individuals are left to reproduce. The removal of weak and sick individuals also tends to conserve food supplies so that there is more available for healthy organisms. By thinning out populations, there is less stress placed on the ecosystem. A healthy ecosystem is able to provide more energy throughout the community. Most predator-prey relationships are thus beneficial to the prey species.

Even though prey populations seem to be at the mercy of predators, predator populations likewise are affected by the number of prey available. In good years, a large number of prey animals supports a large number of predators. When the prey population is low, there is not enough energy available to meet the needs of the predator population and its numbers fall. There must always be fewer predators than prey or there is a risk of eliminating the entire prey population. The elimination of the prey can lead to the predators' decline. Figure 8-5 illustrates predator-prey relationships.

Within prey populations, the young are more vulnerable to predation than the mature members of the population. Species have strategies to insure the survival of their offspring. Salmon and field mice produce large numbers of young, which insures the survival of at least a few. The young of antelope and deer are born with the ability to run within hours of birth. Herding by large mammals such as antelope and bison also offers the young the protection of large numbers of older individuals.

8.2 Section Review

1. Compare the processes of cellular respiration and photosynthesis.
2. How do ecosystems lose the energy stored by producers?
3. Explain why harmful predator-prey relationships are seldom found in nature.

Figure 8-5 In a predator-prey relationship, the prey are found in greater numbers than the predators. Why?

8.3 PHOTOSYNTHESIS AND THE BIOSPHERE

Green plants and algae are able to produce their own food. Through photosynthesis, they use the energy in sunlight to make carbohydrates, proteins, and fats. All other living things make use of these energy-rich compounds either directly by feeding on green plants, or indirectly by feeding on organisms that eat green plants. Because of this relationship, the green plants in an ecosystem are the food producers. (See Figure 8-6.) They are called **autotrophs**, which means they are self-nourishing organisms. All other living things are consumers, or **heterotrophs**, since they must rely on an outside source of food to meet their energy requirements.

Life as it exists on Earth today would not be possible without an abundance of photosynthetic organisms. Both directly and indirectly, photosynthesis supplies nearly all of the chemical energy and organic compounds needed for the life processes that maintain the biosphere. Photosynthesis is partly responsible for the composition of the atmosphere.

It is thought that Earth's early atmosphere contained little or no molecular oxygen. As a result of the chemical reactions carried out by the first photosynthetic bacteria, oxygen probably appeared in the atmosphere about 2 billion years ago. These organisms used carbon dioxide from the atmosphere to produce carbon compounds. The oxygen released during these chemical reactions drastically

Figure 8-6 Green leaves have the pigment chlorophyll, which is necessary for photosynthesis. The chlorophyll in these fallen leaves has broken down, revealing other pigments that were always present but unseen.

changed the composition of the atmosphere and forever altered Earth's climate.

On a global scale, most photosynthesis is carried out by the leaves of green plants that grow on the ice-free areas of the continents and by phytoplankton that float in the upper layers of the oceans. Photosynthesis helps maintain the ratio of carbon dioxide and oxygen in the atmosphere. Today, 21 percent of the atmosphere is molecular oxygen, while 0.035 percent is carbon dioxide. Aerobic organisms, which rely on oxygen for respiration, depend on green plants to maintain this concentration of oxygen in the atmosphere.

Ecosystems usually get their energy through photosynthesis by producer organisms. **Productivity** is a measure of the amount of biological material that producers are able to make using incoming solar energy. Scientists use productivity as an important tool to help them understand the workings of an ecosystem. Productivity is related to biomass, the total mass of living material in an ecosystem. In turn, biomass determines the amount of energy available to consumers.

Earth's main source of energy is sunlight. Even the fossil fuels—coal, oil, and natural gas—we use to power our industrial societies come from organic materials. These organic materials were produced by green plants millions of years ago and preserved under the ground.

8.3 Section Review

1. How have photosynthetic bacteria changed the composition of the atmosphere?
2. How do ecosystems get their energy?
3. What are fossil fuels?

LABORATORY INVESTIGATION 8

The Role of Sunlight in Photosynthesis

PROBLEM: *What is the role of sunlight in photosynthesis?*

SKILLS: *Observing, manipulating*

MATERIALS: *Geranium plant, denatured alcohol, water, hot plate, 500-mL beaker, 250-mL beaker, petri dish, Lugol's solution or iodine, black paper or aluminum foil, paper clips*

PROCEDURE

1. Place a geranium plant in the dark for at least 24 hours.
2. Remove a leaf from the plant.
3. In this step, you will use alcohol to extract the chlorophyll from the leaf. For safety, put the leaf in the smaller beaker, half filled with denatured alcohol. Place this beaker into the larger beaker, half full of water. Then place the nested beakers on the hot plate and heat them, much like a double boiler. Heat until the leaf loses its color. (CAUTION: Care must be taken in the extraction of chlorophyll since alcohol vapors are highly flammable.)
4. When the leaf has lost its color, rinse it with water and place it on a petri dish. Test the leaf for starch by applying a drop or two of Lugol's solution or iodine. If starch is present, the leaf will turn blue-black where the Lugol's solution or iodine touches it.
5. Cover part of a leaf on the plant with black paper or aluminum foil. Fasten the paper or foil with a clip so that both sides of the leaf are covered. Expose the geranium plant to sunlight for a day or two or to a 100-watt lamp for 24 hours.
6. Remove the covered leaf from the plant. Remove the paper or foil from the covered leaf.
7. Repeat step 3 to extract the chlorophyll from the covered leaf. Test the leaf for starch with Lugol's solution or iodine.

OBSERVATIONS AND ANALYSES

1. Was starch present in the first leaf from which the chlorophyll was removed?
2. Was starch present in the part of the leaf that was not covered?
3. Was starch present in the part of the leaf that was covered?
4. Account for any difference in the amount of starch in the uncovered part of the leaf.

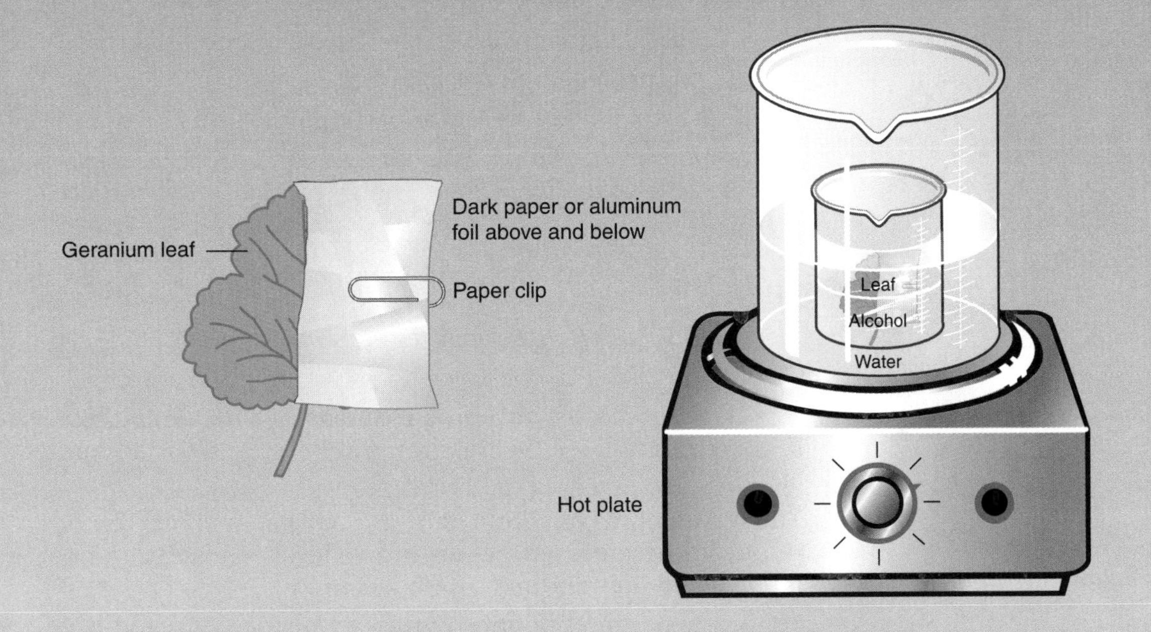

Chapter 8 Review

Answer these questions on a separate sheet of paper.

Vocabulary

The following list contains all the boldfaced terms in this chapter.

aerobic, anaerobic, autotrophs, carbohydrates, chlorophyll, chloroplasts, fermentation, heterotrophs, phytoplankton, plankton, predator, prey, productivity, proteins, recycled, stomata, transpiration, zooplankton

Fill In

Use one of the vocabulary terms listed above to complete each of the following sentences.

1. Pores in leaves that allow the exchange of gases are ______.
2. ______ is the pigment that traps the energy in sunlight.
3. ______ is the anaerobic release of energy.
4. An animal that kills and eats another animal is called a(an) ______.
5. The measure of the amount of energy producers get from the sun is ______.

Multiple Choice

Choose the response that best completes the sentence or answers the question.

6. The stages of photosynthesis are called *a.* light and dark. *b.* slow and fast. *c.* cellular respiration. *d.* lactic acid and alcohol.
7. The wavelengths of visible light used by green plants during photosynthesis are *a.* green and yellow. *b.* violet and blue. *c.* yellow and red. *d.* blue and red.
8. The waste products of photosynthesis are *a.* CO_2 and H_2O. *b.* $C_6H_{12}O_6$ and H_2O. *c.* C_2H_5OH and O_2. *d.* O_2 and H_2O.
9. The carbon used in photosynthesis comes from *a.* coal. *b.* plant tissue. *c.* carbon dioxide. *d.* water.
10. Which of the following is not a product of cellular respiration? *a.* $C_6H_{12}O_6$ *b.* H_2O *c.* CO_2 *d.* energy
11. Which of the following is an example of a predator-prey relationship? *a.* moose and squirrel *b.* robin and worm *c.* bat and cow *d.* leopard and tiger
12. What effect does a decrease in the prey population have on the predator population? *a.* It increases. *b.* It decreases. *c.* It remains the same. *d.* It increases, then decreases.

13. Which elements make up carbohydrates? *a.* carbon, nitrogen, sulfur *b.* carbon, hydrogen, nitrogen *c.* hydrogen, nitrogen, oxygen *d.* carbon, hydrogen, oxygen
14. Anaerobic means without *a.* carbon. *b.* water. *c.* oxygen. *d.* hydrogen.
15. Zooplankton feed on *a.* phytoplankton. *b.* insect larvae. *c.* frogs. *d.* hawks.

Short Answer (Constructed Response)

Use the information you learned in this chapter to respond to the following items.

16. What nutrients make up carbohydrates?
17. Explain the function of chlorophyll in the process of photosynthesis.
18. What are chloroplasts?
19. Describe the process of transpiration.
20. What chemicals are the building blocks of proteins?
21. Describe the process of fermentation.
22. Explain the difference between autotrophs and heterotrophs.
23. Describe some of the differences between phytoplankton and zooplankton.
24. What does the productivity of an ecosystem measure?
25. Why are predator-prey relationships considered beneficial for both species?

Essay (Extended Response)

Use the information you learned in the chapter to respond to the following items.

26. Explain how the predation of deer by mountain lions can benefit the deer population.
27. Describe the relationship between photosynthesis and fossil fuels.
28. Are humans heterotrophs or autotrophs? Explain your answer.

UNIT THREE

NATURE'S CYCLES

The Space Shuttle carries with it into orbit enough food and oxygen to last through its mission. In addition it carries some extra, in case its return to Earth is delayed. After being used, its supply of water and air is purified and recycled.

In many ways, Earth is like a spaceship. Natural cycles purify our air and water. But minerals, such as iron and gold, cannot be replenished by natural cycles. When these substances are used up, they are gone for good. Unlike the Space Shuttle, there is no place Earth can go to get more supplies.

CHAPTER 9
Matter and Energy Move Through the Biosphere

When you have completed this chapter, you should be able to:

Define food chain and food web.

Distinguish between the pyramid of numbers and the pyramid of mass.

Diagram food chains and food webs.

Energy from the sun is the only factor in the biosphere that can be resupplied. Everything else must be recycled to be reused. Just as energy cycles through the biosphere, so does matter. In this chapter you will track the sun's energy as it flows from producers to consumers.

9.1 FOOD CHAINS AND WEBS

All living things within the biosphere are made from a limited supply of raw materials. The raw materials are made up of the atoms of elements, such as hydrogen, carbon, nitrogen, oxygen, phosphorus, calcium, and iron. These elements are the chemical building blocks of the molecules of all living things. They must be recycled. In the billions of years since life began, the same atoms have been used to make molecules that are part of algae, jellyfish, trees, insects, dinosaurs, and humans. The atoms that make up the living things in today's biosphere will be used again and again in the future. There is nothing new under the sun.

Ecologists study how energy enters and is transferred through an ecosystem. To do this, they must discover the feeding relationships among the organisms in the ecosystem. Energy is transferred through organic compounds produced by plants and eaten by animals in recurring cycles of eating and being eaten.

Consumers

Within an ecosystem, the producers trap and store energy from the sun. These plants synthesize, or make, organic compounds to build new tissues and also use them as a source of energy for carrying out their life functions. Herbivores are primary consumers, which means they eat green plants. Herbivores make direct use of the energy stored in the organic matter manufactured by the producers. Deer and rabbits eat grasses, leaves, and twigs; some birds eat seeds and buds; many insects consume a variety of plant parts. They all use the nutrients stored in plant tissues for energy and as a source of materials to build new tissues.

Matter and energy are passed farther along through the ecosystem when secondary consumers eat the herbivores. Secondary consumers include predatory birds, fish, small mammals, and some insects. **Tertiary consumers** prey on the secondary consumers and continue the flow of energy and matter through the ecosystem. Tertiary consumers include large predators such as lions, wolves, sharks, and eagles.

Animals that feed on dead organisms they did not kill are called **scavengers**. Scavengers are the clean-up crews within the biosphere. They may be secondary or tertiary consumers, depending on the type of organisms they eat. Scavengers make use of animal remains that would otherwise litter the environment. Some scavengers can digest bones and hides. In terrestrial ecosystems, coyotes, buzzards, carrion beetles, hyenas, vultures, ravens, and many types of ants, worms, and flies are scavengers. In marine ecosystems, lobsters, crabs, and various isopods, amphipods, and worms are the chief scavengers.

Decomposers feed on organic wastes and the remains of dead plants and animals. They decompose, or break down, organic compounds and recycle them within the ecosystem. The elements that make up living things are borrowed from the biosphere, and after the death of an organism, these elements are returned. Decomposers take the final step by extracting energy from dead tissues and organic waste. In doing so, complex organic molecules are broken down into simple molecules and returned to the biosphere where they can be used again. Bacteria, fungi, and protozoa are the main decomposers in most ecosystems. Decomposers are the link in the natural cycle of life and death.

Food Chains and Webs

The stepwise flow of matter and energy through a community is called a **food chain**. Figure 9-1 illustrates a food chain. The organisms within food chains occupy different feeding stages, called **trophic levels.** Each food chain begins with a producer at the first

Figure 9-1 A food chain illustrates the flow of matter and energy.

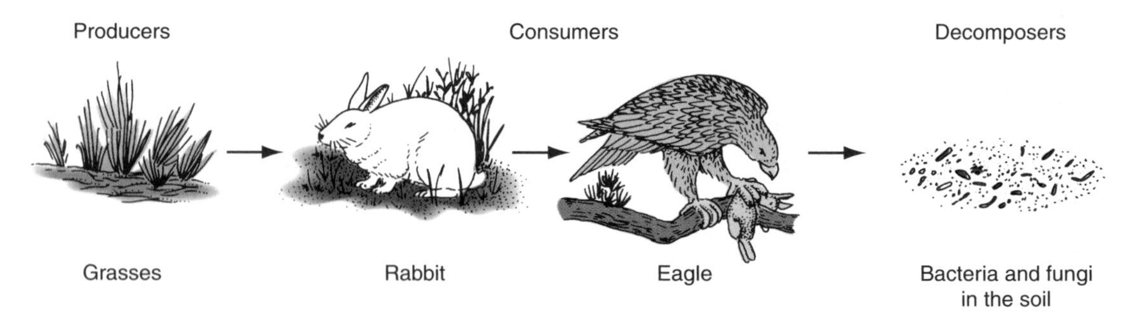

trophic level and identifies the various consumers at succeeding levels. All food chains end with the decomposers. At each stage in the food chain, some energy is passed on to the next trophic level. The biosphere represents a banquet to which all living things are invited. Each organism starts off being a guest and ends up being the meal.

Within an ecosystem, a food chain is a useful way of showing which organisms feed on which other organisms. Any one producer may be fed on by many primary consumers. These in turn are preyed on by many secondary consumers, which then may prey on each other. Every animal in a food chain is caught up in a game of eat and be eaten. Very few animals feed on just one kind of animal or plant. The risks of being dependent on a single food source are too great. Therefore, animals often are part of several food chains.

If you were to diagram all the food chains that begin with the apple tree as the producer, you would find that they crisscross each other many times. This produces a complicated food web rather than a simple food chain. A **food web** is a collection of food chains that shows all the interactions occurring among the producers and consumers within an ecosystem. Figure 9-2 on page 144 illustrates a food web.

Terrestrial Food Chains

Most food chains are classified as **grazing food chains**, since the energy stored by green plants is first transferred to a grazing herbivore. Insects such as moths, caterpillars, beetles, and flies feed on the leaves and fruit of apple trees. These animals are primary consumers. Many small mammals (squirrels, raccoons, and field mice) and birds (jays, grosbeaks, and robins) eat insects, fruits, and seeds. These are **omnivores**, creatures that eat plants and animals. Therefore, they are both primary and secondary consumers. Opossums, foxes, hawks, and snakes, which feed on smaller consumers, can be both secondary and tertiary consumers.

An alternate method of energy transfer is the **detritus food chain. Detritus** is composed of the organic remains of plants and animals, animal droppings, and partially decomposed materials. The remains of green plants and animals are first fed upon by microorganisms

Figure 9-2 **A food web shows the relationships that exist among different organisms.**

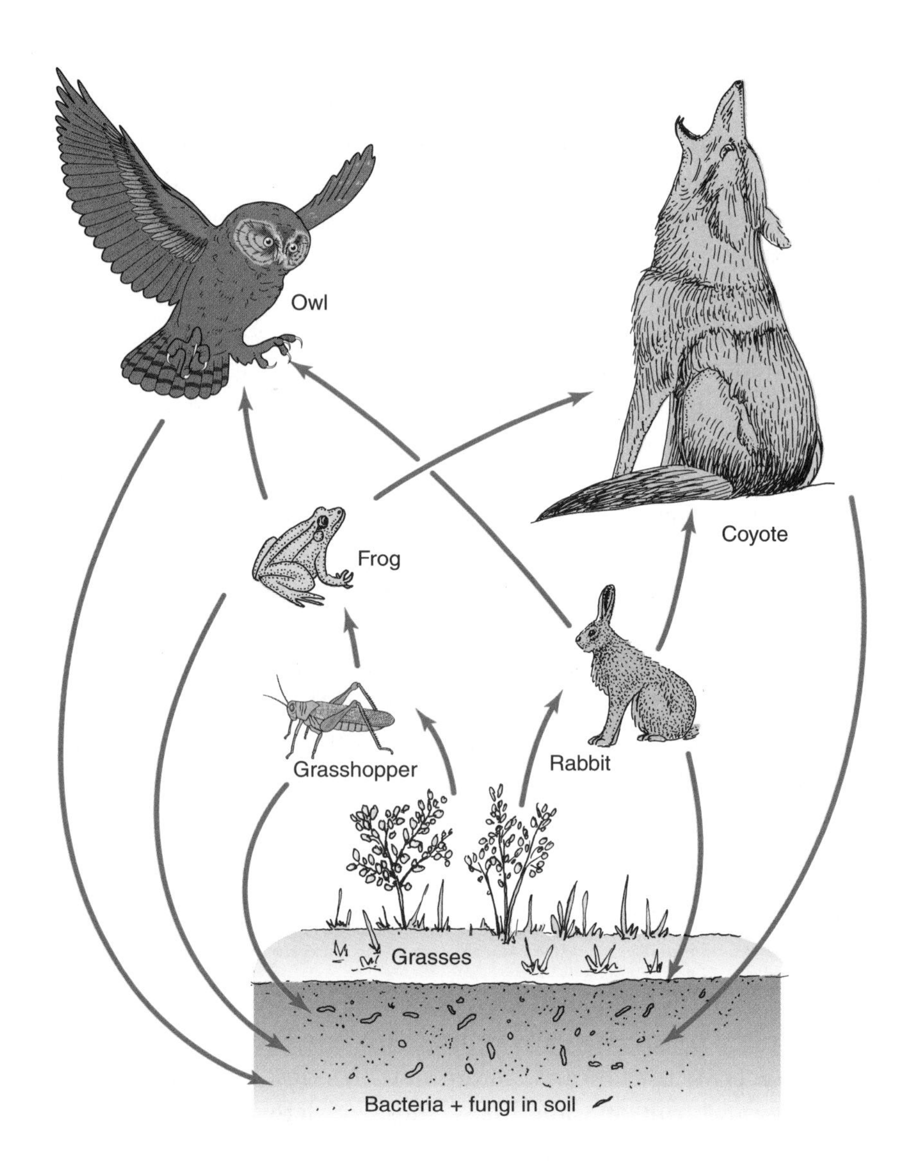

and broken down. The microorganisms reproduce and become a source of food for larger organisms. Eventually, these are fed upon by predators. Energy is still passed from producer to consumer, but the primary consumer is a decomposer microorganism.

Earthworms, wood lice, millipedes, snails, and slugs are **detritivores**, organisms that feed directly on detritus and thus begin the detritus food chain. Detritivores are scavengers. These organisms are able to tackle large pieces of detritus and turn it into their droppings.

The droppings are then more easily broken down by bacteria and fungi. As earthworms eat their way through the soil, they mix the organic and inorganic portions together, enriching the soil for later use by plants. As earthworms move through the soil, they also aerate it. The presence of earthworms therefore increases the overall fertility of soil. Fungi, bacteria, and other decomposer organisms take part in natural cycles that enrich the soil. They break down organic materials and thus return important minerals to the ecosystem.

Marine Food Chain

Phytoplankton are the producers in most marine ecosystems. They live near the surface of the water where sunlight is available for photosynthesis. The phytoplankton are eaten by zooplankton, which are the primary consumers in this ecosystem. Phytoplankton and zooplankton are not true swimmers; they drift slowly through the waters. As a group, these tiny organisms are called plankton, the Greek term for "wanderer."

Plankton are a source of food for small fish, which are preyed upon by larger and larger fish along the food chain. The **top predators**, those animals at the end of the food chain, are large organisms such as the killer whale, the great white shark, the tuna, and the swordfish. Figure 9-3 illustrates a marine food chain.

Some of the largest marine organisms feed directly on plankton. These giants swim along slowly with their mouth open to filter plankton from the water with their gills or baleen. Baleen is a hornlike material that grows down from the upper jaw of some whales, which enables them to strain plankton from the water. These plankton consumers include the blue whale, the largest organism that

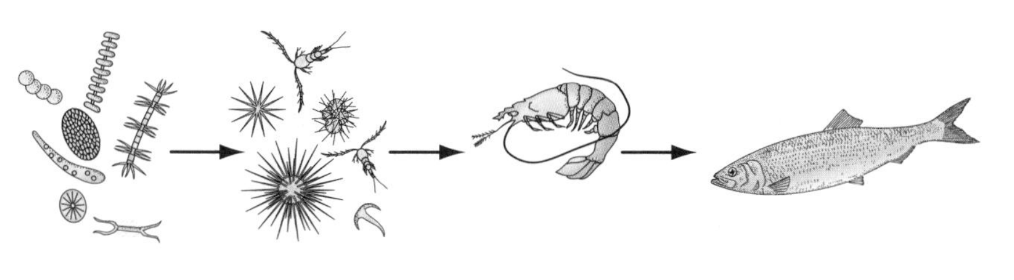

Figure 9-3 Follow along from producers to consumers in this marine food chain.

INTERACTIONS

The Top "Top Predator"

Ever since humans first developed weapons to help them hunt prey, we have been the top predator in any food chain. No other predator can compete with an armed human. Armed hunters can easily prey on any species that fulfills their food needs. Even large carnivores cannot escape from the ever-growing human population and its demand for food. Carnivores that are not considered to be edible are often hunted because they compete with humans for food sources.

People have had a major impact on the biosphere. Because of overhunting, many species have been reduced in numbers or have disappeared completely. The near extermination of the American bison and the sea otter shows the effect of excessive hunting on a species. In fact, the passenger pigeon, the great auk, and the Steller's sea cow were all hunted to extinction within the last few centuries.

Until about 10 thousand years ago, a community of large animals existed throughout North America. Then these animals seem to have disappeared suddenly. Mammoths, mastodons, cave bears, giant sloths, giant bison, and saber-toothed cats were all part of the North American megafauna. (The prefix *mega-* means "large," and *fauna* means "animals"; therefore, *megafauna* means "large animals.") Evidence indicates that human hunters had arrived in the Americas from Asia shortly before these animals began to disappear. It is thought that overhunting by these early Americans led to the extinction of the megafauna.

Figure 9-4 The manta ray (left) and the whale shark (right) are two fish that are plankton-feeders.

ever has lived; the whale shark, the largest living fish; and the manta ray, the largest member of the family that includes skates and rays. Some plankton-feeders are shown in Figure 9-4.

Coastal Food Chain

Almost 10 percent of the world's food production occurs in estuaries and coastal marshes. An estuary is an area where freshwater and salt water meet and mix. Estuaries and coastal salt marsh communities get much of their energy from the detritus food chain. These communities use the decomposed remains of *Spartina* grasses and algae to satisfy most of their energy needs. In the early stages of their development, the young of many food fish need these materials to supply nutrients and energy. In some coastal waters, mangrove leaves or kelp may be the main components of detritus. Because of their productivity, the protection of estuaries and coastal marshes is an ongoing conservation issue.

9.1 Section Review

1. What important role do scavengers play in ecosystems?
2. Give two examples of the types of organisms that occupy the various trophic levels in a food chain.
3. How are grazing food chains different from detritus food chains? You may use drawings and/or photographs to illustrate your answer.

9.2 ECOLOGICAL PYRAMIDS

When the food chains that make up a food web are examined, the following facts are observed.

- The farther along, or higher, an animal is on a food chain, the larger it usually is.
- As you move up a food chain, fewer animals occupy each succeeding trophic level.
- As you move up a food chain, the biomass, or total mass of organisms, is less at each succeeding level. embarrassing

Pyramid of Numbers

Let us examine the food chain in which caterpillars feed on leaves, robins feed on caterpillars, and hawks prey on robins. Along this food chain, robins are bigger than caterpillars but smaller than hawks. There are many more leaves than caterpillars, more caterpillars than robins, and more robins than hawks. In any ecosystem, only a small fraction of the plant material produced each year is turned into herbivores, and only a small part of this herbivore material is turned into carnivores. This type of relationship is often shown in the form of a pyramid, with a wide base and a narrow top.

As mentioned earlier, organisms higher up on the food chain usually are larger in size and fewer in number than organisms that are lower down. Therefore, it takes a huge number of producers to support just a few tertiary consumers. The pyramid of numbers represents this fact. The **pyramid of numbers** shows the number of organisms at each trophic level in an ecosystem. (See Figure 9-5.)

The amount of organic material at each trophic level is known as the **standing crop**. This represents the energy available to the next trophic level. Because there is a limited amount of energy available to support life, predators must always be fewer in number than their prey. If predators increase in number beyond their prey, starvation may result, bringing the number of predators back into balance.

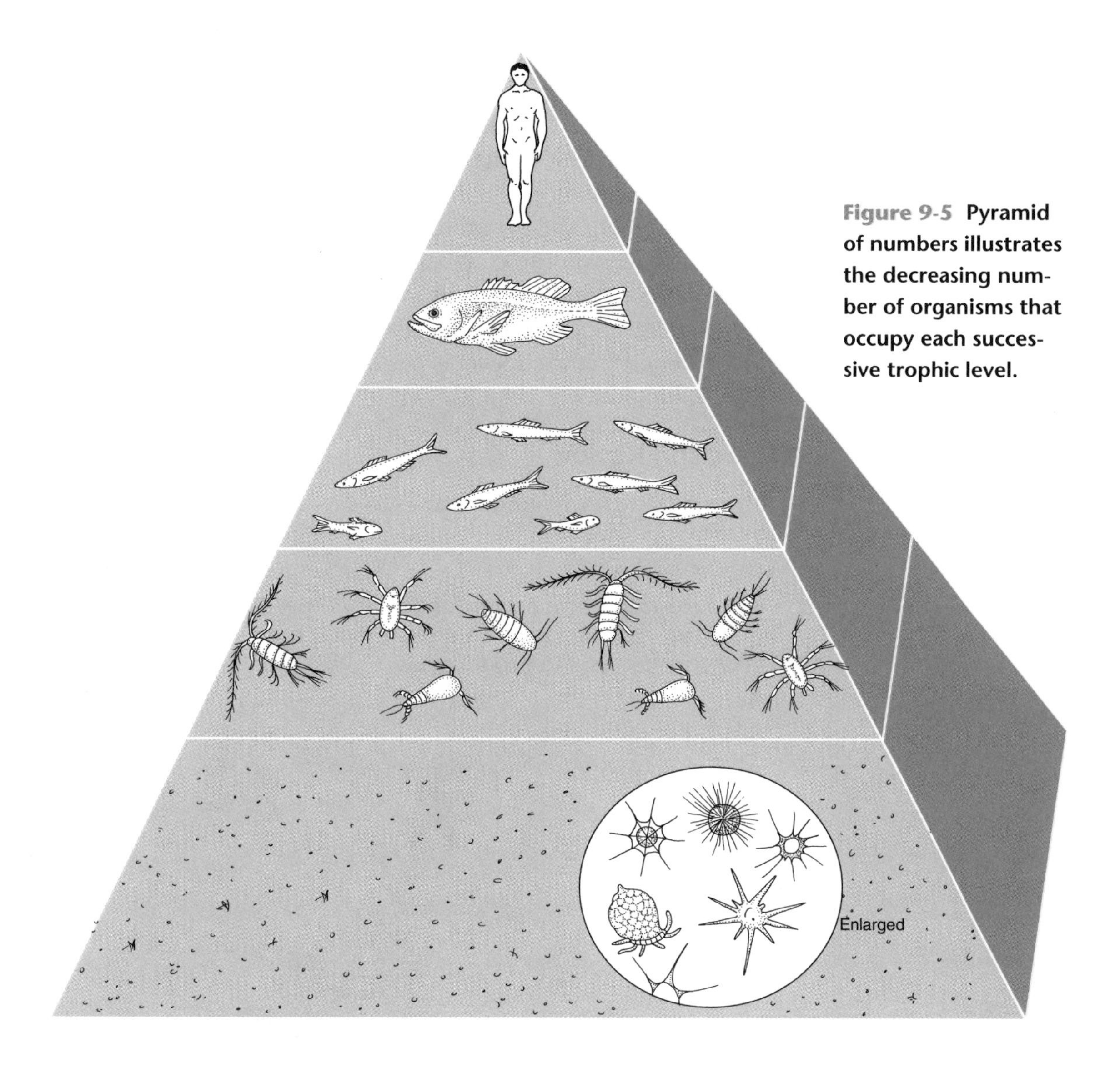

Figure 9-5 Pyramid of numbers illustrates the decreasing number of organisms that occupy each successive trophic level.

Pyramid of Mass

In the leaves-caterpillar-robin-hawk food chain, the total mass of leaves is greater than the total mass of caterpillars. The total mass of caterpillars is greater than that of robins. And the mass of robins is greater than that of hawks. These relationships describe the flow of matter and energy through an ecosystem. In a food chain, each individual usually is larger than the organisms it eats. However, the lower down on the food chain, the greater the number of organisms. Therefore, the total mass of organisms, or biomass, at each lower level is greater than the mass at the higher level. The **pyramid of mass** shows the biomass available at each trophic level. (See Figure 9-6.) The producers form the greatest mass, while large carnivores at the top of the pyramid form the smallest total mass. The total mass of the standing crop can be expressed as the biomass of that trophic level. The organisms at each level carry out activities that use or release most of the energy they obtain. Only 10 percent of the total energy at each level is passed along any food chain.

9.2 Section Review

1. Why are there fewer animals at each succeeding level of a food chain?
2. What is biomass?
3. On a field trip around your school or home, identify some organisms that are part of a typical food chain or pyramid. Label consumers and producers.

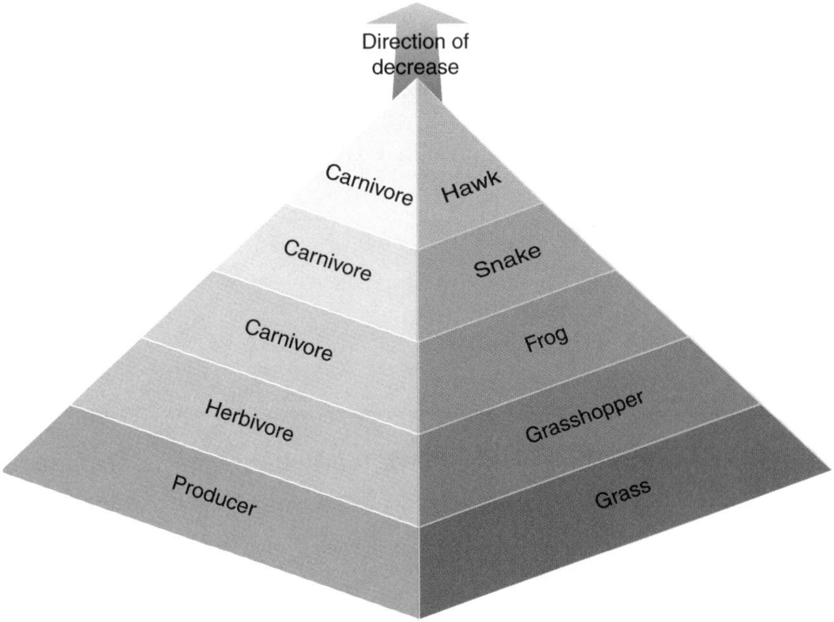

Figure 9-6 Pyramid of mass illustrates that at each successive trophic level the biomass decreases.

9.3 PRODUCTIVITY

Energy from the sun is taken and converted to chemical energy by plants. The productivity of an ecosystem is a measure of photosynthesis and energy storage by producers. It is, in effect, the rate at which green plants capture and store energy. Under similar conditions, different plants have different rates of production. The amount of energy stored depends on the number of plants present and the balance between photosynthesis and cellular respiration. Other environmental factors that affect productivity are the temperature range, the yearly rainfall pattern, and latitude, which determines the number of hours of sunlight.

The productivity of an ecosystem determines the population of consumers it can support. Different biomes store energy in the plant materials at different rates. As shown in Table 9-1, the tropical rain forest is the most productive biome, and the desert is the least productive.

Plants store energy in carbohydrates within a variety of structures. In carrots and beets, carbohydrate storage is in the swollen taproot.

TABLE 9-1 PRODUCTIVITY OF EARTH'S BIOMES

Most Productive	Tropical rain forest
↓	Coral reefs
	Tropical wetlands
	Temperate rain forest
	Temperate deciduous forest
	Tropical savanna
	Boreal coniferous forest
	Aquatic ecosystems (lakes, rivers, and streams)
	Temperate grasslands
	Marine ecosystems (continental shelf)
	Arctic tundra
	Marine ecosystems (open oceans)
	Chaparral
Least Productive	Desert

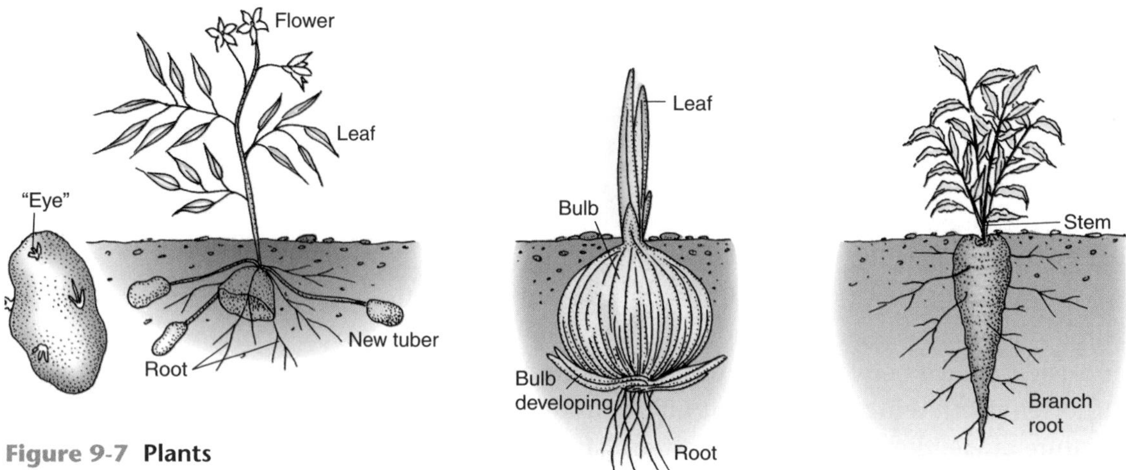

Figure 9-7 **Plants store carbohydrates in tubers (potatoes), bulbs (onions), and roots (carrots).**

In white potatoes, carbohydrates are stored in a tuber, an underground stem. Sweet potatoes store carbohydrates in the root. Onions store carbohydrates in a bulb, an enlarged stem. In artichokes, carbohydrates are stored in the fleshy leaves. Other plants store carbohydrates in fruits, nuts, and seeds, which serve as foods for animals. In turn, the animals aid in seed dispersal. Figure 9-7 illustrates some of the structures in which plants store carbohydrates.

People learned long ago how to identify and cultivate those plants that are highly productive and useful as food. An understanding of the factors that affect productivity has led to the use of chemical fertilizers and irrigation systems to maximize production of these crops. However, extensive farming of ecosystems that have high rates of productivity has led in some areas to the formation of deserts and a loss of biodiversity.

9.3 Section Review

1. Explain why productivity in green plants determines the consumer population in an ecosystem.
2. How do green plants store energy from the sun?
3. Construct a food chain that links the following organisms: pine marten (a weasel), pine nut, red squirrel, and lynx. Label each trophic level. You may use words, photographs, or drawings to describe the food chain.

LABORATORY INVESTIGATION 9

Releasing the Energy Stored in Food

PROBLEM: *Determine the amount of energy stored in different foods*

SKILLS: *Measuring, observing, calculating*

MATERIALS: *Shelled peanut, margarine, white bread, water, 3 test tubes, dissecting pin, Bunsen burner, triple-beam balance, stand, clamp, thermometer, 2 evaporating dishes, rubber stopper*

PROCEDURE

1. Use the balance to find the mass of one of the evaporating dishes.
2. Put a peanut in the evaporating dish and determine the mass of the evaporating dish with the peanut. Calculate the mass of the peanut by subtracting the mass of the empty evaporating dish from the mass of the dish with the peanut. Remove the peanut.
3. With the evaporating dish on the balance, place a small amount of margarine in the dish. Determine the mass of the dish and the margarine. Add or remove margarine until the mass of the dish and the margarine is equal to the mass of the dish and the peanut. Leave the margarine in the dish.
4. Determine the mass of the other evaporating dish. Follow the procedure in step 3 to determine a mass of white bread equal to the mass of the peanut. Leave the bread in the dish.
5. Place 10 mL of water into each of three test tubes.
6. Clamp one of the test tubes to the ring stand. Measure the temperature of the water.
7. Carefully force the blunt end of a pin into a rubber stopper. Secure the peanut on the sharp end. Place the rubber stopper under the test tube filled with water. Adjust the test tube so that its bottom is 3 centimeters above the peanut. Use the flame of the Bunsen burner to ignite the peanut. Let the peanut burn until the fire goes out.
8. Measure the temperature of the water in that test tube.

9. Repeat steps 6 through 8 with the margarine in the evaporating dish in place of the peanut and a different test tube of water. In this case, the pin is not needed. It is important to have the test tube of water the same 3 centimeters from the burning margarine.
10. Repeat steps 6 through 8 with the white bread.

OBSERVATIONS AND ANALYSES

1. Mass of empty evaporating dishes: ________ and ________
2. Mass of peanut: ________
   Mass of margarine: ________
   Mass of bread: ________
3. Temperature of water heated by burning peanut: ________
   Increase in temperature: ________
   Temperature of water heated by burning margarine: ________
   Increase in temperature: ________
   Temperature of water heated by burning white bread: ________
   Increase in temperature: ________
4. Which sample contained the most stored energy? ________
5. Which sample contained the least stored energy? ________

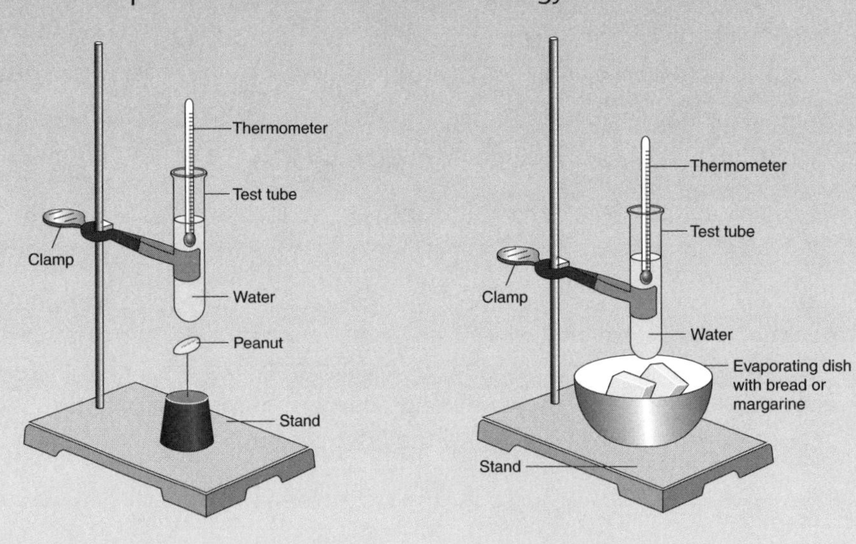

Chapter 9
Review

Answer these questions on a separate sheet of paper.

Vocabulary

The following list contains all the boldfaced terms in this chapter.

detritivores, detritus, detritus food chain, food chain, food web, grazing food chains, omnivores, pyramid of mass, pyramid of numbers, scavengers, standing crop, tertiary consumers, top predators, trophic levels

Fill In

Use one of the vocabulary terms listed above to complete each of the following sentences.

1. Animals that feed on dead organisms that they did not kill are ________.
2. Cattle and grasses are part of ________.
3. The feeding stages within a food chain are called ________.
4. A(an) ________ describes the flow of matter and energy through an ecosystem.
5. The detritus food chain begins with ________.

Multiple Choice

Choose the response that best completes the sentence or answers the question.

6. Which organism is an example of a primary consumer? *a.* tiger *b.* mule deer *c.* mole *d.* garter snake
7. Which organism is an example of a tertiary consumer? *a.* mountain lion *b.* armadillo *c.* raven *d.* moose
8. Which of the following is *not* a detritivore? *a.* wood louse *b.* snail *c.* millipede *d.* robin
9. Which animal is usually called a scavenger? *a.* elephant *b.* whale shark *c.* hyena *d.* grasshopper
10. In a marine ecosystem, which organism is *not* a scavenger? *a.* lobster *b.* isopod *c.* crab *d.* tuna
11. In most marine ecosystems, the primary producers are *a.* baleen whales. *b.* zooplankton. *c.* phytoplankton. *d.* isopods.
12. Which of the following factors does *not* affect the productivity of an ecosystem? *a.* rainfall *b.* temperature *c.* latitude *d.* longitude

13. The final step in every food chain is accomplished by *a.* predators. *b.* decomposers *c.* scavengers *d.* producers.
14. What is a primary source of energy in a detritus food chain? *a.* a decomposer *b.* a carnivore *c.* remains of plants and animals *d.* an omnivore
15. Which animal feeds on plankton? *a.* tuna *b.* killer whale *c.* swordfish *d.* blue whale

Short Answer (Constructed Response)

Use the information you learned in this chapter to respond to the following items.

16. What function do scavengers play in terrestrial ecosystems?
17. Describe the trophic levels that make up most food chains.
18. What is an omnivore?
19. Explain the differences between a food chain and a food web.
20. What is a detritus food chain?
21. What is detritus composed of?
22. What is shown in a pyramid of mass?
23. What factors control the productivity of an ecosystem?

Essay (Extended Response)

Use the information in the chapter to respond to these items.

24. Explain how a raccoon can be both a primary and a secondary consumer within the same ecosystem.
25. Describe the detritus food chains one might find in a mangrove salt-marsh community.
26. How have human activities affected regions of high productivity?
27. Determine your place in the food chain and food web by making a list of all the foods you consume in one day. Include breakfast, lunch, dinner, and snacks. For each food on your list, construct a food chain that ends with you as the final consumer. In each chain, label producers, consumers, and the feeding levels of the consumers. Where you can, make connections between feeding levels to form a food web.
    a. In which chains are you a primary consumer?
    b. In which chains are you a secondary consumer?
    c. In which chains are you a tertiary consumer?

CHAPTER 10
Oxygen and Carbon Dioxide

When you have completed this chapter, you should be able to:

Explain how the atmosphere was formed.

Discuss the greenhouse effect.

Describe what is meant by carbon sink.

Although our moon, as does Earth, receives energy from the sun, there are no living things native to the moon. Scientists explain that there is no life on the moon because the moon has no atmosphere. No atmosphere means no oxygen and no carbon dioxide. Oxygen is the gas in our atmosphere that supports Earth's many and varied forms of life. In this chapter, you will learn about the cycles that replenish oxygen and carbon dioxide in our atmosphere.

10.1 OXYGEN AND THE BIOSPHERE

At room temperature, oxygen is a colorless, odorless, and tasteless gas that occurs naturally in the atmosphere as a diatomic (two-atom) molecule with the molecular formula O_2. Oxygen is soluble in water. A substance is **soluble** when it dissolves in another substance. The amount of oxygen dissolved in water varies with the temperature and depth of the water. Cold water can hold more dissolved oxygen than warm water can.

Oxygen is the most abundant element on Earth. It reacts with many other elements to form compounds called oxides. Some of these compounds are basic to life, for example, water (H_2O) and carbon dioxide (CO_2). Reactions in which oxygen combines with other elements or compounds are called **oxidation reactions**. These reactions are often accompanied by the release of energy. Combustion, or burning, and aerobic respiration are oxidation reactions in which carbon is oxidized to form carbon dioxide. The following equations illustrate some oxidation reactions:

$$C + O_2 \xrightarrow{\text{COMBUSTION}} CO_2 + \text{energy}$$

CARBON + OXYGEN → CARBON DIOXIDE + ENERGY

$$2\,H_2 + O_2 \xrightarrow{\text{COMBUSTION}} 2\,H_2O + \text{energy}$$

HYDROGEN + OXYGEN → WATER + ENERGY

$$C_6H_{12}O_6 + 6\,O_2 \xrightarrow{\text{RESPIRATION}} 6\,H_2O + 6\,CO_2 + \text{energy}$$

GLUCOSE + OXYGEN → WATER + CARBON DIOXIDE + ENERGY

Formation of the Atmosphere

Before life appeared on Earth, there was little oxygen gas free in the atmosphere. Earth's oxygen was combined with other elements to form compounds such as water, carbon dioxide, iron oxide, and silicon dioxide. The early atmosphere was composed mainly of carbon dioxide (CO_2), methane (CH_4), and ammonia (NH_3). There were also small amounts of hydrogen (H_2) and helium (He). This mixture of gases would be poisonous to most organisms that now live in the biosphere. Chemical and physical reactions changed the composition of the original atmosphere. Some gases were removed through chemical reactions that created the minerals of Earth's crust. The lighter gases, such as hydrogen and helium, could not be held by Earth's gravity and escaped into space.

The appearance of the first primitive cells more than 3 billion years ago led to another change in atmospheric composition. These ancient organisms were photosynthetic anaerobes that did not require oxygen for respiration. They used carbon dioxide and hydrogen sulfide (H_2S) to synthesize nutrients and release energy. During photosynthesis, the ancient organisms produced oxygen as a byproduct and released it into the atmosphere. Molecules of oxygen were kept from escaping into space by the pull of Earth's gravity. This caused the concentration of oxygen within the atmosphere to increase. The stage was set for the evolution and survival of new forms of life.

As the concentration of oxygen gas increased in the atmosphere, it initiated other chemical processes that caused more changes on Earth. Some of the minerals that made up the rocks in Earth's early crust combined with oxygen to form oxides, silicates, and carbonates. This changed the composition of the lithosphere. During geologic processes, chemical changes in the lithosphere formed and released water. Water, which was a necessary medium for the continued evolution of life, became more abundant in the biosphere.

The evolution of the atmosphere and the evolution of the biosphere went hand in hand. The activity of photosynthetic organisms slowly built up the concentration of oxygen in the atmosphere. Today, 21 percent of the atmosphere is oxygen gas and 78 percent is nitrogen gas (N_2).

Figure 10-1 An ozone molecule is made up of three oxygen atoms.

Within the stratosphere, some oxygen molecules are changed to ozone when they absorb ultraviolet radiation from the sun. Ozone (O_3) is the triatomic (three-atom) form of oxygen. (See Figure 10-1.) In the stratosphere, the ozone layer forms a protective screen against ultraviolet radiation. Scientists have found that ultraviolet radiation may inhibit photosynthesis, damage DNA, and kill microorganisms. The formation of the ozone layer in the atmosphere helped life to colonize the land. The ozone layer is illustrated in Figure 4-3 on page 56. Increases in the amount of oxygen available in the atmosphere made possible the evolution of consumer organisms that could use the abundant green plants as a source of energy. Aerobic organisms, which then evolved, were able to feed on the producers and on each other. These were the forerunners of the animals that inhabit Earth today.

The Atmosphere Now

Living things cannot make use of many elements in their elemental, or uncombined, form. Since oxygen has the ability to combine chemically with other elements and form compounds, it is important to life. For example, elemental nitrogen (N_2) in the air becomes useful to living things only after it has reacted with oxygen to form nitrogen oxides, such as NO, NO_2, and NO_3. These compounds are created by oxidation reactions begun by lightning.

Erosion of Earth's crust exposes new surfaces to the air and minerals to oxidation. This ongoing process should use up oxygen in the atmosphere. However, the percent of oxygen remains relatively constant. There must be processes within the biosphere that replace the oxygen as fast as it is removed.

Biological systems, which get energy from the decomposition and oxidation of organic molecules, release carbon dioxide. Photosynthesis reactions, which produce carbohydrates, release oxygen. For the composition of the biosphere to remain steady, aerobic respiration of organic compounds and the photosynthetic generation of carbohydrates must balance each other. An increase in either will shift the balance and alter the composition of the atmosphere.

Both plants and animals use oxygen for respiration. It is readily obtained from the atmosphere by those organisms that live in terrestrial environments. Oxygen is scarce only at high altitudes where the air is thin, deep beneath Earth's surface, and at the bottom of very deep bodies of water. Organisms that live in aquatic ecosystems take dissolved oxygen from the water in which they live.

10.1 Section Review

1. Why is today's biosphere considered to be a product of Earth's first photosynthetic organisms?
2. Explain the importance of atmospheric ozone to the biosphere.
3. Why do you think modern city planners recommend extensive park areas within cities?

10.2 CARBON DIOXIDE AND THE BIOSPHERE

At room temperature, carbon dioxide is a colorless, odorless, and tasteless gas. It is soluble in water. Carbon dioxide makes up only about 0.03 percent of the atmosphere, and its concentration varies from place to place. Carbon dioxide enters the oceans where the water meets the atmosphere. Respiration by marine organisms also adds carbon dioxide to the water of the ocean. As with all gases, carbon dioxide is more soluble in cold water than in warm water. The oceans contain about 60 times as much carbon dioxide as does the atmosphere.

Carbon is the building block of all living things. Molecules of carbohydrates, proteins, and fats contain carbon atoms. Initially, green plants get carbon from carbon dioxide gas, which they take in during photosynthesis. Animals get carbon from the green plants or animals they eat. When animals oxidize food, they release energy and carbon dioxide. The carbon dioxide, the waste product of respiration, enters the ecosystem.

Respiration in living things is not the only process that releases carbon dioxide into the atmosphere. There is a huge reservoir of carbon dioxide bound up in carbonate rocks, for example, **limestone**. These rocks are readily dissolved by groundwater and carried to the oceans by rivers and streams. When fossil fuels, such as coal, oil, and natural gas, are burned, carbon dioxide is released into the atmosphere. The carbohydrates from which plants are composed release carbon dioxide when plant material is burned as fuel or in forest and brush fires.

Carbon Dioxide and Climate

Carbon dioxide plays a key role in moderating Earth's climate. Light energy from the sun penetrates the atmosphere and is absorbed by Earth's surface. Some of this energy is changed to heat, or **infrared radiation**, and sent back toward space. Part of the heat is absorbed by certain atmospheric gases, which causes the atmosphere to warm. The action of these gases is similar to the effect of the glass panels of a greenhouse that let light in and out but trap heat. This trapping of heat in the atmosphere is called the **greenhouse effect**. Carbon dioxide, water vapor, ozone, and methane are some of the **greenhouse gases** found in the atmosphere. The greenhouse gases make Earth 33°C warmer than it would be without these gases in the atmosphere. This natural greenhouse effect makes it warm enough for life as we know it to flourish. Figure 10-2 illustrates the greenhouse effect.

Carbon Sinks

Scientific research indicates that in the past, atmospheric concentrations of carbon dioxide may have been 15 times greater than they are today. Chemical reactions within the biosphere have tied up much of the carbon dioxide that was present in the past. Where has all this carbon dioxide gone? Much of it has been stored by living

Figure 10-2 Why are scientists concerned about the greenhouse effect?

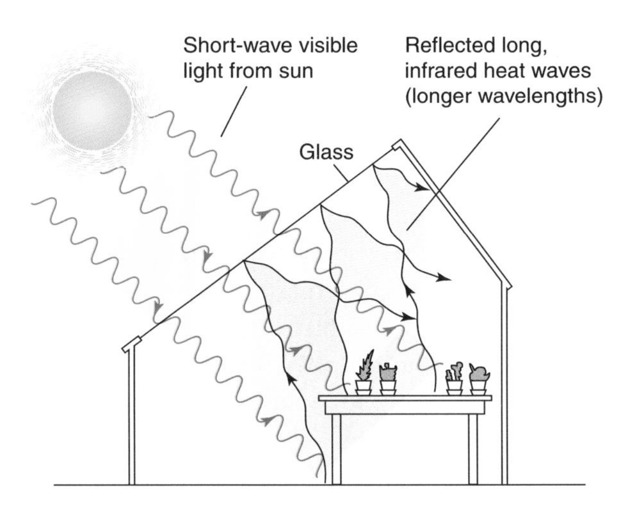

INTERACTIONS

Spring Will Be Early Next Year

Atmospheric observations conducted over the last 30 years indicate that climate changes are altering Earth's carbon dioxide cycle. During this period, rising temperatures have lengthened the growing season in the Northern Hemisphere. The rise in winter and spring temperatures has stimulated plants in this area to begin their growth about a week earlier than they did 20 years ago. The extended growing period has led to a change in the seasonal carbon dioxide cycle.

During the year, atmospheric concentrations of carbon dioxide naturally rise and fall due to the absorption of carbon dioxide by green plants during spring and summer, as part of the process of photosynthesis. During winter, microbes in the soil release carbon dioxide into the atmosphere as they decompose organic materials. Studies indicate that the spring decline in carbon dioxide concentrations has shifted forward by about seven days since the mid-1970s.

The rise in carbon dioxide levels is not only suspected to be a contributor to global warming, but is also expected to increase productivity in green plants. Carbon dioxide is the basic raw material plants use in photosynthesis. They use the energy in sunlight to convert carbon dioxide into carbohydrates and increase total biomass. Plants grow larger and faster under increased levels of carbon dioxide. For many years, agriculturists have added carbon dioxide to their greenhouses to raise the yields of fruits and vegetables.

If the trend continues at the present rate, the concentration of carbon dioxide in the atmosphere will double by 2050. This change will have far-reaching effects. Even though it will not pose a threat to human life, it will alter Earth's geography. As temperatures rise, ice caps and glaciers will melt, sea levels will rise, and climates will change worldwide. A doubling of the carbon dioxide concentration will also increase plant productivity by one-third. Food production would increase, making more food available to the peoples of the world. Since plants are at the bottom of food pyramids, a boost in productivity will increase animal populations and provide more feed for livestock.

We must view the rising carbon dioxide concentration as a two-edged sword. Will we be able to balance increased productivity with the decrease in land area and the reduction in land available for agriculture?

things in environmental "sinks." An **environmental sink** is a feature within the environment that traps a particular substance. Carbon **sinks** are large reservoirs of materials that contain carbon or carbon dioxide.

The living things that make up the biosphere are composed of compounds, such as carbohydrates, proteins, and fats, that were formed from carbon dioxide. Earth's forests, grasslands, and marine biomes are vast reservoirs of carbon dioxide. Deposits of fossil fuels represent another reservoir. Some marine organisms, for example mollusks, use carbon dioxide to form calcium carbonate ($CaCO_3$) for their shells. When these animals die, their shells sink to the ocean bottom. Here they form thick layers of sediment. In time, geologic processes can convert these sediments into limestone, which is formed from calcium carbonate. The chemical processes carried on in the biosphere tend to reduce the atmospheric concentration of carbon dioxide gas.

The interactions of the organisms in the biosphere create a delicate balance that maintains the amount of carbon dioxide available in the environment. These organisms, the biotic components of the atmosphere and oceans, are the storehouse for vast quantities of carbon dioxide. In the future, however, green plants may not be able to use carbon dioxide fast enough to remove the additional amounts people are releasing into the environment through the burning of fossil fuels. This may cause changes in global climate, which most surely will lead to changes in the biosphere.

10.2 Section Review

1. How do animals obtain carbon for building new tissues?
2. Explain how carbon is stored in environmental sinks.
3. How does our reliance on fossil fuels endanger the biosphere?

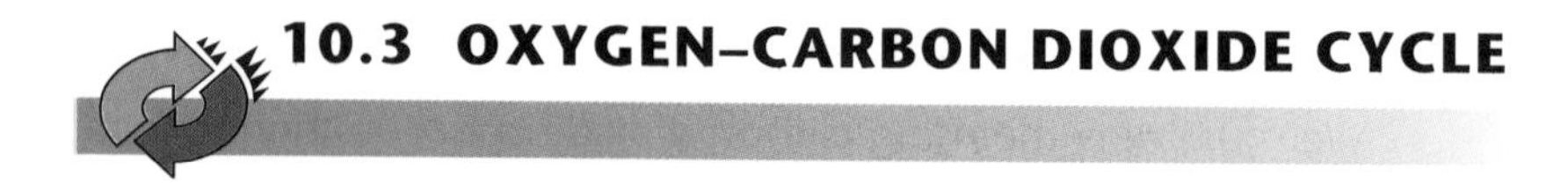

10.3 OXYGEN–CARBON DIOXIDE CYCLE

The processes that occur within ecosystems make use of and at the same time renew the biosphere's supply of oxygen and carbon diox-

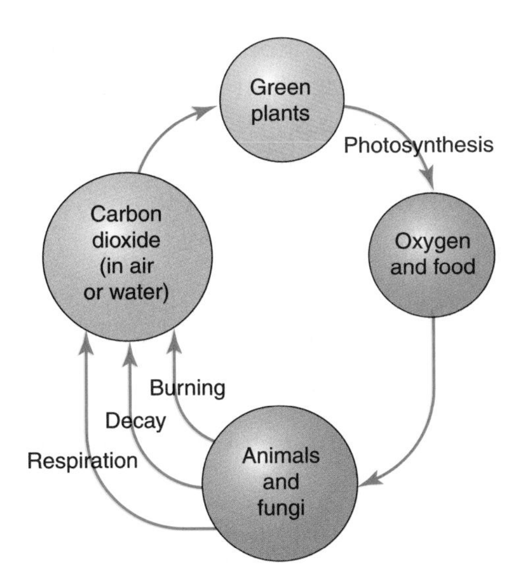

Figure 10-3 Oxygen–carbon dioxide cycle moves oxygen and carbon through the biosphere.

ide. The sum of all these processes is the **oxygen–carbon dioxide cycle**, or the carbon cycle. Carbon serves a dual purpose for all organisms. It is the structural component of organic molecules and the site of energy storage within the chemical bonds between carbon atoms.

In the oxygen–carbon dioxide cycle, photosynthetic organisms take in carbon dioxide and break it down into carbon and oxygen. (See Figure 10-3.) Oxygen, the by-product of photosynthesis, is released into the atmosphere. Carbon atoms are then used to make molecules of sugar, which can be used to form plant tissue or can be combined with oxygen to produce carbon dioxide and energy. Carbon dioxide and oxygen may be locked up in environmental sinks for long periods and not released into the biosphere. All of the processes that release carbon dioxide or oxygen remove materials from these sinks.

Chlorophyll played a major role in the development of the biosphere. This green compound, which gives leaves their color, enables plants to trap the energy in sunlight. Chlorophyll drives the process of photosynthesis. In these unique chemical reactions carried out inside the leaves of plants, all Earth's food is made and the atmosphere renewed with oxygen.

All life depends on a complex series of delicately balanced and interrelated chemical reactions. The oxygen–carbon dioxide cycle is an example of the varied interactions that occur between the abiotic and biotic portions of the biosphere. Just as organisms are storehouses of carbon, other reservoirs of carbon are available to organisms in the form of carbon dioxide gas, which is present in Earth's atmosphere and water. In the unending cycle, photosynthetic organisms remove the carbon dioxide and use it to build the organic molecules necessary for all life.

Carbon's Many Paths

Once synthesized into organic molecules, carbon follows one of several routes:

- It may be cycled up through the various trophic levels within the ecosystem.
- It may be returned to the environment through cellular respiration by plants or animals.
- It may be returned to the environment through the actions of decomposers.
- It may be stored.

Once stored within environmental sinks, in the form of living plant tissues such as wood, as fossil fuels, or as carbonate sediments, carbon dioxide is temporarily removed from the cycle. Figure 10-4 illustrates the carbon paths.

The route followed by an individual carbon atom may be simple or complex, depending on how it is used by an organism. What happens to a carbon atom in a molecule of sugar that is a part of a cookie you eat? Once digested, the sugar is absorbed by your bloodstream and made available to your cells for respiration or to make other organic molecules. If it is used for respiration, the carbon atom that was in the sugar will be exhaled as carbon dioxide within several hours. If the carbon atom is used to synthesize other molecules, it could become a permanent part of your cellular structure. The carbon would remain with you until death. Then decay would release it into the atmosphere as carbon dioxide.

The carbon atoms that make up the wood of trees, the limestone shells of marine animals, or fossil fuels may not be returned to the cycle for a long time. Decay bacteria and molds may eventually convert the wood in fallen trees to atmospheric carbon dioxide. Com-

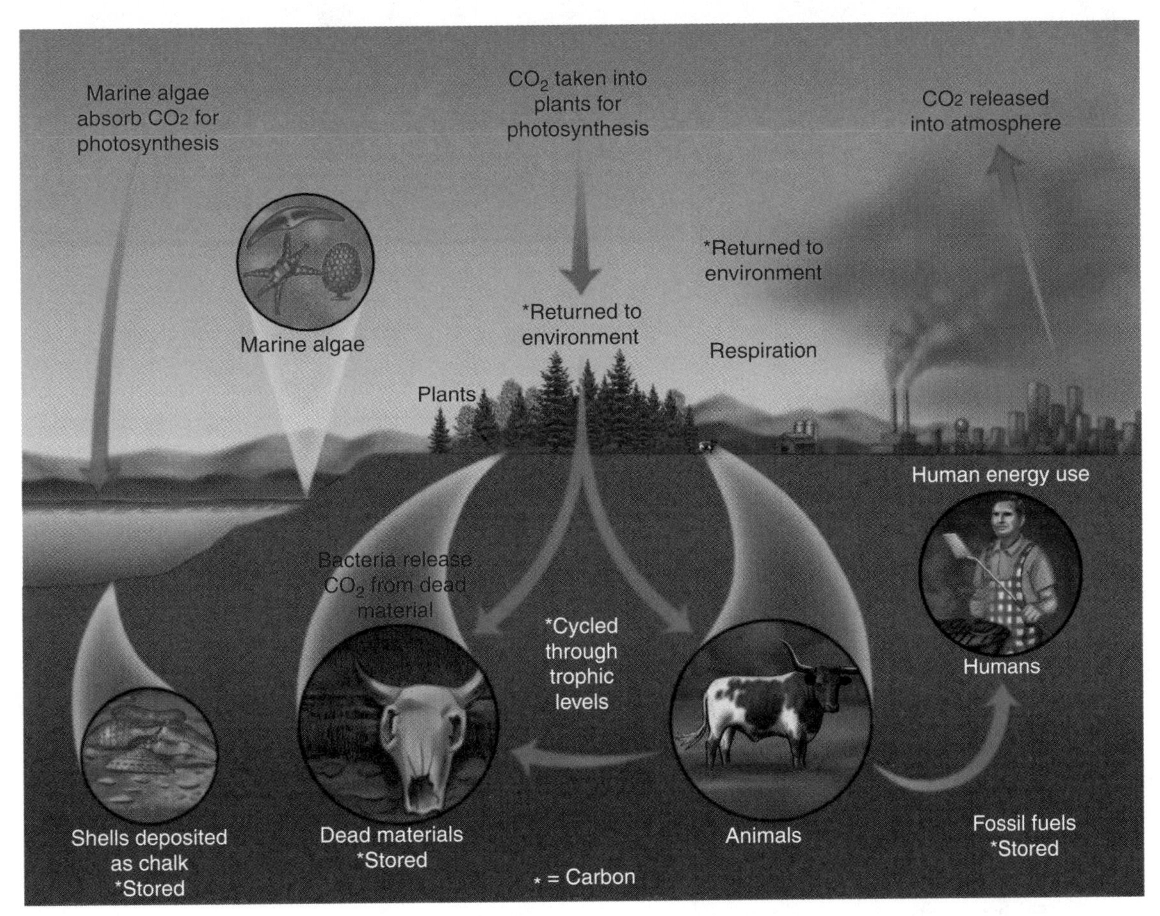

Figure 10-4 This illustration shows the possible paths carbon can follow as it travels through the biosphere.

bustion of the wood in a forest fire may shorten the time it takes for a carbon atom to be returned to the ecosystem. Even the carbon locked away in limestone deposits for thousands of years is eventually recycled. Geologic processes draw sedimentary deposits into the molten portion of the crust, and from there they may be returned to the atmosphere during volcanic eruptions. Burning of fossil fuels for energy sends huge quantities of carbon dioxide back into the biosphere.

10.3 Section Review

1. Describe the routes a carbon atom can follow through the oxygen–carbon dioxide cycle.
2. What two functions does carbon serve for living things?

LABORATORY INVESTIGATION 10

The Oxygen–Carbon Dioxide Cycle

PROBLEM: *To observe the oxygen–carbon dioxide cycle*

SKILLS: *Observing, manipulating*

MATERIALS: *2 liters tap water, 150-mL screw-cap culture tubes (8), labels, bromthymol blue, dropper, 4 elodea sprigs, 4 freshwater snails*

PROCEDURE

1. Let 2 liters of tap water stand uncovered for several days to allow the chlorine to escape. This forms dechlorinated water.
2. Divide the eight culture tubes into two sets of four. Label each set of four tubes from A through D. Fill each set of culture tubes with the dechlorinated water.
3. Add 3 drops of bromthymol blue to all eight tubes. (Bromthymol blue is used as an indicator for carbon dioxide. In the presence of an acid, bromthymol blue turns from blue to yellow. Carbon dioxide forms carbonic acid when it is dissolved in water.)
4. Cap both A tubes, which contain only water and bromthymol blue.
5. Into both B tubes place a 7.5-centimeter sprig of elodea or another aquatic plant. Cap the tubes.
6. Into both C tubes place a small freshwater snail. Cap the tubes.
7. Into each D tube place a 7.5-centimeter sprig of elodea or other aquatic plant and a small freshwater snail. Cap the tubes.
8. Place one set of tubes in artificial light. Place the second set in a closed box, where no light will reach them.
9. After 24 hours, examine both sets of tubes and record your observations.

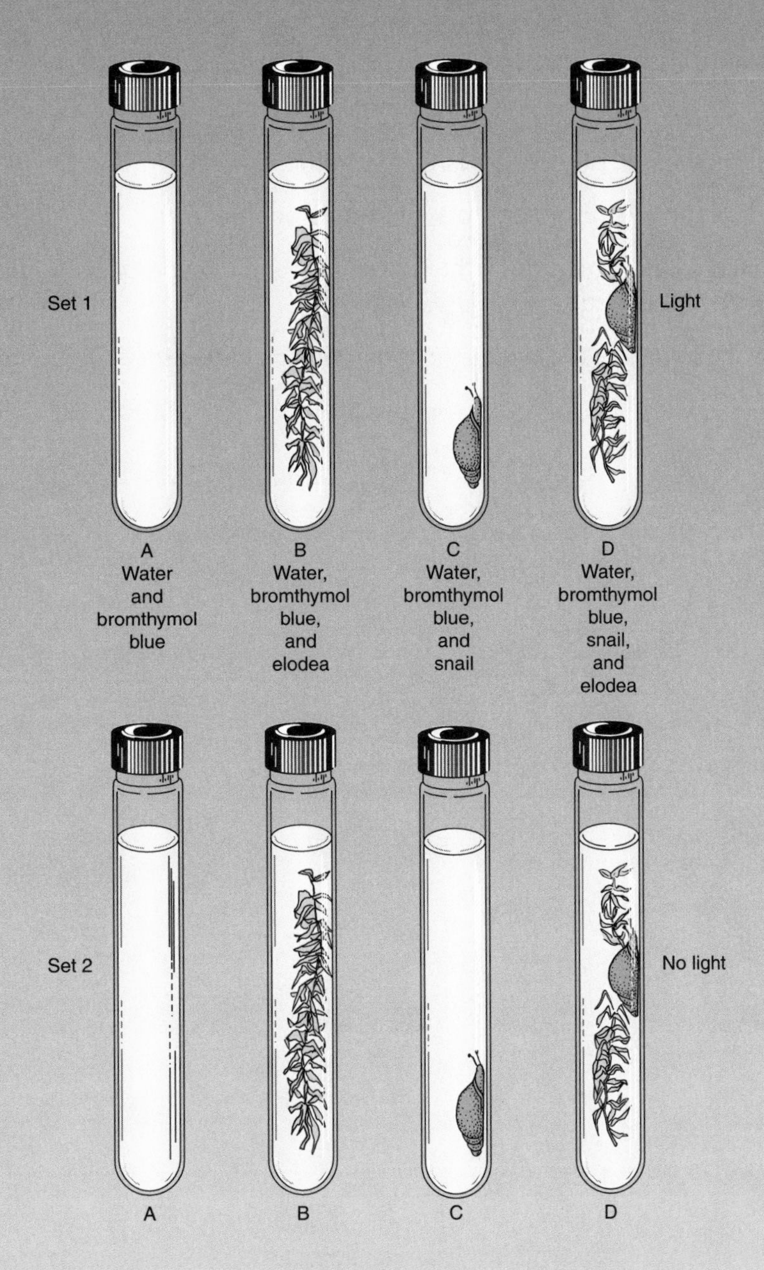
Set 1
Light
A
Water
and
bromthymol
blue
B
Water,
bromthymol
blue,
and
elodea
C
Water,
bromthymol
blue,
and
snail
D
Water,
bromthymol
blue,
snail,
and
elodea
Set 2
No light
A
B
C
D

OBSERVATIONS AND ANALYSES

1. In which tubes did the indicator change color?
2. Describe the role photosynthesis plays in this experiment.
3. Describe the role respiration plays in this experiment.
4. In which tubes do you think the oxygen–carbon dioxide cycle will be able to proceed the longest?
5. Draw a diagram of the oxygen–carbon dioxide cycle as it occurred in the test tubes.

Chapter 10 Review

Answer these questions on a separate sheet of paper.

Vocabulary

The following list contains all the boldfaced terms in this chapter.

carbon sinks, environmental sink, greenhouse effect, greenhouse gases, infrared radiation, limestone, oxidation reactions, oxygen–carbon dioxide cycle, soluble

Fill In

Use one of the vocabulary terms listed above to complete each of the following sentences.

1. Because oxygen gas dissolves in water it is said to be ______ in water.
2. Chemical reactions in which oxygen combines with another element are called ______.
3. Carbon dioxide, water vapor, ozone, and methane are some of the ______.
4. The heating of the atmosphere caused by reradiated energy is called the ______.

Multiple Choice

Choose the response that best completes the sentence or answers the question.

5. Reservoirs of calcium carbonate are *a.* carbon sinks. *b.* greenhouse gases. *c.* carbon cycle. *d.* oxygen cycle.
6. The sum of respiration and photosynthesis is *a.* greenhouse effect. *b.* oxygen–carbon dioxide cycle. *c.* transpiration. *d.* precipitation.
7. A mineral made of calcium carbonate is *a.* shale. *b.* quartz. *c.* halite. *d.* limestone.
8. The atmosphere is warmed as gases absorb *a.* infrared radiation. *b.* ozone. *c.* thermal springs. *d.* ions.

Short Answer (Constructed Response)

Use the information you learned in this chapter to respond to the following items.

9. What is meant by the term "carbon sink"?
10. Describe the role of anaerobic microbes in modifying Earth's primitive atmosphere.

11. Explain how the atmosphere protects Earth's life against the effects of ultraviolet radiation.
12. Describe the oxygen–carbon dioxide cycle.
13. Describe how the atmosphere is heated.
14. What role does limestone play in the oxygen–carbon dioxide cycle?
15. Why are the oceans considered a storehouse for carbon dioxide?

Essay (Extended Response)

Use the information in the chapter to respond to these items.

16. Describe the composition of Earth's early atmosphere.
17. Describe how greenhouse gases heat Earth.
18. Describe two ways the carbon in plant carbohydrates may be returned to the biosphere.
19. Your body contains a huge number of carbon atoms. Explain how it is possible for one of these atoms to have been part of a prehistoric creature.

CHAPTER 11
The Hydrologic Cycle

When you have completed this chapter, you should be able to:

Explain how hydrogen bonding affects the properties of water.

Define capillary action, specific heat, and precipitation.

Diagram the water cycle.

The water we use today has been recycling through the biosphere for billions of years. It rained on the dinosaurs and again on cave people. Where does rain come from? Most of it is water that has evaporated from the oceans. Why isn't it salty? Only water molecules evaporate, leaving the salts behind. In this chapter you will learn about the water cycle.

11.1 FRESHWATER: A LIMITED RESOURCE

Water is essential to life. Almost 75 percent of the mass of most living things is made up of water. It is constantly recycled among the atmosphere, lithosphere, hydrosphere, and biosphere. Thus water creates a variety of environments that can support life. Only 3 percent of Earth's water is freshwater, and 70 percent of that is locked in polar ice caps, permafrost, and mountain glaciers.

Water has unique chemical and physical properties that make it a highly versatile compound. These properties shape life at the molecular and cellular levels as well as determine the structure of the biosphere. A water molecule (H_2O) is composed of two atoms of hydrogen bonded to one atom of oxygen. The atoms in a water molecule are not arranged in a straight line. The molecule is v-shaped with the oxygen at the point and the hydrogen atoms at each end. The shape of the water molecules creates an unequal distribution of electrical charge on each molecule; thus water is a polar molecule. By definition, a **polar molecule** has an unequal distribution of charge. The oxygen end of the molecule is slightly negative, while the end that contains the two hydrogen atoms is slightly positive. (See Figure 11-1.) The polarity of water molecules allows them to interact with one another to form hydrogen bonds, weak electrostatic forces of attraction between adjacent molecules.

The tendency of molecules to attract one another creates internal **cohesion** within water. Cohesion is responsible for surface tension and capillary action. **Surface tension** is the formation of a surface

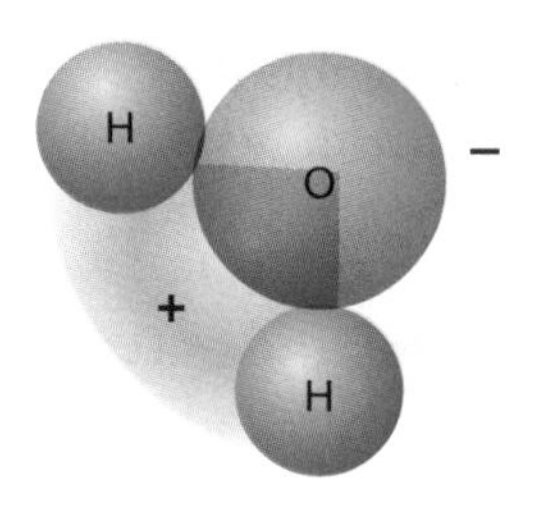

Figure 11-1 The polarity of water molecules is responsible for the attraction of one water molecule to another.

film that is strong enough to support some aquatic organisms, for example the water strider. It is surface tension that makes water droplets spherical. **Capillary action** is the physical property that pulls water up through the vascular tissues of plants.

Water and Living Things

The polarity of water is also responsible for its ability to dissolve a wide variety of substances. Water is called the universal solvent because many different substances dissolve in it. This property affects all living things because almost all the compounds necessary for life must be dissolved before cells can use them. Many mineral salts, sugars, amino acids, and fatty acids are transported throughout organisms in a water-based fluid. Animal blood and plant sap are water-based fluids. Plant roots absorb water that contains soluble nutrients from the soil. Animals rely on water in their respiratory tissues to absorb oxygen from the air.

Water plays a vital role in the biological processes that occur in living things. Within every cell, water is the medium in which all of the chemical reactions that sustain life occur. In addition, water plays an active part in many chemical reactions that take place in cells. Most organisms have a basic physiological need for water. Humans require approximately 2 liters of water each day. In animals, water is an excretory medium for eliminating the waste products of cellular metabolism.

Earth, unlike other planets in the solar system, is blessed with an abundant supply of liquid water. Water exists throughout Earth in three forms, or states: solid water (ice), liquid water, and gaseous water (water vapor). Water's different forms are the result of changes in hydrogen bonding caused by heating or cooling. Water exists as a liquid over a wide temperature range. It freezes to form a solid at 0°C and boils to form a gas at 100°C. Because of this 100-degree (Celsius) temperature range, water is in its liquid form over most portions of Earth throughout the year.

The polarity of water molecules is responsible for its ability to resist changes in temperature. **Specific heat** is the amount of heat energy needed to raise the temperature of 1 gram of a substance by

1°C. The higher the specific heat, the greater the resistance to changes in temperature. The specific heat of water is very high, which allows water to store large amounts of heat without its temperature changing very much. Large bodies of water can absorb and hold vast amounts of heat and thus have a moderating effect on the surrounding environment and, to a large extent, control the climate of an area.

Because of its high specific heat, water also resists freezing by losing energy at a very slow rate. But at or below 0°C, molecules of water become locked together and form solid ice. Water at first contracts when it is cooled. But, unlike most liquids, at temperatures near its freezing point, water expands. Because of this, solid ice is less dense than liquid water, so it floats. The formation of a surface layer of ice insulates underlying layers. This prevents deep lakes from freezing all the way down to their bottom, which would kill the aquatic organisms. As ice is heated, the water molecules absorb energy, and the bonds between them are weakened. When the molecules absorb enough energy to break free of the hydrogen bonds, they form liquid water.

11.1 Section Review

1. Why is water such a precious commodity, even though Earth has a seemingly abundant supply?
2. Describe some of the physical properties of water that are due to the polarity of water molecules.

11.2 EVAPORATION, CONDENSATION, AND PRECIPITATION

Global processes of evaporation and condensation occur night and day, recycling water through the biosphere. These processes are driven by energy from the sun. When liquid water on Earth's surface is heated by sunlight, the molecules absorb energy. This weakens the hydrogen bonds between molecules. Some water molecules at or near the surface gain enough energy to break the hydrogen bonds, become water vapor, and escape into the atmosphere. The process by which liquid water changes into water vapor is **evaporation**. Evaporation can take place over a wide range of temperatures. When water evaporates, any impurities in the water are left behind.

The reverse of evaporation is condensation. In **condensation**, water vapor changes to liquid water. In general, cooling leads to condensation of water vapor and the formation of tiny droplets of water suspended in air; these droplets may accumulate to form clouds. Condensation occurs when moist air cools as it passes over a mountain range, when warm moisture-laden air comes in contact with cool air, and when warm, moist air rises to cooler levels.

The amount of moisture the air can hold is limited and depends on the air temperature and pressure. The warmer the air, the greater is its water-holding capacity. When this capacity is reached, the air becomes saturated. When the amount of moisture in the air exceeds the saturation level, condensation occurs and, depending on the temperature, dew, clouds, mist, fog, snow, or rain may form.

For rain or snow to form, air must contain small particles, called **condensation nuclei**, upon which water vapor can condense. Many droplets must come together to make one drop of rain or one snowflake. Before a drop of rain or a snowflake can fall, it must gather enough mass to allow the pull of gravity to overcome the forces holding it in the atmosphere. Water that evaporates from the oceans may fall on continents that are some distance away.

Gravity plays a role in the recycling of water by bringing water back down to Earth as precipitation. Precipitation is water that falls from the sky as rain, sleet, hail, or snow. Gravity moves the water along in streams and causes it to seep into the ground.

The **water cycle**, or **hydrologic cycle**, describes the circulation of water through the biosphere. The water cycle is illustrated in Figure 11-2 on page 178. Solar energy is the driving force for the water cycle. Heat from the sun evaporates water from the surface of oceans, lakes, and seas.

Plants play an important role in the water cycle by absorbing water from the soil and pumping it into the atmosphere during transpiration, the transport of water through a plant and its evaporation through stomata on leaf surfaces. In tropical rain forests, as

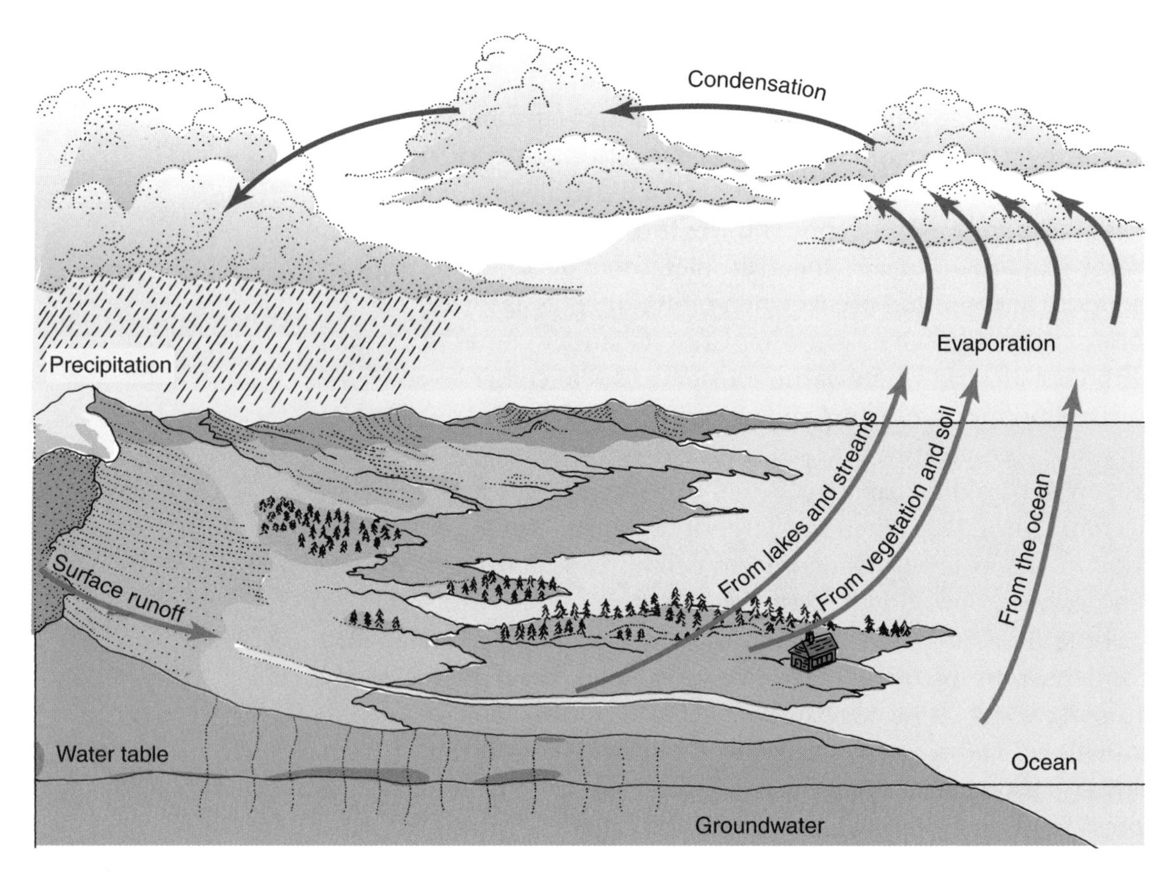

Figure 11-2 **The water cycle.**

much as 75 percent of the annual rainfall is returned to the atmosphere through transpiration by plants. Respiration by plants and animals also releases water into the atmosphere.

Precipitation and the Water Cycle

In cooler regions within the atmosphere, water vapor condenses into droplets or ice crystals. When droplets and ice crystals become heavy enough, precipitation occurs. When precipitation occurs over the oceans, the cycle is completed.

When precipitation occurs over the land, the water may be retained by plants; accumulate as puddles, snow cover, or ice; enter the groundwater supply; or flow along the surface as runoff, emptying into large bodies of water. Ice or snow can eventually melt, which frees the water to rejoin the cycle. The water cycle is part of the larger system of world climate.

Some precipitation never reaches the ground but evaporates on the way down; the water vapor then reforms clouds. Prevailing winds may transport this atmospheric moisture great distances.

From the surface, water sometimes moves underground to saturate the soil or across Earth's surface into rivers, lakes, and seas. This process supplies the biosphere with freshwater, plays a role in maintaining Earth's climate, and moderates worldwide temperatures.

The water cycle that occurs over a particular region is of great interest to hydrologists, civil engineers, and water managers. This is known as the **water budget**, or water balance for that region. The water budget is a measure of precipitation, use, and storage of water. It is determined by the differences between the inflow and the outflow of water for a definite time period. Inflow is based on precipitation and other sources of water that enter the area. Outflow is the total of runoff and evaporation that occurs. If outflow exceeds inflow, there is no storage. The water budget for a municipal reservoir affects the public's use of water supplies. Figure 11-3 compares

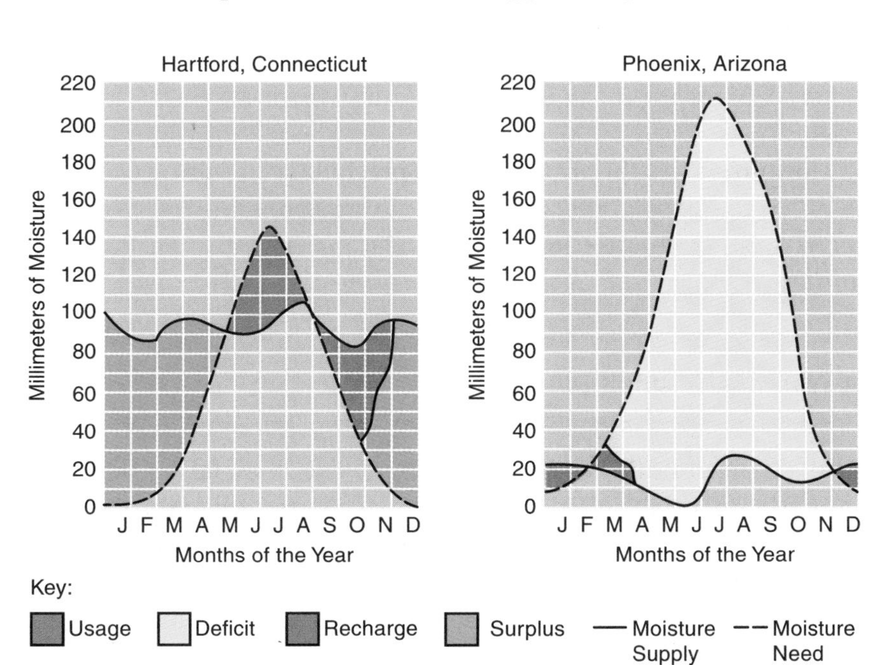

Figure 11-3 The water budget for Phoenix, AZ, and Hartford, CT, is shown in this illustration. Where does the water budget show the greatest deficit?

the water budget for Phoenix, Arizona, which is in the desert, with that of Hartford, Connecticut, which is near the Atlantic coast.

11.2 Section Review

1. Compare the processes of evaporation and condensation.
2. In what way might the burning of fossil fuels lead to increased rates of precipitation?
3. What forces drive the water cycle?
4. Explain why it is important for municipalities to determine a water budget for their local regions.

11.3 WATER STORAGE AND TRANSPORT

There is approximately the same amount of water on Earth today as there was during the ice ages or at the time of the dinosaurs. The oceans play a major role in the storage and transport of Earth's water supplies. Because of the huge surface area of the oceans, large quantities of water evaporate into the atmosphere. In the atmosphere, water cools and condenses to form clouds or precipitation. Clouds move about on the winds, which are fueled by solar energy.

Moisture leaves the atmosphere as precipitation. During transport between the oceans and continents, the atmosphere serves as a huge reservoir of freshwater in the form of water vapor, water droplets, and ice crystals.

A Water Molecule's Journey

In the Amazon rain forest, a water molecule may fall as part of a drop of rain, be absorbed by the roots of a plant, leave through the stomate of a leaf, and return to the atmosphere, all within a single day. But water molecules that are frozen into polar ice may take thousands of years to be recycled back into the atmosphere.

Figure 11-4 The Grand Canyon formed as the Colorado River eroded layer upon layer of rock.

There is a large underground supply of freshwater in the form of groundwater, but much of this is locked deep beneath Earth's surface. Soils near Earth's surface are reservoirs for some groundwater. However, much is lost by evaporation and seepage into deep underground storage areas.

Some of Earth's freshwater is present in the atmosphere as water vapor and droplets, which can be observed when cooling causes the formation of clouds. But most of Earth's freshwater supplies are locked in cold storage, in polar ice caps, glaciers, and permafrost.

Precipitation hitting the ground and the flow of water along Earth's surface are major forces in **erosion**, the geological process that wears away Earth's crust. Figure 11-4 shows the Grand Canyon, which was cut by the Colorado River. Flowing water can dissolve minerals within rocks, which causes them to break apart. The broken parts can be transported by water over Earth's surface to the seas and oceans. Flowing ice in glaciers also breaks up and transports crustal materials. During transport, rocks are broken up into smaller and smaller pieces. **Sedimentation** is the accumulation of small pieces of eroded rock in a body of water, such as a lake or an ocean, and leads to the deposition of large quantities of crustal materials

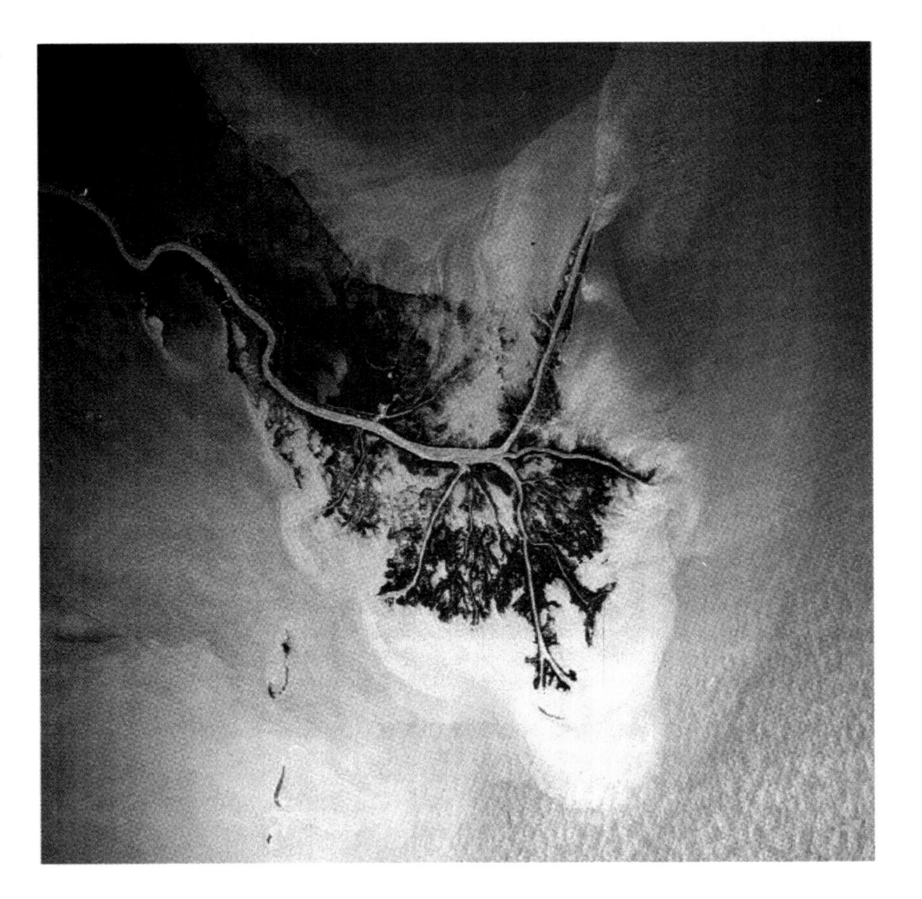

Figure 11-5 The Mississippi River Delta is formed from the sediment carried by the river.

in ocean basins. Figure 11-5 shows the delta formed by the sediments carried by the Mississippi River.

11.3 Section Review

1. Describe some of the processes by which water shapes the land.
2. Explain where Earth's freshwater supplies are stored.
3. Make a diagram to explain the processes by which water is transported between the oceans and the continents. Illustrate your diagram with photos or original drawings.

LABORATORY INVESTIGATION 11

The Water Cycle

PROBLEM: *To observe evaporation and condensation*

SKILL: *Making observations*

MATERIALS: *Small (4-L) aquarium, clear plastic wrap, lamp with 100-watt bulb, food coloring, ice cubes*

PROCEDURE

1. Place about 3 centimeters of water in the bottom of a small aquarium.
2. Completely cover the top of the aquarium with clear plastic wrap.
3. Place the aquarium in the sunlight and observe it. Look for signs of evaporation.
4. Place several ice cubes on top of the plastic wrap and observe.
5. Carefully open the plastic wrap and add a few drops of food coloring. Cover the aquarium and put the ice cubes back on top.
6. Direct the light from a 100-watt bulb onto the water in the aquarium. Note any change in the rate of evaporation.

OBSERVATIONS AND ANALYSES

1. Can you see evaporation taking place in the aquarium?
2. How can you indirectly determine that evaporation is taking place?
3. With the ice cubes in place, could you detect the formation of "clouds" or fog?
4. What effect did the ice cubes have on the rate of evaporation?

5. How did the light of the 100-watt bulb affect the rate of evaporation?
6. Does the food coloring appear in the water that condenses on the plastic?
7. Draw a diagram of this activity and use it to illustrate the various factors in the water cycle.

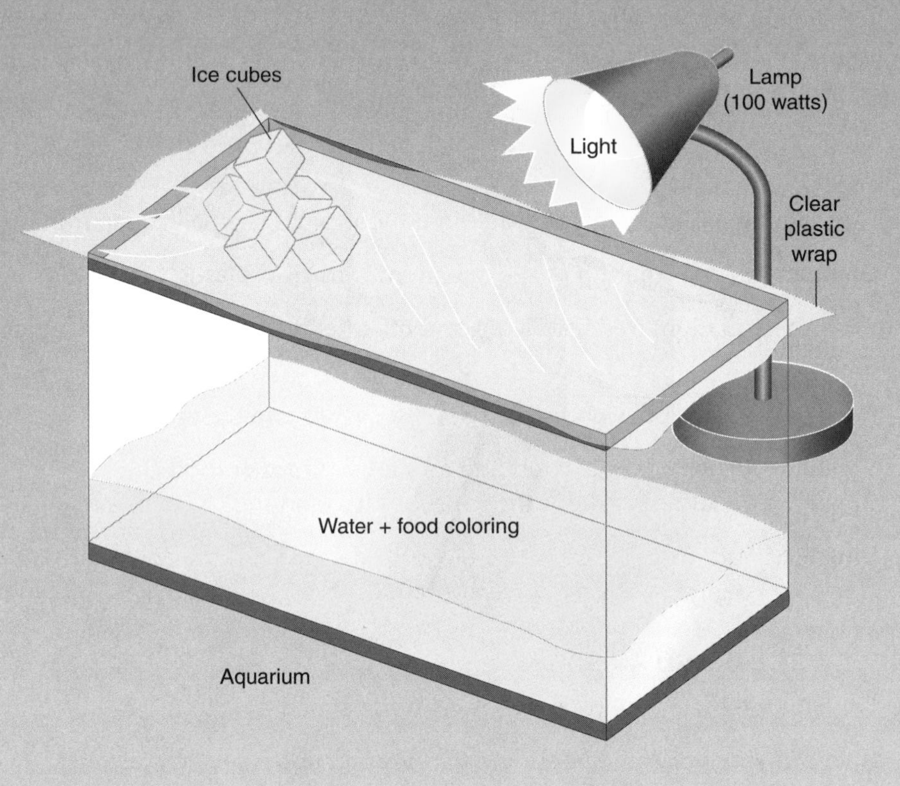

Chapter 11 Review

Answer these questions on a separate sheet of paper.

Vocabulary

The following list contains all the boldfaced terms in this chapter.

capillary action, cohesion, condensation, condensation nuclei, erosion, evaporation, hydrologic cycle, polar molecule, sedimentation, specific heat, surface tension, water budget, water cycle

Fill In

Use one of the vocabulary terms listed above to complete each of the following sentences.

1. The amount of heat needed to rasie the temperature of 1 gram of a substance 1°C is ______.
2. The tendancy of molecules to attract one another is ______.
3. ______ is the accumulation of small pieces of eroded rock in a body of water.
4. ______ is the geological process that wears away the land.
5. The ______ describes the circulation of water through the biosphere.

Multiple Choice

Choose the response that best completes the sentence or answers the question.

6. A molecule of water is composed of *a.* one atom of oxygen and one atom of hydrogen. *b.* one atom of oxygen and two atoms of hydrogen. *c.* two atoms of oxygen and one atom of hydrogen. *d.* two atoms of oxygen and two atoms of hydrogen.
7. One of the factors in the hydrologic cycle that is controlled by gravity is *a.* evaporation. *b.* condensation. *c.* precipitation. *d.* transpiration.
8. Which of the following is *not* a form of precipitation? *a.* rain *b.* snow *c.* water vapor *d.* ice
9. What percent of the mass of most living things is water? *a.* 30 percent *b.* 50 percent *c.* 70 percent *d.* 75 percent
10. The collection, purification, and transport of water through the biosphere is driven by energy from *a.* gravity. *b.* sunlight. *c.* Earth's interior. *d.* Earth's magnetic field.

11. The capacity of air to hold moisture increases as *a.* air is warmed. *b.* air is cooled. *c.* moisture is added. *d.* the wind blows.
12. As water freezes, its density *a.* increases. *b.* decreases. *c.* remains the same. *d.* decreases, then increases.

Short Answer (Constructed Response)

Use the information you learned in this chapter to respond to the following items.

13. Why is water a polar molecule?
14. What is surface tension?
15. How do plants utilize capillary action?
16. Explain what a water budget is.
17. What is meant by evaporation?
18. What are some of the causes of erosion?
19. Describe the water cycle.
20. Explain why sources of freshwater are important resources.
21. Why are bodies of water able to store large quantities of heat?
22. Why are condensation nuclei important to precipitation?

Essay (Extended Response)

Use the information in the chapter to respond to these items.

23. How do the physical properties of water affect life on Earth?
24. How does running or falling water contribute to the erosion of Earth's crust?
25. Describe the path a drop of water might follow through the hydrologic cycle.
26. A warming or cooling of the worldwide climate would affect the hydrologic cycle. Choose either warming or cooling and tell how it would affect people where you live.

CHAPTER 12
Nitrogen and Phosphorus Cycles

When you have completed this chapter, you should be able to:

Describe the role of scavengers and decomposers in the biosphere.

Explain the nitrogen cycle and the phosphorus cycle.

Discuss how detritivores increase the fertility of soil.

Scavengers are part of the biosphere's waste disposal system. They eat the bones and carcasses of dead animals. Microorganisms further decompose what the scavengers leave behind and return the components to the biosphere where they can be used again. This is one aspect of the ecosystem's cleanup effort. In this chapter, you will learn about the importance of the nitrogen and phosphorus cycles.

12.1 OUR SELF-CLEANING ENVIRONMENT

In most natural communities, there are carnivores that do not actually kill animals for food. These carnivores, called scavengers, feed on **carrion**, the remains of already dead animals. Scavengers play a vital role in the community by disposing of the remains of animals that have been killed by predators or died of natural causes. Many predators are part-time scavengers that will feed on carrion when it is available. Depending on the animals on which a scavenger feeds, it may be a secondary or tertiary consumer.

Large, carrion-eating birds, as illustrated in Figure 12-1, show many specialized features that help them feed on animal carcasses. Condors, vultures, and buzzards feed by thrusting their head and neck into a carcass. These birds have no feathers on their head and neck, which helps them stay relatively clean as they feed. Some carrion-eating birds possess large, powerful beaks that can tear flesh and hides. Others can only make use of carcasses already torn open.

Lammergeier vultures take the larger bones from scavenged carcasses, carry the bones to great heights, and drop them onto rocks. The bones crack open and expose the rich marrow inside. The vultures swallow the broken bones. Enzymes in the birds' stomach enable them to digest both marrow and bone.

The spotted hyena is a scavenger found throughout most of Africa. Armed with powerful jaw muscles and strong teeth, hyenas

Figure 12-1 Carrion-eating birds, such as the king vulture, play an important role in recycling materials. The king vulture opens carcasses with its sharp beak, giving access to other scavengers.

can crack the leg bones of large animals such as buffaloes, hippos, and elephants to get at the marrow. This provides them with a food source that other carnivores cannot obtain.

The Role of the Decomposers

Within every community, the food chain cycles energy from producers to consumers. Part of the food material taken in by animals is unusable and is returned to the biosphere as waste matter in the form of droppings, or dung. The droppings are rich sources of energy for other organisms. The death of an organism, either a plant or an animal, does not stop this cycle. Decomposers, such as bacteria and fungi, perform the next step in the cycle. (See Figure 12-2.) They **decay**, or break down, dead tissues and wastes, changing them into simpler compounds that are returned to the community. Some of these inorganic compounds are carbon dioxide (CO_2), ammonia (NH_3), water (H_2O), and minerals such as iron (Fe), magnesium (Mg), phosphorus (P), and sulfur (S).

Figure 12-2 **Fungi, such as mushrooms, decompose organic wastes.**

Decomposers and scavengers are the final links in nature's recycling process. Without them, dead organisms and animal droppings would accumulate and eventually overwhelm the ecosystem. Fortunately for all life, when an animal or plant dies or when an animal deposits waste droppings in the environment, this material is attacked quickly by decomposer organisms. The organic materials are converted into inorganic nutrients that can be used again by producer organisms.

The Role of Fungi

Mushrooms, molds, mildews, rusts, yeast, and smuts are different types of fungi. They are plantlike organisms that reproduce by means of spores, lack chlorophyll, and have no vascular tissue to transport fluids. Refer to Figure 7-5 on page 116, which illustrates the structure of several fungi. Most fungi are decomposers. They grow in soils that contain large amounts of organic matter. Fungi are multicellular organisms usually made up of a cluster of slender threads, called hyphae, though some exist as individual cells.

Fungi have an important role in the decomposition of wood. Mushrooms feed on both living and dead trees, but mainly they attack older, dying, or dead ones. Fungi have visible spore-producing structures, such as mushroom and toadstool caps, puffballs, and shelf fungi. The spore-producing structures attach to the tree by a network of threadlike parts called mycelia, which penetrate deep into the wood. The mycelia produce enzymes that break down the wood into compounds the fungus can absorb.

Because of the chemical reactions fungi carry on, people have found many uses for them. Some fungi are used in the manufacture of cheese, such as Roquefort and Camembert. Others are used to produce antibiotics, for example penicillin. Mushrooms are used as food. Fermentation by yeast is part of the process used in the manufacture of bread, beer, and wine. Some beneficial fungi form relationships with forest trees and orchids, aiding the plants in obtaining nutrients. In return for shelter, some microscopic fungi enter into relationships with ants and termites and help these organisms digest their food.

Not all fungi are beneficial. Harmful fungi are responsible for diseases in humans, for example athlete's foot and ringworm. Others cause plant diseases such as Dutch elm disease, which destroyed many of the elm trees in the United States.

12.1 Section Review

1. How are some carrion-eating birds specialized to feed on dead animals?
2. Explain how decomposers are a final step in the energy cycle.
3. What role do fungi play in the forest?

12.2 CYCLING NITROGEN THROUGH THE BIOSPHERE

A relatively unreactive gas, nitrogen makes up 78 percent of the atmosphere by volume. Nitrogen plays an important role in the biosphere. It is essential for the synthesis of amino acids, which are

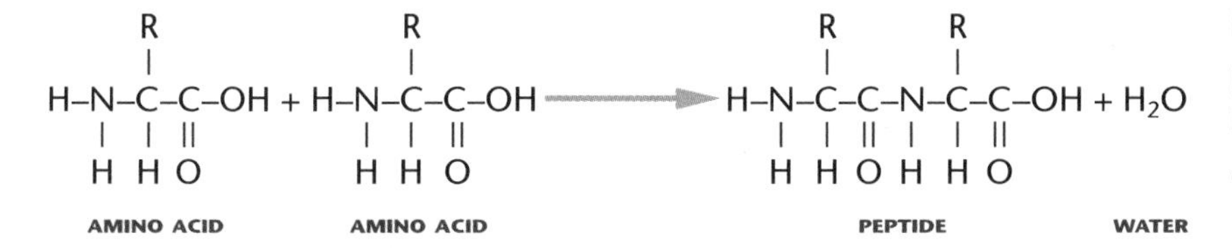

Figure 12-3 The peptides are formed by the joining of amino acids.

the building blocks of proteins and the genetic materials DNA and RNA. The atmosphere is rich in nitrogen gas (N_2). However, because it is nonreactive, it cannot be used by living organisms in its elemental form. Many green plants need nitrogen in the form of **nitrates**, compounds that contain nitrogen, oxygen, and at least one other element. Potassium nitrate (KNO_3) and sodium nitrate ($NaNO_3$) are two such compounds. These substances are water soluble. Therefore, plant roots can absorb them from the soil. Plant cells synthesize amino acids from nitrates and carbohydrates. They use amino acids to build peptides and peptides to build proteins, which are long chains of peptides. (See Figure 12-3.)

At each trophic level, nitrogen compounds are converted into living tissue. Some nitrogen is excreted into the environment as the waste products of protein metabolism. **Nitrogenous wastes**, composed mainly of urea and uric acid, usually are dissolved in urine and sweat. Decomposers break down nitrogenous wastes, as well as the tissues of dead plants and animals, in a series of complex chemical reactions.

The Nitrogen Cycle

The **nitrogen cycle** is a series of natural processes that cycle nitrogen through the biosphere. The nitrogen cycle is illustrated in Figure 12-4 on page 192.

Nitrogen can follow any one of several paths within the cycle. Decay bacteria break down proteins into amino acids and ammonia. **Nitrifying bacteria** take in the ammonia and convert it to nitrites. Another group of bacteria change nitrites to nitrates. The nitrates are released into the soil where they are again available to green plants. **Denitrifying bacteria** act on soil nitrates and nitrites to produce nitrogen gas, which returns to the atmosphere to complete the cycle.

Figure 12-4 The nitrogen cycle shows how this important component of proteins is moved through the biosphere.

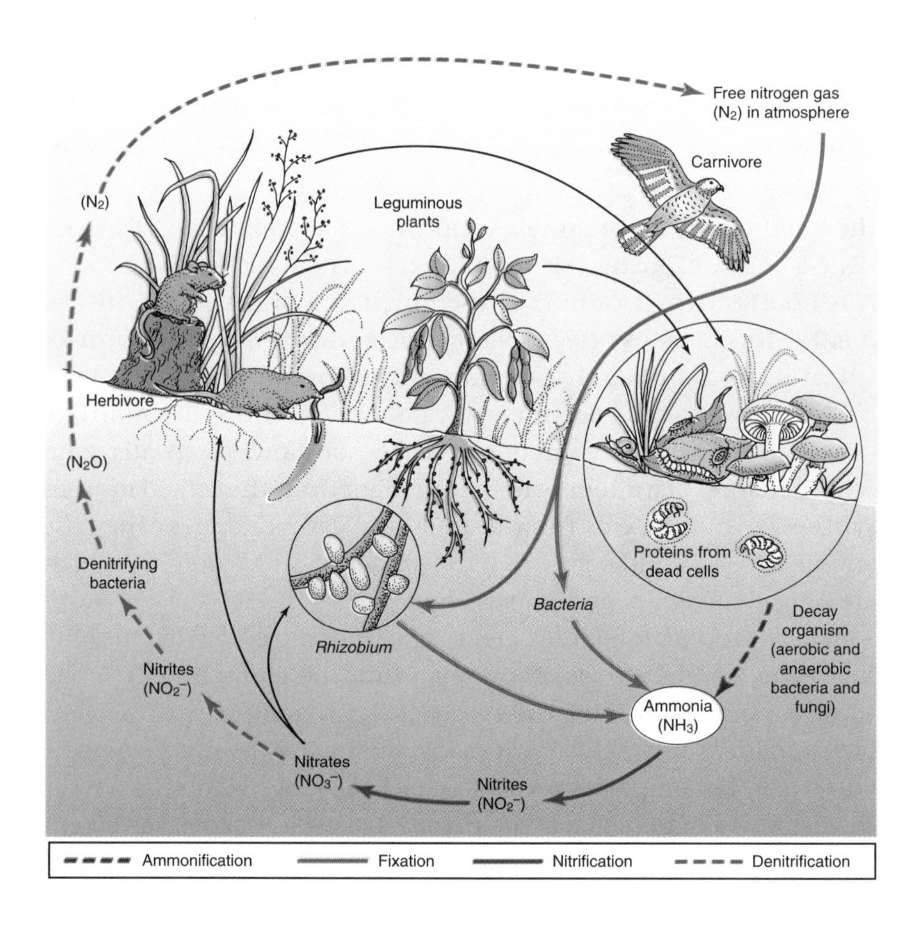

Various processes in the biosphere create alternate pathways for nitrogen to cycle through the ecosystem. During thunderstorms, some atmospheric nitrogen gas is turned into nitrates by lightning. Nitrogen from the lithosphere is recycled to the atmosphere by volcanic processes. Specialized nitrogen-fixing bacteria, called *Rhizobium,* live in nodules on the roots of **legumes**, such as beans, peas, alfalfa, and clover. (See Figure 12-5.) **Nitrogen-fixing bacteria** convert free nitrogen from the air into nitrates. In exchange for the shelter and nutrients they obtain from the root nodules, the bacteria supply the plants with nitrates. Cyanobacteria can also fix nitrogen to produce nitrates and thus make these compounds available to the ecosystem.

The action of nitrogen-fixing, nitrifying, and denitrifying bacteria continually replenish the supply of nitrates in an ecosystem and cycle nitrogen into the atmosphere. But human intervention can create breaks in the nitrogen cycle. Clearing fields for agriculture can remove nitrogen-bearing wastes and destroy those plants that support nitrogen-fixing bacteria. On the cleared land, each new crop of plants removes nitrates from the soil without there being any way to replace them. Eventually, the soil is not able to support the growth of any plants.

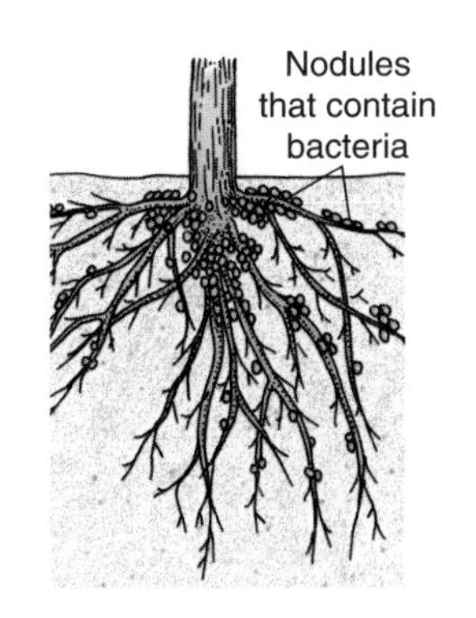

Figure 12-5 The roots of a legume have nodules that contain bacteria, which are able to convert nitrogen from the air to forms that can be used by plants.

The Importance of Nitrogen

Farmers have developed many methods to maintain productive soils. Unused plant materials are plowed back into the soil. In some areas, animal wastes, such as manure, are spread over the soil to replace lost nitrates. Legumes, which can replace nitrates, are planted every two or three years in place of other crops. This practice is called **crop rotation.** Another way to replace lost nitrates is to spread chemicals, called **fertilizers**, over the land. Fertilizers may contain materials in addition to nitrates necessary for plant growth, including potassium and phosphates. Fertilizers play an important role in maintaining the high productivity that is characteristic of modern agriculture.

Water that runs off heavily fertilized cropland contains high concentrations of nitrates and phosphates. The accumulation of fertilizers in groundwater can contaminate drinking water supplies. Fertilizers that get into lakes and ponds stimulate the growth of algae. When the algae use up all the available nutrients, they die. The decomposition of the dead algae uses up the oxygen dissolved in the water, which leads to the death of other organisms in the lake or pond.

12.2 Section Review

1. Describe the role nitrates play in plant metabolism.
2. Explain the relationship between nitrogen-fixing bacteria and legumes.
3. Why are fertilizers used extensively in modern agriculture?

12.3 MINERALS RETURN TO THE SOIL

The thin layer of soil that covers much of Earth's land area is the foundation of the biosphere and an important resource. Not all the soil on Earth's surface is fertile enough to support plants. **Fertility** is a measure of the soil's ability to support a plant community. Soil structure and composition are the most important factors in determining fertility. The structure of fertile soil must be loose enough to allow water to pass through it, dissolving minerals, and to allow plant roots to penetrate deep enough to reach the water.

Naturally fertile soil, called **loam**, contains particles that range, in increasing size, from clay to silt to sand. These particles are derived from rocks that have been weathered, that is, broken down by rainwater, atmospheric gases, ice, and plant roots. The soil is interspersed with cracks, pores, and crevices that contain air or water. Dead and decaying organic matter, called **humus**, is a vital portion of soil. Humus acts like a sponge and retains water and is a source of dissolved minerals. An army of living organisms, composed of bacteria, fungi, and small invertebrates, makes up yet another portion of the soil.

The Role of Living Things

Soil teems with varied forms of life. In 30 grams (about 1 ounce) of fertile soil, there can be one million bacteria, 100 thousand yeast cells, 50 thousand fungi, and thousands of invertebrates. Without these decomposers (bacteria and fungi) and detritivores, materials would not be broken down into nutrients that plants can use. Soil can be thought of as a living organism that supports plant growth through a continuous, self-regenerating process. As soil organisms die, their remains also are broken down to enrich the soil so that it maintains its fertility.

Without the organic materials provided by living organisms, soil would be just a collection of mineral particles. The conversion of organic materials into soil nutrients is an important process in the production of soil. Different types of soil vary in their nutrient content. Loam is a rich, fertile soil that can support a large and varied plant community. Sandy or rocky beaches and desert soils that have a low nutrient content can support few producers.

Topsoil is the upper layer of the soil that contains particles of weathered rock and organic matter. (See Figure 12-6.) An examination of topsoil shows the importance of decomposers to soil production. Leaves, bark, and other debris are found in various states of decomposition. The result is a nutrient-rich humus. The spongy humus traps water. Nutrients from the soil are dissolved in the water, and the water is absorbed by the roots of plants and used in the synthesis of organic materials. Soil rich in humus can support a large plant community and is termed fertile soil.

Earthworms, soil mites, insects, and crustaceans are invertebrate organisms. These detritivores feed directly on organic detritus to produce humus. Detritivores are important agents in the formation of soil. Earthworms eat their way through the soil, mixing the organic and inorganic portions together, thus enriching the soil for use by plants. Worms eat decaying organic material in the soil and

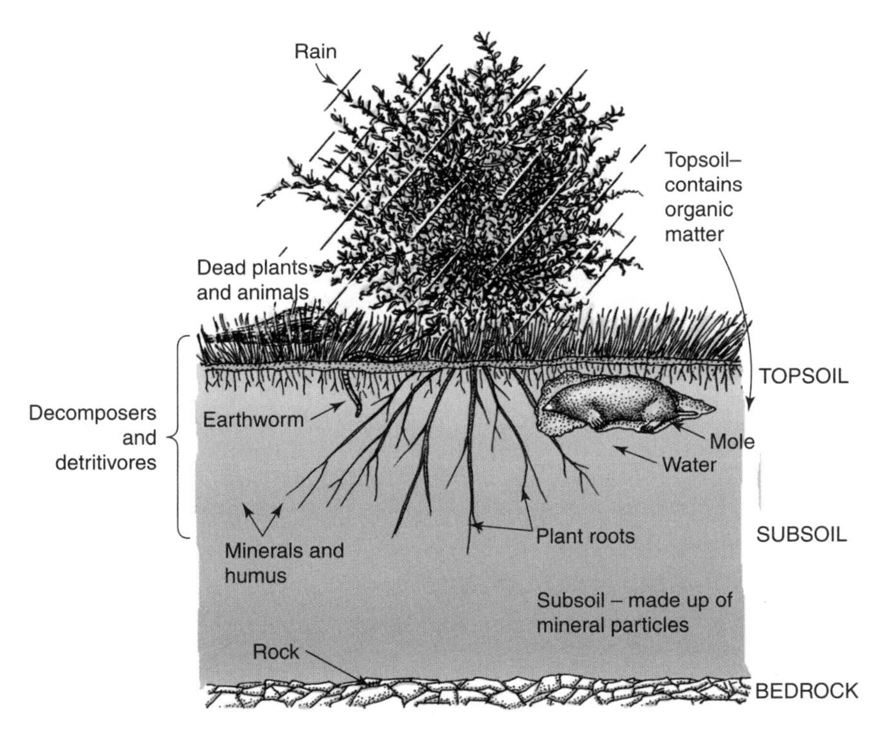

Figure 12-6 This soil profile shows a view of the top few meters of soil. In what horizon is most of the organic material found?

deposit the undigested wastes underground. Bacteria and fungi decompose the droppings, further enriching the soil.

Earthworms also bring material from below ground to the surface, where it accumulates in piles called **casts.** This activity helps to turn over the soil. The casts are a rich source of nutrients. The tunnels that earthworms form aerate the soil and channel rainwater to lower levels, which increases the overall fertility of soil. Snails, slugs, wood lice, and millipedes are some other detritivores that take part in natural cycles that enrich the soil. These detritivores start the breakdown of organic compounds and thus return important minerals to the ecosystem.

The formation of fertile topsoil is a very slow process. It may take up to 100 years or more to form about 2.5 centimeters of topsoil. Natural disturbances and human activities can reverse the process very quickly. Soils can disappear almost overnight. The dust bowl conditions that affected the midwestern United States during the 1930s scoured the rich topsoil from the land in just a few years.

12.3 Section Review

1. What are the characteristics of fertile soils?
2. What are some of the physical factors found in loam?
3. How does the presence of earthworms increase soil fertility?

12.4 PHOSPHORUS IN THE ECOSYSTEM

Phosphorus is one of the most important minerals used for plant and animal growth. Most organisms require phosphorus in relatively large amounts, but only small quantities are available within most ecosystems. This greatly affects the productivity of ecosystems.

The Phosphorus Cycle

The **phosphorus cycle**, illustrated in Figure 12-7, describes the movement of phosphorus through the biosphere. In most soils,

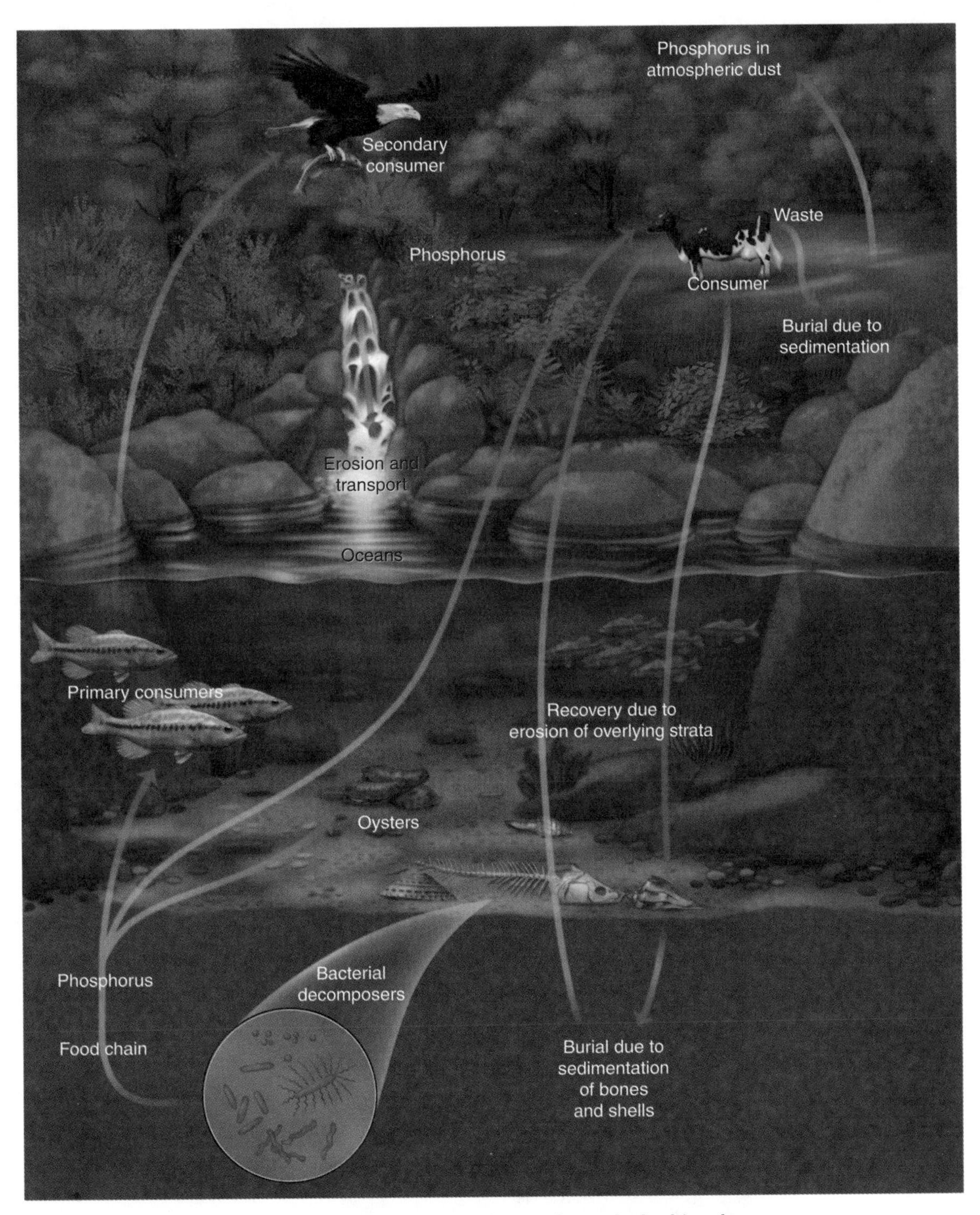

Figure 12-7 **The phosphorus cycle moves phosphorus through the biosphere.**

igneous rocks are the chief source of phosphorus compounds, called **phosphates**. Igneous rocks form as magma cools below Earth's surface. Volcanic eruptions spew phosphate-rich rocks onto Earth's surface, where they are weathered into particles of soil. The most fertile soils are found in regions where there has been past volcanic activity. As part of the phosphorus cycle, water-soluble phosphate minerals are readily absorbed by plant roots and passed through the trophic levels within an ecosystem.

Plants need phosphorus for proper development of roots, stems, flowers, and seeds. In plants and animals, phosphorus is necessary for the production of DNA and RNA. Phosphorus is needed for the production of ADP, adenosine diphosphate, and ATP, adenosine triphosphate. These molecules transfer energy in living cells.

A major function of phosphate in animals is the formation of bone, a material not easily broken down by decomposers. When animals die, phosphate may become locked in their skeletal remains and not be readily available for use by producers. Erosion often removes these materials from the ecosystem and deposits them in the ocean depths. Deep ocean sediments become rich sources of phosphorous. It may take many millions of years for geologic processes to make this phosphate available again to the biosphere.

Some phosphorous from deposits of bird and bat droppings, called **guano**, is directly available for recycling. Deposits of guano are mined and sold as fertilizer. The bones of animals slaughtered for food are ground to make bonemeal, another rich source of phosphate fertilizers. During the 19th century, many buffaloes were slaughtered indiscriminately, and their bones were left to litter the American prairies. These bones were finally gathered and used commercially to produce phosphate fertilizer.

12.4 Section Review

1. How is phosphorus used by organisms?
2. Explain why it may take a long time for phosphorus to cycle through the ecosystem.
3. Draw a diagram of the phosphorus cycle. Illustrate your diagram with original drawings or photographs. What role do people play in the cycle?

LABORATORY INVESTIGATION 12

Testing Soils

PROBLEM: *To measure the pH of several soil samples*

SKILLS: *Measuring, observing*

MATERIALS: *4 zippered, plastic sandwich bags; incubator; marking pencil; mortar and pestle; 100-mL test tubes (5); distilled water; hydrion acid-base indicator paper; glass stirring rod*

PROCEDURE

1. Use the plastic bags to collect soil samples from at least four different environments. Label each as to where it came from and the type of soil it is.
2. Thoroughly dry the samples. This can be done by placing the samples in an incubator or spreading the samples out on trays or by letting them stand for a week or more until they feel dry to the touch.
3. Measure 10 grams of the first soil sample and grind it into a fine powder with a mortar and pestle. Place the measured sample into one of the 100-mL test tubes. Label the test tube. Repeat with the remaining samples, thoroughly cleaning the mortar and pestle after grinding each sample.
4. Add 50 mL of distilled water to all five of the test tubes, stopper them, and thoroughly shake them. Let the sediments settle for about 10 minutes.
5. Place a 3-cm strip of hydrion paper on the table in front of each test tube. Dip a glass stirring rod into the water in the first test tube. Place a drop of water on the hydrion paper in front of that test tube. Repeat with the remaining test tubes, cleaning the stirring rod in distilled water each time.
6. Compare the color of the hydrion paper with the hydrion reference chart.
7. Record the pH of each sample.

OBSERVATIONS AND ANALYSES

1. List those soils that are acid, that is, have a pH less than 7.
2. List those soils are basic, that is, have a pH greater than 7.
3. What was the purpose of the tube that contained only water?
4. Based on the origin of each of the soil samples, what type of pH (acid or basic) has the greatest effect on the growth of plants?

Chapter 12 Review

Answer these questions on a separate sheet of paper.

Vocabulary

The following list contains all the boldfaced terms in this chapter.

carrion, casts, crop rotation, decay, denitrifying bacteria, fertility, fertilizers, guano, humus, legumes, loam, nitrates, nitrifying bacteria, nitrogen cycle, nitrogen-fixing bacteria, nitrogenous wastes, phosphates, phosphorus cycle, topsoil

Fill In

Use one of the vocabulary terms listed above to complete each of the following sentences.

1. Naturally nutrient-rich, fertile topsoil is called ________.
2. Earthworms deposit nutrient-rich material on the soil surface in the form of ________.
3. Green plants can use nitrogen only when it is in compounds called ________.
4. ________ is the measure of the ability of the soil to support a large plant community.
5. ________ act on soil nitrates to return nitrogen to the atmosphere.

Multiple Choice

Choose the response that best completes the sentence or answers the question.

6. Planting clover instead of corn every few years is called *a.* irrigation. *b.* crop rotation. *c.* the nitrogen cycle. *d.* fertilization.
7. Deposits of bird and bat feces are rich in compunds containing *a.* carbon dioxide. *b.* salt. *c.* nitrogen. *d.* pesticides.
8. Which of the following is not a fungus responsible for decomposition? *a.* mushroom *b.* mildew *c.* mold *d.* termite
9. Breaking down dead animal tissues is called *a.* decay. *b.* nitrifying. *c.* nitrogen fixing. *d.* fermentation.
10. Which of the following organisms do not play a role in the nitrogen cycle? *a.* nitrogen-fixing bacteria *b.* nitrifying bacteria *c.* denitrifying bacteria *d.* cyanobacteria

Short Answer (Constructed Response)

Use the information you learned in this chapter to respond to the following items.

11. What is carrion?
12. Describe the composition of topsoil.
13. Explain how earthworms enrich the soil.
14. Why are green plants not able to use the elemental nitrogen in the atmosphere?
15. What are some examples of nitrogenous wastes?
16. What are legumes?
17. Describe some of the methods used by farmers to replace soil nitrogen lost through agriculture.
18. Why is guano useful for agriculture?
19. What is the nitrogen cycle?
20. Why are ocean sediments good sources of phosphorus?

Essay (Extended Response)

Use the information in the chapter to respond to these items.

21. How do scavengers protect the biosphere against the spread of disease?
22. What is the role of decomposers in the various cycles within an ecosystem?
23. Why is soil sometimes considered to be a living thing?
24. Draw a diagram that combines all the cycles you studied in this unit: carbon cycle, oxygen–carbon dioxide cycle, water cycle, nitrogen cycle, and phosphorus cycle. Illustrate your diagram with original drawings and/or photographs. What role do people play in these cycles?

UNIT FOUR

ADAPTING TO THE ENVIRONMENT

We are constantly adapting to changes in our environment without giving it much thought. When we feel cold, we put on a sweater or jacket or turn up the thermostat. In warm weather, we dress in lightweight clothing and turn on the air conditioner or fan.

Other creatures are not as lucky. Animals and plants that live in the desert must be adapted to conditions that include hot days, cool nights, and lack of water. In this unit, you will learn how living things are adapted to their environment and what can happen to them when their environment changes.

CHAPTER 13
The Changing Biosphere

When you have completed this chapter, you should be able to:

Define natural selection.

Explain how geographical isolation and reproductive isolation lead to the evolution of new species.

Discuss how mutations increase diversity.

Offspring that are the result of sexual reproduction receive half their genes from one parent and half from the other. This mixing of genes in each generation produces organisms with a wide range of characteristics. Some characteristics increase an organism's chance of survival in Earth's ever-changing biosphere. The surviving organisms pass on these helpful traits to their offspring. This process is the basis of evolution. Therefore, as the environment changes, species that are present may change; older species adapt or new species move in.

13.1 CLUES FROM THE PAST

The only thing constant about the biosphere is that it is always changing. The movement of continental plates creates changes in the crust; the climate alternates between warm eras and ice ages; dry periods seesaw with wet periods; and mountains rise up only to be eroded away. Everything changes, including the plants and animals that make up the biosphere. These changes are a continuous, modifying force within the biosphere. Every species has the potential to evolve into a new species. The interactions between organisms and their environment have been going on since life began. Whenever we observe the ways living things interact, it is necessary to take into account the history of their environment and the evolutionary path followed by their ancestors.

The Fossil Record

An examination of fossils from around the world reveals an unbroken progression of life that goes back more than 3.5 billion years. The Earth on which life first appeared was quite different from the planet we live on today. Most living things on Earth today would have found the early environment quite inhospitable. On ancient Earth, surface temperatures were very high, thick clouds obscured the sun, and volcanoes spewed poisonous gases and molten rock onto Earth's surface. Scientists think that the ancient atmosphere was composed of methane, ammonia, and water vapor.

Radiation from the sun triggered chemical reactions among the atmospheric gases. These reactions produced nitrogen, hydrogen, and carbon dioxide. This led to the gradual replacement of methane and ammonia, while the concentration of water in the atmosphere increased. Because of hydrogen's low molecular weight, molecules of the gas were able to overcome the pull of gravity and escape into space, thus depleting the atmosphere of this gas. What remained were nitrogen, water vapor, and carbon dioxide.

The atmosphere of primitive Earth most likely contained little oxygen and therefore could not support organisms that carried on aerobic respiration. Without oxygen there could be no ozone, and the lack of an ozone layer exposed Earth's surface to deadly ultraviolet radiation. Oxygen and ozone are the products of photosynthetic bacteria that were the first true living things to evolve. The appearance of these gases in the atmosphere set the stage for living things that are present today. As Earth's atmosphere changed, the biosphere evolved along with it. **Evolution** is the process of gradual change that occurs in living things over time.

Changes in Living Things

Simple organisms first developed in the seas and evolved into more highly organized forms. Photosynthetic cyanobacteria colonized the oceans and enriched the atmosphere with oxygen, which created an environment suitable for aerobic organisms. Unicellular forms evolved into multicellular forms. Eventually, evolutionary changes enabled some organisms to leave the sea to live in freshwater and then, finally, to colonize the land.

Colonization of the land by green plants opened the way for the evolution of land animals that could use the plants as food. Plants and animals spread over the land. New forms of plants and animals developed, which were able to colonize the ever-changing environments within the biosphere. Evolution has been, and still is, one of the driving forces that change the biosphere.

Most of the millions of species that existed in the past were not able to adapt to environmental changes and lost the struggle for survival. These species no longer exist; they became **extinct**. For the most part, new species evolve, flourish for a time, and then become extinct. Species may change over time into better-adapted species, or be replaced by other well-adapted species. Only those species that have evolved rather recently in geologic history and a few older, well-adapted species, such as the cockroach, are still with us today. The continual development of new species in response to changing environmental conditions is part of the process of evolution.

There are some species, truly **living fossils**, which have survived unchanged through millions of years. For example, the tuatara, a reptile found on islands in the seas around New Zealand, has survived relatively unchanged since the time of the dinosaurs. Crocodiles,

Coelacanth

Figure 13-1 **The ginkgo tree (left) and the coelacanth (right) are living fossils, organisms that have not changed much since they first evolved.**

coelacanths (fish that live in the waters around southeast Africa), ginkgo trees, and certain species of shark are other living fossils. (See Figure 13-1.) In contrast, compare these species with the wide variety of dinosaur species that dominated all other life on Earth for many millions of years. Scientists theorize that the dinosaurs were, ultimately, unable to adapt to environmental changes that occurred about 65 million years ago and thus became extinct.

The fossil record clearly shows the ancestry of many of the species that inhabit Earth today. Horse evolution is one of the best examples of how fossil evidence can be used to trace anatomical changes through time. In horses, sequential changes in teeth and in foot structure are seen through their fossil history. In response to a changing environment, horse species evolved from forest browsers to grassland grazers.

13.1 Section Review

1. Describe some of Earth's ongoing changes that affect the abiotic part of the biosphere.
2. Compare the composition of Earth's primitive atmosphere with our present-day atmosphere.
3. How does horse evolution show adaptations to a changing environment?

13.2 NATURAL SELECTION

A species is basically a group of similar organisms that can interbreed and produce fertile offspring. Horses, dogs, raccoons, and jaguars are all examples of animal species. **Variation**, or differences

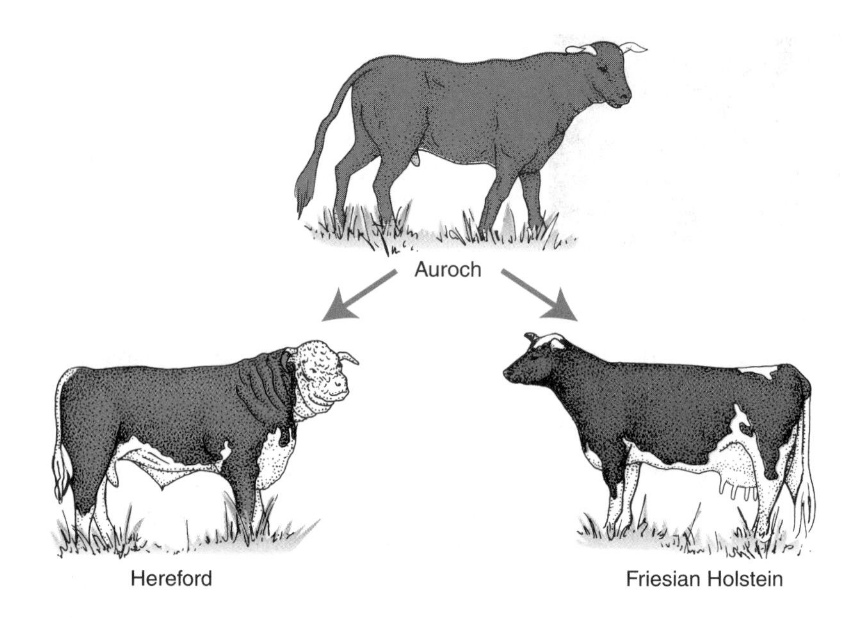

Figure 13-2 **The extinct auroch is regarded as the ancestor of all domestic cattle. There are many varieties of cows, but just one species.**

among individuals, is a basic characteristic of most living things. Figure 13-2 illustrates the varieties found in a species. The raw material of evolution is variation. If you were to examine any animal species, you would find many variations among the individuals. For example, in a group of horses, some are dark colored and others are light colored, some are heavy bodied and others are thin bodied, some have long legs and others have short legs, some run fast and others are slower. You might also find all sorts of variations between these extremes. But whatever variations there are within a species, all the individuals can interbreed to produce fertile offspring. Any differences can be passed along to their offspring.

Variations are transmitted from generation to generation through inherited characteristics. In all living things, **genes** are the carriers of inherited characteristics. Genes are made of DNA (deoxyribonucleic acid), which contains the codes for the production of proteins. Genes are the building blocks of chromosomes. In most cells, chromosomes are found in the nucleus. During sexual reproduction, organisms receive genes from both parents; the traits they possess are a mixture of the parents' traits. Those individuals that, because of their combination of genes, are better adapted to conditions in their environment have the best chance to survive and reproduce.

This process by which some individuals survive to pass their traits along to their offspring is called **natural selection**. As environmental changes create new demands on groups of organisms, the process of natural selection can lead to changes, or adaptations, in organisms. **Adaptations** are traits, or characteristics, an organism inherits that allow it to function better than other individuals in a changing environment. Over time, adaptations can lead to the development of new species. Those species that do not adapt to changing environments become extinct.

Under normal conditions, animals and plants usually produce more offspring than can survive. These offspring vary in characteristics, such as color, size, body shape, or tooth structure. In the struggle for survival, some variations may be more favorable than others. Those offspring that possess favorable variations have an increased chance of living to adulthood and producing offspring of their own. Over many generations, the favorable variations gradually become more common until all members of the species possess them. Inherited features that allow organisms to survive in changing environments are thus passed along to future generations by natural selection.

For a new species to develop, a breeding group, or population, has only to acquire a difference that prevents them from interbreeding with other groups. When this happens, they become isolated from the rest of their species. Once a population stops interbreeding within the species, a new species is born.

Kaibab Squirrel

Abert Squirrel

Figure 13-3 The Abert and Kaibab squirrels evolved from the same species.

Isolation

The most common event that causes new species to evolve is **geographic isolation.** This can occur when a new mountain range or river forms a barrier that separates a population into two groups. Because they are separated and exposed to different environmental conditions, the two groups can evolve in different directions to form two new species. A striking example of geographic isolation is the case of the Abert squirrel and the Kaibab squirrel. (See Figure 13-3.)

Erosion of Kaibab Plateau by the Colorado River led to the

formation of the Grand Canyon. The 1.5-kilometer-deep, 16-kilometer-wide gorge formed a barrier that ran east-west across the plateau. This isolated the populations of Abert squirrels on the northern portion of the plateau from those on the southern portion. Over time, the northern population evolved to form a new species, the Kaibab squirrel.

Characteristics within a species, such as color, thickness of fur, length of extremities, or body size and shape, may vary geographically. In addition, there could be a change in a food preference, mating ritual, mating season, or chemical attractant. These factors can lead to the segregation and eventual isolation of populations within a species. A population that is isolated from the main species interbreeds among its own members and becomes more and more different as time passes. Eventually, this can lead to the formation of a new species. This process is called **reproductive isolation.** Darwin's finches, which were discussed in Chapter 1, are an example of reproductive isolation as well as geographical isolation.

The change that triggers isolation can be caused by a simple change in a gene or chromosome, which leads to a change in a physical or behavioral characteristic. The new population that evolves loses contact with the original species. Ultimately, changes in the DNA that makes up genes and chromosomes are the source of all biological diversity.

Convergent Evolution

In similar ecosystems around the world, different plant and animal species have adapted with similar lifestyles and structures to make use of the available resources. These organisms fill a similar role in different ecosystems. This process is known as **convergent evolution.**

Convergent evolution is common in the animal world. The floor of the world's rain forests provides food in the form of fallen leaves, nuts, and fruit, as well as roots and tubers that grow beneath the soil. In each rain forest around the world, some type of animal has adapted to make use of this resource. These animals all look quite similar, but each belongs to a different species. They are all about

INTERACTIONS

Dog Breeding

Evidence indicates that wolves and jackals were domesticated more than 12,000 years ago. At one time, scientists thought that the Australian dingo was the ancestor of the domestic dog. However, dingoes are now considered to be feral domestic dogs. This means they returned to being wild after having been domesticated. The ability to keep and raise animals was an important survival factor for early humans. By using dogs, early humans extended their hunting abilities.

Once dogs were domesticated, human control was extended to their breeding. By selective breeding, humans were able to manipulate the dogs' genetic makeup to emphasize desired physical characteristics. Factors such as size, strength, speed, or obedience were selected as desirable traits. Breeding for these traits led to the production of totally new breeds that excelled in hunting, transport, and herding. The results of breeding animals for desired traits are similar to those of natural selection. The animals that are most desirable, or "fit," are the ones that are allowed to breed to produce the next generation.

All dogs belong to the same species, *Canis familiaris,* and are still closely related to wolves and jackals. As a result of years of crossbreeding, humans have developed many different types, or breeds, of dogs. Even though they look very different, Great Danes, German shepherds, dachshunds, and all the other breeds of dogs belong to the same species. They still share the same gene pool, and they can (when physically possible) interbreed and produce fertile offspring.

The differences among the various breeds of dog are so great that is almost impossible to make a list of what the original characteristics of the species were. Breeds are classified by the work the dogs perform. The modern breeds include hounds, sporting dogs, spaniels, terriers, nonsporting dogs, sheep dogs, watchdogs, draft dogs, and toy dogs.

Figure 13-4 Penguins (top) and auks (bottom) are examples of convergent evolution.

the size of a large house cat, brownish in color, and have thin legs that end in a tiny hoof or sharp claw. The mouse deer, which lives in the Asian rain forest, is not a deer at all but is related to the pig family. The royal antelope, found in the African rain forest, is the smallest member of the antelope family. The South American agouti is a rodent. Over time, all have become adapted to fill a similar role in similar but widely separated ecosystems.

Penguins and auks are another example of convergent evolution. As shown in Figure 13-4, these birds are similar in appearance but belong to different families. Penguins live in the Southern Hemisphere, while auks live in the Northern Hemisphere. Their two-tone black-and-white coloration is an effective underwater camouflage. Both of these birds feed on small fish and krill. In adapting to a life in the oceans, penguins have lost their ability to fly. Auks have also adapted to a marine environment, but they still can fly short distances. Both hunt and feed by using their wings to fly through the water.

13.2 Section Review

1. Explain how natural selection is similar to people breeding animals.
2. Compare the terms *geographic isolation* and *reproductive isolation.*
3. Write a story in which environmental change stimulates the evolution of a new species of plant or animal, even humans. Describe the adaptations that occur. Draw a picture of your new species.

13.3 DIVERSITY AND MUTATIONS

Each characteristic of an organism is the product of countless generations of natural selection. The many individual variations within a species are reflected in the **gene pool** of the species, which is the total of all of the genetic variation found in a species. The degree of diversity found within a species determines the range of environments it can colonize.

The Peppered Moth

Figure 13-5 The peppered moth adapted to its changing environment. Notice how the light moth is easily visible on the dark tree while the dark moth is harder to see (top). The dark moth is easily visible on the light tree while the light moth is difficult to see (bottom).

The peppered moth, shown in Figure 13-5, is a common insect of England's woodlands. When a population is examined carefully, variations in color are found among the individuals. Some are a light peppered color, and some are dark. In the 1950s, insect collectors were startled to find that there were many more dark moths being collected than light ones. This was strange because collections made 100 years before, in the 1850s, contained many more light individuals than dark ones.

Scientists discovered that during the intervening 100-year period, industrialization in the surrounding area had blackened tree trunks with soot. Dark moths were favored in this environment because they were better adapted to avoid predation by birds. Against the soot-stained tree trunks, the dark moths were almost invisible, while lighter individuals were easily seen and eaten. Prior to industrialization, the opposite had been true because the natural coloring of the bark favored the lighter moths. Natural selection, which had at first favored light-colored moths, now favored dark ones. Interestingly, with air pollution controls in effect, lighter moths are now more prevalent, as soot has been reduced in the area.

Evolution is thought to be the result of spontaneous mutations that occur in individuals. A **mutation** is a change in a gene that controls a specific trait. These genetic changes are passed along to offspring as variations. Mutations make up a small portion of the gene pool of every species. It is not always possible to determine whether a mutation will help an individual adapt to its environment. However, mutations do increase species diversity. If the variation allows the organism to adapt to changes in climate, food supply, or predators, the variation will have a beneficial effect. Many mutations are either damaging or lethal to the organisms they affect. Those mutations that do not kill the organism are passed from one generation to the next and are incorporated into the gene pool. Eventually, the accumulation of these changes in the gene pool may lead to the evolution of a new species.

Individual organisms do not undergo evolutionary changes during their lifetime. Populations may change over a period of time, but evolutionary changes occur over long periods of time. Mutation is a random process. However, fossils clearly show that most of the evolutionary changes through geologic time are not random. Competition between the individuals within a species for food, water, light, territory, and other environmental resources is one of the most important factors in their survival. Which individuals survive and which do not is determined by inherited genetic differences. Mutations are spontaneous changes in gene structure that may be caused by environmental factors. Although the environment may cause changes to occur in genes, it does not in any way direct those changes. The environment works only to preserve those organisms whose traits give them a survival advantage.

Rate of Evolutionary Change

Though most scientists accept the theory of evolution by natural selection, not all agree on the rate at which evolutionary changes occur. Some scientists believe that evolution is a gradual, ongoing process, while others think that evolution occurs in a series of rapid jumps.

Charles Darwin explained evolution as a slow and steady process, later termed **gradualism.** But scientists find few fossil records that indicate gradual changes. The record usually contains many gaps, but this may be explained by the fact that the remains of few organisms are fossilized and many fossils are destroyed with the passage of time.

In 1972, evolutionary biologists Stephen Jay Gould and Niles Eldridge came up with a theory that explained the gaps that exist in the fossil record. They theorized that species might undergo little or no change for long periods of time. This balance is then interrupted by a period of sudden and rapid change over a period of just a few thousand years. They called this process **punctuated equilibrium.** Controversy exists today as to the roles that gradualism and punctuated equilibrium play in evolution.

Preserving Diversity

With more and more people eating fish these days, the diversity of the marine ecosystem is threatened. Some fish populations have become severely depleted. Stocks of prized species such as cod, tuna, and swordfish, which were once plentiful, have declined due to overfishing. It is estimated that more than two-thirds of all commercially important fisheries have been overexploited. Many of these species soon will become endangered, which could progress to their extinction, thus reducing the diversity in the marine ecosystem.

This is an environmental problem on which consumers can have a positive impact every time they buy fish. Species that are in decline, and that should be avoided, include cod, grouper, snapper, Atlantic halibut, monkfish, orange roughy, Chilean sea bass, shark, swordfish, and bluefin tuna. Preferred species with healthy populations are Alaskan salmon, mahi-mahi, squid, tilapia, striped bass, and Pacific halibut.

13.3 Section Review

1. How do mutations affect species diversity?
2. What role do mutations play in natural selection?
3. Describe the main difference between gradualism and punctuated equilibrium.
4. What can you do at the supermarket to preserve the diversity of the marine ecosystem?

LABORATORY INVESTIGATION 13

Variation in Peas

PROBLEM: *What are some of the variations in peas?*

SKILLS: *Observing, measuring, graphing*

MATERIALS: *Peas, micrometer, graph paper*

PROCEDURE

1. Examine a collection of pea seeds from one species of pea plant.
2. Use a micrometer to measure the diameter of the peas and group them by size.
3. Separate peas by wrinkled and smooth seed coat.
4. Separate peas by yellow and green seed coat.

OBSERVATIONS AND ANALYSES

1. How many peas did you examine?
2. List three variations you found in the peas.
3. Construct a bar graph to indicate the numbers of peas that fell within a chosen size range.
4. Construct a pictograph to indicate the numbers of peas of each color.
5. Construct graphs to indicate any other variations you noticed in your peas.
6. What can you determine about the variations found in your population of peas?

Chapter 13 Review

Answer these questions on a separate sheet of paper.

Vocabulary

The following list contains all the boldfaced terms in this chapter.

adaptations, convergent evolution, evolution, extinct, gene pool, genes, geographic isolation, gradualism, living fossils, mutation, natural selection, punctuated equilibrium, reproductive isolation, variation

Fill In

Use one of the vocabulary terms listed above to complete each of the following sentences.

1. ________ describes the changes that occur in living things over time.
2. Species that disappear from the biosphere forever are said to be ________.
3. ________ are traits that an organism inherits that allow it to function better than other individuals in a changing environment.
4. In similar ecosystems around the world, different species have adapted to similar lifestyles to make use of the available resources. This process is called ________.
5. The total of all genetic variation within a species makes up the species' ________.

Multiple Choice

Choose the response that best completes the sentence or answers the question.

6. In the early atmosphere, radiation from the sun triggered the production of *a.* oxygen. *b.* methane. *c.* nitrogen. *d.* ammonia.
7. The accumulation of oxygen in the atmosphere led to *a.* atmospheric warming. *b.* the ozone layer. *c.* anaerobic bacteria. *d.* continental drift.
8. Which of the following is an extinct group of animals? *a.* crocodiles *b.* dinosaurs *c.* horses *d.* Kaibab squirrels
9. Evolution is thought to be the result of all of the following except *a.* random mutations. *b.* geographical isolation. *c.* reproductive isolation. *d.* controlled breeding.
10. Which of the following is *not* a source of competition between individuals within a species? *a.* food *b.* water *c.* space *d.* methane

Short Answer (Constructed Response)

Use the information you learned in this chapter to respond to the following items.

11. Give one example of convergent evolution in birds.
12. Explain the term *living fossil.*
13. What are genes?
14. How did geologic isolation contribute to the success of marsupials in Australia?
15. Give an example of convergent evolution found in rain forest animals.
16. Explain how Darwin's finches illustrate the process of reproductive isolation.
17. How is punctuated equilibrium different from gradualism?
18. What is natural selection?
19. What effects do genetic mutations have on species?
20. Why is a wide range of genetic variation important to the survival of a species?

Essay (Extended Response)

Use the information in the chapter to respond to these items.

21. Why do scientists think that green plants colonized the land before animals did?
22. How has the peppered moth adapted to its changing environment?
23. How does sexual reproduction create variations in a species?
24. Prepare a graph that shows the amount of time each of the following species has existed in the biosphere.
    a. *Homo sapiens* (modern humans)
    b. coelacanth
    c. tuatara
    d. crocodile
    e. ginkgo

Research Project

In your state, investigate the plant and animal species that are on the endangered list. What is currently being done to protect these species from extinction? Choose one species and prepare a plan of action to protect it. In addition, your plan should lead to an increase in the population of the species and an increase in its genetic diversity.

CHAPTER 14
Jobs and Homes in the Biosphere

When you have completed this chapter, you should be able to:

Define niche and habitat.

Explain how an animal's teeth determine the food it eats.

Discuss how food supply affects an animal's home range.

Many times during introductions, people will exchange names, where they live, and what they do for a living. Some people live in large cities while others live in small towns. People can be lawyers, farmers, auto mechanics, or students. Where someone lives and what he or she does for a living tells us a lot about that person. The same is true in the biosphere. Animals and plants also must live somewhere and do something to be able to survive. In this chapter, you will learn about where different organisms live, their habitat, and what they do, their niche.

14.1 NICHE = OCCUPATION

In the day-to-day struggle for survival, every organism carries out activities that use the resources available in its ecosystem. An organism's activities affect the other members of its species and the other species that make up the community. An organism's activities, as well as its general relationships in the community, are called its **ecological niche**, or just its **niche**. You can consider an organism's niche as its occupation, or the way it makes a living, within the ecosystem.

An organism's niche depends on its physical and behavioral traits and how it uses those traits to interact with the environment. Niches are closely related to the food resources available in an ecosystem. Thus a wide variety of niches usually exists within any ecosystem. Over time, organisms evolve to make use of the resources an ecosystem offers.

Organisms can be producers, consumers, decomposers, or scavengers. Consumers are herbivores, carnivores, or omnivores. Some herbivores eat grasses, some eat leaves, some browse on the shoots of young plants, some eat fruit, and some eat seeds. Some carnivores eat insects and small invertebrates, some eat fish, some prey on small mammals and birds, others prey on larger mammals. Omnivores eat a wide variety of plant and animal foods. When feeding, each organism uses a different resource within the ecosystem; thus each occupies a different niche.

Where and When to Eat

Organisms that eat the same food in different parts of the ecosystem or at different times of the day occupy different niches. There are organisms that are **diurnal**; they feed and are active only in the daytime. Others are **nocturnal**; they feed and are active only at night. Some feed and are active only at dawn and dusk, these are **crepuscular**. Some are **arboreal**; they live and feed up in the trees. Some are **aquatic**; they live and feed in the water. Others are **terrestrial**; they live and feed on the ground. Still others are **subterranean**; they live and feed underground. Some organisms, called parasites, make their living within or on the body of another organism. Since each fills a different role in the community, each has its own niche.

Bats and swallows, which both feed exclusively on insects, can be found within the same ecosystem. Because swallows are diurnal and bats are nocturnal, they do not compete for the same food resources and thus fill different niches. (See Figure 14-1.)

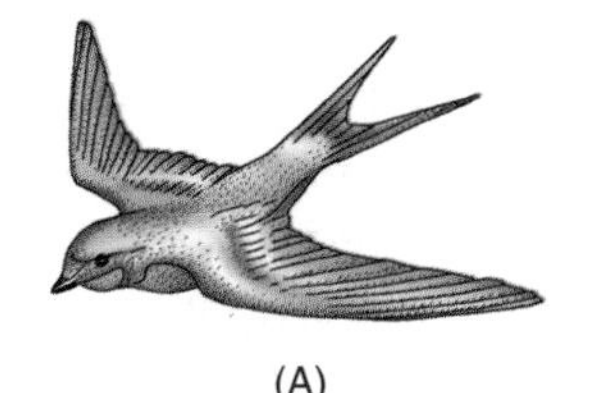
(A)

(B)

Figure 14-1 **The swallow (A) hunts for insects in the day. Some bat species (B) eat night-flying insects. Do these birds and bats occupy the same niche?**

As an ecosystem changes, new niches become available and species evolve to fill these niches. Each species places demands on the environment and contributes to the ecosystem in its own way. Over time, each species evolves physical and behavioral characteristics that define where and how it will live. Woodpeckers or red squirrels do not inhabit grasslands, nor do zebras or wildebeests live in pine forests.

Although mammals share many common physical traits, they show a great diversity in size, color, and appearance. This diversity comes from evolutionary modifications that have allowed each organism to function in a particular environment. For example, mammalian hair evolved into many forms with different functions. The wool of sheep acts as insulation against the cold. The quills of porcupines protect them from predators. The horns of rhinoceroses are used for protection and display in mating. Through embryological studies, scientists determined that all these structures have evolved from hair.

Job Opportunities

In any ecosystem, two different species cannot occupy the same niche at the same time or in the same place. Eventually, the species that is more efficient at making use of the available resources will displace the other. Though it sometimes appears as though two neighboring species occupy the same niche, they in fact use the same resources in different ways.

On the island of Trinidad in the Caribbean, eight species of tanager coexist. The tanagers are closely related, brightly colored songbirds. Three species, the speckled tanager, the bay-headed tanager, and the turquoise tanager, all live in the same bushes and trees and feed on the same insects and fruit. These three species seem to fill the same niche in the ecosystem, but observations of these birds in the wild show that differences exist in their niches.

Figure 14-2 The speckled tanager eats insects it finds on the underside of leaves. The turquoise tanager eats insects it finds on fine twigs. The bay-headed tanager eats insects it finds on branches. Do these three species compete with one another?

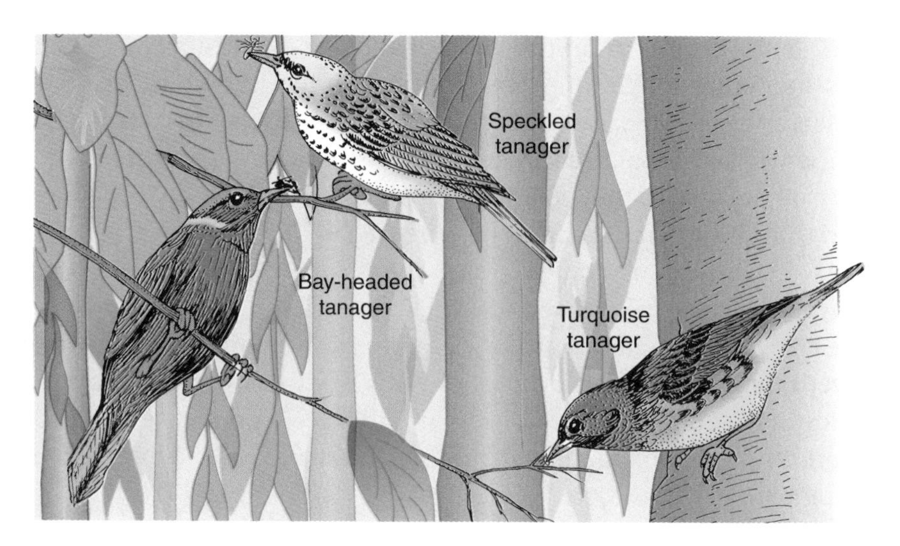

The speckled tanager searches the leaves for insect prey and feeds only from the underside of foliage. The bay-headed tanager takes insects only from large branches. The turquoise tanager feeds exclusively from small twigs and dead branches. These birds divide up the food resources in such a way that they do not compete with one another. Each tanager has a different niche even though they share the same food resources. (See Figure 14-2.)

The Coral Reef

The coral reef provides a variety of environmental niches for marine life. The reef contains a variety of places to **forage**, or look for food. Sea urchins graze on marine algae. Sea slugs devour coral polyps. Anemones filter tiny organisms out of the water. Butterfly fish, with their long beaklike mouth, pick organisms off the coral. Parrot fish crunch up the coral with their sharp beak. Certain wrasses pick bits of dead skin and parasites from the bodies of other reef fish. Barracuda prey on small fish. And some reef sharks prey on large fish. Each is adapted to make use of a different food resource that the reef offers, and each fills a different environmental niche.

Every reef species possesses a specialized pattern of behavior that determines where and how it lives. This creates further variety in the niches present around the reef. Some reef dwellers prefer deeper waters, while others swim near the surface. Some never stray far

INTERACTIONS

Loss of Biodiversity

Biodiversity is a measure of the numbers of different species of plants and animals present in an area. Loss of biodiversity occurs by extinction of a species or loss of a habitat's population of the species. The problems that arise from a loss in biodiversity may be irreversible. As people destroy habitats in the pursuit of resources, they are decreasing the biosphere's biological diversity.

Throughout history, extinctions of species have occurred naturally. However, human activity has greatly accelerated the rate of extinctions and reduced many populations to the point where they are in danger of becoming extinct. It has been estimated that extinctions of plant and animal species are now occurring at a rate 1000 times faster than at any other time in the past 65 million years.

The rain forests, the most biologically diverse ecosystems on Earth, face the greatest losses. The wild varieties of many of the world's agricultural crops, such as yams and rice, exist only in the rain forest, where they originated. Plants protect themselves by producing compounds that fight microorganisms; these compounds often become the basis for medicines. For example, quinine, which is used to treat malaria, is derived from the bark of a rain forest tree. Rubber, certain oils, lumber, and petroleum substitutes also originated in the rain forest. There are, no doubt, many other potential products waiting to be discovered.

Many of today's crops have a limited genetic makeup. These crops are vulnerable to disease and insect pests because they lack the genetic variations that would make them resistant. To solve this problem, plant breeders obtain genetic material from closely related wild plant varieties and add it to crops to make them more resistant to disease or insects.

Since most rain forest areas exist in developing nations, population pressures and poverty make the continued destruction of rain forests inevitable. These poor nations need the rain forests to supply their needs. It is, therefore, sensible for the more developed nations to help poorer nations. More developed nations also can provide support for developing conservation measures and technologies for obtaining resources from the rain forests, while protecting them. The fate of the rain forests will, to some degree, determine the fate of the biosphere.

Figure 14-3 A coral reef community includes a number of different species.

from the vicinity of the reef itself. Some fish are active during the day, while others hunt at night. Some prefer the cooler, ocean side of the reef, while others prefer the warmer, landward side. Figure 14-3 shows a coral reef community.

14.1 Section Review

1. Why is an organism's niche often described as its occupation?
2. Describe how the niches of animals that feed on the same food within an ecosystem may vary.
3. Take a field trip to identify animals living on the school grounds. Describe the niche of each organism identified.

14.2 SURVIVAL ADAPTATIONS

To meet the demands of a predatory lifestyle, teeth evolved in fish. An animal's teeth indicate its food preferences, and the food preferences determine its niche. In animal species, the evolution of differences in tooth structure has allowed animals to adapt to a wide variety of diets. Since teeth fossilize very well, fragments of teeth and jaws are widely used by paleontologists to draw inferences about the diet, behavior patterns, and appearance of the prehistoric animals from which they came.

Mammalian Teeth

In number, shape, and function, mammalian teeth vary from species to species. Most teeth are set firmly into the jawbones. The shape of the teeth determines their function. Mammals have four types of upper and lower teeth, which match up and work together. Incisors are the chisel-shaped, sharp, flat teeth in the front of the mouth. These teeth are best adapted to probing, gnawing, nibbling, and cutting. The pointed teeth next to the incisors are called canines. The canines pierce prey and tear flesh. The large teeth in the back of the mouth that have rounded bumps, called cusps, and sharp edges are premolars and molars. These teeth crush, grind, or shred food, to make it easier for the digestive system to process the food. The types of teeth are illustrated in Figure 14-4.

Rodents, such as rats, mice, squirrels, and beavers, have large incisors that grow throughout their lives. This is an important factor in rodent survival since these teeth are worn down by use. The incisors of rodents are specially adapted for gnawing tough plant materials. Most grazing animals, such as antelope and cattle, have short, flat incisors for clipping vegetation. They have large premolars and molars to grind up the food material so it can be processed by their digestive system. Carnivores, such as dogs and cats, usually

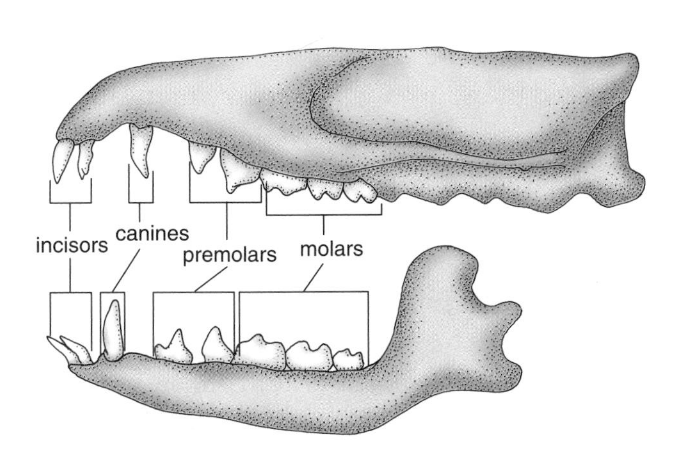

Figure 14-4
The number and shape of a mammal's teeth are adaptations to a particular way of life.

have small incisors and large, sharp canines for catching and killing prey. They have sharp premolars and molars, called carnassial teeth, which cut meat into pieces that can be swallowed.

Animals such as the anteater, aardvark, aardwolf, and echidna feed exclusively on ants and termites. All have weak jaws and few or no teeth. Some have a long sticky tongue that they can snake into insect nests to capture their prey. All have powerful forelegs and sharp claws, which enable them to rip open ant or termite nests. These animals have adapted to a point where they cannot eat any other food.

Fish and Reptile Teeth

The teeth of fish, as well as those of reptiles, are uniform pointed spikes that are embedded in the jawbones. The upper and lower teeth interlock to form an effective cutting or gripping tool. These teeth cut large pieces out of a carcass. Since these animals cannot chew or grind up their food, they are limited to swallowing it whole or in large chunks. Some reptiles have modified teeth, called fangs. These hollow teeth, which act like hypodermic needles, inject poison into prey animals. The poison kills the prey and starts the digestive process.

14.2 Section Review

1. How is the shape of teeth related to a mammal's food preference?
2. Explain the differences between mammalian teeth and the teeth of reptiles and fish.
3. Why are reptiles limited to swallowing prey whole or in large chunks?

14.3 HABITAT = ADDRESS

Animals and plants do not live in separate worlds. They are part of a **habitat**, a specific environment that contains an interacting community. Examples of habitats are a pine forest, a mountain meadow, or a coral reef. A habitat is a special place in an ecosystem where an organism lives. If an organism's niche can be considered its occupation, then the organism's habitat can be considered its address. A habitat can be a particular place that an organism needs for survival, such as a riverbank—the habitat of the river otter. An organism lives in a particular habitat because the members of its species have adapted over the years to function best in that environment. Within this habitat they have developed those traits that best help them eat and avoid being eaten. This gives them an edge over animals or plants that are not as well adapted to live there.

Figure 14-5 This photograph shows various habitats.

Habitats differ in physical factors, such as the yearly temperature range and precipitation, soil type, amount of daily and yearly sunlight, and geology and topography of the terrain. A habitat supplies an animal with its basic requirements of food, water, and shelter, as well as being a home, a playground, a hiding place, a nursery, and a place where it can get together with others of its species to mate. In many habitats, plants and animals must be able to adjust to changing conditions during annual cold or dry periods. (See Figure 14-5.)

All habitats must contain areas that provide cover for animals. Habitat **cover** includes places that offer protection from predators, shelter from the weather, and sites for nests, dens, or homes. Cover consists of vegetation used for concealment by deer and as nesting places by birds; soil in which ground squirrels, prairie dogs, and moles dig their burrows; or rocky terrain used by mountain goats, wild sheep, and marmots.

Plants are an important part of every animal's habitat. The local climate determines the variety and structure of plants in an area. Factors that determine where plant species live include sufficient light to carry on photosynthesis; proper temperature and moisture; the availability of minerals and nutrients in the soil; the soil composition; the presence in the soil of algae, fungi, and invertebrates; and the presence of pollinators, such as insects, birds, or bats. Plants form the framework of every major habitat type, or **biome.** Examples of biomes are the coniferous forest biome, which is dominated by conifers, or cone-bearing trees; the deciduous forest biome, which is dominated by deciduous plants (plants that shed their leaves); and the temperate grassland biome, which is dominated by grasses.

Natural grasslands are usually maintained by a combination of fire and grazing. If these processes are disrupted, the grassland changes in a series of distinct stages until it is replaced by a forest. The grassland species migrate or die, and forest species occupy the new habitats that have formed. An ecosystem dominated by grasses, bison, pronghorn antelope, wolves, and prairie dogs might be replaced by one dominated by deciduous plants, white-tailed deer, black bear, and squirrels.

Extending the Meaning of Habitat

The term *habitat* can be used in several different ways. In broad terms, it can refer to a major type of environment, such as a coral reef habitat or a temperate rain forest habitat. It can refer more specifically to an environment in a particular place, such as an alpine meadow habitat or an oxbow lake habitat. It may also refer very specifically to the home environment of a particular plant or animal and the environmental conditions the organism needs for survival. Examples are the habitat of the beaver, which is a mountain stream; the habitat of the sea otter, which is a marine kelp forest; and the habitat of the bison, which is a grassland or prairie.

The major habitat types, or biomes, contain a wide variety of environments with many different habitats. For example, if you climbed through a mountainous area of a coniferous forest biome, you would encounter many different forest habitats, each dominated by a distinct type of vegetation and with its own kinds of animal life suited to the conditions there. Live oaks, buckeyes, and gray pines—trees that are adapted to grow under dry conditions—dominate the lower slopes of the montane coniferous forest. The middle slopes contain stands of lodgepole pines, ponderosa pines, and Jeffrey pines—trees that grow best when supplied with sufficient rainfall and sunlight. Areas on the higher slopes are dominated by hardy trees such as juniper, blue spruce, fir, larch, and bristlecone pine, which are able to tolerate the low temperatures and limited moisture found there.

Minihabitats make up these larger habitat areas. Fallen logs provide a habitat for salamanders, insect larvae, and small mammals. Slopes covered with weathered rock, called talus, are the preferred habitat for golden marmots and picas. Flying squirrels nest in cavities in dead, standing trees and forage for nuts and seeds in the surrounding area.

Coping With Changing Temperature

Organisms cope with changes in their habitat in many ways. Each organism develops a survival strategy best suited for life in its habitat. To escape winter's freeze, deciduous trees shed their leaves and spend the cold months in a dormant, or inactive, state. Annual plants, which die off in the fall, produce seeds that survive the winter's cold and germinate in the spring, bringing new life to the land.

One way animals avoid daytime heat is to search for food only during the cooler periods of dawn and dusk. These animals are called crepuscular. They include reptiles, such as desert tortoises and horned lizards, which are cold-blooded and can overheat and die if they get too warm. They dig into the sand or soil, or retreat to their burrows to escape the heat of the midday sun.

As a means of conserving body heat, some mammals that live in cold climates, for example the arctic hare and arctic fox, have evolved stout bodies, short legs, and short ears. In warmer climates, mammals of the same families, for example the jackrabbit and kit fox, have evolved thinner bodies, longer legs, and longer ears to facilitate heat loss. Each animal functions best in the habitat it has adapted to, and each has a definite survival edge in its preferred habitat. (See Figure 14-6.)

Arctic hare (tundra)

Jackrabbit (warm deserts and semi-deserts)

Figure 14-6 In what ways are the different structures of these two rabbit species adaptations to their environment?

To escape unfavorable periods that occur in their preferred habitat, some animals enter a state of **hibernation**, a kind of suspended animation when the animal's respiration, metabolism, and heartbeat are slowed. Black bears, chipmunks, frogs, and salamanders hibernate during cold periods, remaining in a den, under the soil, or buried in mud until the arrival of the warmth of spring or the rains that signal the coming of the wet season.

Coping With a Shortage of Water

Deserts often experience long periods of drought when all available surface water evaporates. Some animals minimize water loss through a process similar to hibernation, called **estivation**, a state of suspended animation below ground. When ponds in the North American desert dry up, spadefoot toads burrow into the soil, envelop themselves in a cocoonlike layer of dead skin, and await the next rains. African lungfish can weather several years of drought, buried in a coat of dried mud.

In terrestrial habitats, organisms get water from lakes, ponds, rivers, springs, groundwater, dew that collects on the leaves of vegetation, or from the tissues of the organisms they feed on. In some desert habitats, plants make use of fog that condenses on their leaves and then drips onto the ground. Some insects allow fog to condense on their bodies and drink the water that collects. Some organisms regulate their activities to coincide with rainy seasons, even though these seasons may occur at intervals of several years or more. The seeds of many desert plants lie dormant in the soil for years and germinate only following heavy rains.

The Advantage of Adaptability

Some species show great adaptability in their requirements. They can readily adapt to new habitats when conditions in the environment change. Examples of animals that live in a wide range of habitats are coyotes, crows, and raccoons. Because of their adaptability, these species are present throughout the temperate regions of North America. One characteristic these animals have in common is the ability to exist on a varied diet. They will eat any food that is available, be it fresh meat, vegetable, or carrion.

Species have evolved many adaptations to help them make a living in ecosystems that are constantly changing. As one type of vegetation is replaced by a different type, original habitats disappear and new ones spring up. Species must evolve and adapt to these changes or they will be replaced by better-adapted species. If the displaced species cannot find suitable new habitats, they may be doomed to extinction, gone forever from Earth.

Most natural habitats have been modified in some relatively harmless way by human activity. However, many habitats are now under threat of catastrophic changes caused by people. As habitats are lost, the species that depend on them will disappear, too. Even in the United States, where conservation efforts are aimed at protecting natural habitats, many species are in danger of extinction due to habitat loss.

14.3 Section Review

1. How do habitats differ from one another?
2. What must a habitat supply to the creatures that live there?
3. Based on what you have learned about habitats in nature, explain how a city street is a habitat.

14.4 HOME RANGE

The area over which an individual animal or family group travels is called its **home range.** The knowledge an organism acquires about its home range allows it to find refuge or escape routes in times of danger and recognize when and where food will be available. The size of the home range is determined by the animal's needs and by the availability of food and shelter. The home range of a field mouse may be no more than 1000 square meters, while that of an Alaskan grizzly bear may be 260 million square meters. The total population of the Devil's Hole pupfish can be found in a single desert spring in Nevada. Other animals, such as wildebeests and zebras of the African savanna, are constantly on the move in search of food; they occupy a wide home range. The manatee's home range extends from Florida's inland rivers to the Atlantic and Gulf coasts.

Sometimes temperature extremes and droughts can make an animal's habitat unlivable for part of the year. To solve this problem, some animals **migrate**, or travel between different areas to find food, water, or safe areas to raise their young. These animals have become adapted to living in different habitats for different parts of the year and thus occupy a wide home range. For example, in Jackson Hole, Wyoming, elk herds spend the summer in the high forests and the winter in the lower grasslands. Other animals migrate very long distances to reach suitable habitats.

Migrating Birds

One bird, the Arctic tern, follows a 40,000-kilometer annual migration route between its summer Arctic breeding sites and its winter Antarctic feeding grounds. In addition to the Arctic tern, many North American birds spend their summers in the Arctic tundra where there is an abundant supply of food. During the short growing season, tundra plants quickly flower and produce seeds and fruits. The spring thaw melts the surface ice and snow. This provides favorable conditions for hatching insect eggs deposited the previous summer. In the Arctic, the birds have few predators. When the cold weather returns and insects and plants are no longer available, the birds migrate back to the southern United States and the Gulf of Mexico where food is abundant. Each year, they follow **flyways**, well-defined migration routes. Waterfowl, including ducks and geese, and shorebirds such as plovers and sandpipers travel in large flocks during their spring and fall migrations.

Migrating Wildebeest and Zebra

In East Africa's Serengeti National Park, large herds of wildebeest and zebra migrate during the alternating wet and dry seasons. The herds follow a yearly migration route that is determined by the emergence of new grasses on which they graze. These herds have been following the same migration routes for thousands of years and occupy a huge home range.

Migrating Caribou

Throughout the Arctic tundra, caribou, or reindeer, are found in herds that number many hundreds of thousands. Each year, the herds migrate as far as 9700 kilometers back and forth between the tundra and the northern coniferous forests. The caribou spend the winter in the forests, where the tall trees provide shelter, and food is present beneath the powdery snow. The caribous' broad, flat feet help them travel over deep snow and scrape away snow cover to get at the vegetation beneath. During the winter, the herds move about through the forests, seeking new food sources as old ones are depleted.

In the spring, the caribou migrate north to the tundra, where new crops of grasses and lichens are plentiful. The females give birth during the spring migration north. Within hours of their birth, young caribou are able to run with the rest of the herd. This ability gives them protection from the packs of wolves that follow the herd to prey on stragglers. In the tundra, the caribou can fatten up and then mate before they begin their fall migration southward to the coniferous forests.

The Lapps, nomads who live in the northern European tundra, are totally dependent on the caribou herds. The animals provide the Lapps with the essentials of life: milk and meat for food, thick furs for clothing, skins for tents and ropes, sinews for sewing, and antlers and bone for tools. Since the Lapps cannot control the wanderings of the herds, they follow the caribou on their year-round migrations.

Loss of Home Range

The spotted owl population in the Pacific Northwest is decreasing. This decrease is linked directly to the reduction of the owl's home range due to logging operations in old-growth coniferous forests. Populations of migratory birds, which summer in the United States, are also on the decline due to habitat destruction of forests in the Caribbean region and South America. The destruction of these birds' habitats has led to a loss of part of their home range and disrupted their migration patterns. The fate of many species lies in the success of efforts to maintain existing natural habitats throughout the world.

14.4 Section Review

1. What factor has the greatest impact on the size of the home range of a species?
2. Why has there been a decrease in the spotted owl population in the Pacific Northwest?
3. Why do some animals migrate long distances each year?

LABORATORY INVESTIGATION 14

Animal Habitats

PROBLEM: *To identify habitats around your school*

SKILLS: *Observing, identifying, classifying*

PROCEDURE

1. Take a field trip around your school. Make a list of the various plant and animal habitats you can identify. These areas might include woods or clumps of trees, mowed lawns, clumps of shrubs, concrete walks, asphalt lots, the school building, sources of water (ponds, streams, drainage), and fields overgrown with weeds.
2. List and describe the plants and animals you see in each habitat area.
3. In terms of availability of food and shelter, describe the interactions that occur in each habitat area.

OBSERVATIONS AND ANALYSES

1. What factors in each habitat affect the types of plants found there?
2. What factors in each habitat affect the types of animals found there?
3. Do any of the animals occur in more than one habitat? Which ones?
4. Which plants and animals have benefited most from the environmental changes made by humans? Explain.

Chapter 14 Review

Answer these questions on a separate sheet of paper.

Vocabulary

The following list contains all the boldface terms in this chapter.

aquatic, arboreal, biome, cover, crepuscular, diurnal, ecological niche, estivation, flyways, forage, habitat, hibernation, home range, migrate, niche, nocturnal, subterranean, terrestrial

Fill In

Use one of the vocabulary terms listed above to complete each of the following sentences.

1. Organisms that are active during the day are called ________.
2. The way an organism makes its living is called its ________.
3. The state of reduced metabolism, used by some animals to escape the cold season, is called ________.
4. In order to escape conditions of extreme heat or lack of water, some organisms go into a state of suspended animation called ________.
5. Some animals travel far, or ________, through different habitats in search of food.

Short Answer (Constructed Response)

Use the information you learned in this chapter to respond to the following items.

6. How are nocturnal organisms different from diurnal organisms?
7. How are mammals adapted to feed on ants and termites?
8. How is the niche of swallows different from the niche of bats?
9. What is a subterranean habitat?
10. Why is vegetation an important factor in the environment of most mammals?
11. List two ways that organisms escape cold or dry seasonal changes in their environment.
12. How is a home range different from a habitat?
13. What is a biome?
14. Why do many North American bird species spend the summer in the Arctic tundra?
15. Name three physical requirements supplied to an organism by its habitat.

Essay (Extended Response)

Use the information you learned in the chapter to respond to the following items.

16. Why are two species usually not found occupying the same niche within an ecosystem?
17. Choose an animal and describe the adaptations that make it able to survive in its natural habitat.
18. Discuss habitat protection and habitat restoration as ways of protecting endangered species.
19. Identify any habitats in your area that are endangered. (*a*) How are they endangered? (*b*) List the plants and animals that live there and tell whether they are endangered.

Environmental Issue

Over the past 100 years, Earth has lost more than half of its natural wetlands. These include coastal and inland swamps, bogs, and marshes. Wetlands are fragile ecosystems that are saturated with water for most of the year. Wetlands regulate water flow, recharge groundwater, provide flood control, are nurseries for many species, and provide a diverse habitat for a large number of biological communities. The main causes of wetland destruction are human activities associated with agriculture or urban development.

The loss of wetland habitat has led to a reduction of our planet's biodiversity. Many species are disappearing and many others are endangered. Biodiversity is a measure of the number of species inhabiting an ecosystem, the degree of genetic variation within the population of each species, and the variety of niches, habitats, and trophic levels available within each ecosystem.

Scientists have found that the genes of many organisms provide blueprints for the production of lifesaving pharmaceuticals and other compounds that can benefit society. The loss of biodiversity can mean the loss of a valuable product before we have a chance to discover it.

- What can be done to prevent habitat loss?
- How can human activities be better directed to protect Earth's biodiversity?

CHAPTER 15
The Competition for Living Space

When you have completed this chapter, you should be able to:

Distinguish between intraspecific and interspecific competition.

Explain how an animal benefits from territoriality.

Describe the relationship between predator and prey populations.

Relate carrying capacity to limiting factors.

We tend to think of competition in human terms. Students compete for the teacher's attention in class, for a position on a team, and for good grades. Humans also cooperate with one another. Much of the same is true for other living things. In the animal kingdom, there is competition for food and mates, among other things. Plants compete for sunlight and room to grow. There is also cooperation in nature. In this chapter, you will learn more about different types of competition in nature.

15.1 TYPES OF COMPETITION

Within every ecosystem, there are constant battles over living space and food. These battles are an example of **competition**, a relationship in which organisms struggle with one another to obtain the essentials of life from their environment. To survive, an organism must not only compete with other organisms for food, water, and space, but it must also adapt to the environment or modify it to meet its own needs. All the interactions in an ecosystem are not necessarily competitive. Some organisms share in cooperative relationships with individuals of their own species or with members of other species.

Within a community, two species of plants or animals cannot occupy the same niche for an extended period of time. In their competition for the same food, water, or shelter, the species that is better adapted to use the resources available in that habitat always displaces the less well-adapted species. Charles Darwin described this as "survival of the fittest through natural selection." Individuals from one species will forage, hide, and reproduce more efficiently and thus benefit most from all the habitat has to offer. The members of this species will win the battle for survival, possibly without any blood being spilled during the process. One species will dominate, and the other will disappear from that community. For example, in some areas, mallard ducks displace black ducks. This selective force is known as the principle of **competitive exclusion.**

Intraspecific and Interspecific Competition

Plants compete for living space beneath the soil and above the ground. There is a competition for sunlight, soil nutrients, and water. Animals compete for food, water, cover, mates, and nesting and feeding sites. Competition among members of the same species is called **intraspecific competition**, while competition with organisms that belong to different species is called **interspecific competition.** Some forms of intraspecific and interspecific competition are illustrated in Figure 15-1.

Figure 15-1 On the African savanna, there are many examples of intraspecific and interspecific competition.

Intraspecific competition is often quite intense, because all members of one species have the same needs for food and space. To cope with this form of competition, some plant species have adaptations that favor mature reproductive plants over immature seedlings. To reduce competition, seeds produced by these plants cannot germinate in the shade of the parent plants. But these seeds are adapted for **seed dispersal**, to be carried to distant locations, by wind, water, or animals. For example, the seeds of dandelions and maples are

Howler monkey

Three-toed sloth

Figure 15-2 The howler monkey and the three-toed sloth both live in the tropical rain-forest biome.

dispersed by the wind; cockleburs and beggar-ticks are dispersed by animals; and coconuts are dispersed by oceans' waters.

In the life cycle of amphibians and some insects, the **juvenile**, or immature, stage and adult stage require different habitats or foods. These animals undergo **metamorphosis**, or change in form, accomplished through a series of stages. This survival strategy reduces intraspecific competition between adults and juveniles. Juvenile frogs are herbivorous, aquatic tadpoles. Their needs are quite different from those of the adult frogs, which are carnivores. Insect metamorphosis, which reduces competition, can be readily observed in the case of juvenile leaf-eating caterpillars and adult nectar-sipping butterflies.

Organisms have evolved strategies to reduce interspecific competition. Organisms adapt to different diets to remove competition. For example, howler monkeys and sloths are herbivores that live in the canopy of the South American rain forest. (See Figure 15-2.) Howler monkeys eat a wide variety of leaves but not the leaves of the cecropia tree. The sloth eats only the leaves of the cecropia tree. These differences in diet enable both species to exist side by side.

Among many herbivores that populate the African plains, there is little competition for the available food supplies. Wildebeests graze on tender young grasses. Zebras graze on coarser grasses. Some antelopes browse on leaves and young twigs; others feed exclusively from the tops or bottoms of shrubs. The giraffes browse on the leaves growing on the upper branches of trees. Each species has adapted to feed on a specific food source.

The food preferences of African carnivores also are quite different. Lions prey on the larger grassland herbivores, such as wildebeests and zebras, while the cheetah's great speed enables it to chase down the smaller antelopes and gazelles. The leopard also feeds on the smaller antelopes, but it does so by ambushing them in or at the edge of the forest rather than on the open plains.

15.1 Section Review

1. For which environmental factors do plants compete?
2. Why is intraspecific competition often more intense than interspecific competition?
3. Explain how the food preferences of African carnivores reduce competition.

15.2 TERRITORY

One method some animals have adopted to cope with intraspecific and interspecific competition is to establish a home area, or territory. A **territory** is the area an organism defends against intruders to protect food sources or living space. Territory is different from home range in that a territory is defended against intruders while the home range is not.

In some species, intrusion into a territory by a member of the same species leads to aggressive actions that stop only when one individual is driven off. For example, the stickleback is a species of fish in which the male defends the area around his nest against intruders. This strategy, called territoriality, helps to allocate resources throughout a habitat area. Territorial behavior spreads out the organisms in a population as well as scattering some to other locations.

Individuals of many animal species instinctively occupy and defend a living space, or territory, from members of the same species and from members of different species. Within the ecosystem, this behavior provides survival benefits to the organism and the species. Territoriality is a strong force that regulates the interactions within a community. It shapes the boundaries that allow organisms to coexist within an ecosystem.

As used here, the term *territory* should not be confused with human notions of property. You cannot know exactly how an animal perceives its geographic range. However, it is possible to see how an organism behaves in relation to other organisms in its region. You can easily observe where it wanders, how it makes its presence known, and what it does when it meets intruders. Most territorial behavior does not seem to be a conscious action, but rather is guided by **instinct**, an inborn response.

Marking a Territory

Some animals define their territory with threats in the form of sounds and visual displays. Aggressiveness is usually a clear display

Figure 15-3 **Visual displays by wolves are a way these animals communicate with each other.**

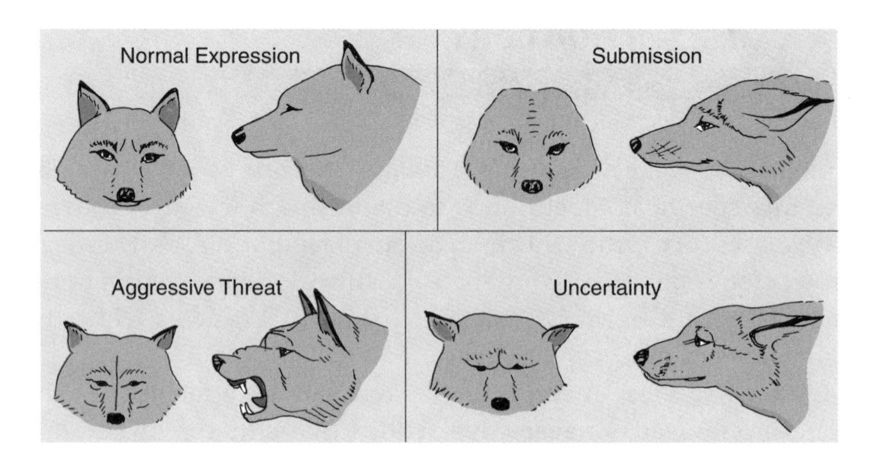

of territorial behavior, warnings that tell others to move on. As shown in Figure 15-3, the snarl of a wolf or dog is a clear visual and vocal threat to indicate that the animal's territory is being intruded upon. The roaring of a troop of howler monkeys announces their location and serves as a warning to other nearby troops not to enter into their feeding territory.

Another way some animals mark their territory is with scent. **Scent marking** may be in the form of a spray of urine, a pile of dung, or a deposit of musk. These chemical messages give information about the animal that left the mark, such as the animal's identity, health, gender, sexual state, when the scent was left, and when the animal is likely to return. These signs mark the territory within the home range that the animal or family will defend in the search for food or mates. Have you ever noticed dogs or cats sniffing out scent messages during their daily walks?

Many animals have special scent glands that produce a strongly scented secretion, called **musk**. Members of the weasel family have anal musk glands. The secretions of some weasels, such as skunks and polecats, have a very unpleasant odor. Many antelope have musk glands at the base of their hooves or on their head. Rodents, such as beavers and muskrats, have musk glands along their back. These animals rub the area around their scent glands against trees, shrubs, or rocks to deposit the musk. As they make their daily rounds looking for prey, tigers and leopards mark their territories by spraying urine on trees and shrubs.

Temporary Territories

Some species of mammals and birds set up temporary territories, which last only through the mating season. The males of these species set up individual display areas from which they try to attract mates. They defend the area against intrusion by other males of the same species. Though fighting does occur, visual or vocal displays and threats minimize it, which usually discourage challenges.

Sea lions, walruses, impala, and sage grouse set up temporary group territories during their mating season. In this form of intraspecific competition, the strongest, healthiest individuals "beat the competition" and get to pass their genes on to the next generation.

During their breeding season, many birds become intolerant of other members of the same species. The area around the nest becomes the exclusive property of the nesting pair until the young have left the nest. The size of the territory varies with the species. Penguins nest in dense colonies. Their nesting territory extends only a short distance from the nest. Eagles and hawks, which need a large hunting area to supply food for their young, may patrol a nesting territory that covers many square kilometers.

15.2 Section Review

1. How does territorial behavior affect the population of a species?
2. What are some ways that animals mark their territories?

15.3 ENVIRONMENTAL LIMITS

The resources, or the amount of matter and energy, available limit the size of a population within an ecosystem. The maximum number of organisms that can be supported by the resources in an ecosystem is called the **carrying capacity.** The carrying capacity of an ecosystem or a habitat is not constant. It can vary during the year as seasons change or from year to year due to climate factors.

The ecological conditions that control population size are called **limiting factors.** For plants, limiting factors include hours of light per day, availability of light, availability of water, presence of nitrogen compounds in the soil, and the presence of essential minerals. These factors affect the efficiency of photosynthesis and the amount of energy available to the plants. For animals, the limiting factors include plants or prey animals (food), water, and shelter or cover.

Within a stable ecosystem, when a population increases, the resources available to each individual decrease. Once the carrying capacity is reached, fewer offspring are produced or more organisms in the population die. Births are then balanced by deaths. Once the carrying capacity is exceeded, overcrowding occurs, and the environmental resources are "stretched." The stretching of resources causes such factors as intraspecific competition and predation to have a greater impact than they would under normal conditions. The population decreases because more animals die and some individuals leave the home range. If these individuals find new resources available, it leads to changes in the home range of the species. Under these conditions, individuals that have beneficial variations can adapt to new lifestyles and may eventually evolve into a new species.

Impact of Predators

Predator and prey populations interact in various ways. They can cause recurring cycles in population size for both the predators and the prey, or they can stabilize populations of both species. When predators keep the prey population from reaching the carrying capacity of the ecosystem, the result is population stability. When predators do not reproduce as fast as their prey, the prey exceeds the carrying capacity of the ecosystem. The predator population grows due to the increased availability of prey, thus increasing predation and causing the prey population to drop. A drop in predator population then follows the drop in prey population.

Throughout the North American coniferous forests, lynx feed on snowshoe hares. Fluctuations in each population affect the other. As shown in Figure 15-4, an increase in the hare population is followed by an increase in the lynx population. This leads to a decrease

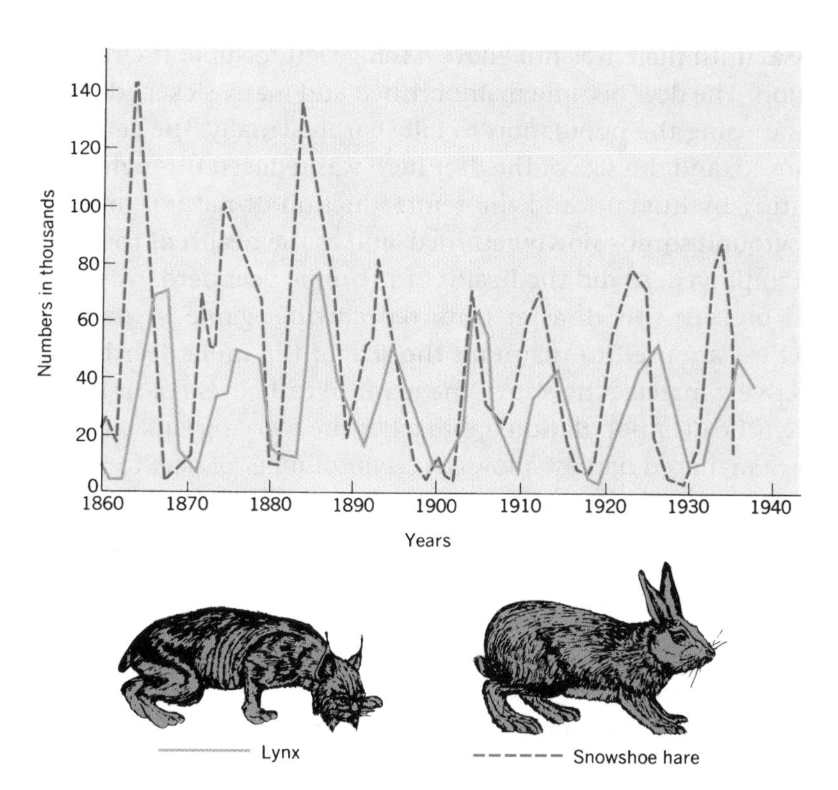

Figure 15-4 **This graph shows changes that occur over time in the population of lynx and snowshoe hares.**

in both populations. The decreased lynx population then allows the hare population to rise. Each species is a natural limiting factor on the population of the other.

A reduction in the population of predators can tip the balance within an ecosystem since it removes a natural limiting factor. In 1906, when the Kaibab Plateau in northern Arizona was declared a wildlife refuge, it supported a large mule deer population. Even though many ranchers grazed cattle on the Kaibab, there was little competition for food. The diets of the deer and cattle were quite different. The deer browsed on ground shrubs and leaves while the cattle grazed on grasses.

To protect the mule deer and cattle populations, all predators were eliminated from the ecosystem. Wolves, coyotes, bobcats, and mountain lions were exterminated or relocated. In response to decreased predation, the mule deer population exploded and exceeded the carrying capacity of the area. Soon there were not enough ground shrubs, seedlings, or leaves to feed the increased deer population. The deer began to compete with the cattle for grasses until there was not enough forage left to support either population. The deer became malnourished and many sickened or died, thus causing the population to fall sharply. Finally, the cattle were removed, and the size of the deer herd was reduced through limited hunting by humans and the reintroduction of natural predators. The ground shrubs slowly returned, and as the health of the ecosystem improved, so did the health of the mule deer herd.

To prevent this disaster from reoccurring, game management practices are used to maintain the size of the mule deer herd. To effectively manage the size of the herd, a yearly deer count is made. When the deer population approaches the carrying capacity of the area, a managed harvest allows a certain number of deer to be killed by hunters or game wardens in order to reduce the deer population.

15.3 Section Review

1. How are the limiting factors of plant populations different from those of animal populations?
2. Describe what happens when a population exceeds the carrying capacity of an ecosystem.

15.4 SHARING RESOURCES

Forest biomes are **stratified**; that is, they have a layered structure. As plants compete for living space, they create levels, and each level contains a variety of habitats and niches for organisms to occupy. The strata, or layers, of the forest ecosystem are the emergent, canopy, understory, and forest floor layers. The forest animals form subcommunities, which occupy a specific level where they carry out most of their activities.

Stratification

Due to the wide variety of habitats found in the tropical rain forests, they are thought to contain almost half of Earth's animal and plant

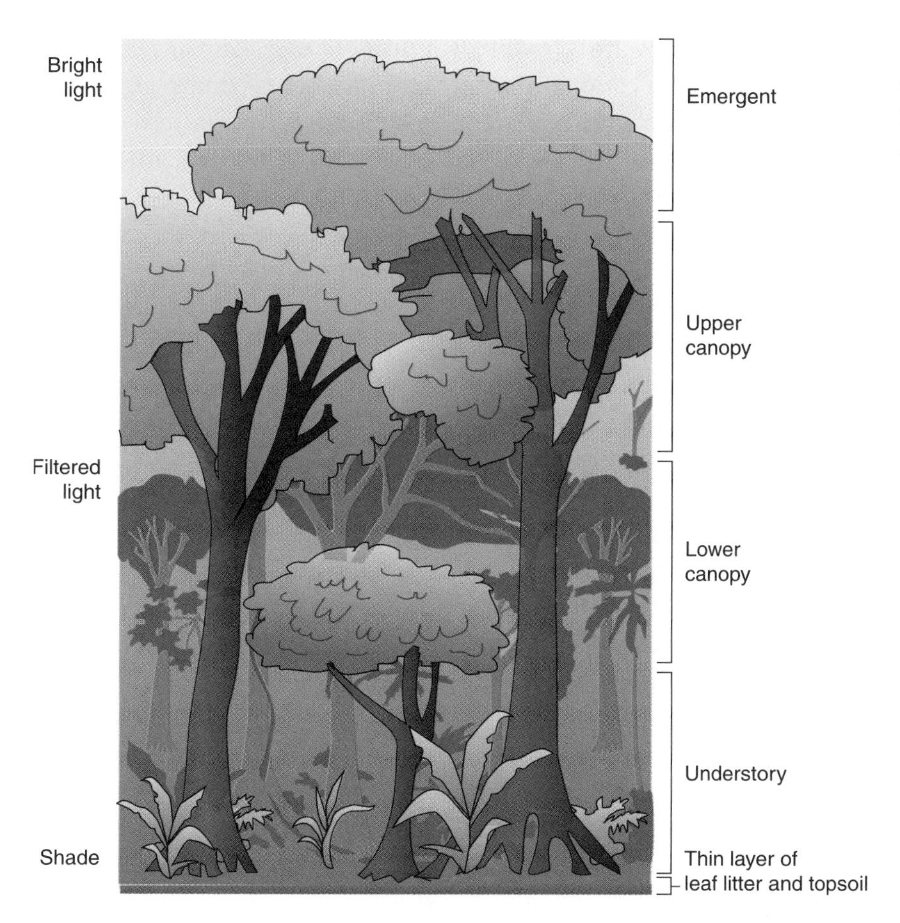

Figure 15-5 Stratification in tropical rain forests produces diverse habitats.

species. Howler monkeys and spider monkeys live in the canopy. Macaws and parrots fly through the understory. Peccaries, a kind of wild pig, and tapirs search for food on the forest floor. These animals spend most of their lives in one specific rain forest stratum. They live on a diet of leaves and fruit, and since each feeds within a different layer of the rain forest, there is little competition among them. (See Figure 15-5.)

Stratification in other ecosystems also creates varied habitats. In the grasslands, some animals make their living on the surface of the ground. Others carry out the same activities while burrowing though the soil. Within this ecosystem, shrews, woodchucks, and king snakes feed on the ground while moles, badgers, and hognose snakes feed beneath the surface of the ground.

In the temperate and evergreen forests, some animals feed on the forest floor, others are active in the understory shrubs, while still others feed in the canopy among the branches of the trees. Porcupines, martens, and warblers feed in the canopy; woodpeckers and nuthatches feed on the trunks of trees; and lynx, deer mice, and ravens feed on the ground.

Day Shift and Night Shift

Eyes designed for hunting at night

Talons for catching and carrying prey

Figure 15-6 Most owl species are adapted to a nocturnal lifestyle.

By conducting their activities on either a day shift or a night shift schedule, organisms have adapted to share resources and eliminate competition. Within the same ecosystems, nocturnal owls and diurnal hawks prey on small mammals. Hawks have acute vision that enables them, even when flying at great heights, to detect the slightest movement on the ground. Owls have a three-dimensional sense of hearing that allows them to pinpoint the position of their prey by the sounds it makes. The owls' feathers are designed to eliminate noise so that they can silently swoop down on their prey. (See Figure 15-6.)

Nocturnal animals have adapted in various ways to cope with hunting or foraging at night. They usually have large eyes, wide pupils, and retinas that are densely packed with light-sensitive cells, called rods. These adaptations allow them to take advantage of the limited light available at night. Owls, tarsiers, and ring-tailed cats are examples of nocturnal hunters that have large eyes.

Many nocturnal animals have developed an opaque layer behind the retina in the eye, called the **tapetum**. This layer reflects light back through the retina for greater efficiency in seeing. The tapetum produces the eyeshine seen when these animals are caught in a car's headlights. Domestic cats, alligators, and hyenas exhibit eyeshine.

Many nocturnal animals, such as the fennec fox, have enlarged ears and sensitive hearing. Their acute sense of hearing helps them to detect and track prey even in total darkness. Fennecs live in the deserts of northern Africa and Arabia, where their large ears also help to radiate excess body heat.

Bats have evolved echolocation to find prey in the dark. In **echolocation**, the animals emit high-pitched sounds. Their sensitive hearing detects the echoes of these sounds bouncing off surrounding objects. Bats make use of sound waves in much the same way as a ship's sonar does. This gives them an accurate picture of their environment and aids them in locating the insects on which they feed. Bats can see, but their eyesight is poor. They depend mainly on echolocation to guide them through their nocturnal activities. Dolphins, whales, and oilbirds also rely on echolocation as a means of gathering information about their surroundings.

The Benefits of Specialized Food Preferences

No matter how remote the living space or how bad we think a food tastes or how uncomfortable the environmental conditions, there is a species that has evolved traits so that it can exploit that resource. The more food available in an ecosystem, the larger the community it can support. But as the size of a community grows, there is more competition for the available food. Thus species often evolve to make use of a particular food. By specializing their food preferences, competition with other species is reduced. But this narrows the organism's niche and ties the existence of the species to the availability of one food resource. When a species limits its food preferences to a readily available food resource, there is a positive effect on population growth. But natural disasters can drastically reduce the amount of food available, which may lead to starvation and a rapid population decline or extermination of the species.

A good example of a species that has a specialized food source is the giant panda, which lives only in China. The panda is a large mammal that eats only bamboo. The tallest member of the grass family, bamboo is found throughout the forests of central China. Bamboo usually reproduces asexually, by means of rhizomes, or underground stems. But at 120-year intervals, all the individuals in a certain species of bamboo reproduce sexually by flowering. After the bamboo has flowered, all the plants wither and die. The bamboo must regenerate from seeds or surviving rhizomes. During the regeneration period, few plants are available as food for the pandas. This happened recently within the panda's range and caused the panda population to drop to a dangerously low level.

Figure 15-7 The panda (left) and the koala (right) are animals that have narrow food preferences. What would happen to these animals if their food supply was reduced or eliminated?

Some other organisms with narrow food preferences are the three-toed sloth of South America, which lives exclusively on the leaves of the cecropia tree; the koala, an Australian marsupial, which feeds only on eucalyptus leaves; and the proboscis monkey from the islands of Malaysia, which feeds mainly on mangrove leaves. (See Figure 15-7.)

15.4 Section Review

1. Describe the stratification of a grassland habitat.
2. How are nocturnal predators adapted to hunting in the dark?
3. Some animal populations have made use of human garbage in and around cities as a food source. What effects has this had on the animal and human populations?

LABORATORY INVESTIGATION 15

The Effects of Competition on Plant Growth

PROBLEM: *How does competition for space affect the growth of radish seeds?*

SKILLS: *Measuring, observing, calculating*

MATERIALS: *4 planting boxes or trays (approximately 15 cm wide × 30 cm long), soil, radish seeds, grass seeds, metric ruler*

PROCEDURE

1. Fill four planting boxes or trays with equal amounts of soil.
2. In the first box, make holes 1 centimeter deep at evenly spaced, 7.5-centimeter intervals; plant one radish seed in each hole. Label this box A.
3. In the second box, make holes 1 centimeter deep at 7.5-centimeter intervals and place three seeds in each hole. Label this box B.
4. In the third box, make holes 1 centimeter deep at 2.5-centimeter intervals; plant one seed in each hole. Label this box C.
5. In the fourth box, make holes 1 centimeter deep at evenly spaced 7.5-centimeter intervals; plant one radish seed in each hole. Sow grass seeds over the surface of the soil and cover them with a thin layer of soil. Label this box D.
6. Place the planting boxes where they will all be exposed to the same amount of light. On a regular schedule, add an equal amount of water to each box. Add just enough water to keep the soil moist.

TABLE 15-1 EFFECTS OF COMPETITION ON PLANT GROWTH

Box	Number of Plants	Height of Each Plant	Total Height	Average Height
A				
B				
C				
D				

OBSERVATIONS AND ANALYSES

1. Copy Table 15-1 into your notebook.
2. Observe the boxes daily. Measure and record the growth of each plant in each box.
3. When the largest plants are about 8 centimeters tall, one box at a time, clip all the radish plants at the ground line. Measure the height of each plant in the box. Use the formula below to calculate the average height of the radish plants in each box:

$$\text{average height of plants} = \frac{\text{total height of all plants}}{\text{number of seeds planted}}$$

4. What were the effects of intraspecific competition among the plants?
5. What were the effects of interspecific competition among the plants?

Chapter 15
Review

Answer these questions on a separate sheet of paper.

Vocabulary

The following list contains all the boldface terms in this chapter.

carrying capacity, competition, competitive exclusion, interspecific competition, intraspecific competition, echolocation, instinct, juvenile, limiting factors, metamorphosis, musk, seed dispersal, scent marking, stratified, tapetum, territory

Fill In

Use one of the vocabulary terms listed above to complete each of the following sentences.

1. The relationship in which organisms struggle with one another and with the environment to obtain the essentials for life is ______.
2. A plant adaptation that opens new environments for colonization is ______.
3. A(an) ______ is the area an organism defends against intruders to protect food sources or living space.
4. The replacement of one species by another within a community illustrates ______.
5. ______ are ecological conditions that control population size.

Multiple Choice

Choose the response that best completes the sentence or answers the question.

6. Which animal sets up a temporary territory for mating? *a.* tiger *b.* sea lion *c.* polecat *d.* howler monkey
7. Which is an example of a plant seed that is dispersed by animals? *a.* coconut *b.* maple *c.* beggar-tick *d.* bamboo
8. An African predator that feeds along the edges of forest is the *a.* lion. *b.* cheetah. *c.* leopard. *d.* hyena.
9. An animal that has anal musk glands is the *a.* polecat. *b.* tiger. *c.* musk deer. *d.* beaver.
10. Which bird has a small nesting territory? *a.* bald eagle *b.* osprey *c.* rockhopper penguin *d.* red-tailed hawk
11. A common mammal that inhabits the canopy of the rain forest is the *a.* peccary. *b.* tapir. *c.* howler monkey. *d.* macaw.

12. Which of the following is *not* a nocturnal feeder? *a.* tarsier *b.* barn owl *c.* ring-tailed cat *d.* golden eagle
13. Which of the following animals relies on echolocation to gather information about its environment? *a.* dolphin *b.* screech owl *c.* hyena *d.* woodpecker
14. Which is the food of the koala, an Australian marsupial? *a.* bamboo *b.* cecropia leaves *c.* eucalyptus leaves *d.* mangrove leaves
15. The seeds of dandelions and maples are dispersed by *a.* water. *b.* wind. *c.* animals. *d.* insects.

Short Answer (Constructed Response)

Use the information you learned in this chapter to respond to the following items.

16. What is intraspecific competition?
17. Describe one method by which seeds are spread to distant locations.
18. What is metamorphosis?
19. Describe two ways animals stake out a territory.
20. Explain what is meant by carrying capacity.
21. Describe two adaptations of nocturnal animals.
22. What are instincts?
23. Describe an animal that has a specialized food source.
24. How does stratification of rain forests provide for a wide range of habitats?

Essay (Extended Response)

Use the information in the chapter to respond to these items.

25. Why does insect metamorphosis reduce intraspecific competition?
26. Explain how an organism's territory is different from its home range.
27. What factors can change an ecosystem's carrying capacity?
28. Describe the interspecific competition in an ecosystem located near your school.

Research Project

Of the approximately 10,000 species of birds in the world, more than 10 percent are threatened with extinction. More than 100 bird species have become extinct over the past 200 years. Some species, such as the whooping crane and the California condor, have been brought back from near extinction in recent years. However, their populations are still small, and a natural disaster or the rapid spread of disease could wipe them out.

Choose one of the following questions and research its answer.

- Why are many seabird species being listed as endangered?
- How have logging operations caused a decline in rain forest species?
- Why is the pet trade dangerous for many bird species?

CHAPTER 16
Partnerships in Nature

When you have completed this chapter, you should be able to:

Discuss the various forms of symbiosis.

Distinguish between parasitism and commensalism.

Explain the significance of keystone species.

Describe the advantages of forming a social group.

Along with the competition between and among species, there is also cooperation. These cooperative relationships may be between individuals of the same species or between members of different species. Some relationships benefit both partners, as shown above by the tickbirds on the back of a Cape buffalo. Other relationships benefit only one participant. In this chapter, you will learn about the partnerships in nature.

16.1 SYMBIOSIS

Symbiosis, or living together, means a close and prolonged physical relationship between two or more species. The host is the organism on which another organism, called a **symbiont**, lives. The relationship may be harmful, beneficial, or neutral for one or both species. The organisms in a relationship may be in close contact, as with the fungus and alga that together make up a lichen. However, in some relationships the organisms come into contact only occasionally, for example the cleaning shrimps and the larger fish that they rid of harmful pests.

Mutualism

Mutualism is a symbiotic relationship that benefits the two species involved. Cleaning symbiosis is a mutualistic relationship that is commonly found in ocean reef communities. Small fish such as wrasses, butterfly fish, and gobies, as well as crustaceans such as the peppermint shrimp set up cleaning stations around the reef. Larger fish come to these stations to have harmful pests and dead or diseased tissue removed from their skin and gills. The cleaners, or symbionts, get an easy meal, while their hosts have potentially health-threatening conditions tended to.

Bioluminescent bacteria live in pouches beneath the eyes of flashlight fish. In these pouches, the bacteria are supplied with food and oxygen. The bacteria, which are able to produce a cold chemical light, enable the flashlight fish to feed in the dark ocean depths and to signal other members of its species. Bioluminescent bacteria are also found living in mutualistic relationships with many other sea creatures. (See Figure 16-1 on page 258.)

There are many similar relationships in terrestrial environments. Various species of birds remove insect pests from the hide of large animals. The tickbird grooms rhinoceroses, large antelopes, and water buffalo; the Egyptian plover cleans the mouth of crocodiles; and the cattle egret lives in a relationship with both cattle and large antelopes.

Figure 16-1 **Flashlight fish have a mutualistic relationship with bioluminescent bacteria that live in structures beneath their eyes.**

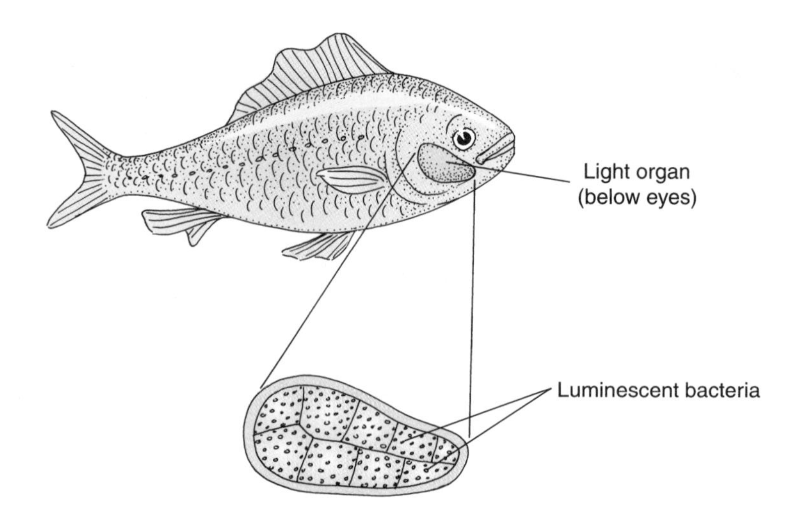

To make use of food sources that otherwise would be unavailable to them, some organisms form mutualistic relationships with microorganisms. Legumes and the nitrogen-fixing bacteria that live in nodules on their roots have a mutualistic relationship. In return for making nitrogen compounds available to the legumes, the bacteria receive nourishment and a place to live from the plants. Symbiotic protozoa that live in the digestive tract of termites aid in the digestion of cellulose. The protozoa provide the termites with usable carbohydrates and in turn receive shelter and a food supply. Ruminants, such as cows and deer, are herbivorous mammals that chew their cud. As is true of many animals, ruminants cannot digest cellulose. Bacteria that live in their digestive system break down the cellulose for them. Thus the ruminants obtain nutrients, and the bacteria are given food and shelter. Algae live in the cells of coral polyps. In return for the protection provided by the polyp, the algae supply the coral with carbohydrates produced during photosynthesis.

Commensalism

Commensalism is a form of symbiosis in which one species benefits from the association without harming or benefiting the other. One partner may receive food, shelter, or transportation from the other, but each could survive on its own.

The remora is a fish that has a suction disk on top of its head. As

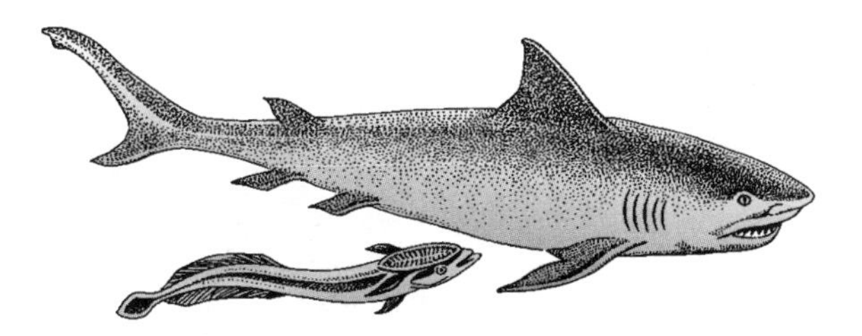

Figure 16-2 The symbiotic relationship between the shark and the remora is an example of commensalism.

shown in Figure 16-2, the remora can attach itself by means of the disk to sharks or other large fish. The shark gains nothing from this commensal relationship. However, the remora receives free transportation. In addition, as the shark feeds, the remora is in a position to pick up fragments of food.

Other examples of commensalism are bluebirds that use woodpecker holes as nesting sites, the protection gained by damselfish that live among the stinging tentacles of sea anemones, and the protective camouflage some crabs achieve when they attach living sponges to their shells.

Parasitism

Parasitism is an extreme form of symbiosis in which the symbiont, called a **parasite**, is totally dependent on the host as a source of nutrients. The activities of the parasite often weaken the host but usually do not kill it. A successful parasite does not kill its host, because the death of the host may kill the parasite as well. However, a weakened host is more susceptible to diseases and attack from predators and, therefore, more likely to die. As the host species evolves defenses against the parasite, the parasite species evolves to overcome the defenses, which produces a nonfatal relationship.

Most organisms are miniature ecosystems that contain a wide variety of symbiotic plants, animals, and microorganisms. Lice, fleas, mites, and ticks are parasitic arthropods that infect many animals. Since these creatures live on the skin or fur of their hosts, they are called **ectoparasites.** The prefix *ecto-* means "outside." Species of roundworms, flatworms, protozoa, fungi, and bacteria may also be parasitic. These usually live within their host and are called **endoparasites.** The prefix *endo-* means "inside." (See Figure 16-3 on page 260.)

Figure 16-3 **The hookworm and trichina worm are endoparasites, while the leech is an ectoparasite.**

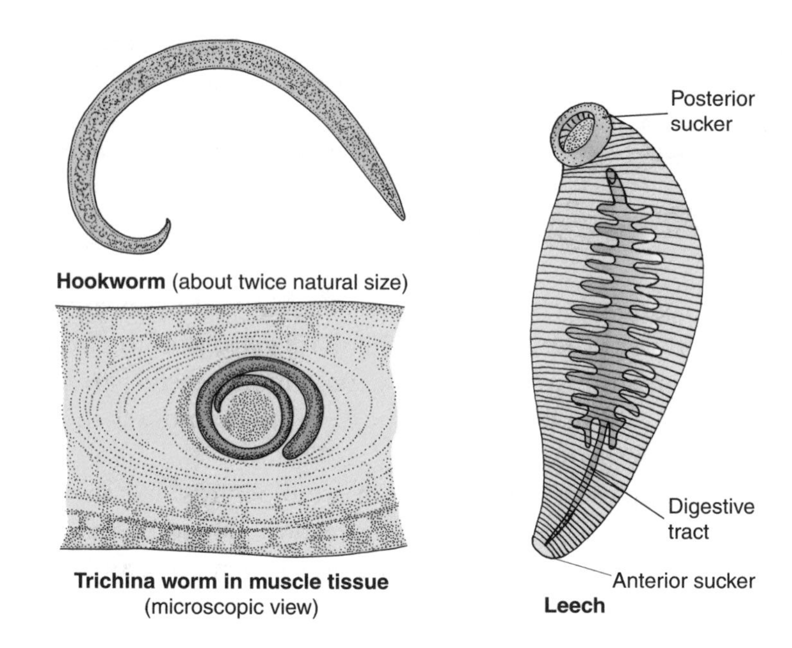

Most parasites infect a particular host and specific areas on or in the host's body. The human malaria parasite can infect humans and several closely related primates. A certain feather mite is found only on the outer portion of the flight feathers of a green parrot. The human head louse lives only among the hairs of the head. Dog fleas do not usually infect people, nor do human fleas infect dogs. The foreheads of humans are home to two kinds of mites. One lives in the hair follicles and the other in the sebaceous glands.

Each species of plant and animal is host to its own collection of parasitic creatures. A single host is able to support a community made up of large populations of many different species. Parasites such as fleas and ticks may harbor their own endo- and ectoparasites. To a parasite, its host's body becomes the whole world.

Mistletoe is a parasitic plant that grows on the branches of trees and feeds on the nutrients produced by the tree. The activities of the plant limit the growth and productivity of the forests of the western United States. There is even a type of mistletoe that attacks the large cacti in the southwest United States.

Birds such as cuckoos and cowbirds practice **brood parasitism.** These birds lay their eggs in the nests of other birds. When these foreign chicks hatch, they are larger than the other chicks. Because they are larger, they get more food. Sometimes they push the other nestlings out of the nest. The host parents unknowingly raise the foreign chicks as their own.

16.1 Section Review

1. Describe the differences between commensalism and mutualism.
2. Choose an ecosystem in your area and draw a diagram that shows examples of commensalism that exist among the creatures that live there.
3. Explain how parasitism differs from mutualism.
4. Why are parasites considered harmful, even when they do not kill their host?

16.2 SPECIALIZED INTERACTIONS

The stability of a community is a measure of the balance created by the interactions within the ecosystem. Many cooperative interactions that occur in nature are a result of the coevolution of two species. In **coevolution**, as one species evolves, the changes in that species affect the selection pressures on another species, causing it to evolve, too. The process by which plants evolved to attract beneficial insects led to adaptations that resulted in the evolution of their insect pollinators.

Coevolution of Plants and Pollinators

In flowering plants, **pollination**—the transfer of pollen from the stamens of a flower to the pistil—is an important step in the reproductive process. In some plant species, there are flowers that have only pistils and others that have only stamens. The flowers of other plant species have both stamens and pistils. The transfer of pollen can be carried out by wind or by insects, birds, or mammals. Over time, plants have adopted a wide range of strategies to accomplish pollen transfer. The diversity of flower types is related to the various methods by which pollen transfer is accomplished.

Flowers pollinated by wind have small sepals (the structures that surround the bud) and petals or lack these structures completely. These flowers, which are located high on the plant, produce large quantities of pollen. The tip of the pistil (see Figure 16-4, left), or stigma, is covered with a sticky fluid, which increases its chance of catching pollen. In plants pollinated by wind, flowers with stamens and those with pistils often are found on the same plant.

Plants pollinated by biological agents have large, brightly colored flowers to attract pollinators. These flowers usually produce a sweet, energy-rich nectar, which is a source of nutrients for the pollinators. The pollen is picked up as the pollinators brush against the stamens that surround the nectar's source. The pollinators leave with a meal and a dusting of pollen grains. The pollen is transferred, by contact, to the pistil of the next plant on which the pollinator feeds. (See Figure 16-4, right.)

Insects, such as bees, moths, and butterflies, are the chief pollinators for the many types of plants they visit. Some plants, though, have adapted over time to form a relationship with only one type of pollinator. This dependency on one pollinator means that the plant cannot reproduce without it, so the extinction of the pollinator would in turn lead to the extinction of that plant species.

The yucca plant, a native of the southwestern United States, is

Figure 16-4 **Plants pollinated by wind (left) have small flowers located high on the plant. Plants pollinated by insects (right) usually have large, brightly colored flowers.**

pollinated by only the yucca moth. The yucca flowers open in the evening when the moths are active. The moths are attracted by the fragrance of the flowers. Male and female moths meet in the flowers, where they mate. After mating, the female collects pollen from the flower, rolls it into a ball, and carries it to another flower. There she bores into the flower's ovary, located at the base of the pistil, and deposits an egg and some pollen into the chamber. In doing this, the female pollinates the flower and initiates fertilization. Some of the developing seeds will serve as food for the moth larvae, but many others will mature and in time germinate and form new plants. Neither moth nor yucca can complete its life cycle without the other.

Seed Dispersal

All living things tend to disperse, or spread out, from their point of origin. In this way, new environments are exploited and competition for living space is kept to a minimum. Animals are **motile**: they are able to move easily from place to place by walking, crawling, flying, or swimming. But plants are **nonmotile**: they are not able to move to new environments on their own. To be dispersed, plants require outside agents, such as wind, water, insects, mammals, or birds.

Many seeds have adaptations for dispersal over a wide area. These seeds are easily dislodged and stick to the coats of birds and mammals that brush against them. Animals transport the seeds to new locations where the seeds are free to germinate. Burdock and cocklebur, for example, have hooks on their seeds that aid them in hitching rides on mammals and birds. Some plants, such as the sticktight, have a sticky secretion that coats their seeds, causing them to cling to anything that brushes against them. The maple's samaras and the dandelion's seeds are dispersed by wind. (See Figure 16-5.)

Plants that produce fruits or berries usually depend on mammals or birds for dispersal. Cherries, raspberries, and grapes contain seeds with tough, thick, indigestible coverings. When a bird or mammal swallows these fruits, the seeds escape digestion and are eliminated with the feces. When birds are involved as the dispersing agents, the seeds may be carried a great distance from the parent plant. In

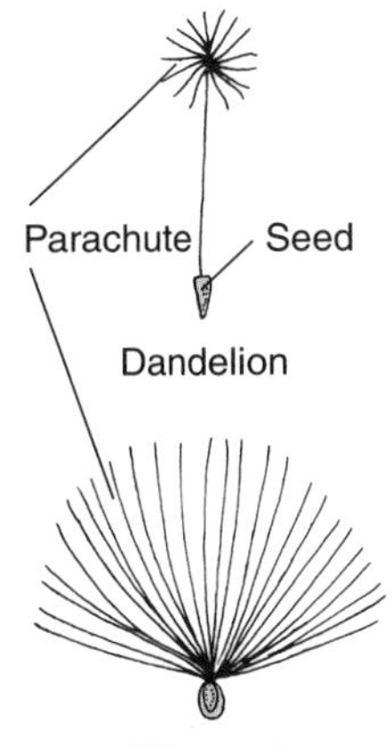

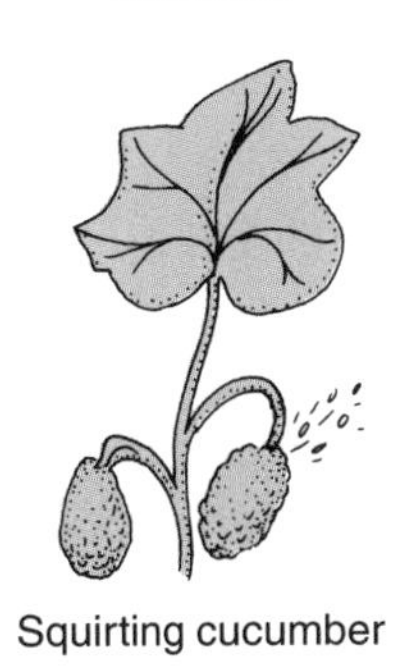

Figure 16-5 Dandelion and milkweed seeds are dispersed by the wind. The squirting cucumber ejects its seeds. The cocklebur sticks to an animal's fur or to people's clothing and is carried away from the parent plant.

fact, birds are responsible for the dispersal of many plants into isolated island ecosystems. The bird's nutrient-rich feces often serve as fertilizer for the germinating seeds. Many seeds cannot germinate unless they are first subjected to digestive fluids or are supplied with the nutrients from feces.

Large seeds and nuts contain high-energy food. Rodents, gnawing mammals such as squirrels, chipmunks, and wood rats, are well adapted to make use of this food source. Since seeds and nuts are not produced year-round, animals often bury them when they are plentiful so there will be food later on. Many of the buried seeds and nuts are overlooked, and germinate when proper conditions arise.

Keystone Species

In many communities, there is one species that is linked to most of the other species, either through the food web or by some other interaction. This species is called a **keystone species**, because its removal adversely affects a substantial part of the remaining community. When a keystone species is restored, the community usually returns to its original state.

Often, a keystone species is not recognized until it is removed from the community. At that point, the ecosystem reacts, and species are affected one after the other, like a row of falling dominoes. To protect the biosphere, people must try to understand all the interactions that occur among its living things, since the extinction of any one species may have an effect on all. Sea otters, beavers, elephants, alligators, and wolves are keystone species.

SEA OTTERS

The story of the sea otters is a good example of the effect of a keystone species. Sea otters are marine mammals that live in the waters off the Pacific coast of North America. (See Figure 16-6.) These animals feed on sea urchins and other shellfish. In turn, the sea urchins feed on kelp and other seaweed. The huge undersea forests of kelp and seaweed support a diverse community of crabs, shellfish, fish, and sea lions. By eating sea urchins, the sea otter naturally helped to maintain the health of the kelp forests. During the latter part of the

Figure 16-6 **Sea otters, a keystone species, keep the sea urchin population in check.**

nineteenth century, the sea otter was hunted nearly to extinction. With most of the otters gone, the population of sea urchins increased dramatically, and they destroyed the kelp and other seaweed. Once the kelp forest was gone, the ocean floor had no vegetation. The changed habitat caused the community to change. When the sea otter population was restored, the kelp forests and the community they supported also returned.

Now the sea otter is protected under the Marine Mammal Protection Act of 1972. But the otters are under attack from commercial abalone fishers. Fishers blame the otters for a decline in the population of abalone, a previously abundant mollusk. Environmentalists blame the decline in the abalone population on overfishing by commercial interests. The controversy, which pits environmentalists against industry, is an example of the types of battles that sometimes occur between these two segments of society.

BEAVERS

The beaver pond is a unique ecosystem created by one of nature's most remarkable engineers. (See Figure 16-7 on page 266.) Beavers often construct a series of dams across a shallow stream to slow the flow of water and create a chain of ponds. Dams are built of fallen trees, branches, and mud. Beavers, which are large rodents, are able to cut down trees with their sharp, chisellike incisor teeth.

Beavers live and raise their young in a lodge, which they construct in the deep waters of the pond. To protect the beavers from predators, the lodge is built solidly out of branches and mud. The entrance to the lodge is hidden underwater. The beavers also construct systems of canals. Beavers are clumsy on land, and the canals give them easy access to trees and shrubs in the surrounding area.

Figure 16-7 **A beaver dam causes a pond to form. The pond provides homes for other species.**

This access is important, since their food is the leaves, young twigs, and bark of trees such as aspen, poplar, and willow.

The beaver changes the environment to suit its needs, and in so doing creates a unique ecosystem. When the beavers cut down large trees, light can reach saplings in the understory, stimulating their growth. The still or slow-moving pond waters allow water plants to take root, which creates sources of food and homes for insects, fish, otters, muskrats, and waterfowl.

The history of the beaver in North America can be traced back more than a million years. In fact, much of the productive farmland in North America developed when beaver ponds filled with silt and became grasslands and forests. Trappers who hunted beaver pelts for the European fur trade spurred the westward expansion of the population in the United States and Canada by European settlers. Their travels opened routes westward.

Mimicry

The pressures of predation have favored the evolution of species that are either bad-tasting, poisonous, or equipped with weapons such as stingers or spines. These species usually display gaudy colors or bold patterns, which serve as a warning to predators. Would-be predators quickly learn to associate these colors and patterns with bad taste or pain. Some weaponless prey species have evolved colors that mimic, or imitate, the warning colors of the more dangerous species. This adaptation of an edible species to resemble an inedible one is called **mimicry**. (See Figure 16-8.)

Figure 16-8 Through mimicry, defenseless prey species, such as the viceroy butterfly (bottom), evolve to imitate the warning colors of a more dangerous species, such as the poisonous monarch butterfly (top).

The black swallowtail and spicebush swallowtail butterflies mimic the unpalatable pipevine swallowtail butterfly. The harmless black-and-yellow striped hoverfly has evolved a pattern similar to that of the yellow jacket, a wasp with a nasty sting. The harmless scarlet king snake mimics the color and pattern of the venomous coral snake. Mimicking a dangerous species confers survival benefits to the harmless species.

16.2 Section Review

1. How do the flowers of wind-pollinated plants differ from those pollinated by insects?
2. Explain how mammals and birds aid in seed dispersal.
3. Explain the relationship between sea otters and kelp forests.

16.3 SOCIAL LIFE IN A SPECIES

Communication holds organisms together within social groups and encourages their cooperation. There are a variety of names given to the social groups found in nature. Examples of animals and the social groups in which they are found are bison and zebra in herds, wolves and wild dogs in packs, lions in prides, prairie dogs and mole rats in colonies, tuna and bluefish in schools, whales and dolphins in pods, baboons and macaques in troops, and herring and anchovy in shoals.

Advantages of Forming Groups

Social living offers many benefits in terms of an organism's ability to find a mate and reproduce, passing its genes along to a new generation. However, in doing so the individual sometimes loses out to the larger group.

A nesting colony of weaverbirds gains protection through numbers, since there are more eyes available to detect predators as well as the combined defensive weapons that a group contains. The formation of a reindeer herd assures protection from predators for those animals located toward the center of the herd. In this way, females and juveniles can be protected. A pride of lions is able to attack and bring down very large prey due to the cooperative efforts of the group. A school of reef fish is able to deter predators through actions that make the group appear to be one large organism. The cooperation that exists among the members of bee, ant, and termite colonies creates complex insect societies. However, a large population in one place can deplete the available food resources and make the group more vulnerable to disease.

Relationships Within Social Groups

Many social groups are extended families. The population of these groups increases as young organisms mature and remain with their parents. But in many societies the young are forced out or leave to join other groups or start groups of their own. In some animal social groups, only one pair, the dominant male and female, mates and produces offspring. Therefore, all members of the group are offspring of that pair. In other groups, a single male or female dominates the reproductive process. The other members that can reproduce are either not allowed to mate or are driven from the group.

This insures that the genes of the fittest animals are passed on to the next generation.

PRAIRIE DOG SOCIETY

Prairie dogs are not dogs at all but rodents related to ground squirrels. They got the name from their high-pitched bark, which the animals use to communicate with one another. Prairie dogs play an important role in the North America short-grass prairie ecosystem. They live in a cooperative society that affords increased protection for its members as well as creating living space for many other species. (See Figure 16-9.)

Prairie dogs live in colonies, called towns, which may cover hundreds of acres and have several thousand inhabitants. The burrows they dig form an underground network throughout the town. Low mounds of soil, brought to the surface by the animals, surround the entrance of each burrow. The digging of burrows helps grassland ecosystems because it opens up air passages in the soil and brings fresh nutrients to the surface.

To insure a clear view when watching for predators, prairie dogs clip and eat the taller grasses and shrubs for great distances around the mouth of their burrows. This activity maintains the shorter grasses by removing competing tall grasses and shrubs. The prairie dogs' decomposing feces add important nutrients to the soil. The activities of the colony stimulate the growth of grass, which attracts larger grazing animals. Abandoned prairie dog burrows become the homes of burrowing owls, kangaroo rats, jackrabbits, tiger salamanders, box turtles, and black-footed ferrets.

Ranchers once considered the prairie dog a pest that competed with their livestock for the available grass. This often led to the

Figure 16-9 Prairie dogs live in colonies.

destruction of whole prairie dog towns with poisons or guns. The delicate balance of this ecosystem was destroyed and the health of the grasslands impaired. Today, ranchers are moving away from these practices and encouraging the return of the prairie dog in hopes of improving their rangelands.

Destruction of prairie dog towns almost led to the extinction of the black-footed ferret, a grassland predator. The prairie dog was the main prey of the ferret. As prairie dog towns disappeared, so did the ferret. By the mid-twentieth century, black-footed ferrets were considered extinct. Then, by chance, a small group of black-footed ferrets was found in Wyoming. To protect the species, all remaining ferrets were captured and their breeding controlled. The number of black-footed ferrets has risen to a point where some have been returned to the wild.

16.3 Section Review

1. What are some benefits of living in a social group?
2. How do prairie dogs maintain the short-grass prairies?
3. Write a short story that tells what it might be like to be the last black-footed ferret.

LABORATORY INVESTIGATION 16

Studying Lichens

PROBLEM: *What do lichens look like, and what do they do?*

SKILLS: *Observing, manipulating, using a microscope*

MATERIALS: *Lichen-covered rock and bark, hand lens, microscope (low-power objective), pick*

PROCEDURE

1. Collect some rocks and bark on which lichens are growing. Examine the lichens with the unaided eye.
2. With a pick, scrape a bit of the lichen off the rock. Examine the exposed surface of the rock.
3. Examine the lichens with a hand lens.
4. Examine the lichens under the low-power (100 ×) objective of a microscope.
5. Draw a diagram of the lichen as you observed it under the microscope. Label alga cells, fungus cells, and fungal hyphae.

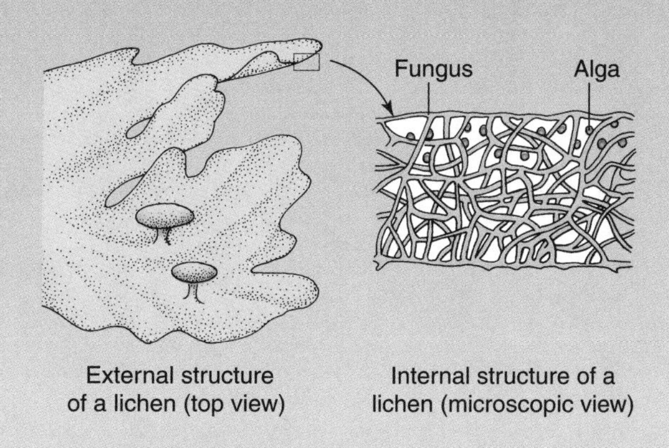

External structure of a lichen (top view)

Internal structure of a lichen (microscopic view)

OBSERVATIONS AND ANALYSES

1. Describe the surface of the rock from which the lichens were removed.
2. What do you think caused the rock to change?
3. Describe how the lichens appear to the unaided eye. Are the lichens crustose (branched) or foliose (leafy)?
4. Describe how the lichens appear under the low-power (100 ×) objective.
5. Describe the relationship that exists within the lichen.

GOING FURTHER

Set up a terrarium that simulates the natural habitat of the lichens you collected. A relatively dry environment is required for lichens to grow. Observe and record the growth rate of the lichens.

Lichens are indicators of air pollution. Most lichens, especially the crustose variety, will not grow in polluted air. Devise and describe an experiment that uses lichens to detect the presence of pollutants in the air.

Look for lichens on walls, stones, and trees outside your home or school. Determine the degree of pollution in the area by the relative abundance of these indicators of air quality.

Chapter 16 Review

Answer these questions on a separate sheet of paper.

Vocabulary

The following list contains all the boldfaced terms in this chapter.

brood parasitism, coevolution, commensalism, ectoparasites, endoparasites, keystone species, mimicry, motile, mutualism, nonmotile, parasite, parasitism, pollination, symbiont, symbiosis

Fill In

Use one of the vocabulary terms listed above to complete each of the following sentences.

1. The close physical relationship that exists between members of two different species is called ________.
2. Organisms that cannot move to new surroundings on their own are ________.
3. The adaptation of an edible species to resemble an inedible species is ________.
4. In ________, as one species evolves it affects the selection pressures on another species, causing it to evolve, too.
5. Sea otters, elephants, and beavers are examples of ________.

Multiple Choice

Choose the response that best completes each sentence.

6. An example of a cleaning symbiont is a *a.* tickbird. *b.* bluebird. *c.* hummingbird. *d.* sunbird.
7. An organism that depends on a cooperative relationship with algae is *a.* a flashlight fish. *b.* kelp. *c.* coral. *d.* a sea anemone.
8. A commensal relationship exists between *a.* termites and ruminants. *b.* tapeworms and dogs. *c.* algae and fungi. *d.* sharks and remoras.
9. An example of a keystone species is the *a.* hummingbird. *b.* kelp. *c.* alligators. *d.* scarlet king snake.
10. A parasitic plant that grows on oaks and other trees is the *a.* mistletoe. *b.* yucca. *c.* aspen. *d.* poplar.
11. A plant whose seeds are dispersed by hitching a ride on the fur of mammals and the feathers of birds is the *a.* cherry. *b.* grape. *c.* burdock. *d.* pine.
12. Plants that have small sepals and petals are usually pollinated by *a.* insects. *b.* birds. *c.* mammals. *d.* wind.

Short Answer (Constructed Response)

Use the information you learned in this chapter to respond to the following items.

13. How is commensalism different from mutualism?
14. What are the two differences between ectoparasites and endoparasites?
15. Explain brood parasitism.
16. List three examples of human ectoparasites.
17. Name three human endoparasites.
18. Explain how mimicry protects the viceroy butterfly from predators.
19. Describe an example of coevolution.
20. Why are beavers considered to be a keystone species?
21. Describe the relationship between legumes and nitrogen-fixing bacteria.
22. Name three adaptations that seeds possess that aid in their dispersal by animals.

Essay (Extended Response)

Use the information you learned in the chapter to respond to the following items.

23. What benefits do cleaning symbionts derive from their hosts?
24. What are the similarities and differences between parasitism and commensalism?
25. Describe how the survival of the yucca plant is tied to the survival of the yucca moth.
26. Determine the keystone species in your area. Describe how changes in its population affect the populations of other species.

CHAPTER 17
How Environments Change

When you complete this chapter, you should be able to:

Distinguish between primary and secondary succession.

Describe the life cycle of a lake.

Compare the process of natural eutrophication with cultural eutrophication.

Contrast the characteristics of an immature river and a mature river.

Discuss how waves and tides affect the shoreline.

Changes in the abiotic environment prompt adaptations in the biotic environment. In addition, living things can cause changes in the nonliving environment. The interactions between the biotic and abiotic environments often follow a predictable pattern that allows ecologists to anticipate how an ecosystem will respond.

17.1 TERRESTRIAL ECOSYSTEMS CHANGE

As the abiotic factors of an ecosystem change, stress is placed on the biotic factors. The living things are forced to move or adapt to make use of the new resources the changed environment has to offer. Whole ecosystems evolve under these environmental pressures. A knowledge of the processes involved provides a basis for understanding and predicting how organisms and ecosystems change following various types of disturbances. It also explains the variety of biomes that make up the biosphere.

From year to year, a forest ecosystem does not seem to change much. Some plants and animals die and are replaced by others, but for the most part you observe little change overall. Fires or severe storms cause major changes, but the forest usually appears to survive these crises. However, when scientists observe forests over an extended period, they can see the forests undergo a variety of changes. The original plants are replaced gradually by different types of plants, until one type of plant dominates the ecosystem. At this point, stability is achieved and the forest may remain unchanged for many years. This gradual replacement of one group of organisms with another group is called **ecological succession** (also called biological succession).

Ecological succession begins when a barren area begins to support plant life or when an ecosystem is destroyed or disturbed by fire or drought. The first plant species to colonize a barren or disturbed area are called **pioneer organisms.** These hardy organisms are able to thrive under hostile conditions. As the pioneer organisms grow, they change the environment so that it attracts other organisms, which initiates the process of succession.

The complex interactions that develop among the plants, animals, and their physical environment create a unique ecosystem. As this ecosystem develops, it is influenced by changes made to the environment caused by living things. Such environmental changes lead to the gradual replacement of the original organisms. One replacement follows another in the orderly process of ecological succession, and the result is the development of a long-term, stable ecosystem, called a **climax community.** Climax communities remain stable as long as the climate and other environmental conditions do not change. Climax communities are more complex and have a greater species diversity than the earlier communities they replace.

Primary Succession

Primary succession occurs on surfaces or in areas that never before have been occupied by living organisms. It may begin on bare rock. This rock may have been exposed by a retreating glacier or formed as lava cooled. The most common pioneer organisms to colonize these areas are lichens. Lichens can live and grow in nutrient-deficient areas under the harsh effects of wind, sun, and precipitation. The ability to carry out photosynthesis and extract nutrients from their surroundings makes them perfectly adapted for the niche they occupy. Lichens form a crusty growth on rock surfaces. These growths attract other microscopic plants and animals. The plants and animals cause changes in the environment that provide new opportunities for consumers. (See Figure 17-1 on page 278.)

The secretions and waste products produced by lichens dissolve the surface of rocks they grow on, softening the rocks and making them susceptible to weathering and erosion. As the rocks crumble, particles accumulate to form soil. The soil also collects particles of dust and debris that settle from the air. Eventually, portions of the lichens die, and the organic materials they contain decay and become part of the soil. These physical and chemical processes lead to the formation of a thin layer of nutrient-rich soil.

Soil builds up and attracts other organisms, such as bacteria, fungi, mosses, worms, and insects. Waste products and decayed remains from these new organisms contribute to a richer, deeper soil. Mosses replace the lichens and attract different types of organisms whose activities contribute to the continuing accumulation of rich soil. Eventually, the nutrient content of the soil is able to support the growth of herbaceous plants, which are small, nonwoody plants, including weeds. Many herbaceous plants are **annuals**, plants that die after one growing season. The herbaceous plants crowd out and replace the moss and pioneer communities.

The death and decomposition of the herbaceous plants adds nutrients and fibrous materials to the soil. As these materials

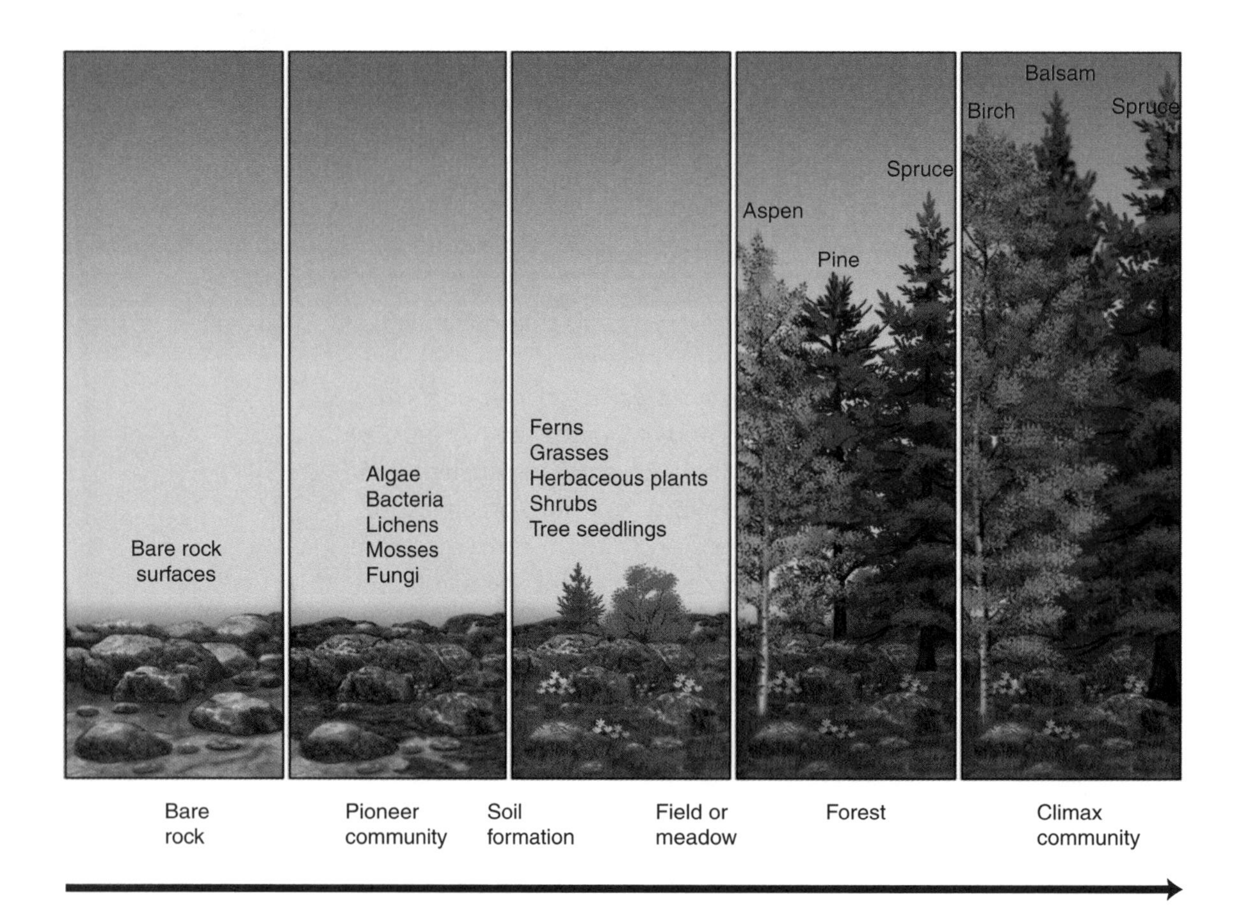

Figure 17-1 **Primary succession takes place in an area that has never supported an ecosystem.**

accumulate, the soil layer becomes thicker and richer in nutrients. These conditions favor the growth of grasses, woody shrubs, and **perennials**, plants that live for many years. The shade cast by woody shrubs prevents the growth of the grasses and encourages the growth of the shrubs. Finally, trees take root and their growth crowds out the shrubs. A forest now stands where once there was bare rock. The process of primary succession takes many years to complete.

Secondary Succession

The process of succession can also begin in an area where an ecosystem existed before but was destroyed or disturbed. This is called **secondary succession**, which is quite different from primary succession. Some disturbances are due to natural causes such as fire, flood, erosion, and earth movements. The actions of people may also cause the destruction of an ecosystem. Clearing land for agriculture, grazing livestock, or logging can result in the devastation of the natural ecosystem.

An abandoned field or farm is an excellent site to observe the process of secondary succession. To open the land for farming, a forest was cleared by cutting or burning. For many years, the land was devoted to agricultural use. Years of cultivation and harvesting crops removed many nutrients from the soil. This finally required the use of fertilizers, herbicides, and pesticides to improve crop yields. The added costs of these chemicals may have made farming unprofitable, which led to the abandonment of the farm.

Annual plants, such as ragweed or fireweed, are the first to appear on the uncultivated soil of an abandoned farm. These are the pioneer plants in secondary succession. They are able to grow quickly in the nutrient-poor soil, even though they are supplied with little water and are exposed to full sunlight. The weeds produce many seeds and spread over the field. When they die, they form a thick mat of vegetation. This prevents the evaporation of water from the soil and allows decomposition of the dead vegetation by soil bacteria. As bacteria decompose the vegetation, they produce humus. Humus, composed of small particles of organic matter, further enriches the soil and helps it absorb and hold water.

Conditions now favor the growth of perennials, such as goldenrod, Queen Anne's lace, and asters, as well as tall grasses. Insects, birds, and mammals feed on and among the many plants. Their droppings help to enrich the soil. Many woody shrubs start to germinate from seeds introduced by birds, other animals, and the wind. As these plants grow, they crowd out the smaller plants, eventually producing conditions that favor the growth of deciduous trees, trees that shed their leaves. When the trees mature, their fallen leaves cover the low grasses, which prevents further growth and adds to the thick layer of nutrient-rich humus. Trees provide cover for large herbivores such as deer and elk, which browse on the low vegetation, promoting further growth of the trees. Eventually, a mature forest stands where once there was an abandoned field. (See Figure 17-2 on page 280.)

The rapid growth of weeds is a constant problem faced by farmers

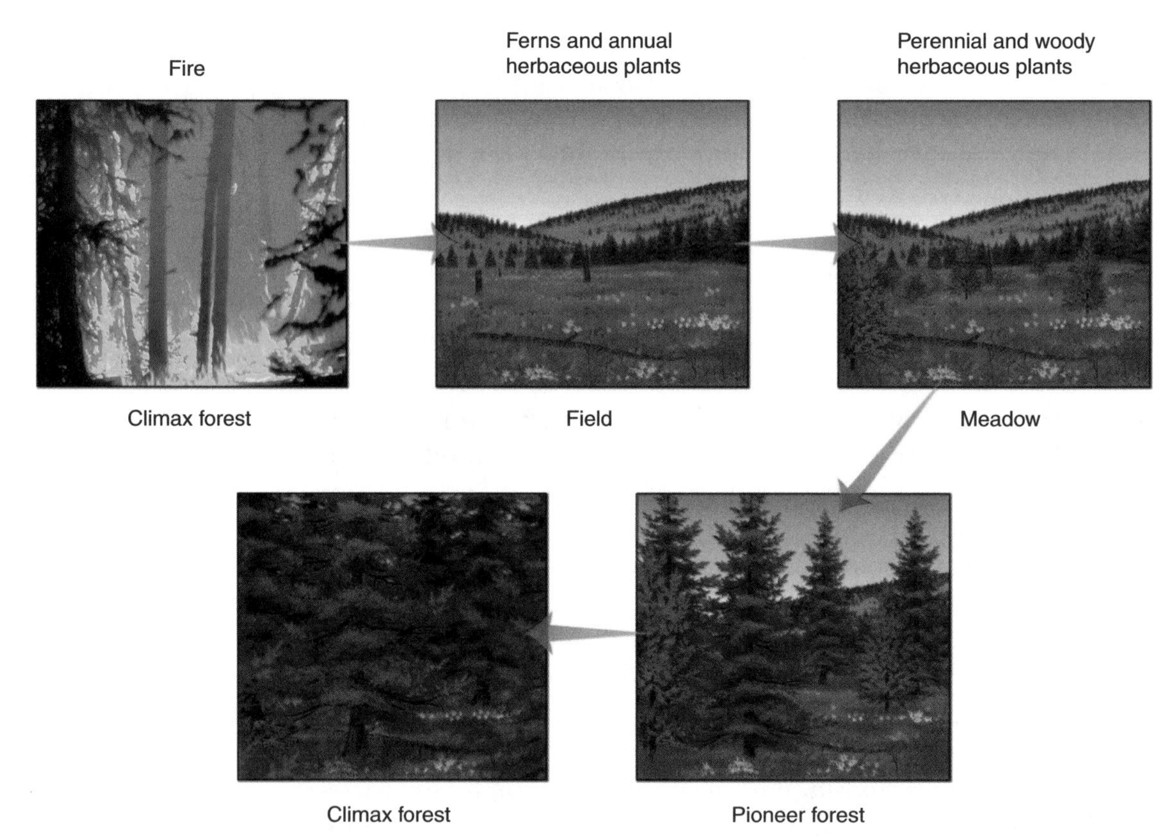

Figure 17-2 Secondary succession occurs after the destruction of an ecosystem.

and gardeners. Weeds are pioneer plants that thrive in the open ground, freshly turned soil, and abundant water available in a garden or farm. These plants are better adapted to colonizing the bare soil of the garden than are cultivated plants. Weeds are masters of initiating the process of secondary succession. If weeds are not removed, their rapid growth interferes with crops. Herbicides, plastic coverings for the soil, and mechanical cultivators are some of the methods used to discourage the growth and spread of weeds.

Climax Communities

Succession in most land environments eventually leads to the establishment of a stable, complex, self-sustaining forest ecosystem—the climax community. Two to three types of trees usually dominate the climax community. Climate plays a major role in determining the type of trees found in a climax community. As long as the climate does not change, the climax community remains relatively unchanged for many years. These communities usually are able to regenerate themselves following natural disasters, such as fires and droughts. Climax communities contain a wide variety of habitats and thus the greatest biodiversity.

Because the climate varies across the country, climax communities differ throughout the United States. This leads to forest ecosystems that contain trees best adapted for each specific area. In the Pacific Northwest, there are evergreen forests composed of spruce, fir, and hemlock. In the southeastern United States, deciduous forests of oak and hickory dominate. And in the Southwest, the forests are piñon and juniper. Beech, maple, and hemlock dominate in the forests of the Northeast.

17.1 Section Review

1. How do the life activities of lichens initiate primary succession?
2. Explain the differences between primary succession and secondary succession.
3. Draw a diagram that illustrates some of the agricultural practices used to deter the growth of weeds on cultivated lands.

17.2 AQUATIC ECOSYSTEMS CHANGE

Lakes and ponds form in depressions on Earth's surface. Some lakes formed when glaciers gouged out their basin. Others formed as rivers were dammed by rocky debris. Still others formed when shifts in Earth's crust left depressions.

Cold water is more dense than warm water. Density differences between warmer and colder waters stratify lakes, and this results in layers that create a diversity of habitats. (See Figure 17-3 on page 282.) Deeper water is usually cooler than surface water. The cooler water is able to dissolve more oxygen than the warmer water can. Nitrogenous wastes accumulate in the bottom layers where decomposition occurs. Since sunlight does not penetrate very deeply, only

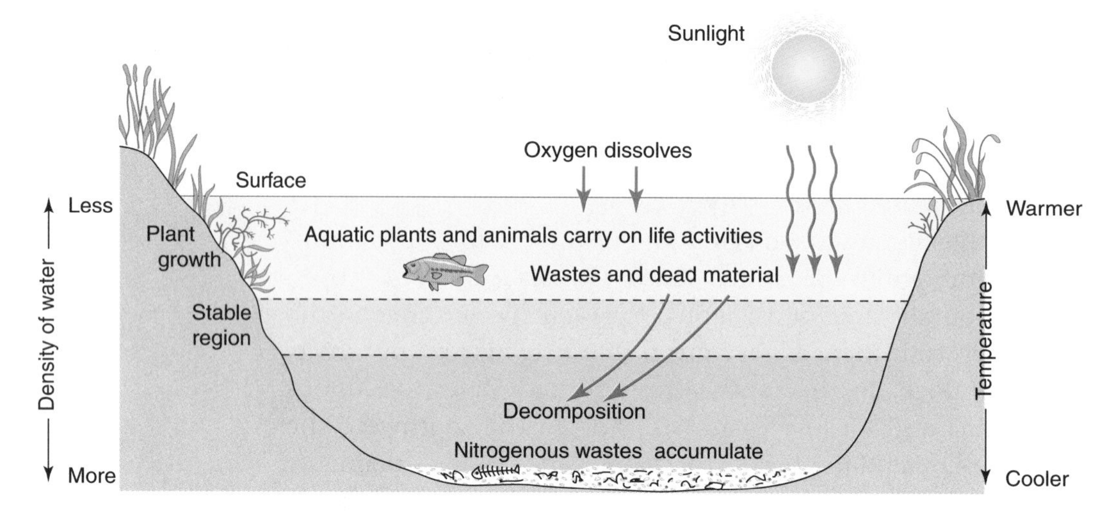

Figure 17-3 Stratification of a lake.

the surface waters can support a community of photosynthetic organisms. If the ecosystem is to be sustained, periods of mixing are critical for redistribution of nutrients (such as oxygen and nitrogen) within the lake.

Lakes and ponds undergo a cycle of youth, maturity, old age, and eventual death. Lakes that are young, in geologic terms, are called **oligotrophic lakes.** These lakes have low levels of nutrients and, therefore, support only a small number of organisms. Food chains in these lakes are quite simple. Because oligotrophic lakes have little organic material, they are usually crystal clear. Examples of oligotrophic lakes are Lake Tahoe, between California and Nevada, Lake Superior, between the United States and Canada, and Crater Lake in Oregon.

Middle-aged lakes, called **mesotrophic lakes**, have a larger amount of nutrients in their waters than oligotrophic lakes. Mesotrophic lakes are biologically more productive; they support a diverse community. Lake Erie and Lake Ontario, between the United States and Canada, and Yellowstone Lake in Wyoming are examples of mesotrophic lakes.

Eutrophic lakes are older lakes that have a very high nutrient content and high productivity. Water quality in these lakes can deteriorate quickly, and this can result in massive fish kills, weed-choked shorelines, and the production of hydrogen sulfide gas by

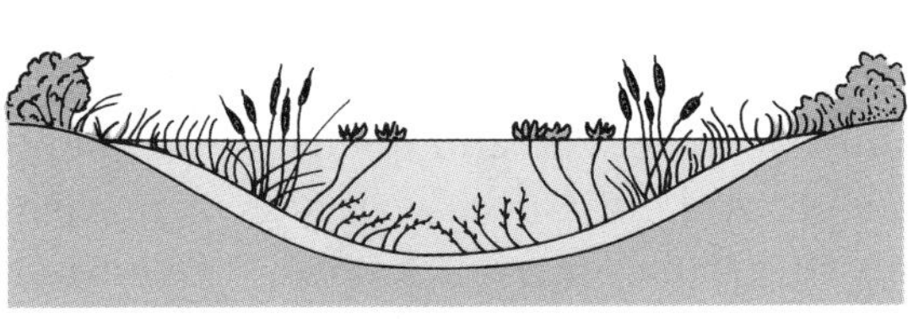

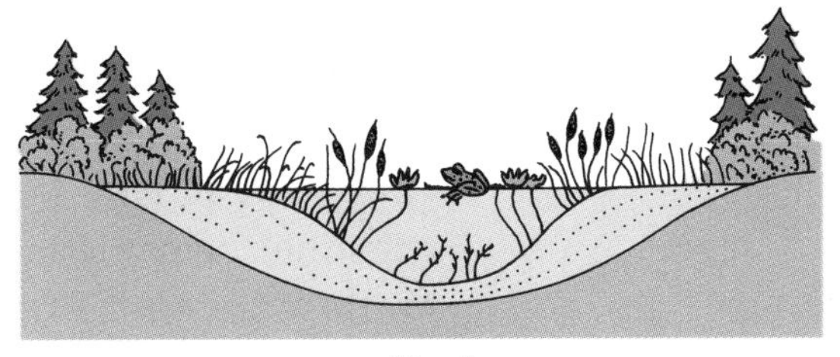

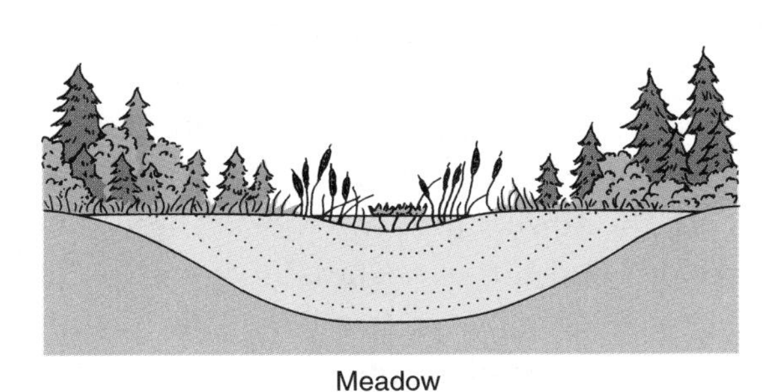

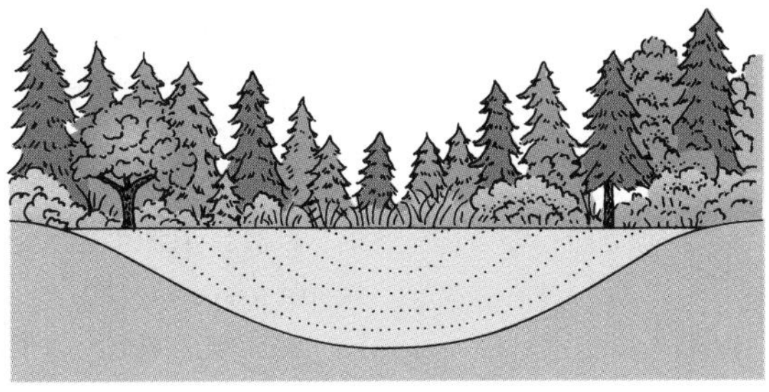

Figure 17-4 Ecological succession in a pond.

decaying organic matter. In recent years, many small lakes and ponds throughout the United States have become eutrophic due to urban encroachment.

Although lakes persist for many years, they eventually disappear in a series of stages known as aquatic succession. Usually, young lakes and ponds contain clear water that is free of particles and debris. But the rivers and streams that drain into the lakes carry sediments such as silt, sand, and gravel. Slowly, these sediments build up, first at the river inlets and later over the entire lake bottom. The buildup of sediments, called sedimentation, gradually fills the basin of the pond or lake.

As silt builds up, lakes become marshy around their edges. (See Figure 17-4.) Each year the marshy area slowly advances toward the center of the lake, eventually filling it in. The lake becomes a marsh or swamp. Pioneer land plants now invade the marshy areas and build up and enrich the soil. Eventually, grasses and woody plants take over. The marsh becomes a meadow, and the meadow finally becomes a forest. The rates at which lakes and ponds undergo succession differ. But sediments slowly fill these depressions until the bottom is raised above the water table. Deep mountain lakes and cold glacial lakes change at a very slow rate, while warm, shallow lakes or swamps fill in more rapidly.

The accumulation of sediments causes a lush growth of plants on the banks and in the shallow areas of lakes and ponds. As the plants die, their remains accumulate and enrich the watery environment with nutrients. This leads to natural **eutrophication**, the enrichment of a body of water with nutrients derived from the remains of plants. These nutrients provide nourishment for plankton, bacteria, and other microorganisms. These in turn serve as a source of food for larger organisms, such as fish, crustaceans, insects, and birds. A complex, ever-changing ecosystem thus exists within the lake.

In the related process of cultural eutrophication, runoff derived from agricultural land and feedlots that surround a pond or stream carry silt, animal wastes, and fertilizers into the water. These substances are rich in phosphates and nitrates. These nutrients build up, and the waters become overenriched with them. This in turn causes dense growths of algae, called **algal blooms**, and bacterial surface scums. The decay of the algae and bacteria uses up most of the oxygen in the water, which makes living conditions unsuitable for most other organisms. Cultural eutrophication leads to the death of the lake and the destruction of the ecosystem.

17.2 Section Review

1. Why does stratification occur in lakes?
2. Describe the process by which a lake fills and becomes a terrestrial habitat.
3. Compare the process of natural eutrophication with cultural eutrophication.

17.3 A RIVER MATURES

The headwaters, where rivers begin, are in mountains or hills. The rivers are fed by melting ice and snow, rain, and runoff from the surrounding land. The slope of the land is steep. Gravity drives the water rapidly downhill to reach a base level, which is usually at

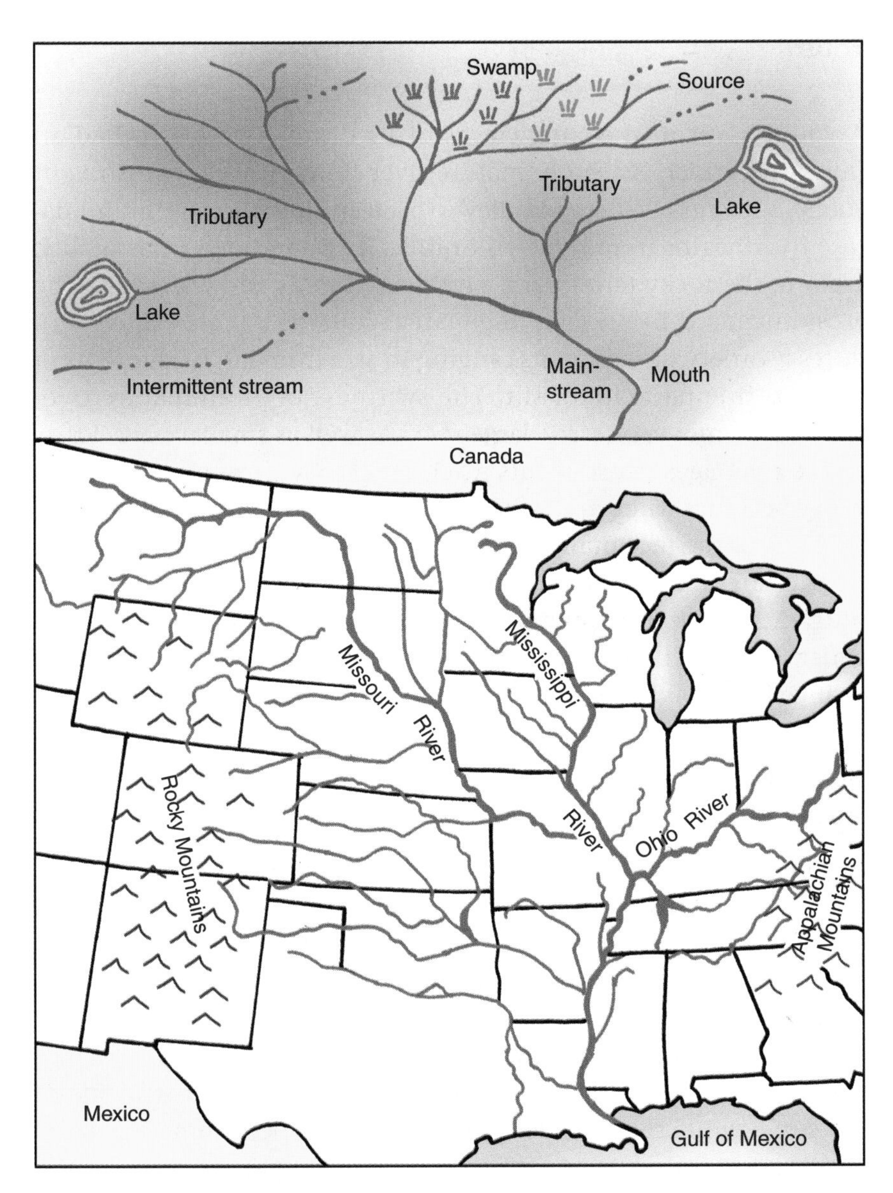

Figure 17-5 The top illustration shows some components of a river system. The map shows the Mississippi River drainage system.

the ocean or sea. However, some rivers empty into lakes. A network of tributaries, made up of streams, creeks, and brooks, forms a branching drainage system that eventually merges to form a river. The area drained by the main river and its network of tributaries is the **watershed**, or **drainage basin.** (See Figure 17-5.) In their journey to the sea, rivers, like lakes, go through stages of development.

Youth

A young river in the early stages of its development is called an **immature river.** As the immature river flows, it follows a straight course and cuts a V-shaped valley with steep sides. Because an immature river begins in mountains or hills, its water flows very rapidly, eroding the rocky terrain and producing rapids and waterfalls. The erosion caused by the moving water scours the land, which produces a variety of sediments ranging in size from large fragments of rocks to fine particles of silt. The swiftness of an immature river allows it to move, or carry, large rocks as well as smaller particles.

As a river ages, erosion cuts its channel to a lower elevation. The valley becomes wider and wider, as moving water removes rock from the sides of the mountains. More streams merge, which drains water from a larger area into the river valley. The waters contain more nutrients and can support a larger number of living things. The wider, shallower river now moves more slowly. The river tends to change from following a straight course to forming a series of curves.

Maturity

As time passes, a river eventually becomes a **mature river.** Along its length, waterfalls and rapids have been eroded away. Erosion has also made the slope gentle and the river valley broad and flat. The slowly moving waters cause the river to form many curves and loops, called **meanders**. The slow-moving river causes little erosion, and the waters can carry only the lighter particles of silt.

Where rivers meander, sediments are deposited on the inside of the curve while the outside of the curve is eroded. Sometimes, the meanders of a river form U-shaped bends that get cut off from the river and form lakes. These are called **oxbow lakes**. They mature as do other standing bodies of water and eventually become terrestrial environments.

Where the river water slows, sediments are deposited. Sediments that settle on the river bottom may eventually divert the river channel or form islands in the river that are often sites for primary succession. The river valley undergoes constant change. Sediments deposited along the riverbank build up the surrounding land. A large river can be immature at its headwaters and mature farther along its course.

On both sides of mature rivers, there are flat areas covered with sediment. These areas are called floodplains. Heavy rains and spring thaws can cause a river to overflow its banks and cover the floodplain. When the water recedes, sediment is left behind. Repeated flooding causes sediment to build up and form thick, fertile soils. These areas support lush vegetation and attract wildlife. They are preferred areas for agriculture and raising livestock. (See Figure 17-6.)

The movement of water in rivers and streams creates conditions that are quite different from those found in lakes. The swiftly flowing water near the head of a river is highly oxygenated because its surface churns and mixes with air. Streams and rivers contain creatures that are able to withstand the movement of the water. Because of the plentiful supply of oxygen, these creatures can breathe well, despite having inefficient respiratory organs. These organisms would suffocate in the still water farther down the river. Many insects and their larvae live under and among the rocks on the river bottom. These organisms feed larger organisms.

Along a river's length, conditions faced by creatures vary greatly. The headwaters of a mountain stream contain cold, swift-flowing, clear waters that are rich in oxygen. Near the river's mouth, the waters are warmer, slow-moving, muddied with sediments, and of a lower oxygen content. Thus the river is composed of a series of diverse ecosystems that stretch along its length. Where streams are

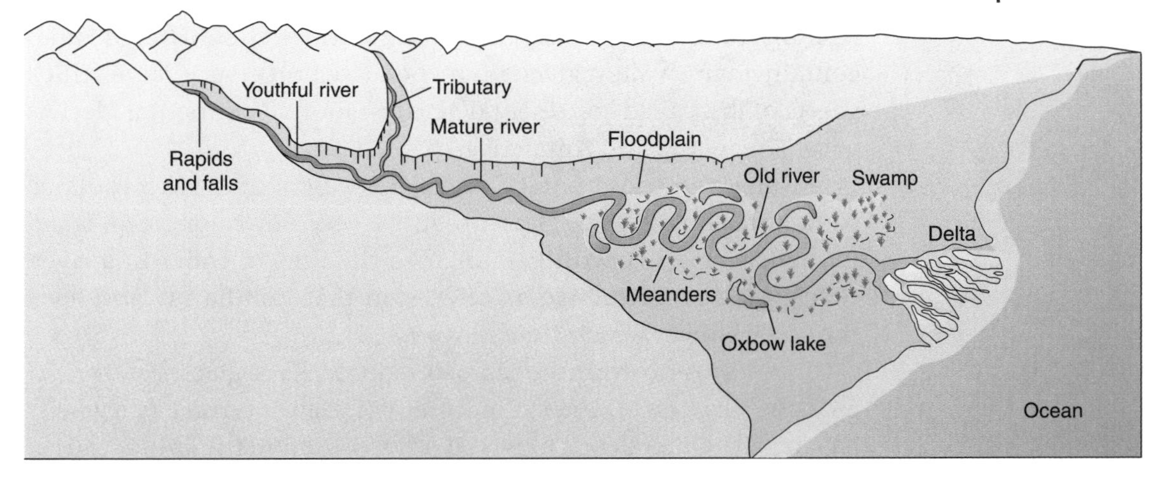

Figure 17-6 Along its length, a river may show different stages of development.

shaded, usually at their headwaters, there is limited plant growth in the water. Primary consumers are limited by the amount of leaf litter that enters the water. Rivers with slow, relatively clear water and pools support many plants and the animals they attract.

17.3 Section Review

1. Explain why a river flows faster at its headwaters than on the floodplain.
2. Why are floodplains preferred areas for agriculture?
3. How is the environment of a river different from that of a lake?

17.4 OTHER ENVIRONMENTAL CHANGES

Rivers are not the only waters to cause changes in the environment. Glaciers were responsible for many features found in North America. In addition, waves and tides in the oceans and seas change the face of the land.

Glaciers

A **glacier** is a large, permanent mass of ice or snow. Glaciers form in areas where more snow accumulates than melts. Valley glaciers form in high mountains, where low temperatures do not permit snow to melt in summer. The Rocky Mountains in the western United States contain many valley glaciers. In polar regions, very large, thick sheets of ice, called continental glaciers, form. Continental glaciers cover Greenland and Antarctica.

In valley and continental glaciers, the weight of the accumulated snow and ice causes the glacier to move very slowly, or creep. Gravity pulls glaciers downhill, similar to the flow of water in a river. Glaciers are powerful agents of erosion that grind away and level the surfaces over which they move.

Glacial erosion forms U-shaped valleys. As a glacier moves, it pushes rocks, gravel, and silt in front of it. Debris carried by a glacier is called till. Other debris loosened or scoured from the land is carried within the ice and snow. Glaciers can dislodge and carry huge boulders for great distances. Rocks carried within the glacial ice scrape against the bedrock, producing scratches called glacial striations.

The glaciers that covered the continents during past ice ages have had a profound effect on Earth's surface. During the last ice age, which ended about 11,000 years ago, glaciers covered most of the northern United States and Canada. Glacial erosion formed many of the surface features seen today. The basins of the Great Lakes were scoured out by glaciers; the Appalachian Mountains were ground down to their present height; and the canyon of the Hudson River was cut by glaciers.

When the advancing end of a glacier reaches a warmer climate, it begins to melt. Debris carried in the glacial ice or that is pushed in front of the glacier is deposited. When a glacier melts and retreats, deposits are left behind in the form of a mound or ridge, called a terminal moraine. Long Island, New York, is an example of a terminal moraine deposited at the end of the last ice age. Because of their rich sediment, moraines and the areas around them are well suited for growing plants. The rich soils of the midwest prairies are made up of debris the glaciers scoured from the northern part of the continent and deposited as they retreated. The land newly exposed by a retreating glacier is a site for primary succession.

Waves and Tides

Wave erosion is a constant source of change along the oceans' shores. Waves cause erosion in several ways. When waves reach shallow water near the shore, they break. The force of the breaking waves against the shore knocks fragments off rock formations. The pebbles and sand carried by breaking waves further erode rocky shorelines. Breaking waves force water into cracks in rocky cliffs along the shore. The pressure of the water can widen the cracks and eventually split the rock. Cliffs are worn away, beaches are created and destroyed, tide pools are hollowed out, and caves are cut into the rocks. These places are colonized by pioneer organisms and develop into habitats that create unique ecosystems.

Waves can carry large quantities of sand, bits of rock, and pieces of shells and deposit them along the shoreline to form beaches,

Figure 17-7 Shoreline features form habitats for marine creatures.

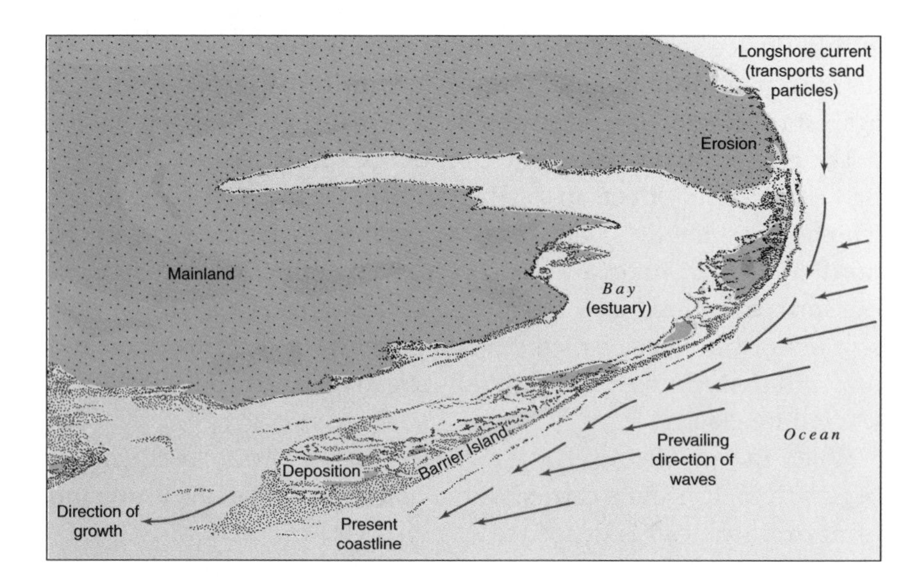

which are habitats for marine organisms. (See Figure 17-7.) Sand and other sediments are carried along the shoreline, constantly changing its shape. New beaches are created as others are carried away. The composition of beaches varies greatly. Some beaches are made up of fine sand, some are coarse sand, some are rocky, and some are composed of silt or mud. On coasts with straight shorelines, long offshore sandbars, called barrier islands, are created by wave action. Barrier islands attract diverse communities of plants and animals.

Tides affect the types of plants and animals that live along a seashore, because they cause shoreline ecosystems to be alternately wet and dry. Tidal currents bring a new supply of water, which carries nutrients and oxygen into coastal areas. And tidal action can multiply the effects of erosion caused by wave action. To be part of a shoreline community, species must be adapted to tidal changes.

17.4 Section Review

1. Compare valley glaciers with continental glaciers.
2. Explain why beaches are constantly changing.
3. Draw a map that shows the extent to which glaciers covered North America during the last ice age. Describe the evidence that exists to support your findings. A high school earth science or geology textbook can be useful in finding this information.

LABORATORY INVESTIGATION 17

Glacial Erosion

PROBLEM: *How does moving ice affect the surface over which it travels?*

SKILLS: *Observing, manipulating, inferring*

MATERIALS: *1 sheet of cardboard approximately 20 cm × 28 cm, soft clay, sandpaper, 125 mL of sand, 2 ice cubes*

PROCEDURE

1. Cover a sheet of cardboard with a layer of soft clay about 0.5 centimeter thick. Rub sandpaper across the top of the clay from left to right.
2. Smooth the surface of the clay. Place a layer of sand on the top of the clay. Slide an ice cube through the sand. Note what happens to the sand.
3. Shake the excess sand off the clay, so that some remains on the surface. Place an ice cube on top of the clay and let it remain there for 2 minutes. Examine the bottom of the ice cube.

OBSERVATIONS AND ANALYSES

1. When you rubbed the sandpaper over the clay, what did you see on the surface of the clay?
2. When you slid the ice cube through the sand, what happened to the sand in front of the ice cube? What did you see on the bottom of the ice cube?
3. Which part of the experiment demonstrated the formation of a moraine?
4. Which part of the experiment demonstrated glacial striations?
5. Which part of the experiment demonstrated the transport of materials by glaciers?

Chapter 17 Review

Answer these questions on a separate sheet of paper.

Vocabulary

The following list contains all the boldfaced terms in this chapter.

algal blooms, annuals, climax community, drainage basin, ecological succession, eutrophic lakes, eutrophication, glacier, immature river, mature river, meanders, mesotrophic lakes, oligotrophic lakes, oxbow lakes, perennials, pioneer organisms, primary succession, secondary succession, watershed

Fill In

Use one of the vocabulary terms listed above to complete each of the following sentences.

1. Agricultural and livestock runoff into lakes and ponds can lead to the growth of ________.
2. The end product of ecological succession is the ________.
3. The area drained by the main river and its channels is called the ________, or ________.
4. ________ occurs in areas that never have been inhabited.
5. ________ is the enrichment of a body of water with nutrients from the remains of plants.

Multiple Choice

Choose the response that best completes the sentence or answers the question.

6. A nutrient that builds up in eutrophic lakes to cause algal blooms is *a.* salt. *b.* phosphate. *c.* sulfate. *d.* hydrogen sulfide.
7. Which of the following is an example of an oligotrophic lake? *a.* Lake Ontario *b.* Lake Erie *c.* Lake Tahoe *d.* Great Salt Lake
8. Which of the following is an example of a mesotrophic lake? *a.* Lake Superior *b.* Crater Lake *c.* Mono Lake *d.* Yellowstone Lake
9. A river that cuts a V-shaped valley is *a.* a mature river. *b.* an immature river. *c.* a meander. *d.* a delta.
10. One characteristic of a rapidly flowing river is that it *a.* has a plentiful supply of oxygen. *b.* has a poor oxygen content. *c.* can carry little sediment. *d.* forms a U-shaped valley.
11. A mound of debris left in front of a retreating glacier is called *a.* an alluvial fan. *b.* a terminal moraine. *c.* glacial striation. *d.* a sandbar.

Short Answer (Constructed Response)

Use the information you learned in this chapter to respond to the following items.

12. How is an annual plant different from a perennial plant?
13. List two examples of pioneer organisms.
14. Give three examples of climax communities.
15. What is a watershed?
16. How does an immature river differ from a mature river?
17. Over time, why do ponds become forests?
18. What is eutrophication?
19. How are valleys formed by glaciers different from valleys formed by rivers?
20. Describe the formation of an oxbow lake.
21. Define ecological succession.

Essay (Extended Response)

Use the information in the chapter to respond to these items.

22. How do annuals change the environment to encourage the growth of perennials?
23. What are some disturbances that open areas to secondary succession?
24. How do tides affect the organisms that live in beach environments?
25. Look for pioneer communities in your local region. Describe the criteria that led to your identification of these ecosystems. Make a diagram or take photographs to illustrate your criteria.

Research Projects

- What environmental changes have occurred in your geographic area during the past 20,000 years? What were the major causes of these changes?
- Has cultural eutrophication affected any lakes in your region? Trace the causes of eutrophication to their sources. What are some steps that can be taken to reverse the process?
- Conduct a study of lichens, a pioneer organism. What factors are necessary for their growth? Where are they usually found? Investigate the mutualistic relationship that exists in these organisms.

UNIT FIVE

ECOSYSTEMS AND BIOMES

Earth contains a wide variety of environments. At one extreme, there are tropical rain forests where it rains every day. These areas teem with many kinds of plants and animals. At the other extreme, there are deserts so dry that few plants and animals can survive there. You probably live in an area in which the conditions fall somewhere between a tropical rain forest and a desert. The characteristics of your environment make it different from other places. Friar's Glenn in Killarney National Park, County Kerry, Ireland, is a forest biome people can enjoy.

In this unit, you will explore the variety of areas on Earth and will discover why these areas differ from one another in such profound ways.

CHAPTER 18
Networks of Life

When you have completed this chapter, you should be able to:

Define biota, flora, and fauna.

Describe how ecosystems function.

Identify the biotic and abiotic factors in an ecosystem.

Although humans manage to survive under a wide variety of conditions, there are creatures that can outdo us. These creatures survive in deserts, oceans, and even in Earth's extremely cold regions, shown here. Can you guess what they are? The answer may surprise you. These hardy creatures are bacteria. Bacteria can survive in places that are inhospitable to other organisms. In this chapter, you will study the factors that make places hospitable to living things.

18.1 HOW ECOSYSTEMS FUNCTION

The Sonoran Desert in Arizona and the Great Barrier Reef off the coast of Australia are examples of large environmental systems, called ecosystems. Ecosystems are made up of the plants, animals, fungi, and microorganisms that live in an area, along with the nonliving environmental factors with which they interact. The nonliving factors include air, water, soil, rock, light, heat, and climate. The living organisms in an ecosystem are called the biota, while the nonliving are collectively called the abiota. The Sonoran Desert is an example of a land, or **terrestrial ecosystem**; the Great Barrier Reef is an example of an oceanic, or **marine ecosystem**.

Matter and Energy Flow

The plant life in an ecosystem is known as the **flora**, and the animal life is known as the **fauna**. In a typical ecosystem, green plants use energy from the sun to produce nutrients. Herbivores eat green plants to obtain the energy stored in their nutrients. Carnivores eat the herbivores, making further use of the energy. Decomposers break down dead organisms and return organic materials to the ecosystem. The decomposers in most ecosystems are bacteria and fungi. The activities of an ecosystem's biota affect the composition of the abiotic environment and cycle matter and energy through the system.

Ecosystems are considered self-sufficient even though energy, air, and water generally enter and leave. Radiant energy, in the form of sunlight, provides the initial energy for most ecosystems. Light energy is changed to chemical energy within the cells of the biota. Chemical energy is stored in the chemical bonds that hold atoms together in the molecules of living things. When used by the organisms in an ecosystem, some of the chemical energy is converted into waste heat that is radiated away and lost to the system. Thus incoming energy is always needed to offset the energy lost as waste heat by the ecosystem's biota.

Although ecosystems require a constant source of energy, usually in the form of sunlight, matter is recycled, or reused. Matter is composed of elements or groups of elements that form compounds. All living matter in an ecosystem contains the element carbon, which is the building block of life. Carbon compounds, also called organic compounds, are synthesized by organisms to create new, living tissue. Carbohydrates, fats, and proteins, the chemical compounds necessary for energy and growth in all living things, are organic compounds. The carbon in these compounds is cycled through the ecosystem from plants, to animals, to decomposers, and back to plants. Thus, organic matter is recycled through an ecosystem.

18.1 Section Review

1. Give some examples of the biotic and abiotic factors found in a desert ecosystem.
2. Describe the role of decomposers in cycling matter and energy through ecosystems.
3. Explain the process by which energy moves through ecosystems.

18.2 TYPES OF ECOSYSTEMS

All ecosystems are made up of a complex network of living and nonliving things. Ecosystems may be as small and simple as a decaying log on the forest floor, a lichen-encrusted rock on a mountaintop, or a water-filled, rocky tide pool along the seashore. Ecosystems may also be as large and complex as the South American rain forest or the Arctic tundra; each contains a great variety of organisms. Together, all Earth's ecosystems form the biosphere. (See Figure 18-1.)

Many areas that at first seem inhospitable to life can support hardy ecosystems. From the icebound Antarctic continent to the scorching deserts of the African Sahara, communities of organisms can be found that use whatever resources the environment has to offer. Ecosystems can also be found in sunless ocean depths clustered around hydrothermal vents at the mid-ocean ridges. There are ecosystems in pockets within basalt rocks deep beneath Earth's surface. The producers in these ecosystems do not carry on photosynthesis, since there is no sunlight to use as a source of energy. Instead,

Figure 18-1 Ecosystems vary greatly. A desert ecosystem (left) and a coral reef (right) are pictured here.

energy requirements are usually supplied through **chemosynthesis**, which uses energy stored in chemical bonds to make food.

A unique ecosystem exists in the boiling, mineral-rich waters of the hot springs in Yellowstone National Park. The producers in these springs are many types of bacteria and algae. Some survive the boiling water; others inhabit the cooler waters that surround the springs. The consumers are ephydrid flies and dolichopodid flies. The ephydrid flies eat the bacteria and algae. Dolichopodid flies eat the eggs and larvae of the ephydrid flies. The hot springs are miniature, self-sustaining ecosystems that contain their own special biota.

Miniecosystems

One organism's body can be the whole world for another organism. Miniature ecosystems often exist on and in the bodies of plants and animals. These miniature ecosystems are symbiotic, life-sharing relationships that exist between organisms and the microorganisms that live in association with them. Ectoparasites and endoparasites spend their whole lives in or on specific hosts. Without their hosts, the parasites cannot survive. A bird, for example, may carry a variety of ectoparasites, including fleas, ticks, mites, lice, and fungi, all making a home on its skin and feathers. Endoparasites live within the host's body. Tapeworms, flukes, ameba, flagellates, and bacteria may live in the bird's digestive and circulatory systems; viruses can live in its cells.

A single tree can be an ecosystem as it supports a community of organisms. Tropical plants and ferns often grow on the branches

Figure 18-2 The three-toed sloth is a "living ecosystem"; algae and moths live in its fur.

and trunks of trees; lichens and mosses may grow on the bark. Insects live on the leaves or under the bark of trees. Squirrels, birds, and frogs nest and search for food among the branches of trees. These organisms interact with one another and with the environment to maintain the miniecosystem.

The three-toed sloth is a mammal that lives in the tropical rain forests of South America. (See Figure 18-2.) The sloth spends most of its life hanging upside down by its claws from the branches of cecropia trees. The major source of food for the sloth is cecropia leaves, which are plentiful throughout the rain forest. The sloth moves very slowly through the branches of the trees and comes down to the ground once every seven days to defecate. Because of the animal's slow behavior, it was named sloth, which means "lazy."

The sloth is an ecosystem in itself. A specialized community lives in the sloth's fur. The producers in this ecosystem are green algae that grow in grooves in the hair of the sloth's fur. The green algae provide **camouflage** to hide the naturally gray or tan sloth from predators.

A moth lives in the sloth's fur. The moth is the primary consumer in this ecosystem; it grazes on the algae. These moths, which live nowhere else on Earth, time their reproductive cycles to be in step with the sloth's weekly trips to the ground to defecate. Female moths deposit their eggs in the sloth's dung, which serves as food for the moth larvae. When the larvae mature, the adult moths take to the trees in search of a sloth of their own to call home.

Bromeliads and orchids are plants that can be miniature ecosystems, too. (See Figure 18-3.) The roots of these plants absorb moisture from the humid air; the debris that collects around the roots forms a source of the plants' nutrients. Rainwater collects in the hollow cup formed by the bromeliads' leaves. These water-filled hollows, as well as the pockets of soil that collect around their roots, often support a diverse community of microorganisms, insects, worms, algae, and frogs.

18.2 Section Review

1. Compare the biotic and abiotic factors within a large ecosystem, such as a South American rain forest, with those of a miniecosystem, such as a bromeliad.

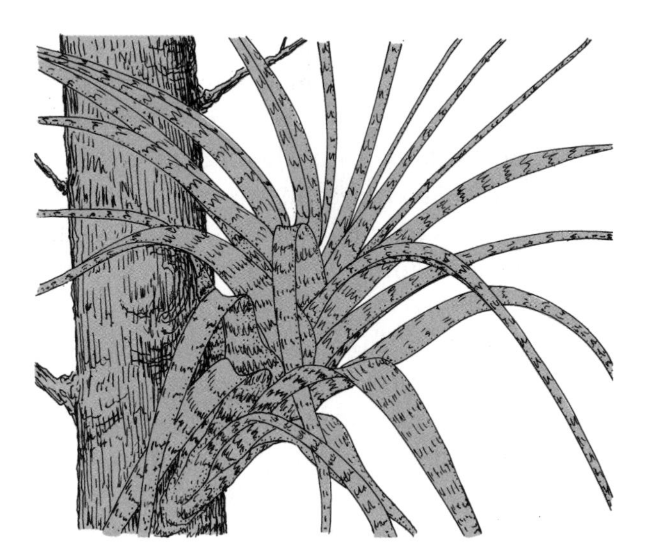

Figure 18-3 The water found in the bowl formed by the leaves of a bromeliad may attract a diverse community of organisms.

2. Why is the ecosystem that exists around Yellowstone National Park's hot springs unique?

18.3 BIOTIC INTERACTIONS WITHIN ECOSYSTEMS

If you examine a drop of pond water with a microscope, you will see thousands of individual organisms of many varieties. There might be unicellular, or single-celled organisms, as well as multicellular organisms. Some of these organisms produce their own food, for example, the green algae *Spirogyra.* Others, such as the protozoan *Ameba,* prey on the producers. Each organism has its own particular way of life. However, each lives in a delicate balance with every other organism as well as with the abiotic environment.

Divisions in Ecosystems

In a drop of pond water, each variety present represents a separate species that contains many similar individuals. Organisms, or individuals within a species, may differ in appearance. However, all organisms in a species have a similar genetic makeup and similar characteristics, as in the members of a human family. The total number of ameba in a sample is referred to as the sample's **population** of ameba. A population contains all of the organisms of the same species that live in a particular ecosystem or sample. It can be compared with a family. The drop of pond water also may contain populations of paramecia, euglena, rotifers, and other species, both seen and unseen.

All the interacting populations within the drop of pond water make up a **community**. A community is like a neighborhood. It consists of all the populations of organisms that live in a particular place at the same time. Every ecosystem can be thought of as a city, composed of one or more communities that interact in specific ways. Similar ecosystems located in a single geographical zone that share the same climate are called biomes. Biomes are like large countries that contain many cities. All Earth's biomes taken together form the biosphere.

In a garden or an overgrown vacant lot, you can see the interactions within an ecosystem. Soil, air, rainwater, and a community of plants, animals, and microorganisms exist together. Each of these factors affects the others, either directly or indirectly. Their various interactions are constantly changing the physical environment around them.

Flowering plants, shrubs, weeds, and trees remove carbon dioxide, oxygen, water, and nitrates from the air and soil. In the presence of sunlight, the plants produce organic materials necessary for their growth. As part of this process, oxygen, carbon dioxide, and water are released into the air. Insects and birds eat plants to obtain nutrients; they also use the oxygen the plants release into the air. Other birds feed on insects to obtain nutrients. Some birds eat plants and insects. The fauna eliminate wastes onto the soil. In the soil, bacteria break down the nitrogenous wastes and change them into nitrates that are needed by the flora. The fauna release carbon dioxide into the air. Plants use the carbon dioxide to make carbohydrates, the sugars and starches that are found in plant tissues. The biota play an important role in the natural cycles that maintain the balance within every ecosystem. Ecologists view the living world as a vast community in which each species plays a specific role in maintaining the biosphere.

18.3 Section Review

1. Describe the interactions that occur in a garden.
2. Describe how the activities of plants change their environment.
3. Explain how the activities of animals affect the environment.

18.4 THE ABIOTIC ENVIRONMENT

Abiotic factors are the nonliving physical and chemical parts of an ecosystem. Abiotic factors include the molecules of compounds in the air, water, and soil; the climate; and the level and variability of energy sources. The primary energy source in most ecosystems is the sun. However, some ecosystems have been discovered that use other energy sources, such as geothermal or chemical.

The Importance of Water

Water is an important compound for the organisms in every ecosystem, since water is the most abundant compound that makes up all living things. Without a constant supply of freshwater, life cannot exist. Water enters most ecosystems in the form of precipitation. The molecules of other useful substances, such as oxygen and carbon dioxide, are in the air or are dissolved in water. Organisms can extract these gases directly from the environment. Animals use respiratory organs, such as lungs, gills, or a moist skin, to obtain oxygen. The stomata of green plants aid in the intake and removal of gases. Nitrogen, the most abundant gas in the air, is also vital to life. However, most organisms cannot make use of gaseous nitrogen. Nitrogen-fixing bacteria convert nitrogen to nitrates that plants absorb from the soil. Animals get nitrogen compounds when they eat plants or feed on other animals that ate plants.

Ecosystems are affected by climate, the long-range weather pattern of a particular area. Weather is the short-term condition of the atmosphere. The average yearly precipitation and temperature determine the climate of an area. In turn, temperature and precipitation

Figure 18-4
The amount of freshwater on Earth is limited. The water cycle shows how freshwater moves through the biosphere.

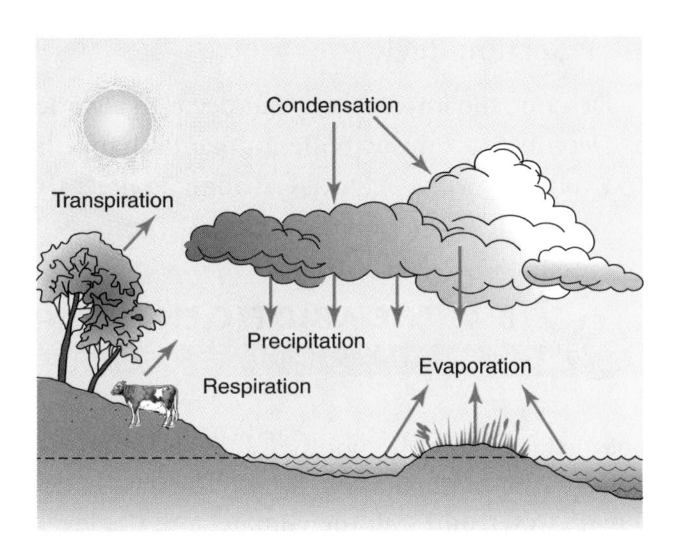

are affected by the number of days and hours of sunlight, the direction and prevalence of winds, and nearness to large bodies of water and ocean currents. The changing period of daylight encountered as you move from the equator to the poles leads to the formation of a variety of ecosystems.

Energy from the sun heats Earth's surface and causes water to evaporate into the atmosphere. The heat from Earth's surface warms the atmosphere, forming convection systems. Convection is the transfer of heat by a moving liquid or gas. Convection systems create huge moving air masses within the atmosphere. Generally, air masses move in an easterly direction over the United States. Moving air masses transports moisture, which enters the atmosphere above the oceans. As warm, moist air rises and cools, the moisture condenses and forms clouds. The clouds eventually return the moisture to Earth as precipitation. Wind systems help to distribute precipitation over Earth's surface. (See Figure 18-4.)

The Importance of Soil

Soil is a complex mixture of organic materials and inorganic minerals that forms a thin layer over many parts of Earth's surface. The organic materials consist of humus, dead and decaying organic matter, along with a variety of living microorganisms. Without humus, there would be no soil. Soil also contains varying amounts of air and water. Climate has an effect on many of the chemical and physical interactions that help form humus and thus create soil. Soil nourishes and supports plant growth. Therefore, soil is an important factor in determining which organisms are found in an ecosystem.

Moist climates promote weathering of underlying rocks and, thus, add mineral particles to the soil. These minerals are in the form of fine particles of silt and clay and larger particles of sand, pebbles, and rocks. Warm, moist climates also promote the growth of soil microorganisms that form humus-rich soils, which support abundant plant life. The soil microorganisms break down organic matter and produce humus. The humus in the soil absorbs water from precipitation. In turn, the minerals and nutrients, which plants require, are dissolved in the water. The water and dissolved minerals are absorbed from the soil by the roots of plants. Dry climates inhibit the growth of microorganisms, thus producing humus-poor soils. These soils cannot retain much moisture and so support limited plant life.

Since the climate of a particular region affects the formation of soil, it directly determines the flora that can grow there and indirectly affects the type of fauna that is present. The plants determine the type of ecosystems, and ultimately the biomes, that form.

Earth's Biomes

The major terrestrial ecosystems are classified into units called biomes, large regions that have their own typical flora and fauna. Each biome has a characteristic climate that directly influences the type of flora it supports. Since animals depend on plants for food, the climate also influences the type of fauna found in an area. For example, seasonal drought restricts the growth of trees in a grassland biome. These large plains support grass eaters, or grazers, such as bison and pronghorn antelope. Deer and elk are browsers that feed on twigs, buds, and leaves. These animals are found in forests where year-round moisture promotes the growth of shrubs and

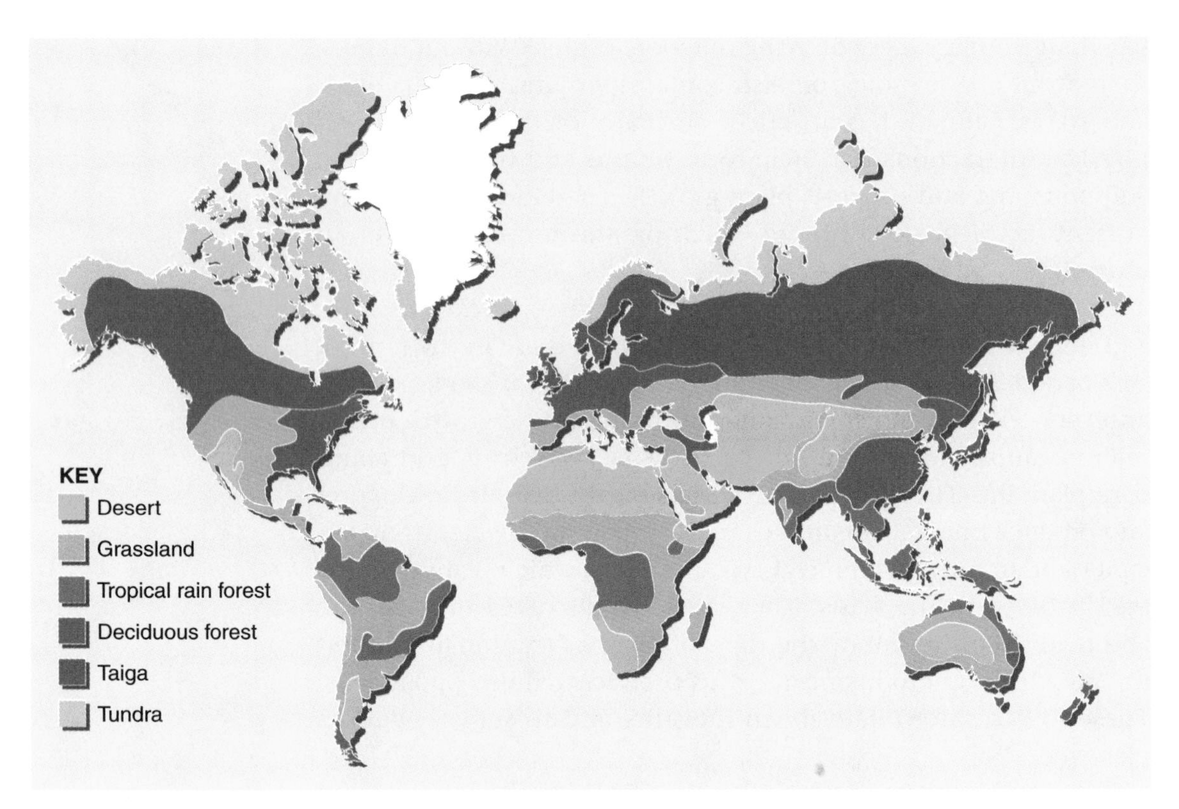

Figure 18-5 Earth's major biomes are shown on this world map.

trees. Biomes are large areas composed of many smaller overlapping ecosystems, which may extend across several continents.

Each biome is a superecosystem, defined by its climate and physical condition, which contains a unique biota. Types of biomes range from the Arctic tundra to the deserts and include various kinds of forests and grasslands. The map in Figure 18-5 indicates Earth's major biomes.

18.4 Section Review

1. How does climate affect the biota within an ecosystem?
2. Explain the role of water in the formation of ecosystems.
3. Select the biome in which you would like to live and tell why.

LABORATORY INVESTIGATION 18

Miniature Ecosystems

PROBLEM: *Can you create a miniature terrestrial and aquatic ecosystem?*

SKILLS: *Manipulating, observing, measuring.*

MATERIALS: *1-L jar with lid, gravel, ½-L jar with lid, activated charcoal, topsoil, sand, rooted plant cuttings (ferns or arrowleaf), land snail or salamander, water plants (*vallisneria, sagittaria, *elodea, cabomba, or duckweed), mealworms, guppies, water snail, water*

PROCEDURE

Aquatic Ecosystem

1. Place about 2.5 centimeters of sand or gravel on the bottom of a ½-liter jar. Fill the jar with water. Firmly root several water plants, such as *vallisneria* or *sagittaria,* in the sand or gravel; or add floating plants such as elodea, cabomba, or duckweed. Add several guppies and a water snail.
2. Cap the jar and place it where it will receive daily sunlight. Make sure the temperature within the ecosystem does not rise too high and kill the biota.
3. Observe your ecosystems daily to make sure the biota have the necessities for survival. If the environmental factors are not correctly balanced, it may be necessary to add new plants or food to maintain the ecosystem.

Terrestrial Ecosystem

1. Cover the bottom of a 1-liter jar with a 2.5-centimeter layer of gravel into which you have mixed 2 grams of activated charcoal. Prepare a mixture of equal parts topsoil and sand. Place a 7.5-centimeter-deep layer of this mixture on top of the layer of gravel.
2. Place several small plants or cuttings, such as ferns or arrowleaf, in the soil layer. Place a land snail or a terrestrial salamander into the jar. If you use a salamander, add several mealworms to the biota in the ecosystem. (See Figure 18-6 on page 308.) Add enough water to dampen the soil. Cap the jar.
3. Place your ecosystem where it will receive daily sunlight. Make sure the temperature within the ecosystem does not rise too high and kill your biota.

OBSERVATIONS AND ANALYSES

1. What are the living components of your ecosystems?
2. What are the nonliving components of your ecosystems?
3. After several days, moisture should appear on the sides of the jar that contains the terrestrial ecosystem. What natural process that occurs in Earth's biosphere does this illustrate?
4. Why do these ecosystems sustain themselves even though they are sealed?
5. Using diagrams, show how the following materials are cycled within your ecosystems: gases, water, nutrients.
6. What is the source of oxygen for the animals in the aquarium?
7. Prepare a progress chart to record your daily observations of the environmental conditions in your ecosystem.
8. What environmental problems arose over time? How did you solve each of these problems?

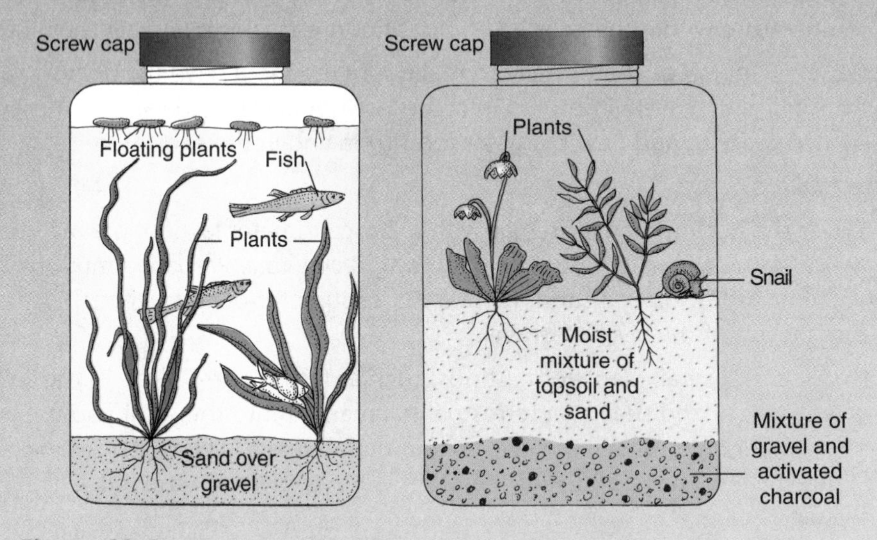

Figure 18-6

Chapter 18 Review

Answer these questions on a separate sheet of paper.

Vocabulary

The following list contains all the boldfaced terms in this chapter.

camouflage, chemosynthesis, community, fauna, flora, marine ecosystem, population, soil, terrestrial ecosystem

Fill In

Use one of the vocabulary terms listed above to complete each of the following sentences.

1. The ______ consists of all the animal life in an ecosystem.
2. Producers that live deep in the ocean or under the ground acquire their energy through ______.
3. The plant life in an ecosystem is known as the ______.
4. All the interacting populations within an ecosystem make up a(an) ______.
5. A ______ is made up of all the organisms of the same species that live in a particular area.

Multiple Choice

Choose the response that best completes the sentence or answers the question.

6. Which of the following is an example of a bird ectoparasite?
   a. ameba *b.* fluke *c.* louse *d.* bacterium
7. Which of the following is an example of a bird endoparasite?
   a. tick *b.* tapeworm *c.* mite *d.* flea
8. Which of the following are the respiratory organs in plants?
   a. gills *b.* lungs *c.* trachea *d.* stomata
9. Deer and elk are browsers; they feed mainly on *a.* grasses. *b.* twigs and buds. *c.* ripe fruits *d.* nuts.

Short Answer (Constructed Response)

Use the information you learned in this chapter to respond to the following items.

10. Name one marine and one terrestrial ecosystem.
11. How is chemosynthesis different from photosynthesis?
12. Why is carbon called the building block of life?
13. Name two abiotic factors found in most ecosystems.

14. In the grassland biome, what is the predominant type of fauna?
15. What is Earth's northernmost biome?

Essay (Extended Response)

Use the information in the chapter to respond to these items.

16. Explain how energy is lost by an ecosystem.
17. How is moisture transported through the atmosphere?
18. What is a biome?

Reading: The Everglades

At the beginning of the twentieth century, South Florida's warm climate and rich soil began to attract people to the area. The increasing population needed dry farmland and freshwater for agriculture and the growing urban areas. In the 1950s, 1600 kilometers (1000 miles) of canals and levees were built around Lake Okeechobee and the Everglades. They were constructed to provide flood control and freshwater to the area. However, they also disrupted the flow of freshwater through the Everglades and drastically changed the ecosystem.

The changes in the Everglades have affected the flora and fauna. The populations of wading birds have decreased by almost 90 percent while the alligators have flourished. Exotic plants introduced by humans, such as Brazilian pepper, malaleuca tree, and water hyacinth have crowded out native species.

It is not possible to return the entire ecosystem to its original state, because at least 50 percent of the land has been converted to urban areas or farms. The future appears grim as the population of Florida is expected to double in the next 50 years. It is hoped that what remains of the Everglades can be maintained through projects that will protect this unique ecosystem.

Base your answers to the following questions on this passage and your knowledge environmental science.

1. Why were canals built around Lake Okeechobee and the Everglades?
2. What has happened to the flora and fauna?
3. Why is it not possible to return the ecosystem to its original state?

CHAPTER 19
Forest Biomes

When you have completed this chapter, you should be able to:

Describe the taiga and deciduous forest biomes.

Distinguish between the tropical and temperate rain forests.

A forest is more than just a collection of trees. Forests are special places that cannot be replaced quickly once they have been cut down. The kinds of trees in the forest create an environment that determines the kinds of mammals, large and small, reptiles, birds, insects, other plants, and microorganisms that can live in the forest. The roots of trees and other plants hold the soil in place and keep it from washing away with the rains. The leaves of the trees shade the soil. Forests are more complex than people realize. In this chapter, you will learn about the different types of forest biomes.

19.1 CONIFEROUS FORESTS

Biomes are large ecosystems that are made up of many smaller, overlapping ecosystems. Each biome is defined by its climate and physical condition. Remember that climate determines which plants grow in an area, and the plants in an area determine the animals that live there. The northern coniferous forest biome is also called the **taiga**, or **boreal forest**. The taiga extends across the northern reaches of the temperate zone through North America, Europe, and Asia. There is no taiga in the Southern Hemisphere, because at corresponding latitudes there is little land.

Coniferous forests are also found at lower latitudes in the Rocky, Wasatch, Sierra Nevada, and Cascade mountains. Another type of coniferous forest—the southern pine forest—is located along the coastal plains of the South Atlantic and Gulf Coast states.

Climate

The taiga is a region that has short summers and long, cold winters. There is also a brief spring and fall. The average yearly precipitation (rain and snow) is 35 to 75 centimeters, and the average yearly temperature range is –10°C to 14°C. Because of the tilt of Earth's axis, the taiga receives more hours of daylight during the summer than areas farther south. The longer period of daylight and warmer temperature of the taiga extend the growing season for plants. The taiga is commonly known as the spruce-moose forest.

Plants

Coniferous forests dominate the taiga biome. **Conifers** are trees that have stiff needles instead of broad leaves, and most species produce cone-shaped seed cases. Cedar, fir, hemlock, pine, larch, and spruce are examples of conifers. (See Figure 19-1.) All conifers shed their needles. Some, such as the larch, shed all their needles at the

Figure 19-1 **Conifers have needles instead of broad, flat leaves. Most species produce seeds in cones.**

same time. However, for most conifers, each needle has a lifetime of several years. Fallen needles are continually being replaced by new ones. This keeps the trees green all year, thus the common name "evergreen."

The long, thin shape of conifers' needles makes these trees well adapted to the climate of the taiga. The needles are able to withstand the weight of snow and ice that coat them through the winter. In addition, the needles resist freezing. Conifer needles have few stomata, or pores, and they have a thick cuticle covering, so they lose only small amounts of water through transpiration.

The dark green color of these specialized leaves indicates that they have a high concentration of chlorophyll, the green pigment necessary for photosynthesis. Recall that during photosynthesis, plants use light energy from the sun to make food (carbohydrates) from water and carbon dioxide. The high concentration of chlorophyll permits the conifer leaves to absorb a maximum amount of light from the sun. Since evergreens do not lose all their needles in the fall, they can carry out photosynthesis whenever there is light. Thus they can start growing at the first sign of spring. This gives them a longer growing season than trees that shed their leaves.

The floor of the coniferous forest is always covered with a thick layer of fallen needles, called the litter layer. There is little undergrowth in the coniferous forest because the branches of the trees block the sunlight, shading the forest floor. Where older conifers have fallen, gaps allow sunlight to penetrate to the forest floor. Sunlight also reaches the banks of streams and lakes. In these areas, you also will find shrubs and broad-leafed deciduous trees, such as birch, willow, cottonwood, aspen, and poplar. Deciduous trees shed their leaves in the fall or during times of drought.

Forest Fires

Wildfires are a natural part of the healthy development of coniferous forest ecosystems. Lightning from summer thunderstorms often ignites the layer of dry needles that carpets the forest floor. The result is frequent wildfires in coniferous forests.

Before the early nineteenth century, forest fires were common in the coniferous forests of North America. About 100 years ago, the National Forest Service began to put out forest fires as quickly as possible. However, these fires were the principal means by which the coniferous forests evolved and changed. Fires kill older, less vigorous trees and open areas to sunlight. This helps the growth of young trees, since many conifer seeds cannot germinate and grow in shade. Without these fires, the forests would become populated with older trees. Where humans have tried to eliminate forest fires from the environment, the ecosystem has been changed drastically. The absence of forest fires leads to deciduous trees crowding out the conifers. This, in turn, affects the animals associated with the forests.

Even though wildfires may destroy thousands of hectares of forest each year, some types of conifers escape destruction because their bark covers and protects the trees' growth layer. Other conifers, such as jack pine and black spruce, produce cones that will not release their seeds until they have been through a forest fire. If it were not for forest fires, these conifers could not produce new trees. In addition, forest fires exterminate insect pests and competing plants. Fires also release needed nutrients into the soil.

Since 1963, the National Forest Service's policy is to let natural fires burn unless they threaten people, property, or endangered species. In addition, in many forests, small controlled fires are set to clear away the underbrush and reduce the occurrences of disastrous fires.

Soil

In the taiga, the long, cold winter slows decomposition of the litter of conifer needles that covers the forest floor. These needles contain resins that make them acidic. Resins are poisonous secretions produced by conifers that discourage primary consumers from eating the needles. Since the recycling of materials is very slow, the result is a nutrient-poor, highly acidic soil. The acid soil, together with the ever-present litter of needles, makes this forest floor an unsuitable environment for many plants and animals.

Many kinds of bacteria cannot survive the acidic soil conditions and cold winters of the taiga. Therefore, fungi are the main decomposers in the soil of this area. **Mycorrhizal fungi** live a symbiotic relationship with the roots of many types of green plants. Because soil nutrients are not released through bacterial decay, conifers depend on mycorrhizal fungi to reclaim nutrients from the fallen needles.

Animals

In the taiga, animals do not live near one another. Therefore, there is less competition for scarce resources. The conifers' resinous, waxy needles are difficult for most herbivores to digest. The most important food source produced by conifers is their seeds. Some birds (crossbills) and mammals (red squirrels) are adapted to extract seeds from pinecones. (See Figure 19-2.) Mice, lemmings, and ground squirrels cannot extract seeds from cones, but they can eat loose seeds from fallen cones. Large herbivores, such as moose and deer, feed on moss, lichens, and the deciduous trees and shrubs that grow in the open areas of the coniferous forest. Moose also eat plants that grow in lakes and ponds. Carnivores, such as lynx and wolves, must range over large areas of forest to track their scarce prey.

Few insects inhabit the litter of needles that covers the floor of the coniferous forest. However, many insects can be found on the trees.

Figure 19-2 Some birds and mammals have adaptations that permit them to extract seeds from pinecones.

INTERACTIONS

Mycorrhizal Fungi and Forests

Some mycorrhizal fungi are the largest living things on Earth. The network formed by the fungi's filaments can extend through the soil of an entire forest ecosystem, and may weigh thousands of kilograms. DNA testing proved that some of these networks form a single, giant organism larger than dinosaurs or blue whales. Some mushrooms are the fruiting structures of mycorrhizal fungi. The mushrooms contain the spores by which the fungi reproduce.

Though most fungi live on the forest's dead organic material, mycorrhizal fungi live within the root systems of trees and shrubs. Within an ecosystem, many different species of mycorrhizal fungi interact with conifers in a mutualistic relationship. Each is dependent on the other. These fungi are called mycorrhizae, which is a general term for all the fungi living in mutualistic relationships with trees. The fungi increase the amount of nutrients available to the plants by extending the volume of soil accessible to their roots. The mycorrhizal fungi also provide protection against parasitic fungi and nematode worms.

Conifers have a shallow, yet extensive, root system that radiates from their trunks and through the soil. The mycorrhizae consist of strands of mycelia, which penetrate into a tree's roots. The fungi benefit because the mycelia absorb carbohydrates produced by the tree. The large surface area of the fungal hyphae enables it to absorb large quantities of mineral ions from the soil. When minerals become scarce in the soil, the fungi release them to the trees. The tree benefits because it gets the minerals that the fungi take from the soil. The mycorrhizae are particularly helpful in absorbing phosphorus. Ultimately, mycorrhizae are highly susceptible to being damaged by acid rain. Mycorrhizal fungi transport and store carbon from the plant roots to other soil organisms, aiding the decomposition process within the food web. The hyphae, rhizomes, and fruiting structures are important habitats and food sources for soil invertebrates.

Some insects lay their eggs beneath the bark of trees. Others bore into the conifers' soft wood to lay their eggs. Woodpeckers are well adapted to locate and extract insects and their larvae from beneath the bark. With their sharp beak, woodpeckers hammer holes into insect tunnels under the bark. These birds then use their long, sharp-pointed tongue to extract the insects. During spring and summer, large numbers of insect larvae feed on the younger, more tender, and less resinous conifer needles. In spring, many songbirds migrate north in search of the plentiful larvae as a food source for their young.

As winter approaches, some animals, such as chipmunks, bats, and snakes, retire to dens to hibernate. These animals escape harsh winter conditions in a state of inactivity, or suspended animation. They live off fat reserves they built up by eating well during spring and summer. Some insects become dormant, or inactive, during the winter. When warm weather arrives, these animals become active once again. Other insect species survive the cold temperatures by laying eggs in protected spots.

To avoid the harsh winter, some animals migrate, or leave the area. In fall, many birds, such as woodpeckers, hawks, and flycatchers, migrate south to warmer climates where food is more readily available. They return to the taiga in spring when food is available again.

19.1 Section Review

1. Describe conditions on the taiga.
2. How are conifers adapted to conditions on the taiga?
3. Explain two ways creatures that live on the taiga escape the harsh winters.

19.2 DECIDUOUS FORESTS

In Earth's temperate zones are areas called the **temperate deciduous forest** biome. Broad-leafed **deciduous trees** that drop their leaves seasonally in preparation for adverse environmental conditions dominate this biome. Temperate deciduous forests are found in only three areas of the world: eastern North America, western

Europe, and eastern Asia. All of these widely separated areas have similar plant life and contain similar species of animals.

Climate

The climate in the temperate deciduous forests is generally much milder than that of the taiga. The average yearly temperature range is 6°C to 28°C. Temperate forests have cold winters, spring and fall seasons, very warm summers, and abundant, year-round precipitation averaging 75 to 125 centimeters. Winter snows may be heavy. The leaves that cover the forest floor in the fall decompose during the warm summer months. This leads to the formation of a thick, nutrient-rich, organic soil. Bacteria, fungi, and many invertebrates, such as beetles, millipedes, mites, springtails, and earthworms, live in the leaf litter. These creatures break down the layer of fallen leaves and work it into the soil. This produces humus and enriches the forest soil with recycled nutrients.

Forest Structure

The deciduous forest is stratified, or layered. The upper layer of branches is called the **canopy layer.** Most deciduous trees are tall, but their leaves are spaced so that sunlight filters between them through the canopy layer and reaches the forest floor below. Because light reaches the forest floor, smaller trees can grow there. These smaller trees form the **understory layer**. Beneath the understory is the **shrub layer**, which contains shrubs, relatively short woody plants that have many stems rather than a single trunk. Below the shrub layer is the **ground cover**, a layer of wildflowers, ferns, and mosses. Decaying leaves and branches along with other vegetation that cover the soil form the bottommost layer, the **litter layer**. A profile of the deciduous forest is shown in Figure 19-3.

Forest fires often destroy deciduous forests. However, the destructive impact on deciduous forests made by humans has been greater. Temperate deciduous forests are located in areas that have nutrient-rich soils and long growing seasons. For these reasons, most of the world's deciduous forests have been cut down and the land used for farming, grazing, or logging.

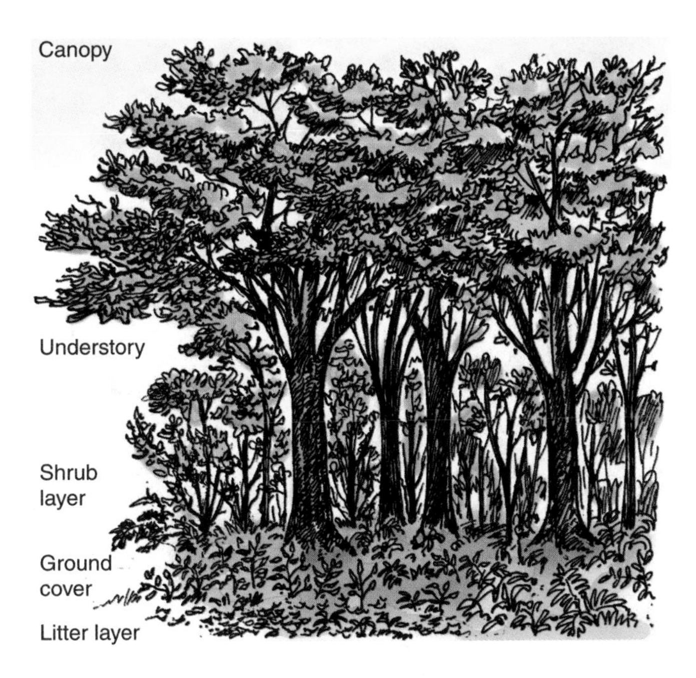

Figure 19-3 What layers are found in the deciduous forest?

Throughout history, humans have modified the world's deciduous forests. The composition of the deciduous forests in the United States has been shaped through controlled burning and cutting. Most of the original forest has been cut down. Our forests no longer resemble those encountered by Native Americans or early settlers. When forests are allowed to regenerate, oak-hickory forests now tend to become established rather than previously dominant fire-sensitive maple forests.

Plants

In this biome, deciduous trees, such as maple, oak, ash, birch, hickory, and beech, are the dominant forms of vegetation. (See Figure 19-4 on page 320.) However, conifers, such as pine, spruce, and hemlock, are also found in scattered locations. Conifers usually spring up in areas where the soil is sandy and has a low nutrient content.

The broad, flattened leaves of deciduous trees have a large surface area that absorbs a maximum amount of sunlight. The bottom of each leaf contains many stomata, or pores, through which carbon dioxide and oxygen enter the leaf and water vapor and oxygen

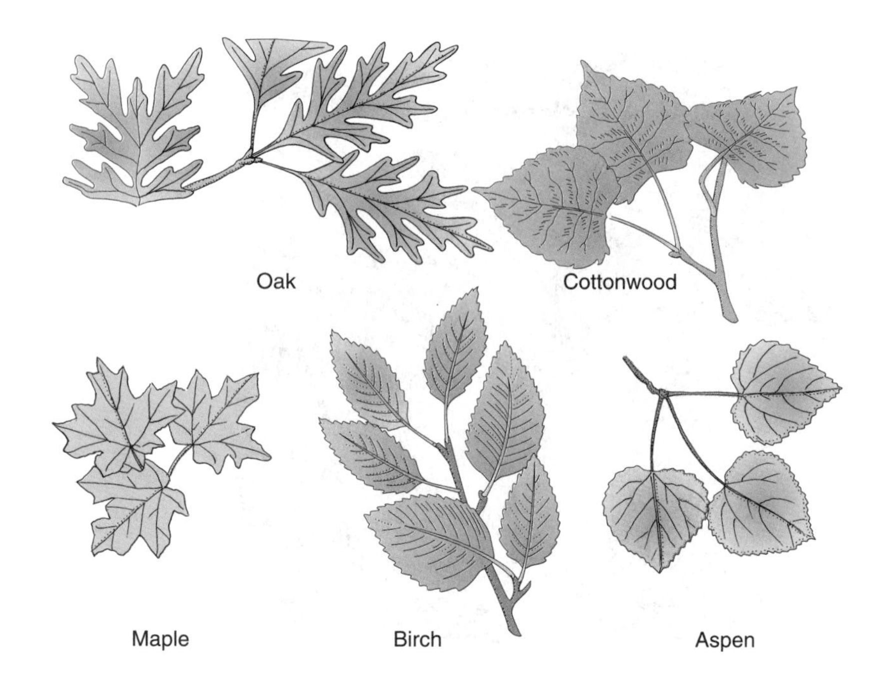

Figure 19-4 **Leaves of some broad-leafed deciduous trees are illustrated here.**

leave. The leaves are highly efficient food-producing factories. During the spring and summer growing season, the leaves manufacture the large quantities of carbohydrates that the tree needs. The carbohydrates are distributed through the tree in the fluid called sap.

As the period of daylight grows shorter and cold weather approaches, deciduous trees begin a period of winter dormancy. In most trees, the delicate leaves cannot last through the cold winter and are shed. First, the green chlorophyll is broken down and withdrawn into the trunk of the tree. Once the green color is removed, other brightly colored compounds—brown, yellow, or red—become visible. Next, the tubes in the stem that carry sap into the leaf are closed off and sealed. The leaf dries, and then the slightest breeze is able to lift it off the branch and send it floating down to the forest floor. Some trees, such as the aspen, keep their dry leaves through the winter and shed them in the spring.

Seed Dispersal

The plants of the deciduous forests produce a wide range of fruits, some in the form of nuts or berries. The fruits contain the plants' seeds. Young deciduous plants cannot grow in the shade of their parents. Therefore, seeds must be widely dispersed, or spread from their source. Some plants produce brightly colored fruit that attracts a wide variety of birds. The birds eat the fruit and spread the seeds in their droppings. The birds act as agents of dispersal by helping the trees spread to new areas.

The seeds of other trees are dispersed by the wind. These trees produce seeds that have wings or parachutes. As winged seeds fall, they are caught by a breeze and carried away from the parent tree. The samara of the maple is this type of seed. The seeds of cottonwoods are attached to a tuft of cottony fibers that act as a parachute. The parachute allows the seeds to be carried great distances with the wind. (See Figure 19-5.)

Trees such as oak, hickory, and beech produce nuts. A nut is a seed encased in a hard shell. Acorns are the nuts produced by oak trees. In late summer and autumn, nuts are an important source of food for many animals, including squirrels and birds. In preparation for the time when food will be scarce, some animals hide nuts beneath the leaf litter layer and under rocks. This also aids in seed dispersal, since many of these nuts are forgotten and thus can germinate when conditions are suitable.

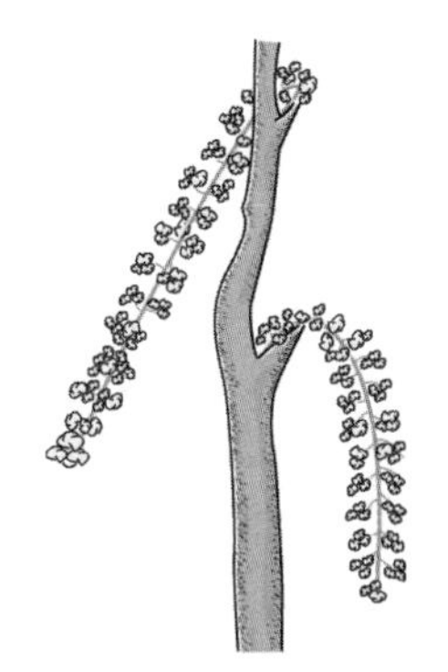

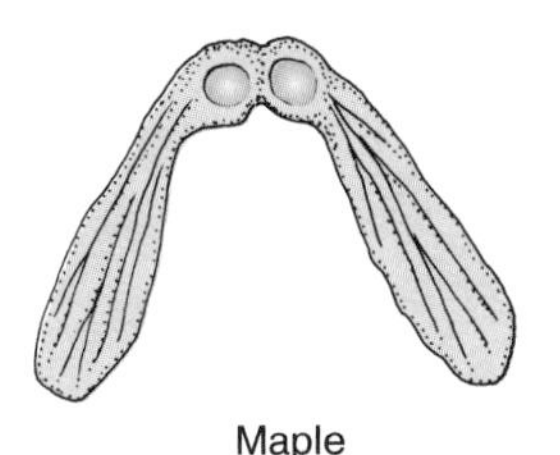

Figure 19-5 Animals eat cherries and spread the undigested seeds in their wastes. Maples produce winged seeds, which are carried by the wind.

Animals

Many species of mammals, birds, and insects live in the temperate deciduous forest. Some animals, such as the gray squirrel, cottontail rabbit, and chipmunk, eat only plants. Some animals eat other animals, for example the bobcat and the woodpecker. Other animals eat both plants and animals, for example, the black bear, raccoon, and nuthatch (a bird).

19.2 Section Review

1. Describe the climate of the temperate deciduous forest biome.
2. Explain how animals aid in the dispersal of seeds of the deciduous trees.
3. What are some examples of deciduous trees?

19.3 TROPICAL RAIN FORESTS

The **tropical rain forest** biome exists in four areas of the world, located along or near the equator. The largest rain forest area is in the Amazon River basin in South America. The next largest is in Indonesia. Another area of rain forest is in Africa in the Zaire River basin. The smallest area of rain forest is in Central America. These areas are commonly, although incorrectly, called jungles. Jungle refers to specific areas of the rain forest that contain dense, tangled masses of vegetation. These areas are usually found only in open spaces where sunlight reaches the ground. Sunlight can reach the ground along riverbanks and where there are gaps in the canopy created by fallen trees.

Climate

In the tropical rain forest, very warm temperatures are the rule. Temperature varies little during the day or year, averaging between 28°C and 35°C. Near the equator there are no seasons, rain falls almost every day, and the humidity is always high. As you travel away from the equator north or south, the temperature is more variable, 21°C to 35°C, and so is the rainfall. Areas of tropical rain forest do not have temperature-related seasons; however, they do have seasons related to rainfall. Rain forests often receive more than 200 to 400 centimeters of rain per year.

Plants

Tropical rain forests are dense forests of broad-leafed, shallow-rooted evergreens, such as cecropia and symphonia. These forests contain the greatest variety of trees and shrubs found on Earth. The **foliage**, or growth of leaves, is very dense and limits the amount of light that reaches the forest floor and restricts the growth of other plants. Some rain forest trees grow to be more than 60 meters tall. Because the trees of the rain forest are shallow-rooted, buttresses support

Figure 19-6 Buttresses support shallow-rooted, rain forest trees.

many tall trees. A **buttress** is a wide, spreading prop root that grows out from the base of the trunk and into the soil to keep the tree from falling over. (See Figure 19-6.)

Epiphytes are plants that have aerial roots that absorb moisture from the humid air. These plants, such as some ferns, mosses, orchids, and bromeliads, grow in great numbers on the branches and trunks of the trees. (See Figure 19-7.) The debris that collects around the epiphytes' roots forms pockets of organic matter from which the roots extract nutrients.

Figure 19-7 Some orchids are epiphytes that grow on trees. Many orchid species can be grown in homes.

Because of the warm temperatures, high humidity, and large numbers of microorganisms in the soil, organic materials on the rain forest floor decompose quickly. However, the soil is shallow and of poor quality. The nutrients released by decomposition are absorbed quickly by the spreading network of roots or washed away by the daily rains. Trees are shallow-rooted to take advantage of these soil conditions.

Stratification in the tropical rain forests creates a wide variety of habitats for animals. The year-round, warm, humid climate produces lush vegetation. The tops of the trees form a dense, green, leafy upper layer, or canopy, which supports a large community of diverse organisms that are adapted to feed on the vast supply of leaves and fruits. Many plants in the rain forest depend on insects and birds to pollinate their flowers. They also rely on fruit-eating birds and mammals for pollination and to disperse their seeds.

The tallest trees, called **emergents**, poke their crowns through the canopy. When seen from above, the emergents resemble islands floating on a sea of leaves. Beneath the canopy is the dark, humid understory layer. Little light penetrates through the canopy. Here the air is very still. Leafless tree trunks and intertwined vines grow everywhere. Where older trees have fallen and created an opening in the canopy, sunlight reaches the ground. In these sunlit areas, new trees can begin their climb up to the canopy. The bottom layer is the bare forest floor. Few plants can survive here due to the low level of sunlight that filters through the canopy. Many popular houseplants come from this area. The conditions in our homes, especially the low level of light, are similar to those on the rain forest floor. Scientists estimate that only 1 percent of the light that shines on the canopy filters through to the rain forest floor.

Stratification creates varied habitats for the rain forest animals. Howler monkeys live in the canopy. Macaws fly through the understory. Tapirs walk along the forest floor. These animals live on a diet of leaves and fruit. They do not compete with one another, because each feeds within a different layer of the rain forest ecosystem. Rain forests are the most diverse of all ecosystems. They are thought to contain almost half of Earth's animal and plant species.

Destruction of the Tropical Rain Forest

The rain forests are shrinking due to the slash-and-burn practices that clear the land for agriculture, ranching, mining, and timber. The roots of rain forest trees anchor the soil and slow the runoff of water into rivers and streams, which prevents erosion. Clearing rain forest areas for agriculture rapidly depletes nutrients in the thin layer of soil. Therefore, farms do not remain productive for very long. Once farming stops, the soil will not support the growth of new plants. There is no cover of vegetation to prevent the daily rains from washing away the soil. This leads to widespread erosion. The cutting down of rain forests has caused many rivers to be clogged with silt from runoff. Cleared rain forest areas do not recover; they remain scars on the landscape. More than 17 million hectares of rain forest are cut down each year. This represents an area the size of Florida.

Ecologists fear that destruction of the rain forests will lead to global warming. Rain forest vegetation absorbs large quantities of carbon dioxide from the atmosphere. Carbon dioxide is one of the greenhouse gases. A greenhouse gas traps heat within the atmosphere in much the same way that glass traps heat in a greenhouse. Therefore, scientists theorize that a buildup of carbon dioxide in the atmosphere will cause the average temperature on Earth to increase, leading to global warming.

Rain forests also play an important role in recycling Earth's water. Most of the water absorbed by the roots of trees is returned to the atmosphere by the leaves in the form of water vapor. This water returns to Earth as precipitation. Thus destruction of rain forests may cause changes in climate on a global scale.

Cloud Forests

Cloud forests are located in mountainous regions within or bordering tropical rain forests in parts of Africa and the New Guinea highlands. These are specialized **montane**, or mountain, ecosystems. Because these forests grow at higher elevations, cooler temperatures prevail. Due to the cooler temperatures, moisture in the air condenses as fog, or mist. Seen from below, these forests seem to be covered in clouds, thus the term *cloud forest.* The mist supplies a great deal of moisture for the growth of vegetation. Cloud forests have less plant diversity than the tropical rain forests found at lower elevations. However, in the cloud forests the density of vegetation is much greater. The canopy of the cloud forest is very low, often near ground level. This makes travel through the cloud forest very difficult.

19.3 Section Review

1. Describe the climate of the tropical rain forests.
2. Compare the physical environment of the cloud forest with that of the tropical rain forest.
3. Why should people who live in the United States be concerned about the destruction of rain forests in South America?

19.4 TEMPERATE RAIN FORESTS

Rain forests are not located exclusively within the tropics. **Temperate rain forest** biomes are found in areas between the tropics and the polar regions. In the United States, the Pacific Northwest supports a temperate rain forest. The Olympic National Forest in Washington State is an example of a temperate rain forest. Other temperate rain forests are located in Norway, Japan, southern Chile, southern New Zealand, southern Australia, and Tasmania.

Climate

Temperate rain forests thrive where there are mild, wet winters and warm, wet summers. The average yearly precipitation is 200 to 400 centimeters, and the average yearly temperature range is 10°C to 20°C. The Pacific Northwest is a region of abundant rain and heavy fog, which produce temperate rain forest conditions. The maritime winds carry warm, moist air from the Pacific Ocean up the slopes of the coastal mountain ranges. On the slopes of these mountains, the water precipitates as rain or fog. The Olympic Peninsula on the coast of Washington State has the wettest climate in the continental United States.

Due to the moist climate, fires do not occur often, and trees grow to be very tall. The canopy is quite dense. This produces low light conditions in the understory and the forest floor. Slugs, snails, and salamanders find that the damp conditions make a perfect home. Many small mammals and birds feed on the abundant seeds. Since temperate rain forests are valued by humans for their timber as well as for recreational uses, conflicts arise regarding their use.

Plants

In terms of biodiversity, the temperate rain forests are second only to the tropical rain forests. Temperate rain forests contain more abundant vegetation than any other ecosystem. The dominant trees

Figure 19-8 The temperate rain forest ecosystem.

here are coniferous evergreens such as redwood, red cedar, Douglas fir, hemlock, and Sitka spruce. The trees in both types of rain forests tend to be giants, with shallow, spreading root systems. You may know that redwoods can be very tall, but specimens of many other rain forest tree species can also grow to heights of more than 90 meters. Epiphytes, such as mosses and lichens, are abundant in the temperate rain forest. Shelf fungi and mushrooms also grow on the trunks and branches of trees. (See Figure 19-8.)

In the temperate rain forest, many trees grow to be very old and have dead branches and decaying trunks. The decaying trunks serve as homes for a variety of animals. Wrens often live in the holes made by woodpeckers. Flying squirrels may nest in rotted-out knot-holes and hollow trunks of living trees. The northern spotted owl is adapted to life in old-growth forests and can live nowhere else.

The dark, moist forest floor is cushioned with a layer of fallen needles, moss, and ferns. Downed tree trunks serve as a nursery for new vegetation. Seedlings root in the rotting logs and take in nutrients for their growth.

19.4 Section Review

1. Compare the flora of the temperate rain forest with that of the tropical rain forest.
2. Why do tall trees dominate the temperate rain forests?

LABORATORY INVESTIGATION 19

Microecosystems

PROBLEM: *Can a pinecone contain an ecosystem?*

SKILLS: *Observing, manipulating, collecting*

MATERIALS: *Several pinecones, magnifying lens, funnel, 500-mL flask, 100 mL of alcohol, water, filter paper, 500-mL beaker*

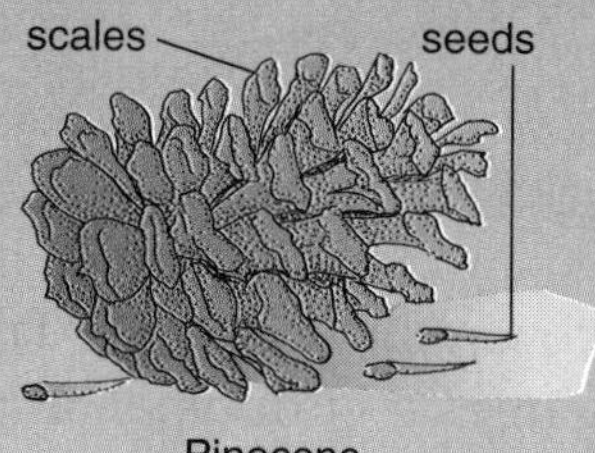

Pinecone

PROCEDURE

1. Observe several pinecones with a magnifying lens. You will notice that a cone is composed of a central axis surrounded by the scales.
2. Examine the scales of a cone and the area above the scales. Can you locate any seeds attached to the top of the scales? If there are no seeds present, look for the impression the seed left on the upper surface of the scale.
3. Place the funnel into the mouth of a 500-mL flask. Place a pinecone in the funnel and slowly pour 100 mL of alcohol over it. Wash the cone with 200 mL of water.
4. Clean the funnel and line it with filter paper. Use this to filter the alcohol-water mixture into the 500-mL beaker.
5. With a magnifying lens, look for spiders and insects that may be trapped on the filter paper.

OBSERVATIONS AND ANALYSES

1. Describe the general appearance of the cones.
2. Draw pictures of the cone specimens you find, so they may be identified.
3. Examine the spiders and insects. See if you can distinguish between those that feed on the conifer and those that feed on the other spiders or insects.

Chapter 19 Review

Answer these questions on a separate sheet of paper.

Vocabulary

The following list contains all the boldfaced terms in this chapter.

boreal forest, buttress, canopy layer, conifers, coniferous forests, deciduous trees, emergents, epiphytes, foliage, ground cover, litter layer, montane, mycorrhizal fungi, shrub layer, taiga, temperate deciduous forest, temperate rain forest, tropical rain forest, understory layer

Fill In

Use one of the vocabulary terms listed above to complete each of the following sentences.

1. ________ are trees that have needles instead of leaves and produce seeds in cones.
2. ________ break down the litter of needles into substances the roots of conifers can absorb.
3. The biome that has cold winters, very warm summers, abundant precipitation, and is dominated by broad-leafed trees that shed their leaves is the ________.
4. The ________ is located along or near the equator, where temperature and humidity are high.
5. The leaves of trees and shrubs are also called ________.

Multiple Choice

Choose the response that best completes the sentence or answers the question.

6. The main decomposers in taiga soils are *a.* bacteria. *b.* squirrels. *c.* fungi. *d.* lichens.
7. Which of the following is an example of a broad-leafed deciduous tree? *a.* pine *b.* birch *c.* spruce *d.* fir
8. The tropical rain forests play an important role in recycling *a.* nitrogen. *b.* carbon dioxide. *c.* air. *d.* water.
9. The floor of the rain forest receives *a.* very little light. *b.* no light for six months of the year. *c.* bright light. *d.* a great deal of light.
10. Bacteria, fungi, and invertebrates decompose leaf litter to form *a.* humus. *b.* nitrogen. *c.* needles. *d.* epiphytes.
11. The taiga is also called the *a.* temperate deciduous forest. *b.* temperate rain forest. *c.* tropical rain forest. *d.* boreal forest.

Short Answer (Constructed Response)

Use the information you learned in this chapter to respond to the following items.

12. What are epiphytes?
13. How are deciduous trees different from coniferous trees?
14. What role do fungi play in the deciduous forest biome?
15. Describe the layers of the deciduous forest biome.
16. What is the dominant vegetation in the taiga?
17. Name three methods of seed dispersal used by deciduous trees.
18. Why do many rain forest trees have buttresses?
19. What is an emergent?

Essay (Extended Response)

Use the information in the chapter to respond to these items.

20. Explain how conifers benefit by not dropping all their needles each year.
21. Identify the factors that contribute to the poor quality of rain forest soils and their shallow depth.
22. How does stratification within the rain forest reduce competition for food?
23. There are only a few scattered areas of old-growth forest left. Old-growth refers to forests that have never been cut. Many of these areas are in national forests. Should the lumber industry be allowed to cut these trees?

Quotations on Earth

Read the following quotations and explain what they mean.

"When we try to pick out anything by itself, we find it hitched to everything else in the universe."

—John Muir, naturalist

"In the end, our society will be defined not only by what we create but by what we refuse to destroy."

—John Sawhill, The Nature Conservancy

CHAPTER 20
Limited Water Biomes

When you have completed this chapter, you should be able to:

Describe the tundra, prairie, desert, and chaparral biomes.

Distinguish among rain shadow, coastal, and interior deserts.

Diagram the life zones on a mountain.

Deserts are not the only biome shaped by a lack of water. A variety of ecosystems, such as the tundra, the prairies in North America, the savanna in Africa and South America, and the chaparral in various parts of the world, are determined by a limited supply of water. These biomes vary in temperature, which creates differences among them. In this chapter, you will learn about the lands of limited water.

20.1 TUNDRA

The **tundra** biome stretches along the Arctic reaches of North America, Europe, and Asia. Because cold air can carry little water vapor, there is low precipitation in these areas. Therefore, the tundra is a vast, treeless, polar desert region that has low year-round temperatures and long months of total darkness. Temperatures high enough to support plant growth occur for only about two months during the year. Few species are hardy enough to endure this harsh environment. This creates a relatively simple community with the least biodiversity of any biome. But life is not rare on the tundra; large numbers of organisms are found there, especially during the brief summer season.

Except for the upper few centimeters, the soil remains frozen for most of the year. This permanently frozen layer of soil is called **permafrost**. The permafrost prevents spring meltwater from soaking into the ground, which causes the formation of many lakes and bogs during the short summer. Bogs are wet, spongy areas that contain partially decayed plant material. The tundra soil is low in nutrient content, since the decomposition of plant material is slow due to the cold.

Plants

Only the hardiest plants can survive tundra conditions. Grasses, sedges, and mats of mosses and lichens cover the ground. Woody plants, such as dwarf willow and dwarf birch, grow close to the ground. The leaves of most plants are either very small or have a hairy covering, which reduces the evaporation of water from their surfaces. During the short summer growing season, flowers appear rapidly, and seeds develop quickly. Though the summer is short, there is a long period of extended daylight during which the sun does not set. This period leads to conditions that favor rapid plant growth.

Since few types of bacteria can tolerate the tundra climate, decomposition of organic material is slow. Undecayed plant material piles up year after year and forms thick mats, called **peat moss**. Peat moss is composed mainly of the partially decayed remains of sphagnum moss, a plant that grows in the cold water of the bogs. As the sphagnum moss grows, the lower parts die and are buried beneath new growth. The dead moss is compacted by the material above it. The lack of oxygen in the cold water retards decomposition and forms peat moss, a dense organic soil that supports the other plant life on the tundra. In some regions, people cut the peat moss into blocks and sell it for fuel or as an additive to enrich garden soil. The harvesting of peat has destroyed many bogs and reduced wildlife areas on the tundra.

Animals

Animal life on the tundra is most active during the short summer. During the long winter, resident animals must cope with extreme cold and scarcity of food. Some animals, such as the arctic ground squirrel, hibernate through the winter months, living on fat stored during the summer. Hibernation is a state of reduced metabolism employed by some animals to escape cold periods. But most tundra animals do not hibernate, because the summer is too short for them to build up adequate fat reserves. During the winter, musk oxen, caribou, and lemmings graze on the vegetation that remains under the snow cover. Arctic foxes, snowy owls, and wolves prey on these herbivores. (See Figure 20-1.)

Animals that live in the tundra are insulated from the cold by thick layers of fat or a dense covering of hair or feathers. Some animals, such as weasels, ptarmigans, and snowshoe hares, change the

Figure 20-1 **Some tundra animals, such as snowy owls and musk oxen, do not hibernate during the winter.**

Snowy owl

Musk ox

Figure 20-2 The protective coloration of the weasel (known as the ermine when its fur is white) and the snowshoe hare lets them blend into the landscape when the seasons change.

color of their coats from brown in the summer to white in winter. Called protective coloration, this adaptation lets them blend into the landscape during summer and winter. (See Figure 20-2.)

As summer approaches, large numbers of migratory birds, such as ducks, geese, sandpipers, and plovers, arrive from the south. These birds travel back and forth during the year, alternating between summer and winter territories. They fly thousands of miles each summer to breed and raise their young in the tundra. Here the migrants find plentiful food, since the lakes and bogs are breeding places for millions of insects, such as mosquitoes. As summer ends and ice returns, these birds head south to their winter feeding grounds.

20.1 Section Review

1. How does peat moss form?
2. Describe the methods used by plants to take best advantage of the short growing season.
3. In what ways are tundra fauna adapted to survive in or escape from cold winter conditions?

20.2 PRAIRIES

Throughout Earth's temperate regions are areas of moderate temperature and seasonal droughts. These relatively flat areas usually are in the interior of continents, especially on the **leeward** side of mountain ranges, the side protected from the wind. In these areas, there is not enough rainfall to support the growth of trees. Grasses, which can subsist on a moderate water supply, are the dominant form of vegetation. These areas are the **temperate grassland** biome or **prairie**. In North America, these areas are also known as the great plains, or prairie, in South America as the pampas, in Africa as the veldt, or savanna, and in Eurasia as the steppes.

Plants

Grasses are the most common of all seed plants. Since grasses grow from the base rather than the tip, as most other plants do, they are resistant to destruction by grazing animals or fires. Grasses are able to withstand long periods of drought and periodic fires because their fibrous root systems often extend several meters into the soil. Even though the blades, or leaves, of the plant are destroyed by fire, drought, or grazing, the roots are able to regenerate new leaves.

The dense fibrous root systems of grasses propagate asexually by sending out underground runners, called **rhizomes**. A clump of grass can send out many rhizomes, each of which can start a new clump at some distance from its source. These new clumps send out more rhizomes until the soil is densely packed with a network of roots, and the ground is covered with a sea of grass. Grasses also reproduce sexually by means of seeds, which form new plants. (See Figure 20-3.)

There are many different types of grasses. Pasture grasses include timothy, fescue, bluegrass, and sorghum. Throughout history, the seeds of some grasses have played an essential role in the development of civilization. Cereal grains are the edible seeds of certain grasses. Wheat has been the preferred cereal grain in Europe, rice in Asia, and corn in the Americas. Barley, thought to be the first cereal grain cultivated, is now fourth in importance worldwide. Rye and oats are important grains that are grown in the poorer soils of the northern regions, and sorghum and millet are favored in hot, dry climates, such as in parts of Africa.

Bamboo is the tallest member of the grass family. It has a hard woody stem from which the leaves grow. There are many species of bamboo throughout the world, and it is an important material used

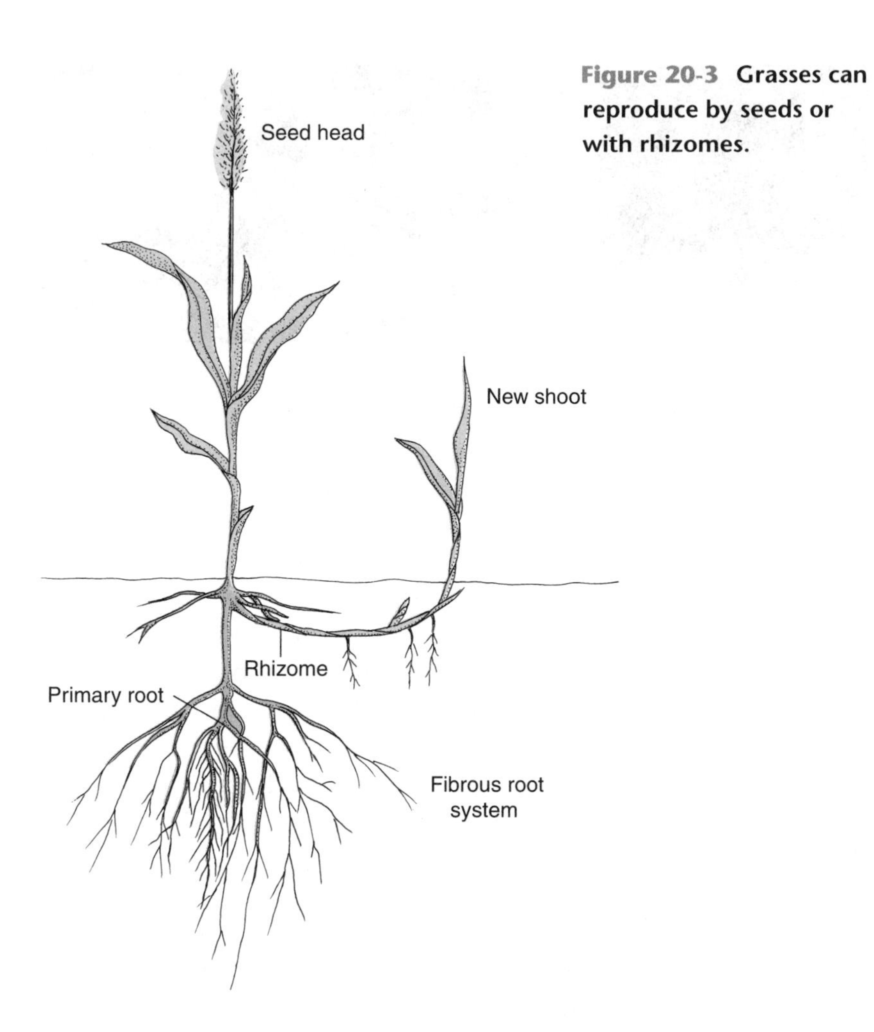

Figure 20-3 Grasses can reproduce by seeds or with rhizomes.

in the construction of homes, furniture, and domestic products. It is, in fact, the most widely used woody building material in the world. The tender young shoots are a source of food to many animals.

Animals

Small burrowing mammals, insects, and invertebrates are abundant in the deep grassland soil. Many other insects live and feed on the grass stalks and seeds. Hoofed herbivores dominate the world's grasslands, except for Australia, where kangaroos are the chief plant-eaters. Accompanying the herbivores are large predators.

Figure 20-4 The ostrich (left), emu (center), and rhea (right) illustrate parallel evolution.

The largest concentrations of wildlife on Earth are found in the grasslands.

Similar types of animals, from different species, can be found in widely separated grasslands. The ostrich in Africa, the rhea in South America, and the emu in Australia look similar but are different species. (See Figure 20-4.) Each continent also has its own species of large herbivores, which graze on the plentiful grasses. Bison and pronghorn roam the North American prairie, antelope and zebra are found throughout the African veldt, and saiga antelopes and wild horses live in the Asian steppes.

Scientists estimate that before the westward expansion of the United States, more than 60 million bison grazed on the grasses of the North American prairie. By 1900, hunting and sport shooting had reduced that number to several hundred. Since then, laws to protect the bison have been passed. The population of bison now numbers more than 20,000. Table 20-1 lists some typical North American grassland animals.

TABLE 20-1 NORTH AMERICAN GRASSLAND FAUNA

Herbivores	Carnivores
Pronghorn antelope	Coyote
Meadowlark	Prairie rattlesnake
Bison	Badger
Dove	Golden eagle
Prairie dog	Swift fox
Deer mouse	Gopher snake
Prairie vole	Prairie falcon
Jackrabbit	Black-footed ferret

Savannas

A **savanna** is a tropical grassland that contains scattered trees. Savannas occur throughout large areas of Africa and South America, where long seasonal droughts and warm temperatures prevail. These conditions favor the growth of grasses but also allow for limited tree growth. Like the temperate grasslands, savannas support large concentrations of wildlife. The African savannas, or veldts, support vast herds of zebra, antelope, and wildebeest. Each herd may number in the hundreds of thousands.

Termites, which are social insects like ants and bees, build castle-like nests of hard soil on the savanna. These huge towers are ventilation systems for the colony, insuring a constant temperature within the nest. Termites collect the dead grasses from the surface and use this material in underground chambers to cultivate the fungus on which they feed. The termites help to recycle important nutrients within the ecosystem, and their nests often serve as home to a variety of savanna animals.

20.2 Section Review

1. What are some of the uses of bamboo?
2. Describe some of the large animals that dominate grassland biomes.
3. How is a prairie different from a savanna?

20.3 DESERTS

Desert biomes are regions where the annual rate of evaporation is greater than the annual rate of precipitation. These arid, or dry, regions experience an extreme range of temperatures throughout each 24-hour day. During the daytime, temperatures are usually high. However, the dry air causes rapid cooling after dark, and nighttime temperatures may be very low. Some tropical deserts, for example the Sahara in Africa, experience very hot daytime temperatures throughout the year. Temperate deserts, such as the Mojave in the western United States, have hot daytime temperatures in summer and cool temperatures in winter. Parts of the Gobi in central Asia are cold in both summer and winter.

Desert biomes usually receive fewer than 25 centimeters of precipitation per year, and some areas may even go several years between rains. The precipitation that does occur is seasonal and is usually brief but heavy. Because the rainfall is intense, much of the water runs off the surface instead of being absorbed by the soil. The runoff causes erosion of the soil and forms **arroyos**, or gullies, throughout the landscape. Since the desert air is warm and dry, what little water the soil does absorb is quickly lost through evaporation. Desert streambeds, called **washes**, are dry most of the time and contain water only following heavy rains.

Rain is not the only source of water in the desert. In some deserts, such as the Namib in Africa and Atacama in South America, fog is an important source of moisture. Underground water may appear at the surface in the form of a spring or seep, which is often surrounded by an oasis. An **oasis** is a fertile area in a desert that supports dense vegetation. Large rivers, such as the Colorado in North America and the Nile in Africa, flow through desert biomes. The banks of these rivers are covered with trees and grasses and can be considered extended oases.

Types of Deserts

Deserts can be classified by the cause of their dryness. Most deserts lie along the Tropic of Cancer and the Tropic of Capricorn and are called subtropical deserts. These deserts form because of circulation patterns in the atmosphere. At the equator, hot, moist air rises, cools, and loses its moisture. The cooler, drier air moves north toward the Tropic of Cancer or south toward the Tropic of Capricorn, where it descends and warms. This air contains little moisture and thus contributes little precipitation to the region. The Sahara and Kalahari in Africa are subtropical deserts located on either side of the equator. (See Figure 20-5.)

Coastal deserts form when cold ocean currents move along the

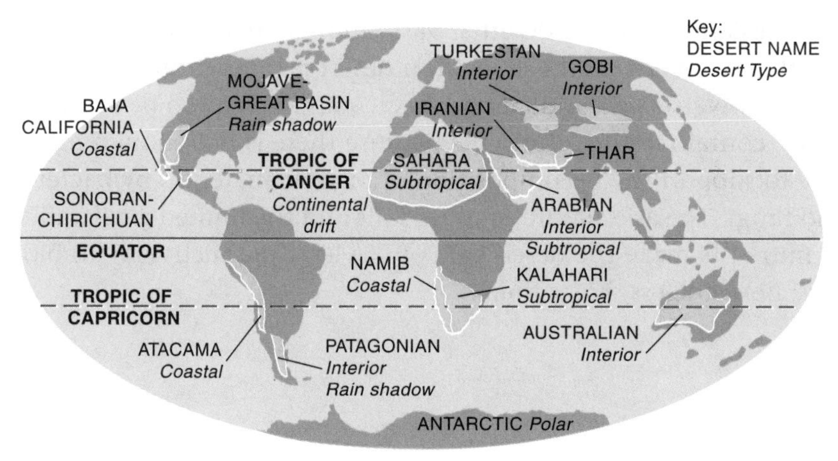

Figure 20-5 Deserts are located worldwide.

continental shoreline. Moist air flowing toward the land is chilled by contact with cold ocean waters, causing thick fog to form. This heavy layer of fog moves inland and blankets the region. Though the humidity is high, there is little precipitation. The Atacama of Chile is a coastal desert.

Interior deserts occur deep within continental regions where moisture-laden winds do not reach. The Gobi in central Asia is an interior desert. Ocean winds have precipitated most of their moisture by the time they reach the Gobi region.

Tall mountain ranges often produce **rain shadows**, or regions of reduced rainfall on the leeward side of the mountains. When moist air moves up and passes over the windward side of a mountain range, it cools and produces rain or snow. This air, drained of most of its moisture, then moves down the leeward side of the range. Since the leeward region receives little or no precipitation, conditions there favor the formation of a desert or a grassland.

Rain shadow deserts form in the dry zone on the leeward sides of mountain ranges. The Mojave, which is flanked to the west by the Sierra Nevada and Cascade ranges of western North America, illustrates this phenomenon. The Pacific Ocean is a source of the weather systems that affect the region. The western slopes of these mountain ranges, which face the ocean, receive adequate rainfall. However, the area to the east receives little rain and is either desert or semidesert.

We usually think of a desert as vast, uninhabited expanse of sand. But this is not always the case. Most desert regions contain soil and support diverse flora and fauna. Desert soils, which can be sandy or rocky, contain limited nutrients because there is not enough moisture to support the decomposition of organic materials by bacteria and fungi. The arid nature of these soils and the limited availability of nutrients make the desert one of the least productive of all biomes; nevertheless, life is there.

Plants

Most desert flora are **xerophytes**, plants adapted to collect and conserve water. These plants have developed various methods of coping with the extremes in temperature, nutrient-poor soil, and lack of moisture found in desert regions. The roots of most desert plants spread out in a shallow network, so when it does rain, they are able to soak up moisture rapidly. This causes the plants to be widely spaced, since their spreading roots limit the growth of other plants nearby. The roots of some other plants, such as the mesquite tree, are very long and reach down more than 30 meters to tap underground sources of water.

To carry them through dry periods, plants such as the barrel cactus, saguaro cactus, and agave are able to store large quantities of water in their spongy tissues. Some desert plants, such as ocotillo, have small leaves to reduce water loss. In other plants, such as many species of cacti, the leaves are modified to form thorns or spines. The leaves of desert plants often have thick, waxy skins that prevent the evaporation of water. Other plants commonly found in the North American deserts are fishhook cactus, prickly pear cactus, paloverde, cholla, desert paintbrush, and sagebrush. (See Figure 20-6.)

Even though drought is the rule rather than the exception, desert vegetation is not always scarce. Following a rainstorm, large numbers of wildflowers and grasses swiftly bloom, produce seeds, and then die. If necessary, the seeds of these plants are able to survive for many years, until the next rainfall occurs.

Figure 20-6 Desert plants have adaptations that help them conserve water.

Animals

Deserts are not barren. In fact, some teem with life. Many mammals, birds, reptiles, and a wide variety of small invertebrates live in these arid regions. (See Figure 20-7 on page 344.) Invertebrates found in the desert include scorpions, tarantulas, beetles, wasps, and ants. Each makes use of the scant moisture present in the ecosystem.

Animals have adapted in many ways to cope with desert conditions. Insects have thick shells and reptiles have thick skins that prevent water loss. To conserve water, some mammals, such as the camel, produce dry feces and concentrated urine. Some animals do not sweat. To cut down water loss through respiration, many small animals live in cool underground burrows. The kangaroo rat never drinks water. However, it is able to survive on water produced during the metabolism of dry plants and seeds that it eats. Reptiles, such as the rattlesnake, horned toad, and Gila monster, meet their need for water by extracting it from the tissues of the animals they eat.

Though some desert animals are diurnal, active during the day, many are nocturnal, active only at night. To avoid the high daytime temperatures, they forage for food when the sun is down. Bats, owls, nighthawks, ringtail cats, and many species of rodents are nocturnal.

Figure 20-7 Animals that live in the desert show many adaptations to reduce water loss.

TABLE 20-2 NORTH AMERICAN DESERT FAUNA

Herbivores	Carnivores
Kangaroo rat	Beaded lizard
Piñon jay	Roadrunner
Desert bighorn	Coyote
Javelina	Pallid bat
Wood rat	Harris's hawk
Desert tortoise	Rattlesnake
Ground squirrel	Elf owl
Chuckwalla	Gila monster

Deserts often go through long periods of drought, when ponds and streams dry up and all the available water evaporates. Some animals escape these conditions through estivation, a process similar to hibernation. During estivation, animals achieve minimum water loss in a state of suspended animation. When ponds in the North American desert dry up, spadefoot toads burrow into the soil, enveloped in a cocoonlike layer of dead skin to await the next rains. African lungfish also estivate and can survive several years of drought conditions buried in a coat of dried mud. Table 20-2 lists some animals that live in the North American deserts.

20.3 Section Review

1. What is the effect of heavy rain on desert landscapes?
2. Explain the relationship between deserts and nearby mountain ranges.
3. Describe some of the adaptations used by desert plants to conserve water.

20.4 CHAPARRAL

The **chaparral** biome is a specialized woodland ecosystem that has cool, wet winters and long, dry, hot summers. It is located in areas of the southwestern United States, around the Mediterranean Sea in Europe, in parts of South Africa, along coastal Chile, and in northern and southern Australia. In the Mediterranean region, this biome is also known as maquis. The terms *chaparral* and *maquis* are also used to characterize the vegetation found in this biome.

Plants

Chaparral vegetation differs from region to region. In the southwestern United States, sage, mesquite, manzanita, small oaks, and junipers are the dominant vegetation. In Australia, eucalyptus trees and shrubs are dominant. Eucalyptus trees have been introduced to areas of California in recent years. These plants thrive there, and they have spread, crowding out and replacing many of the native plant species.

Most chaparral flora are xerophytes. Many of these plants cope with drought by avoiding it. They grow during to the wet season when conditions are favorable. They shed their leaves by early summer, which prevents water loss through their leaves in the dry season. Some trees and shrubs have small, waxy evergreen leaves to reduce water loss during the dry season. Others have leaves covered with tiny hairs, which give them a gray-green appearance. The hairs reflect heat and insulate the leaves, thereby reducing moisture loss.

Chaparral woodlands are composed of low brush, which consists of large shrubs and dwarf trees that have small, waxy, or hairy evergreen leaves. The brush, which is 1 to 3 meters tall, grows in dense clumps that form a low canopy. There is little undergrowth beneath the trees, and the ground is covered with dry plant litter.

In the chaparral, most plant growth occurs during the mild, rainy winter. When the rains stop, the vegetation becomes dry and brittle. A litter of dry branches and leaves soon builds up on the ground. Because of the dry litter, fires caused by lightning are common during the dry summer and fall seasons. These fires spread swiftly through the dry litter, trees, and shrubs. Fire plays a major role in shaping this ecosystem.

The ability of chaparral shrubs to withstand severe seasonal drought and periodic fires is due in part to their specialized root systems. They have long taproots that reach deep sources of water and extensive lateral fibrous roots that absorb water at the soil's surface. The thick underground roots and stems of chaparral vegetation enable these plants to survive the dry summers and frequent fires. Even though fire destroys the shrubs' upper portion, new shoots are able to grow quickly from the surviving underground stems.

Other shrubs have seeds that may lie dormant for years until they are burned or scarred by fire. Following a fire, these seeds sprout and grow, producing new plants that begin the renewal of the chaparral.

Grasses and wildflowers, which usually are rare in the chaparral, spring up at the first rain following a fire. These plants are able to use the nutrients released by the fire and are important to the recycling process in the chaparral. The rapid regeneration of new trees and shrubs soon crowds out the grasses, restoring the chaparral woodland community.

Animals

A great variety of small mammals, birds, reptiles, and insects inhabit the chaparral. Some larger animals migrate in and out between the wet and dry seasons. (See Figure 20-8.) But most large browsing animals, such as mule deer, find passage through the thick vegetation difficult. The waxy or hairy covering on the leaves of many of the plants makes them hard to digest by many chaparral fauna. Due to the abundance of oak trees, acorns are an important food source. Many animals of the chaparral estivate through the hot, dry summer season. Table 20-3 on page 348 lists some animals that live in the chaparral biome.

Figure 20-8 Some chaparral fauna.

TABLE 20-3 CHAPARRAL FAUNA

Herbivores	Carnivores
Rabbit	Gray fox
Pocket mouse	Coyote
Gray squirrel	Gopher snake
Chipmunk	Striped skunk
Scrub jay	Bobcat
Wood rat	Western toad
Quail	Rattlesnake
Acorn woodpecker	Arboreal salamander

20.4 Section Review

1. Describe the climate of the chaparral biome.
2. Explain how periodic fires shape the chaparral vegetation.
3. Why are large browsing animals not abundant in the chaparral?

20.5 MOUNTAIN ZONES

As warm air rises, it cools at a rate of about 2 degrees Celsius for each 150-meter increase in altitude. Therefore, the higher up you go into the atmosphere, the cooler the temperature is. Going 1000 meters up a mountain has the same effect on temperature as traveling many kilometers toward the pole. (See Figure 20-9.)

At higher altitudes in mountainous regions, the winds blow harder, the atmosphere is thinner, and less moisture is present in the air. Because of the thinner atmosphere, the amount of oxygen available for respiration is reduced. At the base of a mountain, the environmental factors may support a rain forest community. Conditions on the top of the same mountain may be able to support only a tundra community, or even be so hostile to living things that few are found.

Variations in climate on a mountain or throughout a mountain range are due mainly to differences in elevation. Changes in vegetation are related to these differences in climate. When climbing a mountain, you encounter a sequence of climate-controlled **life zones**, or belts of vegetation, each of which has its own distinct

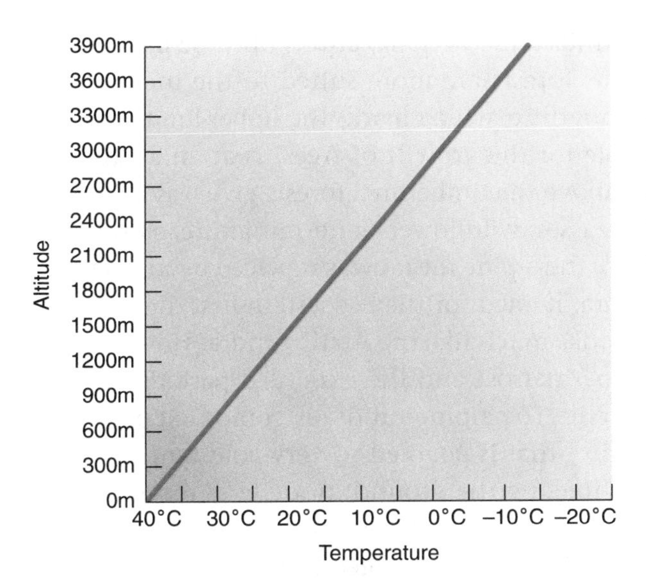

Figure 20-9 When ascending a mountain, the temperature drops 2°C for each 150-meter increase in elevation.

flora and associated fauna. The ecosystems on mountain slopes are called montane.

Life zone boundaries do not follow elevation contours, lines on a map that pass through points of equal elevation. In the Northern Hemisphere, life zones extend to higher elevations on south-facing slopes and drop to lower levels on north-facing slopes. This is because the southern slopes receive a longer period and higher intensity of solar radiation than the northern slopes. Thus the south-facing slopes are warmer and free of snow earlier, which provides a longer growing season for vegetation. Zone boundaries also tend to extend higher on ridges and lower on adjacent valleys, since the ridges generally receive more sun and are free of snow earlier than the valleys.

On some mountain ranges, rain shadow conditions affect life zones. While the windward side of the mountain range facing the ocean receives enough precipitation to support lush forests, the leeward side, which receives little moisture, is a desert or chaparral.

The Rocky Mountains

Changing zones of vegetation are especially evident in many areas of the Rocky Mountains of western North America. Deciduous forests cover the foothills and lower elevations of these mountains. Higher up, coniferous forests are more suited to the increasingly cold climate. The **timberline**, which marks the upper limit at which the climate is suitable for the growth of trees, is at an altitude of about 2450 meters. Above the timberline, forests give way to **alpine meadows**, areas of grasses, wildflowers, and miniature, slow-growing shrubs. Higher up, the alpine meadow is replaced by **alpine tundra**. The alpine tundra, named for the ecosystem first identified in the Alps in Europe, looks much like the Arctic tundra. However, the alpine tundra lacks permafrost and the extended periods of darkness found in the Arctic. The alpine tundra is composed of a plant and animal community that is adapted to very cold temperatures. Plants grow slowly, hugging the ground; none ever reaches more than 1 meter tall. The flora of the alpine tundra have many of the same characteristics as Arctic tundra vegetation. The alpine tundra ends at either bare, rocky scree-covered slopes, or at the snow line. Scree is loose, rocky rubble that has been eroded from the mountain bedrock. (See Figure 20-10.)

Figure 20-10 What determines the life zones on a mountain?

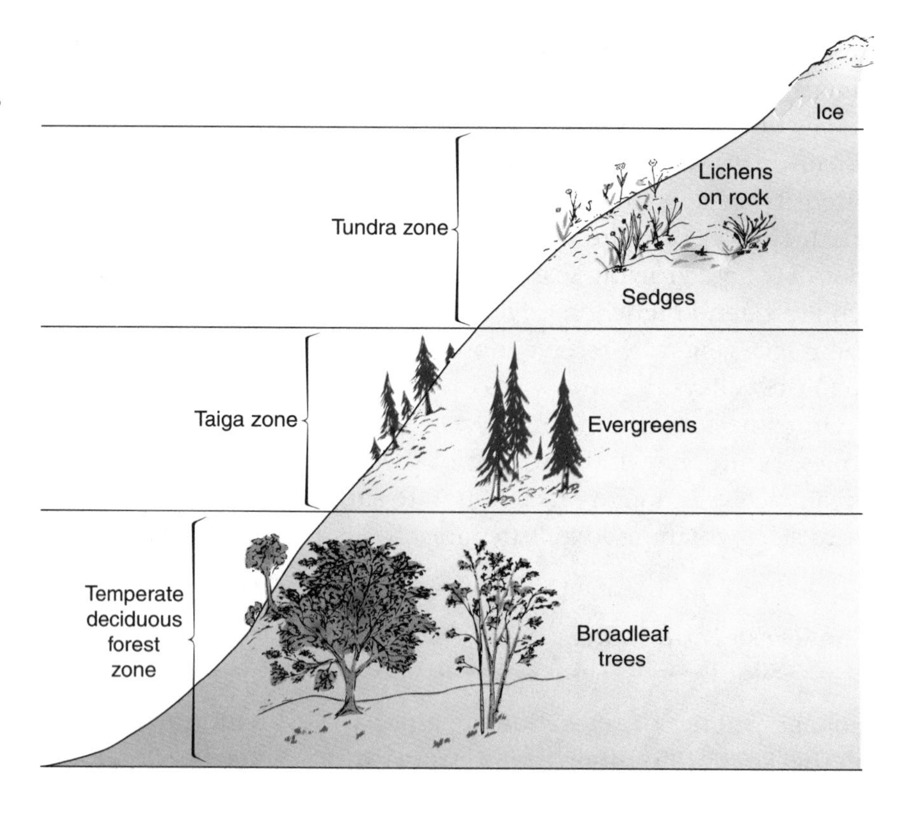

INTERACTIONS

Reduced Sierra Snowpack Affects Climate

The Northern Hemisphere has been warmer in the twentieth century than in any other century during the past 2000 years. The evidence of climatic change is derived from tree ring samples, cores taken from lake and ocean sediments, and examination of ancient ice and coral reefs. Over the past century, the average surface temperature has risen by almost 2°C. The United States Geological Survey has suggested that mountainous areas are particularly vulnerable to the effects of this warming trend. Its effects could impact on montane ecosystems as well as those around them.

Research suggests that the warming trend will severely shrink the snow cover in the Sierra Nevada Range of eastern California. The snow cover, or snowpack, is crucial to the economy of California. The shift in temperature will bring more rain and less snow to the mountainous region. This will affect California's use of the Sierra snowpack to supplement its water supplies. Billions of dollars have been spent on dams, reservoirs, and aqueducts to capture the water released during the annual snowmelt. The snowpack acts to regulate the flow of water. It holds water through the winter in the form of snow and ice. The water is then gradually released during the spring and summer as the snow melts.

When water falls as rain instead of snow, it flows directly into the river system. This can tax the system, leading to flooding, a decline in year-round water supplies, and a decrease in hydroelectric power production. Since the snowpack acts as a giant natural reservoir, its loss is like losing reservoir storage space. To counteract the effects of warming, new reservoirs and dams might have to be built. Since many of the best sites for dams have already been used, new dams might not be able to counteract all the effects due to warming.

The shift in temperature will surely affect the flora and fauna of the region. Whole ecosystems will be changed. California's efforts to restore salmon and other endangered species will likely fail. Supplies of potable water for human consumption will be reduced.

The system of aqueducts and dams helps make southern California one of the nation's richest irrigated agricultural areas. Snowmelt captured in northern reservoirs is carried through a system of aqueducts to southern agricultural areas. These areas usually receive little rainfall. The decrease in water for irrigation would surely create economic hardships for the region.

Large herbivorous animals that live at high altitudes, such as mountain goats and bighorn sheep, have thick, shaggy coats to protect them from winter storms. They have split hooves or pads on the bottom of their toes to make them surefooted on the rocky and snow-covered slopes. These animals have large, efficient lungs to help them breathe the thin air. They usually migrate to the lower elevations in fall and winter to escape the harsh conditions and to find suitable food and shelter. In spring and summer, they return to the higher elevations to feed on the new vegetation. Elk and deer are other large herbivores that live in the mountain forests. The major predators are the mountain lion, wolf, lynx, grizzly bear, and wolverine.

Marmots, golden-mantled ground squirrels, and picas (small mammals related to rabbits) live on the grasses and shrubs that grow on the scree-covered slopes. These animals are most active during the spring and summer. They hibernate through the winter beneath the snow-covered rocks. Foxes, weasels, badgers, hawks, owls, and eagles prey on these animals.

Many large birds, such as eagles, condors, and hawks, nest on the rocky slopes during spring and summer. Here they find a plentiful supply of food to feed their young. The birds use air currents that rise along the steep cliffs to soar great distances using a minimum of energy. They usually migrate to lower elevations or warmer climates in the fall. (See Figure 20-11.)

Figure 20-11 **Some Rocky Mountain fauna.**

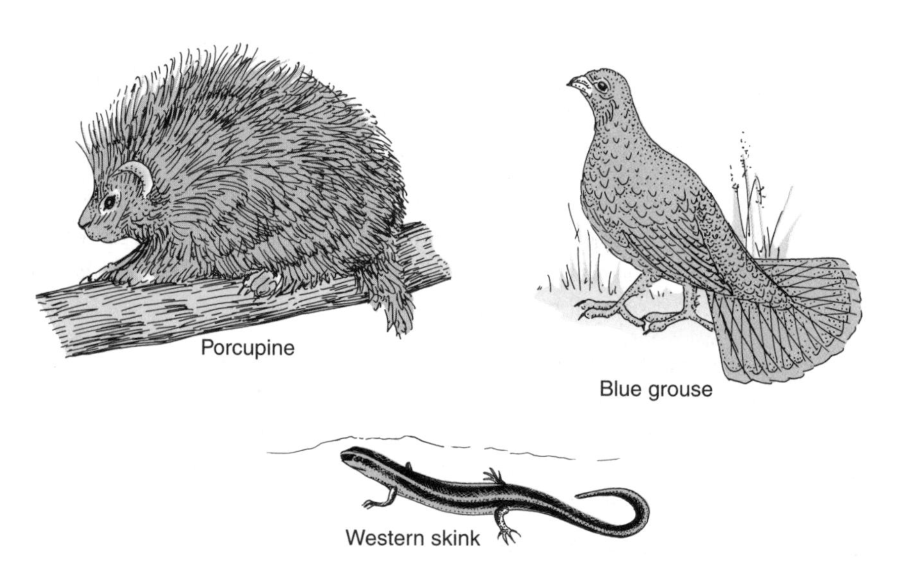

Figure 20-12 Even though Mount Kilimanjaro is located near the equator, its summit is covered with snow.

Mount Kilimanjaro

Africa's Mount Kilimanjaro is located near the equator. In ascending this 5800-meter mountain, the climber encounters the range of biomes that occur from the equator to the North Pole. The climb begins in a tropical, equatorial rain forest. As the elevation increases, conditions grow colder and drier, and winds blow harder. The cold temperatures shorten the growing season for most plants. Climbers can see changes in the flora as they pass from one life zone to another. First, the tropical rain forest of broad-leafed evergreens and ferns is replaced by a temperate evergreen forest. This vegetation is able to withstand the cooler temperatures of higher altitudes. Trees and shrubs reach to the timberline. Beyond the timberline, alpine meadow and tundra communities extend to the snow line. (See Figure 20-12.)

20.5 Section Review

1. Describe the main reasons for the cooler temperatures found at higher mountain altitudes.
2. Why is it harder to breathe at high altitudes?
3. What causes the varied montane life zones?

LABORATORY INVESTIGATION 20

Making a Topographic Map

PROBLEM: *To make a model of a topographic map*

SKILLS: *Measuring, drawing, manipulating*

MATERIALS: *Plastic box (approximately 30 cm long, 23 cm wide, and 15 cm high), ruler, felt-tipped pen, modeling clay, water, sheet of clear plastic wrap, masking tape*

PROCEDURE

1. Along one corner of the box, starting at the bottom and ending at the top, mark off 2-cm intervals.
2. Use modeling clay to construct a mountain that is no higher than the box. Place the mountain in the box.
3. Pour water into the box until it reaches the 2-cm level.
4. Cover the top of the box with a sheet of clear plastic and secure it in place tightly with masking tape.
5. Use a felt-tipped pen to trace onto the clear plastic sheet the line formed where the water and clay mountain meet. This is your first contour line.
6. Remove the plastic sheet from the container and add water to the 4-cm level.
7. Replace the plastic and trace the new line formed by the water and mountain. This is your second contour line.
8. Continue adding water 2 cm at a time and tracing the new lines formed between the water and clay, until you reach the top of the mountain. You now have a completed contour map for your mountain drawn on the plastic sheet.

OBSERVATIONS AND ANALYSES

1. If each contour line on your diagram represents 100 meters, how high is the mountain?
2. Are all the contour lines the same distance apart?
3. Where are the contour lines closest?
4. Where are they farthest apart?
5. Where is the steepest part of the mountain?

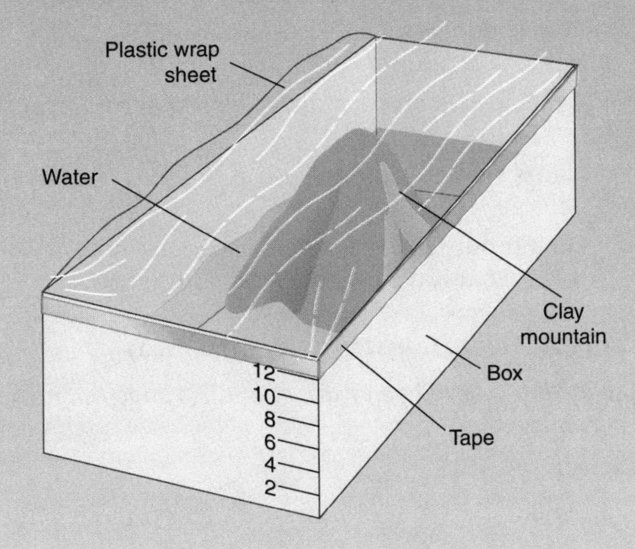

Chapter 20 Review

Answer these questions on a separate sheet of paper.

Vocabulary

The following list contains all the boldfaced terms in this chapter.

alpine meadows, alpine tundra, arroyos, chaparral, desert, leeward, life zones, oasis, peat moss, permafrost, prairie, rain shadows, rhizomes, savanna, temperate grassland, timberline, tundra, washes, xerophytes

Fill In

Use one of the vocabulary terms listed above to complete each of the following sentences.

1. Another name for the temperate grassland is the ________.
2. Rain shadows are often found on the ________ side of tall mountain ranges.
3. A fertile area surrounding a desert spring or river is a(an) ________.
4. Plants that are adapted to obtain and conserve water are called ________.
5. On a mountain, ________ are climate-controlled belts of vegetation that have distinct flora and fauna.

Short Answer (Constructed Response)

Use the information you learned in this chapter to respond to the following items.

6. What is permafrost?
7. Explain why tall mountain ranges often produce rain shadows.
8. Describe the asexual method of reproduction employed by grasses.
9. What is a savanna?
10. Arroyos and washes are usually found in what type of biome?
11. What adaptations allow xerophytes to conserve water?
12. What is the significance of the timberline of a mountain range?
13. How do coastal deserts form?
14. Other than location, how is the Gobi Desert different from the Sahara Desert?
15. What are the dominant forms of vegetation in the southwestern chaparral biome in the United States?

Essay (Extended Response)

Use the information in the chapter to respond to these items.

16. Explain why grasses are not easily destroyed by fires or grazing.
17. Compare the processes that lead to the formation of interior deserts and subtropical deserts.
18. Describe how each of the following adaptations helps plants cope with the chaparral ecosystem.
    a. hairy leaves
    b. waxy coating on leaves
    c. thick underground roots and stems
    d. seeds that sprout only when burned
    e. leaves shed before summer
    f. growth limited to wet season

Quotations on Earth

Read the following quotations and explain what they mean in your own words.

"Humans must understand their universe in order to understand their destiny."

—Neil Armstrong, astronaut

"Consult Nature in everything and write it all down. Whoever thinks he can remember the infinite teachings of Nature flatters himself. Memory is not that huge."

—Leonardo da Vinci, Italian artist and engineer

Research Project: Studying Xerophytes

Cacti and succulents are excellent examples of xerophytes for study. Obtain one or more samples of various cacti or succulents.

- How is the stem of a xerophyte different from that of most plants?
- How are xerophyte leaves modified to retain water?
- In what type of soil do xerophytes grow best?
- Describe the xerophyte root system.
- What are the water requirements for these plants?
- What are the light requirements for these plants?

CHAPTER 21
Freshwater Biomes

21.1
Groundwater

21.2
Standing Water

21.3
Flowing Water

When you have completed this chapter, you should be able to:

Discuss the various sources of freshwater.

Distinguish among lakes, ponds, and marshes.

Compare conditions at the headwaters of a stream with those of the floodplain.

Even though 71 percent of Earth's surface is covered with water, most of it is salt water. Only 3 percent of Earth's water is the freshwater that living things need to survive. Freshwater is present as groundwater, water that is absorbed by Earth's crust, and stored underground; standing water, which includes inland seas, lakes, and ponds; and flowing water, which includes rivers, streams, and brooks. In this chapter, you will learn about Earth's freshwater biomes.

21.1 GROUNDWATER

Precipitation that does not run off the surface into lakes or streams, or evaporate into the atmosphere, is absorbed into Earth's crust. Water flows through and eventually saturates the soil. This water becomes **groundwater**. The boundary between the saturated and unsaturated soil is called the **water table**. Groundwater is stored in underground **aquifers**, layers of permeable rock, gravel, or sand that are saturated with water. (See Figure 21-1.) Aquifers rest on a layer of impermeable rock. Aquifers can be small, localized regions or cover thousands of square kilometers. For example, the Ogallala aquifer stretches between South Dakota and Texas. Groundwater accounts for 98 percent of the world's usable freshwater reserves. In the United States, more than half the water used for drinking and agriculture comes from underground aquifers. During droughts, this water plays an important role in maintaining ecosystems.

Groundwater from aquifers usually appears at the surface through natural springs, seeps, and wells. The water in aquifers located near Earth's surface is not under pressure. Wells that tap these sources

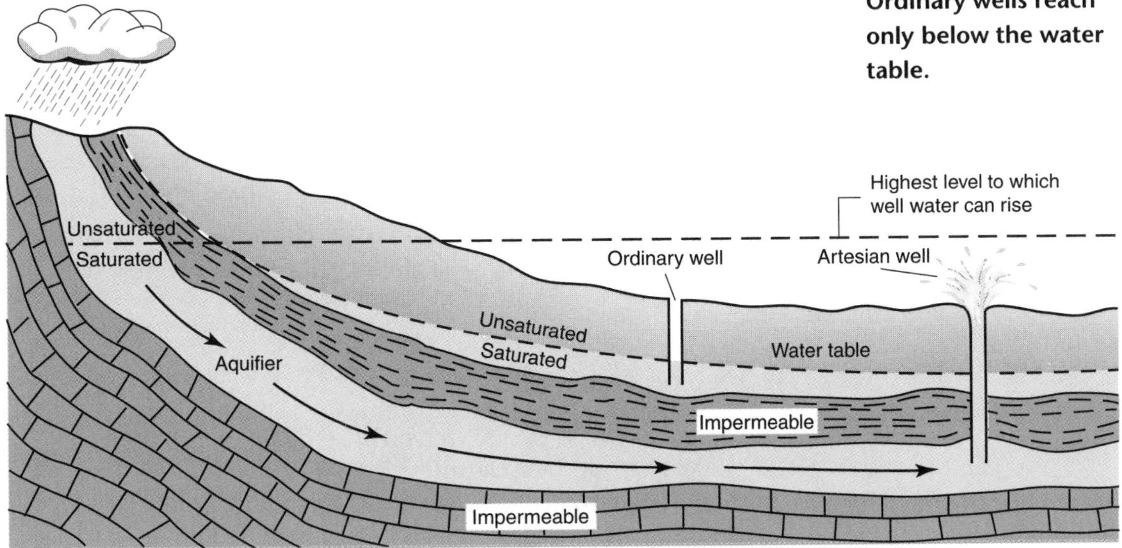

Figure 21-1 Aquifers are places where groundwater is stored. Artesian wells tap into an aquifer. Ordinary wells reach only below the water table.

TABLE 21.1 COMMON MINERALS DISSOLVED IN HARD WATER

Mineral	Common Name
Magnesium sulfate	Epsom salt
Calcium carbonate	Limestone
Calcium sulfate	Gypsum

need pumps to draw the water up to the surface. Aquifers located deep within the soil are usually covered with a layer of impermeable rock. Since the water here is enclosed and under pressure, wells that tap these aquifers can use their natural, internal pressure. When deep aquifers are tapped, the water is often used faster than it can be replaced. New supplies of water can enter these aquifers only in areas where their permeable layers extend to the surface, which makes refilling them a slow process.

Rainwater is soft water. As water passes over and through rocks, minerals in the rocks dissolve. Water that contains dissolved salts of magnesium, calcium, and iron is called **hard water**, because it is hard for soap to form lather in it. The mineral salts also clog plumbing and leave residues in hot-water heating systems. Hard water is often a characteristic of well water and groundwater. Water-softening compounds and detergents usually are used to remove the salts or to lessen their effects. Table 21-1 lists some minerals found in hard water.

Water that is fit for drinking is called **potable water**. Groundwater is the primary source of potable water for most of the population. However, many aquifers have been contaminated by agricultural pesticides and fertilizers, leaking municipal sewer systems and septic tanks, toxic waste from landfills, and leaks in hazardous waste disposal sites. Human activities and technology have severely affected both the quality and quantity of the world's groundwater reserves.

21.1 Section Review

1. Compare the two types of wells: those that tap surface aquifers and those that tap deep aquifers.
2. Why does the extraction of water from deep aquifers often lead to the depletion of this resource?

21.2 STANDING WATER

Water is crucial for the existence of life, because most chemical reactions in living things take place in water. Many living things are found in aquatic, or water-based, ecosystems. Aquatic ecosystems can be freshwater or saltwater. In the following sections, you will learn about freshwater ecosystems.

Water's Response to Changing Temperatures

Most liquids contract, or shrink, as they are cooled. They continue contracting until they freeze and become solid. These solids continue to contract as they are cooled further. As a substance contracts, its density increases. Increased density means a greater mass per unit volume. Since the solid form of a substance is usually more dense than the liquid form, the solid sinks when placed in the liquid. Solid mercury sinks in liquid mercury.

Water does not behave as other liquids when it is cooled. As water cools, it contracts and becomes more dense until it reaches its maximum density at 4°C. If water is cooled further, it expands and becomes less dense. At 0°C, water freezes. It becomes ice and expands even more. Since ice is less dense than liquid water, it floats. Due to this special property of water, rivers and lakes almost never freeze all the way to their bottom. This is very important for the biota of aquatic ecosystems.

In winter, ice that forms on the surface of a lake or river insulates the water below and slows the process of freezing. The densest water, at 4°C, sinks to the bottom and collects there. Within the dense bottom layer, which remains at 4°C, fish and other aquatic organisms live through the winter, protected from freezing.

As rising spring temperatures melt the layer of ice, the water begins to warm. Through contact with the air, the surface water becomes rich in oxygen. When the surface water warms to 4°C, its density is greatest and it slowly sinks to the bottom. The sinking water displaces the less dense water below it, which rises to the surface. The exchange of water replenishes the oxygen that plants and animals removed from the bottom water during the winter. This mixing process is called the **spring overturn** of the lake. The spring overturn aids in recycling nutrients by bringing decomposed organic remains from the lake bottom to the surface.

The heat of summer raises the temperature of the surface water well above 4°C. At these temperatures, the surface water becomes less dense than the water below, and the mixing process stops. The deep bottom water remains cool through the warm summer months because the warmer, less dense surface water insulates it. During this period, organisms that live in the cool lake bottom deplete this water of oxygen.

In the fall as the surface layer cools, it becomes more dense than the water below, and it sinks. The less dense layers below are displaced upward. This is called the **fall overturn**, and it causes a new mixing of the lake's water. The overturn cycle is illustrated in Figure 21-2.

Lakes and Ponds

Lakes and ponds are bodies of standing water. Ponds are usually small and shallow enough to permit light to penetrate to the bottom. Lakes are generally larger than ponds and contain areas that are too deep for light to reach the bottom. Although many lakes and ponds are fed by rivers and streams, they are considered standing water.

In the shallow water of lakes and ponds, sunlight reaches the bottom, and rooted plants such as reeds, sedges, rushes, cattails, and wild rice flourish. These plants, which grow half in and half out of the water, are called **emergents**. In deeper water, water lilies, duckweed, and pondweed float on the surface, where their leaves can be exposed to sunlight. In the deepest water, large plants cannot grow, and microscopic algae drift suspended at or near the surface.

Algae are plantlike organisms that contain chlorophyll. They produce their own food through photosynthesis. These organisms make up a large part of a lake's phytoplankton, free-floating microscopic producers in the lake ecosystem. Microscopic "animals," called zooplankton, eat the algae. Plankton drift through the water and are food for fish and other large aquatic organisms.

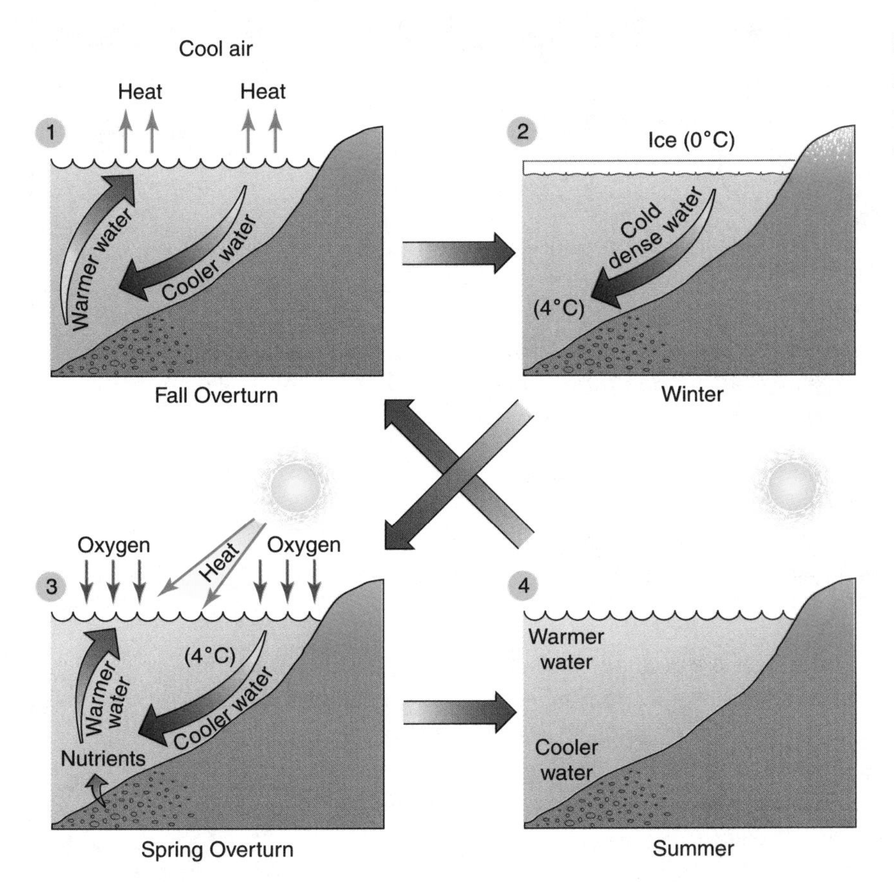

Figure 21-2 The spring and fall overturn in a lake are part of a continuing cycle.

The shore area that surrounds a lake or pond is inhabited by many predators, including mammals, birds, insects, fish, and amphibians. **Amphibians** are vertebrates that begin life in the water and later move to the land, for example, frogs, toads, and salamanders. Many types of migratory birds, such as ducks, geese, herons, and egrets, visit lakes because food usually is plentiful there. (See Figure 21-3.) Water snakes, mink, raccoons, and turtles prowl the water for prey. Muskrats, beavers, and otters are other mammals that make their homes in the water and along the shore.

Figure 21-3 Egrets are predatory birds that inhabit lakes and marshes where food is plentiful.

Marshlands

Marshes, swamps, bayous, potholes, and bogs are examples of freshwater wetlands. All are often referred to as **marshland**, an area that

DISCOVERIES

Mono Lake

Mono Lake, located in the Sierra mountains of western California, is a unique ecosystem. One of the oldest lakes in North America, it covers an area of about 155 square kilometers. Since the lake has no outlet, salts and other minerals that are carried in by mountain streams are left behind as water evaporates. The lake is now two and a half times saltier and 80 times more alkaline than seawater. This gives the water a soapy, or slippery feel.

Many calcium-bearing springs that well up through the lake are also rich in carbonates. The calcium and carbonate ions combine and precipitate as limestone. This forms underwater towers, called tufa, around the mouth of the spring. When the level of the lake drops, the tufa towers are exposed and stop growing.

Some have called Mono Lake dead, but in fact it abounds with life. Although only a few types of organisms can tolerate Mono's salty, alkaline water, those that do are present in great number. Microscopic green algae use sunlight and decayed organic matter from the lake to grow. Brine shrimp and brine flies feed on the algae. The shrimp and flies provide a plentiful food supply for the more than 80 species of migratory birds that visit the lake each spring and summer. The most common birds are grebes, phalaropes, gulls, and plovers. Phalaropes feed on flies from the surface of the lake or from the air above the water. Grebes dive for the brine shrimp that swim through the water. The gulls and plovers nest and rear their chicks along the shores, where food is plentiful.

Many cities in the western United States face severe water supply problems. Since 1941, Los Angeles has been diverting water from some of the streams that feed Mono Lake. California's agricultural areas also tap into this supply. These factors have caused the level of the lake to drop by about 12 meters and doubled its salinity. Concern about the shrinking lake's ecosystem has led to legislation to protect it from further destruction. Due to recent water rights decisions, some of the water that once flowed to the Los Angeles area is now use to keep Mono Lake healthy. The Mono Lake Committee and the city of Los Angeles have instituted long-term water conservation projects to protect the Mono Lake ecosystem and still meet the city's water demands.

Figure 21-4 **Marshes are important breeding grounds for many plants and animals.**

is partially or periodically covered with water during most of the year. Such wetlands serve as important breeding grounds for many plants and animals. Marshlands can form in any depression that holds water long enough for water plants to grow. They often form in aging lakes and ponds, in prairie potholes, and on the floodplains of rivers. (See Figure 21-4.)

In the shallower parts of a marsh, emergents such as grasses, bulrush, cattail, and arrowhead grow half in and half out of the water. Their roots and underground stems form networks that hold the soft mud of the marsh, making it stable. Reeds, wild rice, and water lilies grow in deeper water. Insects breed in the shallow water. In turn, the insects are fed upon by birds and fish. The birds in the marsh area may be migratory or permanent residents. Turtles, frogs, salamanders, and muskrats make permanent homes in the marsh. They escape unfavorable conditions in burrows deep within the mud.

The marshes that dot the North American prairie are important stopovers for many migratory waterfowl. Here they can rest and build up their strength by eating the plentiful plants and insects before they continue on their migration.

21.2 Section Review

1. Describe the factors that lead to the spring and fall overturns in a lake ecosystem.
2. Name the producer organisms in a lake.
3. Explain the importance of emergents to the stability of a marsh.
4. Why are prairie marshes important to migratory waterfowl?

21.3 FLOWING WATER

In any particular area of a river, conditions are usually uniform. But conditions may be quite different in other areas of the same river. At a river's source, called the **headwaters**, it begins as a fast-moving mountain stream fed by melting ice and snow. The icy, churning water is rich in oxygen and carries a large amount of sediment. The many areas of fast-moving water, called **rapids**, are interspersed with pools of quiet water.

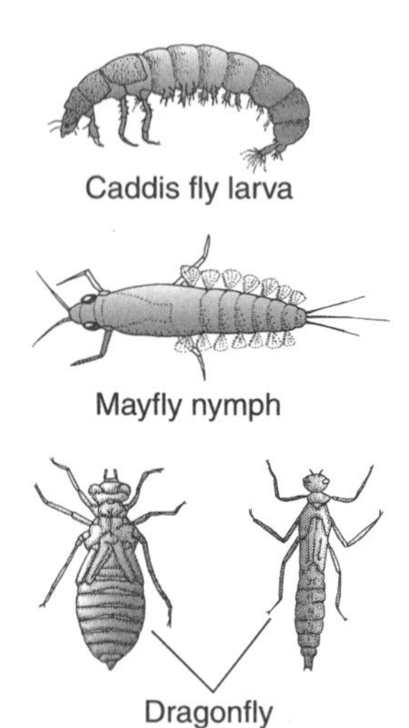

Figure 21-5 The larvae of many insects develop on and under rocks in fast-moving streams.

In the upper reaches of a stream, where there is ample sunlight and oxygen, aquatic organisms must contend with the force of the fast-moving water. Rooted plants are not found in the fast-moving water but are abundant in the slower water of shallow pools and along the shores. Algae and mosses cover the rocks of the stream with a slippery film. Plankton are scarce in cold, fast-moving water, but the larvae of many insects develop attached to or under rocks to avoid the rushing water. These insect larvae include dragonfly, caddis fly, mayfly, and midge. (See Figure 21-5.) Most of the larvae have gills, which enable them to extract oxygen from the water for respiration. They feed on the plentiful debris swept along by the current.

Fish such as trout, which are strong swimmers, frequent the pools between the rapids. They dart into the fast-moving water to snatch passing insects. The dipper, a bird with strong wings, flies through the water to the rocky stream bottom and searches for insects. The fauna here are adapted to life in the rushing water.

On the Floodplain

As the river encounters more level terrain, it slows, spreads out, and forms a wide valley. Here the river erodes its banks into a series of curves, called meanders. Still farther downstream, the river slows to a point where sediments and organic debris settle to the bottom. This forms a wide, flat valley, called a floodplain. Each year, spring floods deposit new sediments that enrich the valley floor. The sun warms the water. Warm water holds less oxygen than cold water. Therefore, there is less oxygen available for aquatic organisms.

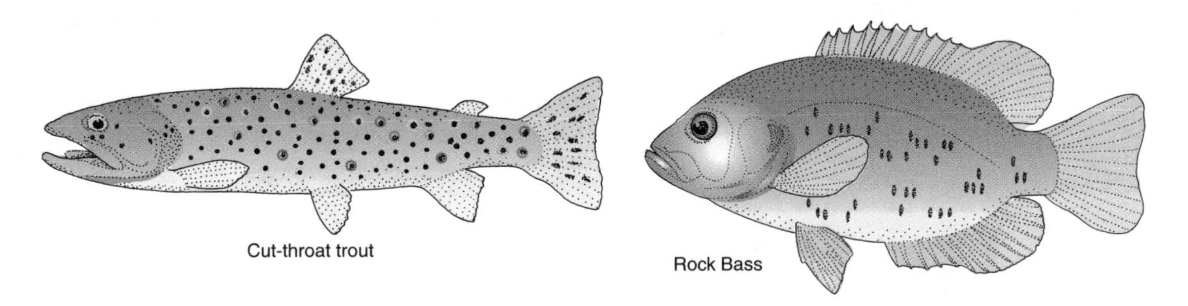

Figure 21-6 **Trout are round-bodied; they live in the headwaters and rapids of rivers. Bass are flat-bodied; they live in the floodplain.**

In slower-moving water, the river bottom contains nutrient-rich sediments. Plants are able to take root here, and many invertebrates burrow into the sediments. The water is warmer and flows slowly enough so that plankton can grow. The biota are similar to those found in lakes. In these waters, flat-bodied fish such as bass and bluegill can swim through the thick growth of plants in search of food. (See Figure 21-6.)

Rivers are important sources of drinking water for wildlife. This brings many animals to the tree-filled banks during the course of a day. Birds and mammals come to hunt for food, to drink, and to hide from predators. The damp riverbanks are the home for a variety of organisms.

Throughout the course of history, people have exploited floodplains for agriculture. The nutrient-rich sediments create fertile farmland, and fresh supplies of drinking water are readily available. This has led to the development of cities on the floodplains.

21.3 Section Review

1. Explain why the headwaters of a river usually contain very cold water.
2. Describe how insect larvae escape the force of a river's rushing water.
3. Why are floodplains used for agriculture?

LABORATORY INVESTIGATION 21

Sedimentation

PROBLEM: *How do sediments settle out of flowing water?*

SKILLS: *Measuring, observing*

MATERIALS: *1-L clear plastic bottle with cap, powdered clay or diatomaceous earth, gravel or pebbles, sand, water*

PROCEDURE

1. To represent river sediments, place a 2-cm layer of each of the following materials in a clear 1-L plastic bottle: powdered clay or diatomaceous earth, gravel or pebbles, sand.
2. Fill the bottle with water and cap it tightly.
3. Turn the bottle over and shake it gently to swirl all the sediments through the water.
4. Turn the bottle right side up, and place it on a table.
5. Allow time for the sediments to settle.

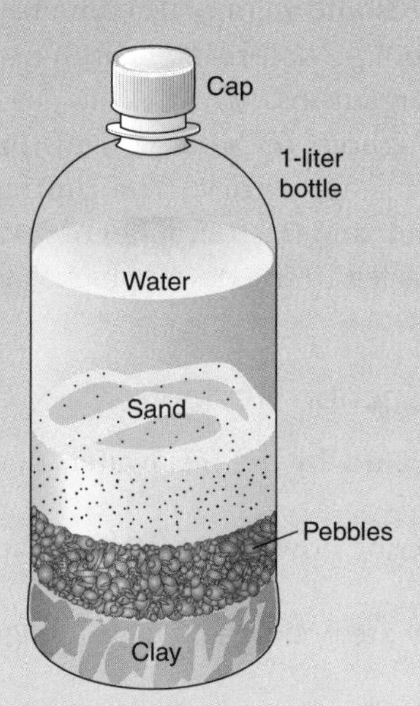

OBSERVATIONS AND ANALYSES

1. List the sediments in order of increasing particle size.
2. List the order in which the sediments settle.
3. Explain why the sediments settle in that order.
4. In what way is this demonstration similar to the sedimentation caused by rivers?
5. Why are fine silts deposited on floodplains?

Chapter 21

Review

Answer these questions on a separate sheet of paper.

Vocabulary

The following list contains all the boldfaced terms in this chapter.

amphibians, aquifers, emergents, fall overturn, groundwater, hard water, headwaters, marshland, potable water, rapids, spring overturn, water table

Fill In

Use one of the vocabulary terms listed above to complete each sentence.

1. _______ is water that is fit for people to drink.
2. The _______ occurs when the surface layer of a lake cools, becomes more dense, and sinks, displacing the less dense layers below it.
3. Vertebrates that begin life in the water and then move to the land are _______.
4. _______ are plants that grow half in and half out of water.
5. The source of a river's water is called its _______.

Multiple Choice

Choose the response that best completes the sentence or answers the question.

6. Which of the following is an example of standing water? *a.* river *b.* mountain stream *c.* inland sea *d.* natural spring
7. Which of the following is not a freshwater wetland? *a.* swamp *b.* bayou *c.* well *d.* bog
8. In a marshland, which plant is not considered an emergent? *a.* cattail *b.* arrowhead *c.* bulrush *d.* willow
9. A fish that is adapted to life in fast-moving mountain streams is the *a.* bass. *b.* trout. *c.* bluegill. *d.* salamander.
10. Which of the following is not an example of a source of freshwater? *a.* glacier *b.* ocean *c.* natural spring *d.* brook

Short Answer (Constructed Response)

Use the information you learned in this chapter to respond to the following items.

11. What is groundwater?
12. Explain how aquifers are able to hold large quantities of water.

13. What is hard water?
14. How do the spring and fall overturns affect the waters in a lake?
15. What is a spring?
16. Why do deep lakes and rivers almost never freeze all the way to the bottom during winter?
17. What are marshlands?
18. What is the difference between the flow of water in a river's headwaters and its floodplain?
19. How do the prairie marshes in North America aid the migration of waterfowl?
20. Describe the Mono Lake ecosystem.

Essay (Extended Response)

Use the information you learned in the chapter to respond to the following items.

21. Where does the water in wells and natural springs come from?
22. How are the flora of the lakeshore different from those of the deeper water?
23. Explain the difference in temperature between the headwaters and the water of the floodplain in a mountain stream.
24. List the sources of freshwater found in your area. How are they endangered? What can citizens do to help protect sources of freshwater?

Research Project

The health of our watersheds is important to all citizens. Many problems are now appearing that can severely curtail their use and impact on the amount of water available to us. You can help by identifying problems, notifying state and local government agencies of your findings, and exploring methods to help solve these problems. Investigate your watershed to identify any problems. Design projects to correct them.

CHAPTER 22
Saltwater Biomes

When you have completed this chapter, you should be able to:

List the factors that determine the type of marine ecosystem found in a part of the ocean.

Distinguish between the pelagic and benthic zones.

Describe sandy and rocky beaches, coral reefs, kelp forests.

Contrast the life zones of the salt marsh.

A day at the beach can be fun, as long as you do not get sunburned. Breathing the salty sea air is refreshing. Floating in the salt water is easier than floating in freshwater. As a matter of fact, it is almost impossible to sink in the Great Salt Lake or the Dead Sea, shown here, because the salt content of their waters is so high. The inhabitants of the saltwater, or marine, environment are different from those of the freshwater environment. The plants and animals of the marine environment are adapted to the special conditions found there.

22.1 THE MARINE ENVIRONMENT

The hydrosphere is 97 percent seawater, or salt water, which contains dissolved mineral salts. Geographically, the seawater part of the hydrosphere is divided into 4 vast interconnected oceans and 21 smaller seas. The four oceans are the Pacific, Atlantic, Indian, and Arctic. Table 22-1 lists the average depth of the oceans. A sea is a division of the ocean that is partly or totally enclosed by land. Some of the major seas are the Caribbean, Mediterranean, Bering, Ross, North, Caspian, Black, and Arabian.

As shown in Figure 22-1, there are two major divisions of the marine ecosystem: the **pelagic zone**, which is composed of the ocean waters, and the **benthic zone**, which is composed of the ocean floor. These zones are subdivided into smaller regions. The waters of the pelagic zone include the coastal waters above the continental shelf, called the **neritic zone**, and the deep waters beyond the continental shelf, called the **oceanic zone**. The benthic zone is composed of the **intertidal zone**, the region between the highest and lowest tide lines, and the **subtidal zone**, the region that begins below the lowest tide line and extends out to the continental shelf.

There are five factors that determine the type of marine ecosystem found in a body of water:

1. The concentration of dissolved salts, or salinity
2. The depth to which sunlight is able to penetrate
3. The amount of dissolved oxygen
4. The availability of essential nutrients
5. The water temperature

TABLE 22-1 AVERAGE DEPTH OF OCEANS

Ocean	Depth *(kilometers)*
Arctic	1.1
Atlantic	3.6
Indian	3.8
Pacific	4.0

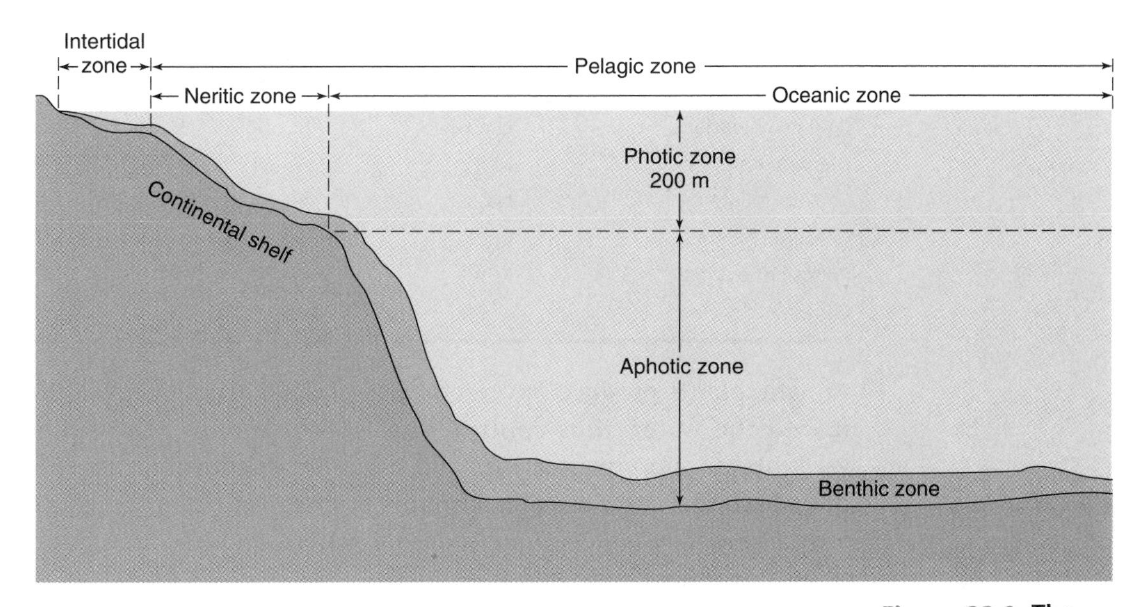

Figure 22-1 The pelagic zone includes the neritic and oceanic zones.

While not necessarily limiting the existence of life, each of these factors has an effect on the types of organisms the environment can support.

What Is Dissolved in the Seas?

The salinity, or concentration of salts, in the water of the open ocean is about 35 parts per thousand (ppt) but varies from one region to another. For example, the Red Sea, which has no source of freshwater, has a salinity of 46 ppt; while the Baltic Sea, due to a large inflow of freshwater from surrounding rivers and streams, has a salinity of only 12 ppt. In the tropics, ocean salinity can reach 40 ppt, due to evaporation. A concentration of 35 ppt tells you that there are 35 grams of dissolved salts in a 1000-gram sample of seawater. Therefore, 3.5 percent of the sample is dissolved salts, while the remaining 96.5 percent is water. The most common salt found in seawater is sodium chloride. Magnesium chloride, sodium sulfate, and calcium chloride are some of the other salts present in much smaller amounts. (See Table 22-2 on page 374.)

Dissolved gases are also found in most aquatic ecosystems. Gases in the atmosphere, such as oxygen and carbon dioxide, dissolve at the water's surface. The wind, waves, and rain aid this process.

TABLE 22-2 RELATIVE PERCENTAGES OF SALTS IN SEAWATER

Salt	Chemical Formula	Percent
Sodium chloride	$NaCl$	67.0%
Magnesium chloride	$MgCl_2$	14.6%
Sodium sulfate	Na_2SO_4	11.6%
Calcium chloride	$CaCl_2$	3.5%
Potassium chloride	KCl	2.2%
Other salts		1.1%

Aquatic plants produce oxygen during photosynthesis and release it into the water, thus contributing to the amount of dissolved gases. Respiration in aquatic plants and animals contributes carbon dioxide to the water. Greater amounts of dissolved gases are usually found in surface waters than in deeper waters, and in colder waters than in warmer waters.

Even though water is transparent, sunlight can penetrate only to a maximum depth of about 100 meters; this region is called the photic zone. Below this level there is total darkness; this region is called the aphotic zone. Only the upper 75 meters receives sufficient light to support photosynthesis in plants. The presence of suspended materials can greatly reduce the depth to which light penetrates. Organisms that inhabit the deep ocean often live in association with bacteria that produce chemical light. It is thought that this light aids in communication between members of the same species and attracts prey organisms.

Water temperature decreases with depth, and since dissolved salts lower the freezing point of water, temperatures in the deep are often a few degrees below freezing. Life can survive even in these frigid waters. Some organisms that live in such cold water, such as the Antarctic sculpin, have a compound in their bloodstream that acts like the antifreeze found in car radiators. This compound keeps their blood from freezing. The cold waters are often rich in nutrients and oxygen and support a diverse biota. In the waters of the Antarctic, seals, whales, penguins, and fish feed on an abundance of smaller organisms.

22.1 Section Review

1. Explain the differences in the divisions of the pelagic zone and benthic zone.

2. What factors determine the type of ecosystem a marine environment will support?
3. Describe the typical composition of seawater.

22.2 SANDY BEACHES

Beaches are the boundary between the land and the sea. They are hostile environments subject to unusually harsh conditions. Waves break against the shore, showering beaches with a salt spray; tides alternately cover the beach with water and expose it to the air; wind and waves continuously assault the beach; and wide fluctuations in temperature occur daily.

Beaches are formed by the action of waves and wind weathering and eroding coastal rocks. Beaches may be rocky or sandy, depending on the composition of the rocks that make up the shoreline and the force of the waves and wind against these rocks. Because of the differences in composition, there is a sharp contrast between sand and rock beaches. Each forms a unique ecosystem with distinctive flora and fauna. Most of the organisms of the sandy beach community live beneath the sand while those of the rocky shore community live attached to the surface of the rocks. In this section, you will learn about sandy beaches.

Supratidal Zone

The rise and fall of the tide along the coast creates three beach zones: the supratidal zone, the intertidal zone, and the subtidal zone. The **supratidal zone** is the beach region above the highest tide line. On most sand beaches, the supratidal zone has a dune area that rises above the high tide line. A **dune** is a mound, or hill, formed as sand is blown inland, away from the beach. Plants that grow on the beach and dune area are exposed to winds, extremes in temperature, and a constant spray of salt water.

Beach plants are specially adapted to withstand such heat, dryness, and exposure to salt spray. To decrease water loss, their leaves are small, contain few stomata, have waxy coverings, and are often silvery and hairy. They have extensive root systems for water storage. The roots also stabilize the sand around them. Many beach plants are **succulents**, plants with thick, fleshy stems and leaves that store water. These plants are able to absorb and store rainwater before it passes through the sand and mixes with the salt water below ground. Beach grass, beach plum, beach rose, sea rocket, and dusty miller are some common dune plants.

The spreading root systems of plants anchor the dunes, keep them from shifting, and prevent beach erosion. The aboveground parts of beach plants act as a windbreak, preventing the sand from blowing away. In those areas where there is enough moisture to prevent them from drying out, invertebrates live in the sand. Insects and crabs patrol the beach, searching for food, and in turn they are preyed on by a variety of shorebirds.

Intertidal Zone

The intertidal zone is the region between the highest and lowest tide lines along a shore. It is alternately covered by water and exposed to air during the course of a day. **Tides** are the rhythmic rise and fall of the ocean's water caused by the gravitational attractions of the moon and sun. Due to the motions of Earth and the moon, tides occur at different times each day. The time between successive high or low tides is 12 hours and 25 minutes. (You may wish to review Chapter 2, Section 3, which discusses the causes of the tides in greater detail.)

On a sandy beach, the strandline marks the upper limit of the intertidal zone. It is clearly defined by piles of natural debris, called **beach wrack**, which litter the area. Shells, seaweed, driftwood, and dead marine life washed in by the tide make up the beach wrack that collects at the strandline. Beach wrack is cool and moist and serves as a source of food and shelter for many beach organisms. Decomposers convert the organic remains into nutrients that enrich this area. Tiny crustaceans called beach fleas are common here. They live in cool, damp burrows in the sand and spend the evenings searching the beach for food. Many other small invertebrates live in the spaces between the grains of sand, while insects and crabs scurry along the beach in search of food.

In the intertidal zone, physical conditions change greatly, since the tides vary during the course of the day, month, and year. At the lowest area, the beach may be covered with water 99 percent of the time, while at the highest it may be exposed to air 99 percent of the time. In the middle of the intertidal zone, the beach is under water as often as it is exposed to the air. This creates differing environments for marine organisms that live within the intertidal zone. (See Figure 22-2.)

Figure 22-2 This illustration shows some characteristic organisms that live in the intertidal zone on a sandy beach.

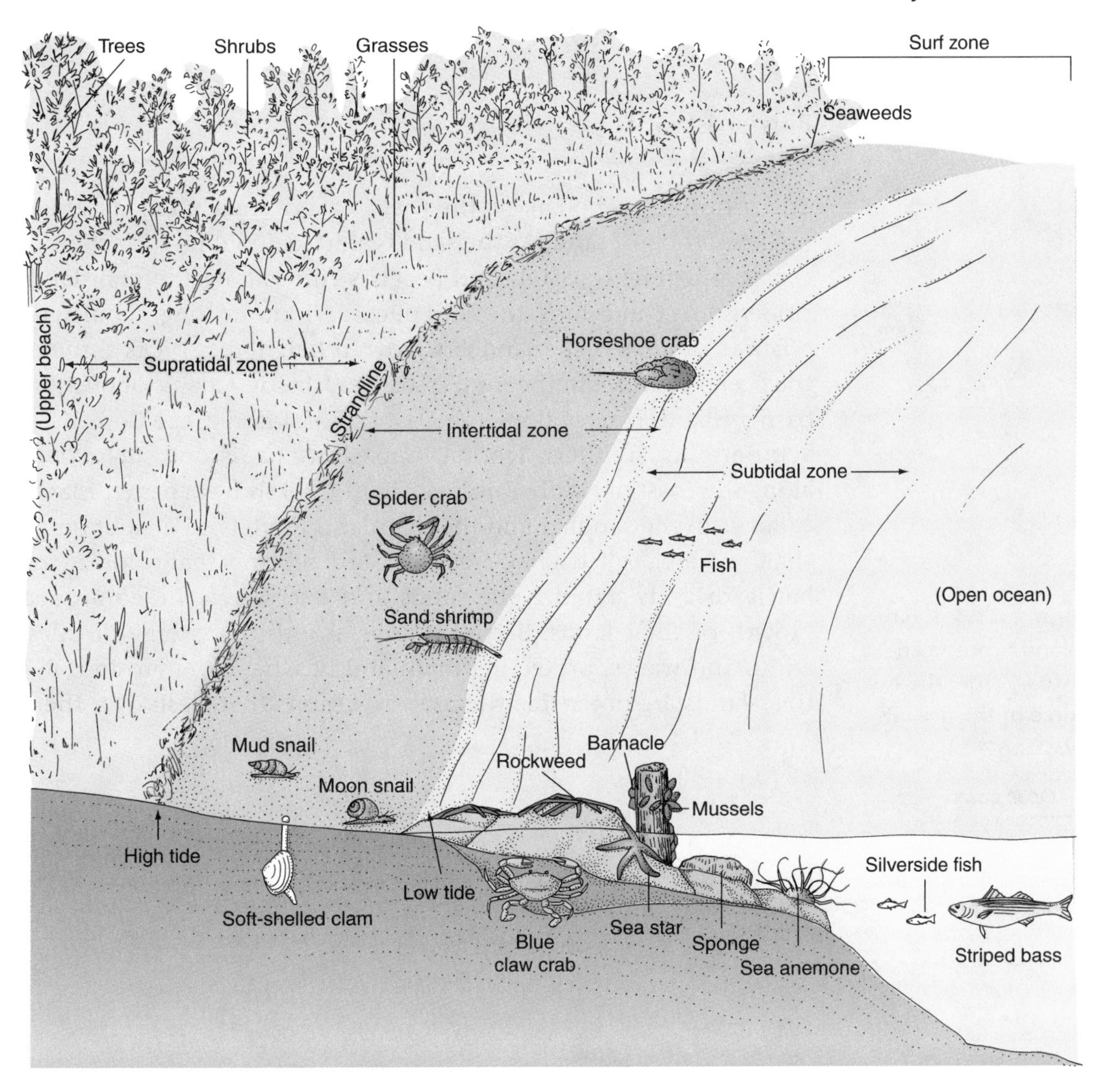

Where the beach remains damp through the day, many tiny organisms live between the sand grains. These include worms, algae, protozoa, and bacteria. The larger animals of the sandy shore escape exposure to sun and air by burrowing, constructing tubes, or digging tunnels in the sand. When the tide goes out, they retreat into their burrows and reemerge when the tide comes back in. The animals here include razor clams, soft-shelled clams, burrowing worms, mole crabs, sand dollars, and fiddler crabs. While the tide is out, birds such as sandpipers, sanderlings, and stilts explore the beach, using their long beaks to probe into the sand for food.

Barrier Islands

The continental shelf along the East Coast of the United States is a huge reservoir of sand. As the sand is shifted about by the action of waves and tides, coastal sand beaches form along the mainland. Under some conditions, the sand piles up, forming long, submerged offshore sandbars. The sandbars grow in height and break through the ocean's surface in the form of barrier islands. Chains of offshore barrier islands extend along the Atlantic coast from Cape Cod, Massachusetts, to southern Florida. Cape Hatteras is one such chain along the coast of North Carolina. Figure 22-3 shows a barrier island.

Between the sandbar and the mainland there is an area of quiet, shallow water, called a bay or lagoon. Beach plants that colonize the barrier islands stabilize the sands and encourage the continued growth of the islands. Barrier islands absorb the energy of tides, wind, and waves, protecting mainland beaches from erosion. But the islands are not stable themselves. Storms that wash over them

Figure 22-3 A barrier island protects an estuary from the full force of the ocean's waves.

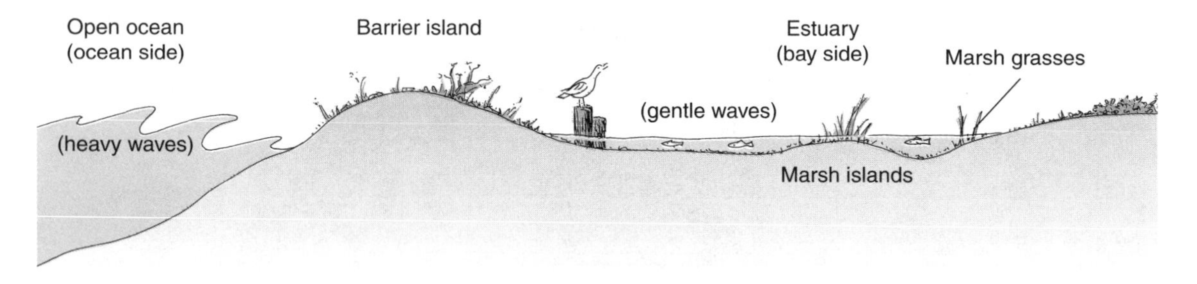

remove sand from the seaward side and transport it to the landward side. This has the effect of moving the islands closer to the mainland and making the lagoon smaller. (See Figure 22-4.) Longshore currents, which result from waves striking the shore at an angle, move sand along the length of the island and deposit it on the opposite end. Thus the entire barrier island migrates in the direction of the longshore currents.

Because of their nearness to water, moderate climate, and availability of recreational activities, barrier islands and beaches are attractive places on which to build homes. Miami Beach, Florida, and Jones Beach and Fire Island, New York, are barrier islands that are used for recreation by large numbers of people. Homes are built on

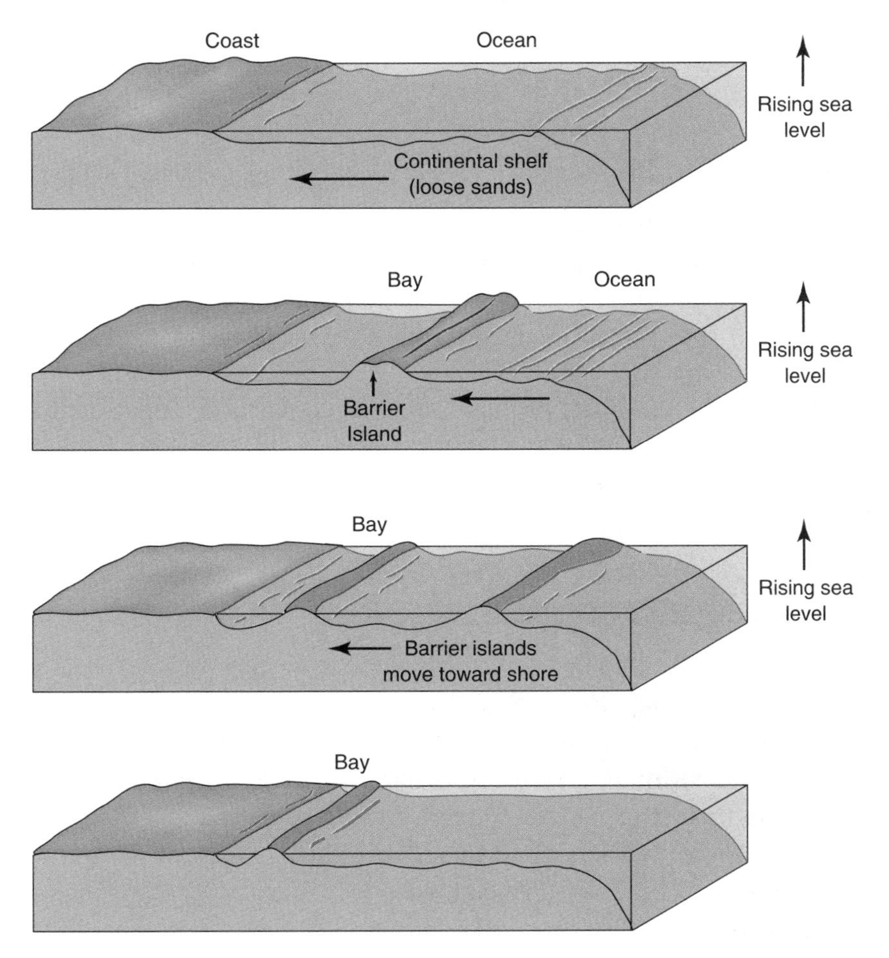

Figure 22-4 The migration of a barrier island.

the sand dunes that form these areas. However, the instability of the beaches and islands leads to a variety of environmental problems.

Erosion caused by severe weather can often drown beachfront property, washing away homes. The construction of breakwaters and jetties to prevent beach erosion only slows the process. The delicate beach grasses that hold sand in place are easily disturbed. Dune areas are quickly destroyed when these plants are trampled on by people or crushed under the tires of recreational vehicles. Without plants to stabilize the beach and dunes, erosion proceeds rapidly. Sand is often dredged from offshore areas to replace sand that has blown or washed away. However, without the stability of a beach plant community, it is of little use. Clearly, if we want to preserve our beach areas for the future, we must develop a strategy that protects them from their instability and the effects of erosion.

22.2 Section Review

1. How are beaches formed?
2. Explain the importance of beach grasses to the stability of dunes.
3. Explain why physical conditions within the intertidal zone vary from day to day.
4. Discuss the potential problems associated with the construction of permanent structures on barrier islands and beaches.

22.3 ROCKY SHORES

On a rocky shore, the area of the supratidal zone just above the high-tide line is subjected regularly to the salt spray from breaking waves. The animals and plants that live here form permanent attachments to the rocks. Moisture from the spray prevents them from drying out. Barnacles, algae, and periwinkles are common inhabitants of this area. (See Figure 22-5.)

The intertidal zone on rocky shores provides living spaces in crevices within and between the rocks. Space is the most prized

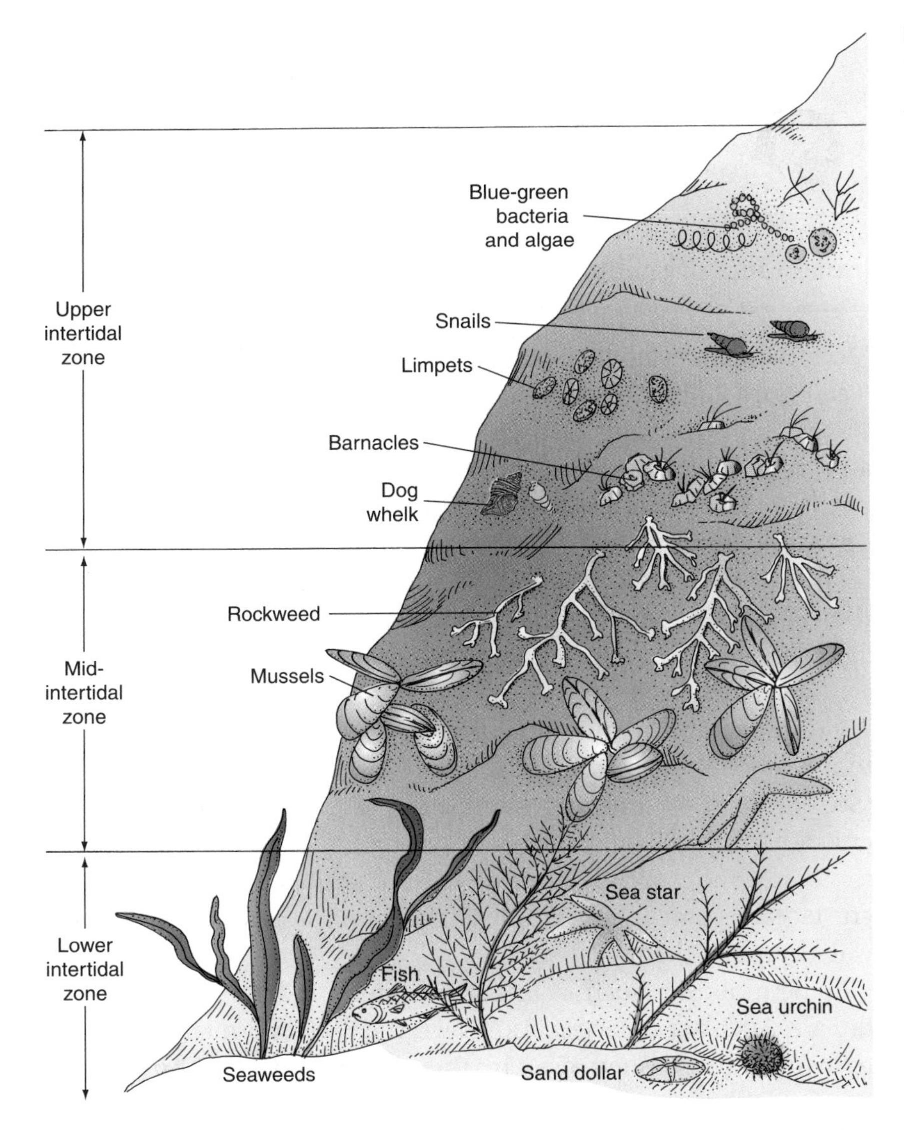

Figure 22-5 **This illustration shows some of the characteristic organisms that live on a rocky shore.**

resource in the rocky intertidal zone. Crevices offer organisms protection from the wind and sun. Organisms that live in the intertidal zone have adaptations that allow them to attach themselves to the rocks, so they will not be washed away by waves or blown away by the wind. Red algae, green algae, rockweed, sea palm, and lichens live where they can find secure attachments. Such attachments are not possible on a sandy beach, where constantly shifting sand prevents the development of permanent living spaces. The rocky shore community is more permanent than that of the sandy beach.

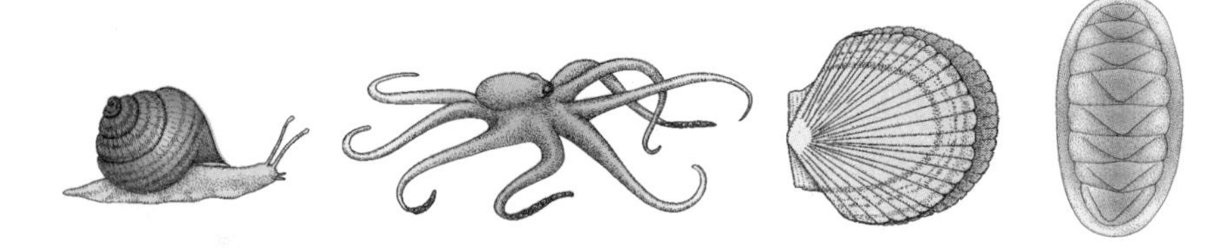

Figure 22-6 Even though they are very different in appearance, the snail, octopus, scallop, and chiton are all mollusks.

Inhabitants of the Rocky Shore

Mollusks are soft-bodied invertebrates, many of which have shells. (See Figure 22-6.) They are divided into several classes. One class is the gastropods. Many gastropods, such as snails, abalone, whelks, and periwinkles, have a single, coiled shell. Another class, the bivalves, such as clams, oysters, mussels, and scallops, have two opposing hinged shells. A third class, the cephalopods, which include the octopus, squid, and cuttlefish, have an internal shell that is greatly reduced in size. Cephalopods do not usually live along rocky shores. Chitons, which have heavily plated shells, form a class by themselves.

Crustaceans are marine arthropods, animals with hard outer skeletons and jointed legs, such as lobsters, shrimp, crabs, and barnacles. (See Figure 22-7.) Amphipods and isopods are two groups of tiny crustaceans that live in the waters along a rocky shore and also on the sandy shore. While the tide is in, many fish, birds, and larger crustaceans come to feed on the smaller crustaceans that live in the waters among the rocks.

Some mollusks (such as mussels, oysters, and chitons) and crustaceans (such as barnacles) also attach themselves firmly to the rocks to avoid being washed away by the pounding waves. Some organisms attach by means of a holdfast, a tough pad of tissue. When the tide goes out, these animals close their shells to avoid being dried by

Figure 22-7 Some typical crustaceans are shown here.

the sun. During low tide, periwinkles and shore crabs graze on the film of algae that covers such organisms and the rocks to which they are attached.

Tide Pools

Along rocky shores, there are **tide pools**, depressions within or between rocks that remain filled with water after the tide goes out. (See Figure 22-8.) Tide pools come in many sizes. Each pool is like an aquarium filled with an assortment of marine life. Anemones, barnacles, sponges, and coralline algae attach themselves permanently to the rocks. These permanent residents are able to adapt to the changing conditions within the tide pool. When the tide is in, sea stars, shrimp, small fish, snails, and crabs wander into the pool to feed. They may be trapped in the pool when the tide goes out and must remain there until the next incoming tide.

The size, depth, and location of a tide pool affect the makeup of the community it supports. The water in a tide pool undergoes changes in salinity, temperature, pH value, and the level of dissolved oxygen each time the tide changes. Those pools farther from the low-tide line will be cut off from the ocean for a longer time than pools that are closer to the low-tide line. Larger and deeper

Figure 22-8 How are some organisms adapted to live in tide pools?

pools warm more slowly when exposed to sunlight. Deeper pools are generally more stable than shallow ones. The many variables create different environments. Thus each pool contains a slightly different community from those around it.

22.3 Section Review

1. How does life on a rocky shore differ from life on a sandy beach?
2. How do organisms of a rocky shore protect themselves from drying when exposed to sun and air?
3. Describe the environment in tide pools.

22.4 WETLANDS

Wetlands are areas where the water table, the upper limit of the zone of saturated soil or rock, is at or near the surface of the land. These areas are covered, or saturated, with water for most of the year. There are two types of wetlands, inland and coastal. Inland wetlands can be found in almost any area, whereas coastal wetlands occur only along the ocean shores.

Coastal wetlands are composed of shallow saltwater pools or lagoons, called salt marshes, and their soft mud beaches, called mudflats. Wetlands form in areas where rivers or bays enter the sea and where the water is sheltered from the effects of the wind and waves. The calm conditions allow fine sediments, such as silt and sand, to settle out of the water and form mudflats. **Silt** is composed of tiny particles of organic and inorganic material. Tidal creeks and freshwater streams often meander through the mudflats, dividing the land into small islands.

Because coastal wetlands are affected by the tides, they also are called tidal marshes. As the tide comes in, it floods the shore. When the tide retreats, it leaves behind layers of muddy sediments. Freshwater, brought in by streams and heavy rain, dilutes the salt water of the marsh area. This reduces the salinity of the marsh water, producing brackish water. **Brackish water** has a lower salinity than seawater.

Coastal wetlands are among nature's most productive and useful ecosystems. Like sponges, they soak up and hold large quantities of water. They are natural filtering systems that maintain water quality by decomposing wastes and regulating the nutrient content of the water. Coastal wetlands reduce erosion of the shoreline and protect coastal areas from being washed away. Marine fish, shellfish, waterfowl, and mammals live in the coastal wetlands. Most commercial and game fish use these areas to spawn, that is, to lay their eggs, and as nurseries for their young.

The shallow, open waters of the salt marsh, which receive abundant sunlight, provide excellent growing conditions for producers such as green plants and algae. Marsh grasses such as cordgrass, salt hay, reed grass, and pickle grass grow in the area along the shore. Algae grow on the surface of the mud. Seaweeds float in the water among the marsh grasses. The plants stabilize the mud, which allows the further accumulation of silt and sand. This expands the marsh area. The incoming tide brings fresh nutrients for the plants, and the outgoing tide carries away the detritus.

Detritus forms as marsh grasses and algae die. Bacteria break down this material, forming a nutrient-rich mixture that is an important food source for microorganisms and the young of many marine organisms. These in turn are preyed on by larger organisms. Filter feeders, such as clams, barnacles, and mussels, strain the nutrients from the water. Crabs, snails, and worms live in the mud along the shore, and many types of fish prowl the waters. All these creatures depend on the producers within the salt marsh for food.

Life Zones in the Marsh

The rise and fall of the tide creates different life zones within the marsh. The part farthest from the water, or upper area of the marsh, is the salt barren. This area usually is covered with water only once each month. Thus the soil is exposed to the sun and air for long periods. The result is a dry soil that has a high salt content. Few plants can grow in this soil, and those that do are stunted. The high marsh is an area that is briefly covered with water each day, while the low marsh is underwater for many hours during the day. Dense plant growth in these areas traps detritus and forms a spongy soil layer. Many animals make their home in the muddy soil, regulating their lives to the rise and fall of the tides.

Mudflats lie closer to the water than the low salt marsh. They appear barren because few plants are able to take root here. Some seaweeds attach themselves to fragments of shells, pebbles, and rocks. They appear as small green clumps that grow on the mud. Microscopic algae grow on the surface of the mud. Clams, worms, and crabs live in burrows beneath the mud's surface. Figure 22-9 illustrates the zones of the salt marsh.

Since there is little wave action in the waters of the salt marsh, the fine particles of silt and sand that make up the soft mud are held close together. Therefore, little or no air can penetrate the mud. Except for the upper 1 centimeter, the oxygen content of the mud is very low. Anaerobic bacteria, which live without oxygen, are abundant in the mud. They decompose organic material to release hydrogen sulfide, which smells like rotten eggs. When the tide is out and the mud is exposed to the air, the odor of hydrogen sulfide is unmistakable.

The lack of oxygen in the mud causes a problem for burrowing animals. They must extend breathing tubes to the surface of the mud and extract oxygen from the water or air. Soft-shell clams, which bury themselves deep in the mud, extend long siphons

Figure 22-9 The zones in a salt marsh provide homes for different creatures.

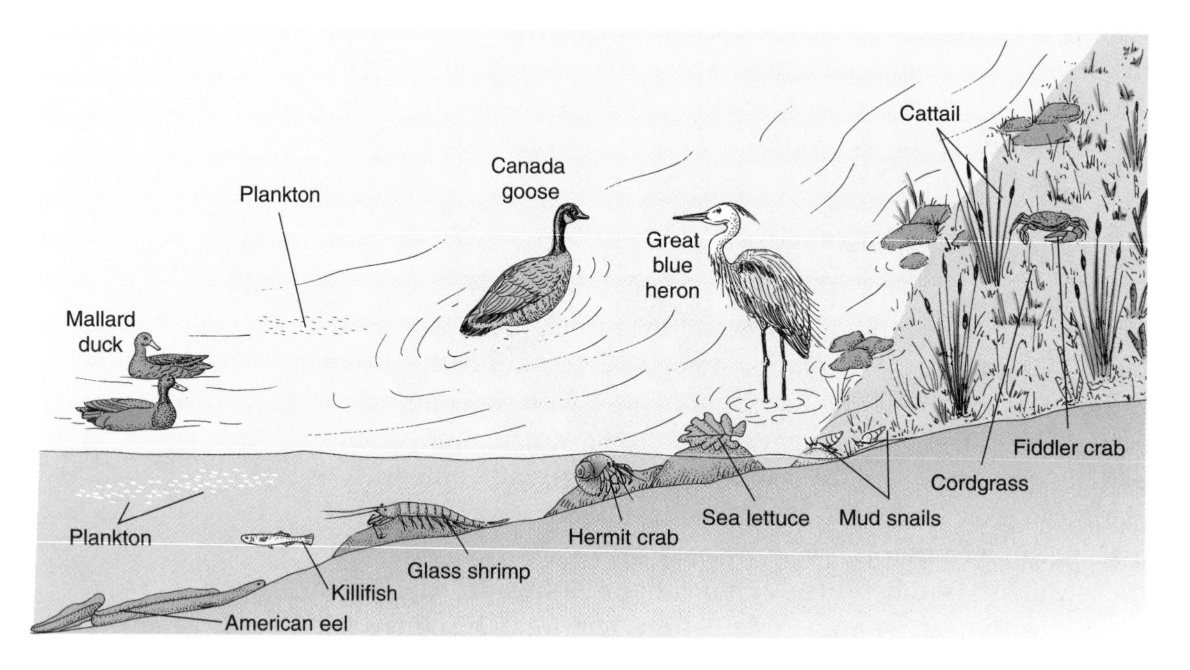

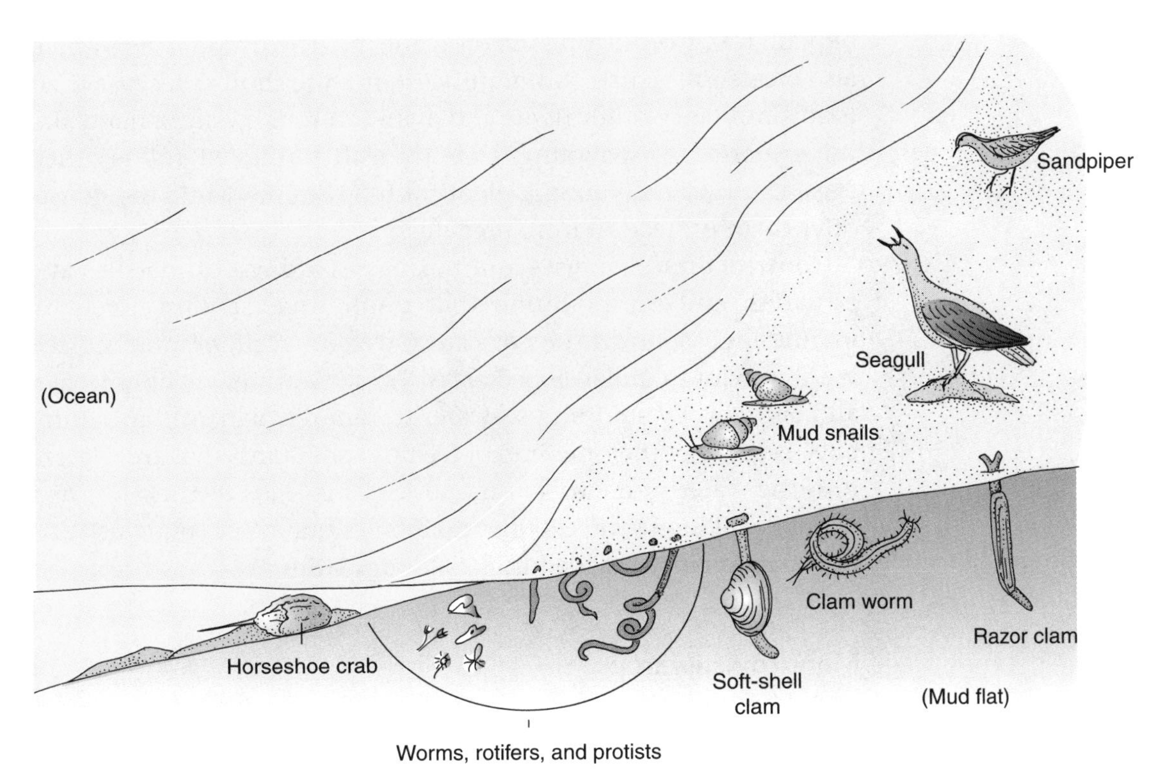

Figure 22-10 Many different kinds of animals live in mud flats.

through the mud into the water. They feed and breathe by pumping water from the siphons through their soft bodies, filtering out plankton and detritus for food and absorbing oxygen. Hard-shell clams and razor clams live buried in the mud but have shorter siphons. Snails, worms, and crabs also live buried under the surface of the mud; they feed on the plentiful microorganisms. (See Figure 22-10.)

Estuaries

An **estuary** is the area where the freshwater of a river joins and mixes with the salt water of the ocean. The water here is usually brackish. Where the waters meet, the freshwater, which is less dense, overrides the salt water, forming two distinct layers. Where the layers meet, they mix and form a middle layer of brackish water.

Estuaries are important areas for many marine organisms. Rivers transport large amounts of suspended particles in the form of silt, sand, and detritus to the estuary. The detritus promotes the growth of plankton, free-floating microorganisms. The nutrient-rich waters also create salt marsh communities along the shores. Because there is an abundance of nutrients and plankton in the waters, many fish use estuaries for spawning. Here the fish can breed and lay their eggs, insuring the young a plentiful food supply while the grasses offer them protection from predators.

Construction of houses and businesses along salt marshes and estuaries, and the pollution that results, has serious effects on marine life. Mollusks such as clams, oysters, scallops, and mussels are abundant in undeveloped areas. These creatures are filter feeders and get food by straining seawater to remove plankton and nutrients. Because of their method of feeding, pollutants that are released into the water often poison mollusks. This leads to their loss as a commercial food source. Other pollutants can affect young fish and thus dangerously reduce future fish populations.

Mangrove Swamps

A mangrove swamp is an ecosystem found along muddy, tropical coasts that is dominated by mangrove trees. (See Figure 22-11.) Mangroves are leathery-leafed trees that grow in salty water. Mangroves have specialized roots that grow upward to absorb oxygen from the air. Beginning above the water line, mangroves have other roots called prop roots, which grow from the trunk. These give the tree stability by growing out to the side and into the soil below the water to form a tripodlike support.

Mangrove seeds start sprouting before they drop from the tree. The seeds are spear shaped so they stick in the mud and take root after they fall and float away from the parent tree. The growth of new trees creates a tangled network of roots that traps sand and soil brought in by the tide. As seeds drop and take root, the mangrove swamp spreads. Mangrove swamps are nurseries for many forms of marine life, since they provide shelter and food among their tangled roots.

Mudskippers are small fish that live on the muddy tidal flats of some mangrove swamps. These fish are able to leave the water to prowl the mud flats for food. Some graze on the algae that cover the mud. Others crawl about the flats or climb mangrove roots to hunt

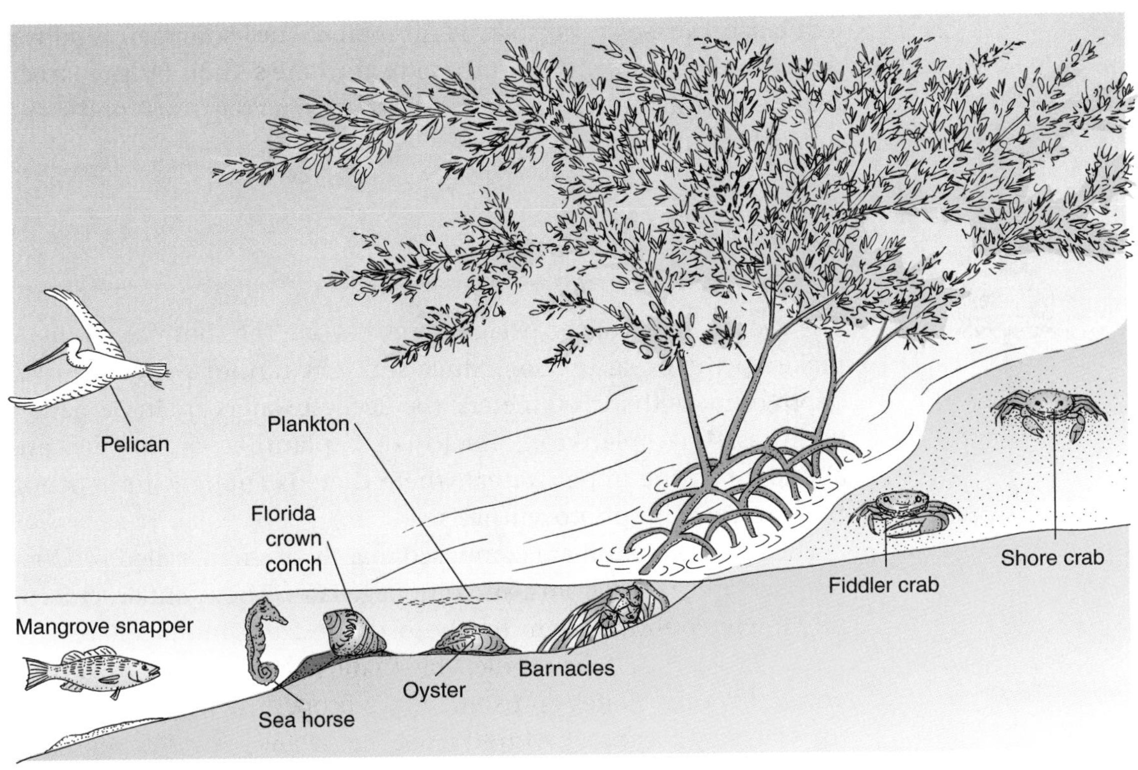

Figure 22-11 The organisms shown here are characteristic of those that live in a mangrove swamp.

insects, worms, and small crustaceans. They must keep their skin and gills moist to aid in respiration.

22.4 Section Review

1. How do burrowing animals breathe in the oxygen-poor mud of the salt marsh?
2. Why are estuaries important breeding places for fish?
3. Why are mangrove swamps important nurseries for marine life?

22.5 OCEAN WATERS

The largest concentration of marine life occurs in the neritic zone, those waters of the pelagic zone that cover the continental shelf. Algae and plant life are abundant in the shallow areas where sunlight can reach the ocean bottom. Plant material is a source of food for many microorganisms, and these in turn are fed on by the larger marine organisms. It is in the neritic zone that "deep-sea" and commercial fishing occurs.

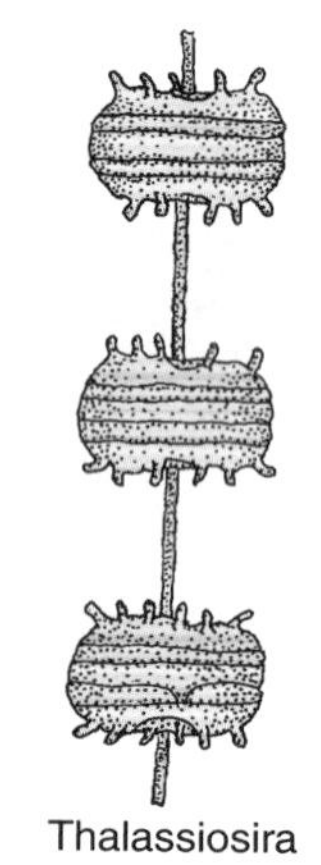

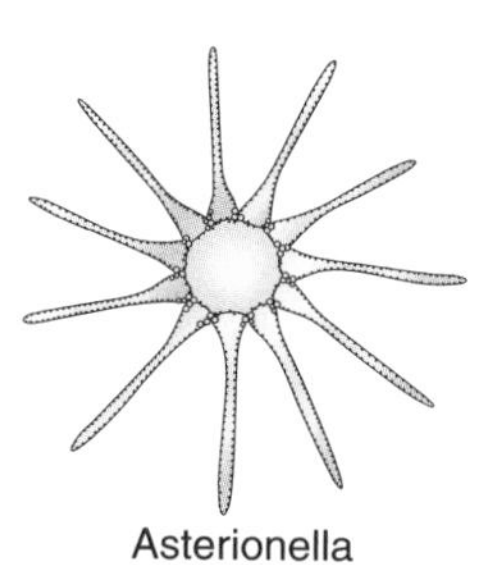

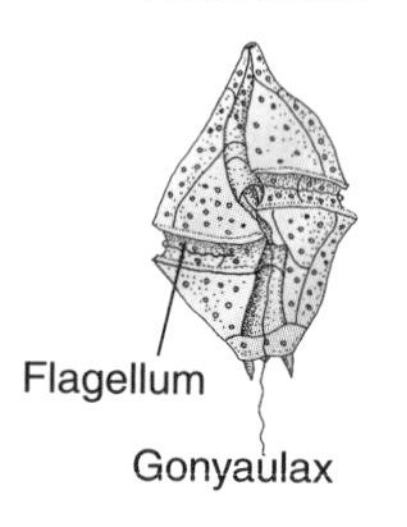

Figure 22-12 Phytoplankton are the dominant food producers in the open ocean.

Life in the Oceanic Zone

The deeper waters of the pelagic zone, beyond the continental shelf, make up the oceanic zone. Since sunlight cannot penetrate to a depth of more than 100 meters, the deepest waters are in perpetual darkness. Phytoplankton, microscopic plantlike organisms, are abundant in the upper waters where there is enough sunlight for them to carry on photosynthesis.

The phytoplankton are composed mainly of single-celled producers, such as algae, diatoms, and dinoflagellates. They contain chlorophyll, which allows them to photosynthesize. Dinoflagellates are phytoplankton that are motile, able to move about under their own power. Dinoflagellates and some algae propel themselves through the water by means of whiplike flagella. Diatoms and other algae are immotile; they drift about, carried by ocean currents.

A diatom is a single-celled alga enclosed in a silica shell. Silica, or silicon dioxide, is a compound found in glass and sand. Diatoms absorb silica from the water and then secrete it in the form of a glassy shell. When diatoms die, their shells sink to the ocean depths, where they accumulate over time to form thick layers of sediment.

Phytoplankton are the dominant food producers in the open ocean. In fact, they carry on more photosynthesis in the oceans than do all the green plants that cover the rest of Earth. Phytoplankton are responsible for the release of huge quantities of oxygen into the atmosphere as well as into the water. (See Figure 22-12.)

Zooplankton, microscopic animals, are the most plentiful fauna in the sea. There may be as many as half a million in 4 liters of seawater. They feed on the phytoplankton. Zooplankton range in size from single-celled protozoans to multicellular jellyfish. Among the microscopic zooplankton are radiolarians and foraminiferans, or forams. These are shelled protozoa related to the ameba.

Radiolarians produce beautiful silica shells. Forams absorb carbon

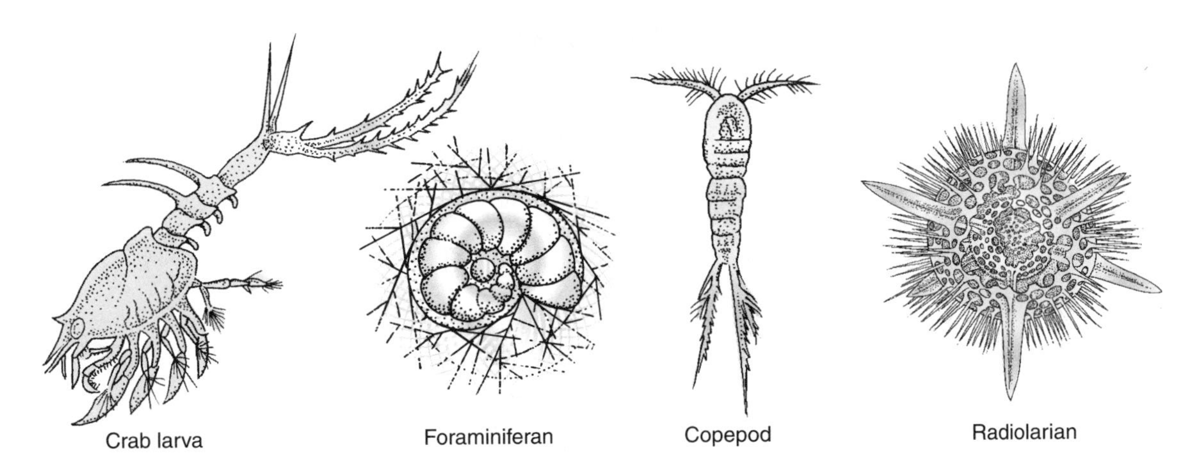

Figure 22-13 Some zooplankton are microscopic. These are color-enhanced.

dioxide from the water and use it to secrete calcium carbonate shells. The White Cliffs of Dover in England are made of limestone, a chalklike rock formed from the countless foram shells that were deposited on the ocean floor over many millions of years. (See Figure 22-13.)

The larvae and eggs of fish, crustaceans, mollusks, worms, and sponges are also part of the oceanic zooplankton. These organisms use the open ocean water as a source of food and a means to disperse their young. The eggs and young are usually produced in large enough numbers to insure that at least some of them will survive to maturity.

Larger zooplankton include copepods and krill, relatives of crabs, shrimp, and lobsters. While both types of crustaceans are found in large numbers, copepods are the most common form of zooplankton. They are among the most important food sources in the oceanic ecosystem and are the main food of small fish such as anchovies and sardines and the young of many larger fish. Huge filter feeders such as the blue whale, right whale, basking shark, and manta ray feed by straining large numbers of krill and copepods from the water.

The organisms that drift about in the open ocean are called plankton. Most cannot swim, and all are too small to resist being carried along by ocean currents. The animals that can swim and control their motions are known as **nekton**. These include fish, whales, squid, seals, penguins, and turtles. They employ various swimming methods to find and catch prey. The leatherback turtle is a slow swimmer that eats drifting jellyfish. The bluefin tuna, with its great

DISCOVERY

Gulf of Mexico Dead Zone

There is a dead zone in the Gulf of Mexico off the coast of Louisiana. This lifeless zone extends from the Mississippi River to the Texas border. In this region of low-oxygen-content (hypoxic) water, most marine life has difficulty surviving. The ecosystem has been severely disrupted. Hypoxic conditions arise when there is not enough oxygen in the water to sustain animal life in the bottom layer of the ocean's water. On the surface, hypoxic water appears normal. However, below the surface, the bottom is covered with dead marine organisms.

The dead zone first appeared in the 1970s. It usually reappeared every two or three years. Recently, it has occurred annually. During the summer of 2001, the zone extended into the waters off the coast of Texas. From the Mississippi River Delta, the zone extended more than 80 kilometers (50 miles) into the Gulf of Mexico. The zone covered more than 20,000 square kilometers (8000 square miles), which is the largest area that it has covered. Hurricanes and tropical storms break up the zone in late August or September.

Scientists think that the main cause of the oxygen depletion is the runoff of nitrogen-rich fertilizer from midwestern farmlands. The runoff drains into the Mississippi River, which in turn carries it to the Gulf of Mexico. Similar dead zones occur worldwide where sewage or fertilizer enters the ocean. Excessive nitrogen leads to eutrophication, the takeover of nutrient-rich surface waters by blooms of phytoplankton or other marine plants. A bloom is an increase in the population of a species that is well above the normal numbers. These organisms use the oxygen in the water, reducing the amount available for the rest of the ecosystem. If the nutrient pollution is not reduced, fish and shellfish will be replaced by anaerobic bacteria, which do not need normal levels of oxygen. Hypoxic conditions cause changes in food chains, loss of biodiversity, and a high rate of death in marine species. Scientists theorize that the dead zone could severely affect the marine fisheries industry in the gulf.

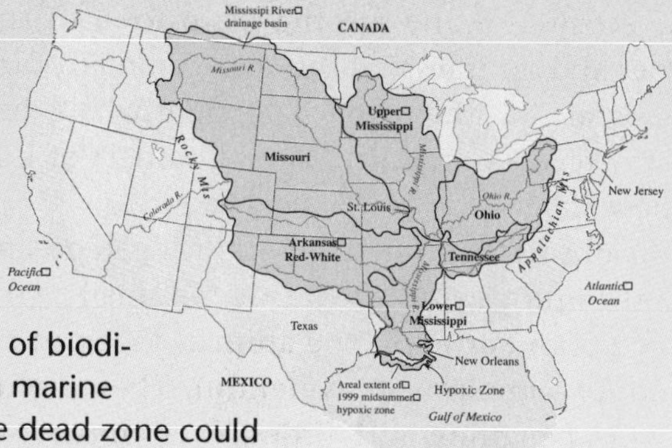

bursts of speed, attacks schools of mackerel and herring. **Benthos** are those organisms that live in or on the ocean bottom. They burrow, crawl, and walk on the ocean floor or live permanently attached to rocks. Benthos include worms, sea stars, lobsters, crabs, anemones, urchins, sponges, and mollusks.

Coral Reefs

A coral reef is a structure composed of the solid remains of countless tiny animals called coral polyps. Coral reefs grow only in the warm, shallow waters of the tropics. A coral polyp has a soft body topped by a mouth surrounded with tentacles. The polyp uses its tentacles to catch plankton and draw them into its mouth. Each polyp is enclosed in a calcium carbonate skeleton. As the coral polyp grows, it continues to secrete material and build up its hard skeleton. When the polyp dies, the soft parts decompose and the hard skeleton is left behind. New polyps grow on this base of coral skeletons. Thus the reef is slowly built up on the accumulated skeletons of dead coral polyps, covered with a thin layer of living polyps.

The Great Barrier Reef off the coast of Australia is more than 2000 kilometers long, but most reefs are much smaller. Corals grow only in warm waters that have rocky bottoms to which they can attach their skeleton. The bottom must be no more than 100 meters deep, and the water must be clear of sediments to allow sunlight to penetrate. Sunlight is necessary because algae live in the cells of coral polyps. The algae need sunlight to carry on photosynthesis and to supply nutrients to the coral polyp.

Hard corals grow in many different shapes. Brain coral is globular, staghorn coral is branched, and lettuce coral forms flat sheets. Growing among the hard corals are soft corals, such as sea fans, sea pens, and whip corals, as well as many types of sponges. The coral reef is honeycombed with crevices, holes, and chambers. These openings provide living spaces for a large variety of marine organisms, such as fish and lobsters. Coral reefs are the most biologically diverse areas in the marine ecosystem and, as such, often are called the rain forests of the sea. (See Figure 22-14 on page 394.)

Many animals feed on the coral polyps. The crown-of-thorns starfish, which eats coral polyps, has caused the devastation of

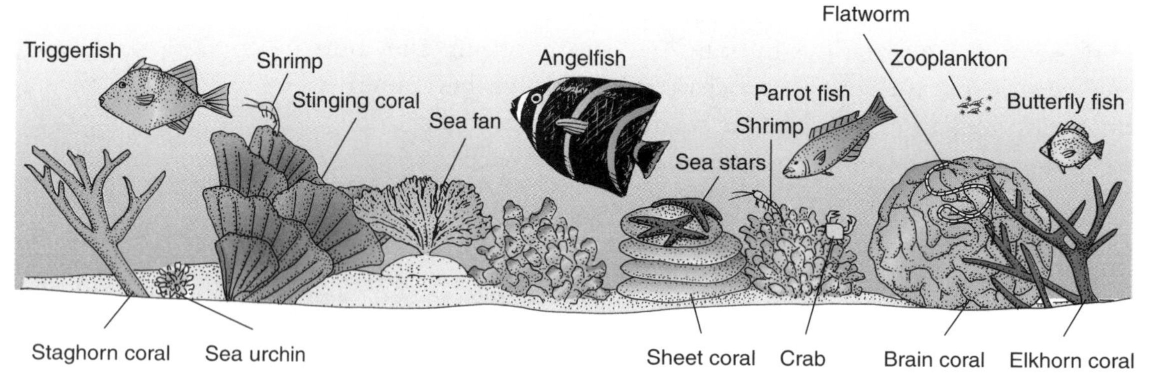

Figure 22-14 Why are coral reefs called the rain forests of the sea?

many Pacific reefs. Parrot fish have a sharp beaklike mouth with which they bite off and chew up chunks of coral. They eliminate this pulverized, sandy material along with their fecal waste. As a result, thick layers of calcium carbonate sand, produced by parrot fish, surround many coral reefs.

Although they are stony and look strong, coral reefs are very delicate. Pollution, sediments, temperature changes, and damage from boats can destroy the thin film of living polyps on the surface of the coral reef. This in turn can have disastrous effects on the whole coral reef ecosystem.

The Kelp Forests

Giant kelp is a brown alga that often grows to a height of 60 meters. It is a tough, leathery seaweed that anchors itself to rocky bottoms with a holdfast. Some species of kelp grow up from the ocean floor to the surface, because they have air-filled sacs, called bladders, located on each leaflike frond. (See Figure 22-15.) Kelp is found off rocky coasts where the water is cold. Giant kelp forms huge underwater forests that provide food, protection, and living space for a great variety of marine organisms including sea otters, seals, and sea lions. They feed on the plentiful fish, sea urchins, abalone, and crabs.

Sargassum and *fucus,* or rockweed, are other important brown algae. Many marine organisms are dependent on these seaweeds. Seaweeds are of economic value since they are used as sources of iodine, potash, vitamins, and algin. Algin is an important ingredient used as a thickener in syrups, salad dressings, and ice cream. Kelp

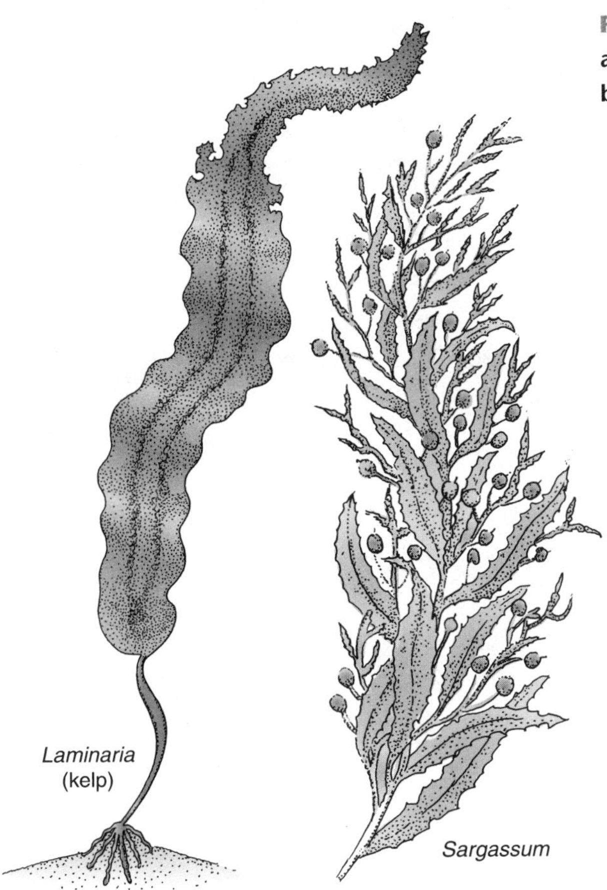

Figure 22-15 Kelp and *Sargassum* are brown algae.

is also used as food for humans and livestock. The harvesting of seaweeds can affect the other organisms within the marine ecosystem. However, kelp is a very fast-growing alga and can be regularly harvested. Other seaweeds are now commercially farmed.

22.5 Section Review

1. How do radiolarians and foraminiferans differ?
2. Why are coral reefs found only in shallow water?
3. Explain why some types of kelp fronds float.

LABORATORY INVESTIGATION 22

The Effect of Salt on Water Density

PROBLEM: *How does the addition of salt affect the density of water?*

SKILLS: *Measuring, observing, inferring*

MATERIALS: *3 unsharpened wooden pencils with erasers, water, sodium chloride (or sea salts), 500-mL beakers (3), china marking pencil, 3 thumbtacks, graduated cylinder, triple-beam balance*

PROCEDURE

1. Stick a thumbtack into the eraser of each pencil. The pencils all should be the same length. The tack will serve as a weight to keep the pencil floating upright in the water.
2. With the marking pencil, label the beakers A, B, and C.
3. Place 400 mL of water in beaker A.
4. Place 400 mL of water and 2 grams of sodium chloride or sea salts into beaker B.
5. Place 400 mL of water and 5 grams of sodium chloride or sea salts into beaker C.
6. Place a pencil in each beaker of liquid so that it floats eraser end down. Observe the level in the beaker to which each pencil sinks. With the marking pencil, mark this level on the side of the beaker.

OBSERVATIONS AND ANALYSES

1. In which beaker did the pencil float highest?
2. Based on the results of this activity, which is more dense, freshwater or seawater?
3. How does the addition of salt affect the density of seawater?
4. Why is it so easy to float in the water of the Dead Sea in Israel and the Great Salt Lake in Utah?

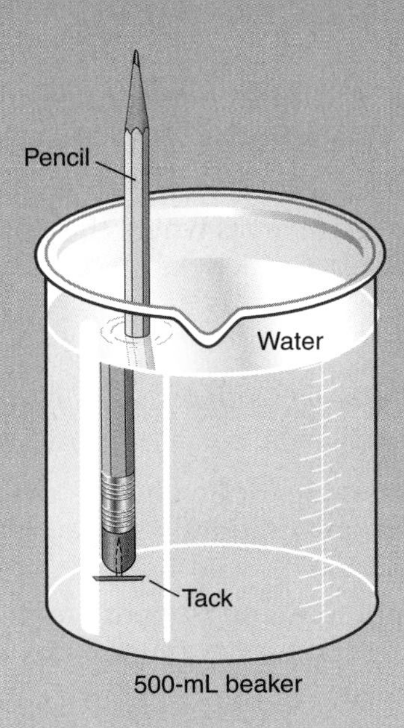

Chapter 22 Review

Answer these questions on a separate sheet of paper.

Vocabulary

The following list contains all the boldfaced terms in this chapter.

beach wrack, benthic zone, benthos, brackish water, dune, estuary, intertidal zone, nekton, neritic zone, oceanic zone, pelagic zone, silt, subtidal zone, succulents, supratidal zone, tide pools, tides, wetlands

Fill In

Use one of the vocabulary terms listed above to complete each sentence.

1. The beach area that is never under water is the ________.
2. The region between the highest and lowest tide lines is the ________.
3. A mound, or hill, of windblown sand is called a(an) ________.
4. The rhythmic rise and fall of the ocean's water are the ________.
5. Productive areas where the water table is at or near the land's surface are called ________.

Multiple Choice

Choose the response that best completes the sentence or answers the question.

6. Beaches are formed by the action of *a.* people and animals. *b.* moon and tides. *c.* waves and wind. *d.* people and tides.
7. Amphipods and isopods are examples of *a.* crustaceans. *b.* copepods. *c.* phytoplankton. *d.* mammals.
8. The water of the salt marsh contains a nutrient-rich mixture of dead and decomposing organic material called *a.* krill. *b.* phytoplankton. *c.* diatoms. *d.* detritus.
9. In the open ocean, the dominant producers are *a.* zooplankton. *b.* phytoplankton. *c.* radiolarians. *d.* shrimp.
10. The larvae of fish, crustaceans, mollusks, and worms are part of the *a.* zooplankton. *b.* phytoplankton. *c.* radiolarians. *d.* foraminiferans.
11. The two major divisions of the marine ecosystem are the *a.* neritic and oceanic zones. *b.* intertidal and subtidal zones. *c.* benthic and neritic zones. *d.* pelagic and benthic zones.

12. Which of the following is the most common salt found in seawater? *a.* magnesium chloride *b.* sodium sulfate *c.* sodium chloride *d.* calcium chloride
13. Which of the following are inhabitants of the rocky shore? *a.* razor clams, mole crabs, sand dollars *b.* sandpipers, sanderlings, stilts *c.* sharks, whales, rays *d.* amphipods, periwinkles, mussels
14. Coral reefs are *a.* rock formations found in the deep sea. *b.* made up of the skeletons of coral polyps. *c.* formed only in cold water. *d.* of volcanic origin.
15. Which statement about giant kelp is *not* true? *a.* It is a type of brown alga. *b.* It grows only in warm water. *c.* It forms underwater forests. *d.* It is a tough, leathery seaweed that anchors itself to rocky bottoms.

Short Answer (Constructed Response)

Use the information you learned in this chapter to respond to the following items.

16. What are the divisions of the benthic zone?
17. What is the salinity of the water in the open ocean?
18. Describe how beaches form.
19. How are beach plants adapted to withstand heat, dryness, and salt spray?
20. What is the cause of tides?
21. Explain how a barrier island forms.
22. How are the organisms of a rocky shore different from the organisms of a sandy shore?
23. What are crustaceans?
24. What is an estuary?

Essay (Extended Response)

Use the information in the chapter to respond to these items.

25. In what ways are beach plants similar to desert plants?
26. Describe the conditions necessary for the formation of a salt marsh.
27. How do coastal wetlands maintain the water quality in ecosystems?
28. List the problems faced by humans in a marine biome. How are oceanic fish and birds adapted for these saltwater conditions?

UNIT SIX

HUMAN POPULATION

In the year 1600, just seven years before the founding of the Jamestown colony in Virginia, scientists estimate that Earth's population was about 500 million people. In 1850, 250 years later, the population had doubled to 1 billion people. It took just 80 years to double again, reaching 2 billion people in 1930. By 1976, only 46 years later, the population had doubled once again, reaching 4 billion people. The next doubling to 8 billion may be reached by 2036, in 60 years. As you can see, the time it takes Earth's population to double decreased, but now appears to be increasing. What does this increasing human population mean to the biosphere?

In this unit, you will learn how humans evolved and how societies developed. You will examine the causes of population's growth. You will explore the effect of the human population on the biosphere.

CHAPTER 23
Early Societies

When you have completed this chapter, you should be able to:

Describe the course of human evolution.

Discuss the role of tools in human evolution.

Explain the significance of the plow in the development of agriculture.

The success of the human species is due to the ability we have to adapt to a wide range of environments through the modification of our physical surroundings. Our early ancestors learned to use fire to keep warm and cook food. They learned to make tools and then built shelters rather than living in caves. From people who gathered food where they found it, they became farmers who grew what they needed close to home. In this chapter you will learn about the variety of physical, intellectual, and social abilities these accomplishments required.

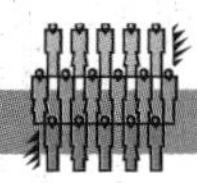

23.1 HUMAN EVOLUTION

The success of humans as a species can be traced directly to the development of culture. **Culture** describes all the human activities that are passed from generation to generation by teaching and learning. It is a special quality of society that is communicated by humans through language. One aspect of culture is **technology**, the inventive ways society solves problems posed by the environment. Using language and tools, humans have been able to bring about technological advances and hasten the spread of these advances from one society to another.

Learning About Early Societies

Historians trace the path of human history by examining the written record found in ancient books. But human history started long before books. **Archaeologists** are scientists who study the remains and artifacts of early civilizations. **Paleontologists** study the fossil record to reconstruct the life-forms that existed in the past. **Anthropologists** study the development and cultures of the human species. The coordinated efforts of these scientists have helped to uncover the evolution of prehistoric human societies.

Our knowledge of human societies comes from the study of fossil remains and **artifacts**, objects made by humans or influenced by human activities. Examples are stone tools, pottery, the foundation of homes, carvings or artwork, and ancient households' trash heaps, called **middens**. The location, size, and depth of a midden allow scientists to estimate the size of the population. Collections of bones from game animals scattered around a midden give clues about climate, tool-making ability, and patterns of hunting and eating.

Hominids are early humans and humanlike species as well as modern humans. Hominid fossils are found scattered throughout many parts of the world. These remains are used to trace the path of human evolution.

Early Hominids

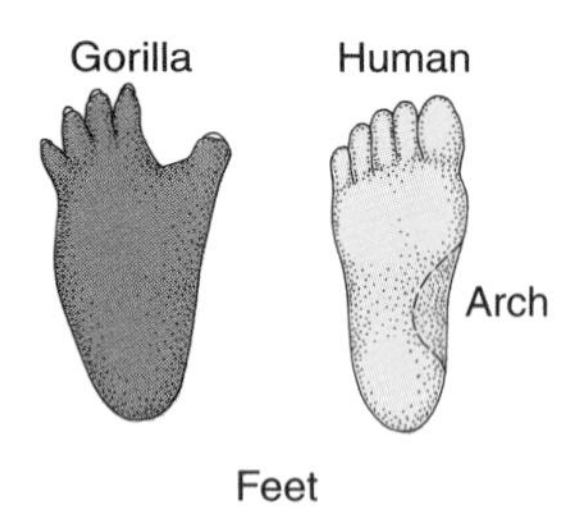

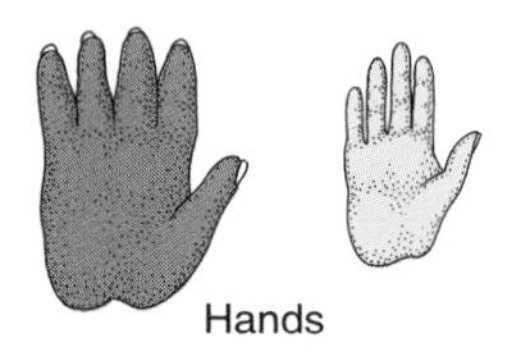

Figure 23-1 The arches in human feet aid in maintaining an upright posture.

The first steps in human evolution were the development of an upright posture and a bipedal gait, both of which are unique among mammals. (Figure 23-1 compares human and gorilla hands and feet.) Evolving a **bipedal gait**, or walking on two feet, freed the hominids' hands. This enabled them to transport infants, tools, and food for long distances. It may have led directly to the establishment of social groups and the division of labor between those who stayed at the campsite and those who foraged for food. From this start about 5 million years ago, the evolution of the human species has been tied to its social habits. The brain of early hominids was no greater than that of modern apes such as gorillas and chimpanzees. The stage was set for the evolutionary advances that were to follow.

Evidence indicates that hominids first appeared in Africa. Ecologically speaking, early hominids were like every other species in the biosphere. They played a role in the various food webs within their ecosystem. They were primary consumers that also may have scavenged at carnivore kill sites and were themselves the prey of predatory animals. Later, with the development of tools, the use of fire, and the evolution of communication skills, hominids became hunters and thus secondary and tertiary consumers. This period, marked by the use of stone tools, is called the **Stone Age.**

The earliest human ancestors included members of the genus *Australopithecus,* which inhabited East Africa about 4 million years ago. *Homo erectus* evolved about 2 million years ago. They were taller, walked with a more upright posture, and had a larger brain than any of the australopithecines. *H. erectus* developed tools, formed complex societies, began cooperative hunting and foraging, and learned to use fire. Their adaptability opened up new environments for habitation and led them to migrate out of Africa. By 750,000 years ago, they had spread to the Middle East, China, Southeast Asia, and Europe.

Our species, *Homo sapiens,* evolved from *H. erectus* in Africa about 250,000 years ago. About 100,000 years ago, *H. sapiens* migrated northward to Europe and Asia, where cooler climates prevailed and where food was not as plentiful. This spurred the development of new tools to aid them in living in their environment. Several types of *H. sapiens* evolved over time. The earliest were the Neanderthals,

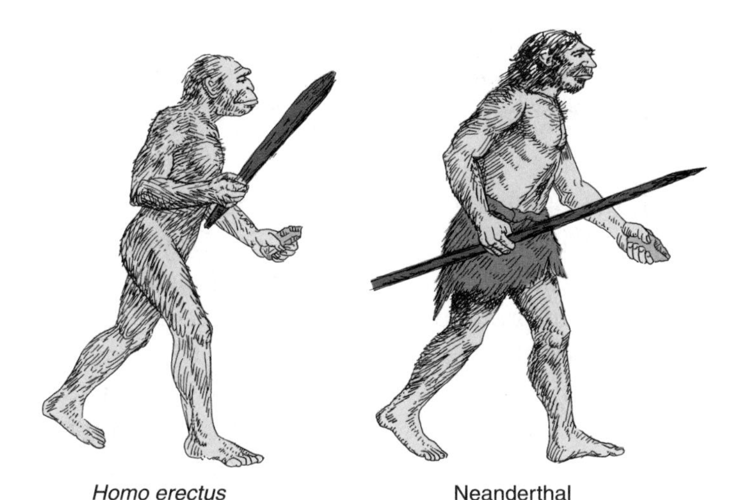

Figure 23-2 Fossils of early hominids show relationships with modern humans.

whose fossils are found scattered through Europe. The Neanderthals had a heavy build, with a protruding nose and jaws, a thick brow ridge above the eyes, and a large brain. They were proficient hunter-gatherers who lived in caves, camps, and rock shelters throughout a wide variety of climates. The Cro-Magnon, who closely resembled modern humans, lived alongside the Neanderthals for much of this time. The Cro-Magnon were highly skilled hunters, tool makers, and artists. (See Figure 23-2.) The Neanderthals disappeared suddenly about 40,000 years ago. From that time to the present, evolutionary trends in *H. sapiens* have been almost completely social and cultural rather than biological.

By 40,000 years ago, *H. sapiens sapiens,* modern humans, had colonized most of Earth, except for North America and some of the Pacific islands. About 25,000 years ago, a shift in climate exposed a land bridge between Siberia and Alaska and opened a route for humans to migrate into North America.

23.1 Section Review

1. How is culture passed on to new generations?
2. Describe the relationship between culture and technology.
3. Make a chart that shows the evolutionary changes that set modern humans apart from their early hominid ancestors. You may illustrate your chart with drawings.

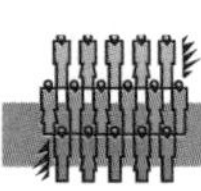

23.2 THE DEVELOPMENT OF SOCIAL GROUPS

The identification and analysis of plant remains and animal bones recovered at archaeological sites can reveal prehistoric hunting and gathering customs, agricultural and livestock practices, and dietary preferences. This information can be used to reconstruct the social structures that existed within ancient communities. A dynamic relationship exists between humans and the environment because agricultural and foraging practices have a measurable impact on an ecosystem. For example, irrigation systems eventually alter the landscape, making it unfit for native plants as well as for further agriculture. In addition, hunting practices can lead to the extinction of some game species, which affects the native plants and other animals.

Hunting and Gathering

From the time of the earliest hominids up to the end of the Pleistocene epoch, about 12,000 years ago, humans made their living by hunting wild animals, gathering wild plants, or scavenging the remains of kills left by predators. These societies were known as **hunter-gatherers**. In fact, they may have been responsible for killing off many species of large prehistoric mammals, the megafauna such as mammoths, mastodons, giant sloths, and cave bears.

Hunter-gatherers were usually **nomads**, people who wandered through a large home territory in bands, clans, or tribes. Their populations never grew very large, though they were knowledgeable about their environment. The primitive weapons they used did not give them much of an advantage over their prey. Hunter-gatherers were experts at using the resources the environment had to offer. They manipulated tools and used fire efficiently. As they moved through their home territory, they set up campsites, where they stayed until the available food was used up. Then they moved on. These people could locate water, game, and edible plants throughout the year and understood the impact of weather and climate on these staples of life. They may even have made use of the medicinal properties of native plants. (There are groups of people who still live this way today, such as the Kalahari bushmen.)

Since cooperation was needed to insure a successful hunt, the reliance on meat as a food source led to social coordination among the members of a tribe or clan. An important advantage of a cooperative social group was the assurance of a steady food supply. The use of fire enabled hominids to cook and eat previously unusable types of food and to preserve meat by smoking it. Fire also offered protection from predators and permitted hominids to migrate to colder climates.

Some hunter-gatherers developed permanent settlements, usually near rich hunting or fishing grounds. Permanent settlements, due to their year-round consumption of resources, were more likely to affect the environment than were nomads.

Tool Making

At some point in time, the australopithecines started using rocks as tools, possibly to dig up roots and tubers or to crack open animal bones to get at the fatty marrow inside. This led to the use of sharp-edged stone flakes as cutting tools. These flakes formed naturally as rocks rolled down hills and collided. *Homo erectus* shaped stone tools to their liking, by striking small flakes from larger rocks. A **tool** is a hand-held device used to perform a specific function. Flint, chert, and obsidian were the preferred stone materials used by early humans to fashion tools.

Hunter-gatherer societies made tools from sticks, stones, animal bones, and horns. The earliest recognizable tools that have been uncovered are chipped stones. Other items that may have been used as tools, such as stones that could be used for battering, twigs used to probe for insects, and wooden sticks used as weapons, are not preserved as artifacts in the fossil record. Early stone tools were used to process animal carcasses for food and clothing. Meat made up only a small portion of the diet of early hominids. The ability to make use of tools to obtain and process meat added a rich nutritional source to the diet of later hominid species.

The manufacture of tools and weapons was an important step in

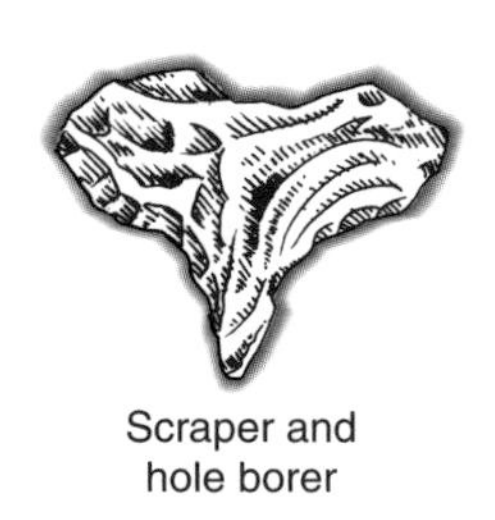

Scraper and hole borer

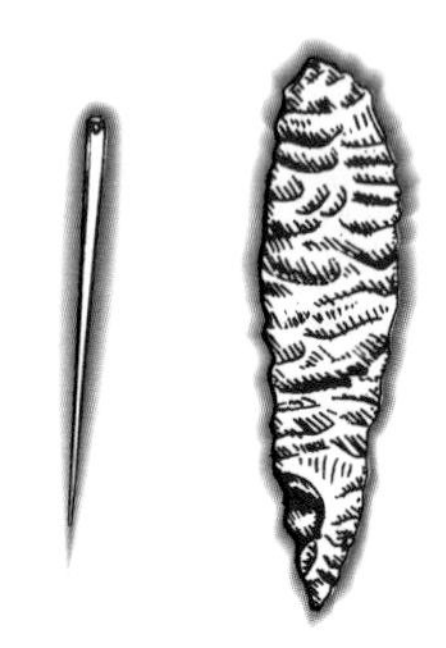

Figure 23-3 Tools used by hunter-gatherers were made from stone and bone.

the development of human society. Stones could be processed to be used as knives, axes, spear points, and bolos. Bones, horns, and antlers were fashioned into buttons, needles, fishhooks, spear throwers, and digging implements. (See Figure 23-3.)

Pottery

People discovered how to make **pottery**, or articles made from baked clay, by observing the effects fire had on clay soil. Pots were first fashioned from hollowed-out lumps of clay. The pottery was hardened by heating it in fire. Later, clay was mixed with straw before shaping to make it less brittle after it was fired. Fire-hardened pottery brought about improvements in cooking and storing food. The use of pottery added items such as soups, bread, and stews to the human diet. Storing food in covered pots protected it from insects and rodents. Humans could now gather foods and preserve them for future use or as items of barter to obtain other goods. Thus the roots of commerce, the buying and selling of goods, began with the manufacture of pottery.

23.2 Section Review

1. Describe life in an early hunter-gatherer society.
2. How did the making of pottery change hominid society?
3. Make a stone tool by splitting off the edge of a squared block of flint or chert. The fragment can be pressure chipped along one edge to form a cutting or scraping tool. Wear goggles to protect your eyes from stone chips.

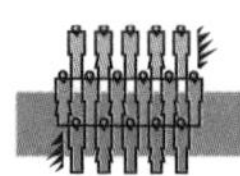

23.3 DOMESTICATION OF PLANTS AND ANIMALS

The origins of agriculture can be traced back about 10,000 years or more to Southwest Asia. As populations began to grow, people discovered the advantages of managing their environment. They were able to increase productivity and reduce the risks associated with a hunting-and-gathering lifestyle.

Within 2000 years, cultivating plants and breeding of animals were practiced throughout the region from Greece to Pakistan. This pattern of mixed farming fueled the economies of the great civilizations of the Old World, including Sumeria, Egypt, Rome, and Greece.

The Development of Agriculture

Seed crops originated in the area between India and China, in northern Africa, and in Central America. Agriculture started with gathering, storing, and sowing wild seeds. Planting was done in areas where water could be controlled and where harvesting could be accomplished easily. Most of the seed crops that primitive humans cultivated were annual grasses, such as wheat, rye, barley, corn, and rice. (See Figure 23-4.) These were well adapted for growing in open areas. The rich store of nutrients within the seeds made them highly useful to the human population. Squash, root crops, and grapes were some of the other plants cultivated.

Figure 23-4 Seeds contain food for germinating plants. Seeds are also eaten by animals.

In rain forests, plots were cleared for planting by cutting down and burning all vegetation. The ashes mixed with the earth formed a fertile soil, and crops were planted and harvested. This practice is called **slash-and-burn agriculture**. Because of the low nutrient content of rain forest soils, after several harvests, new crops failed and the land was abandoned. Once a farmer moved on, the native plants returned, and the rain forest reverted to its original state. The overall impact on the environment was small. Slash-and-burn agriculture is still practiced in some rain forests around the world.

In other areas, woodlands were cleared and their rich soils used for crops. The development of the plow allowed dense grassland soils to be cultivated, which greatly increased productivity. A **plow** cuts, lifts, and turns over the soil. Figure 23-5 on page 410 shows an ancient plow. The plow marked a turning point in cultural evolution and was the beginning of modern technology. Use of the plow resulted in a predictable supply of food that could be stored and used as needed. Humans became less dependent on the natural environment and developed a settled lifestyle. This allowed people to concentrate on other areas, such as trade, arts, and science. Civilizations

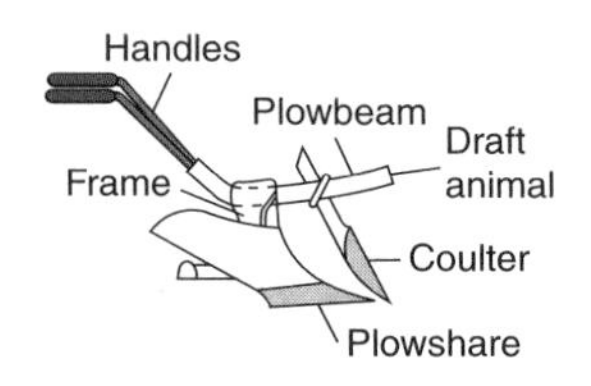

Figure 23-5 The plow enabled people to have a predictable supply of food.

rose up as agricultural techniques improved, and, due to the increased food supply, the population grew. The growth in population due to the plow's success led to increased demands for timber and minerals. Overutilization of such resources seems to have led to the fall of some ancient civilizations.

Domestication is the process of adapting populations of plants and animals to live in association with and to the advantage of people. Often, domesticated varieties depend on humans for survival. In some cases, domestication interferes with the plant's or animal's capacity to reproduce on its own. This requires that the wild varieties be used to maintain the genetic diversity of the species, or that humans help pollinate or breed the species.

In prehistoric times, wolves scavenged kills made by human hunters. Because of their close association, wolves were domesticated by hominids. The wolf is the ancestor of today's dogs. In Southeast Asia, early agricultural societies domesticated animals such as jungle fowl and pigs as sources of food. The domestication of cattle, sheep, and goats expanded the diets and sources of clothing materials for societies in Southwest Asia. (See Figure 23-6.)

The Impact of Domestication

Once animals and plants were domesticated, human influence over them extended beyond their use as sources of food or clothing. Humans were able to manipulate physical characteristics to produce new breeds and varieties of plants and animals by selective breeding. Factors such as length of coat hair, protein and fat content, body size, resistance to disease, strength, and milk productivity could be controlled. However, selective breeding makes the individuals of domesticated species more alike genetically than the populations of their ancestors. This makes them vulnerable to extinction by a disease, when there are no individuals who may be genetically resistant to that disease.

The domestication of plants and animals gave agricultural societies the ability to increase the productivity of the environment for the people's benefit. By controlling the resources of the environment, they could expand beyond the previous limits determined by the natural food supply. As the human population swelled, im-

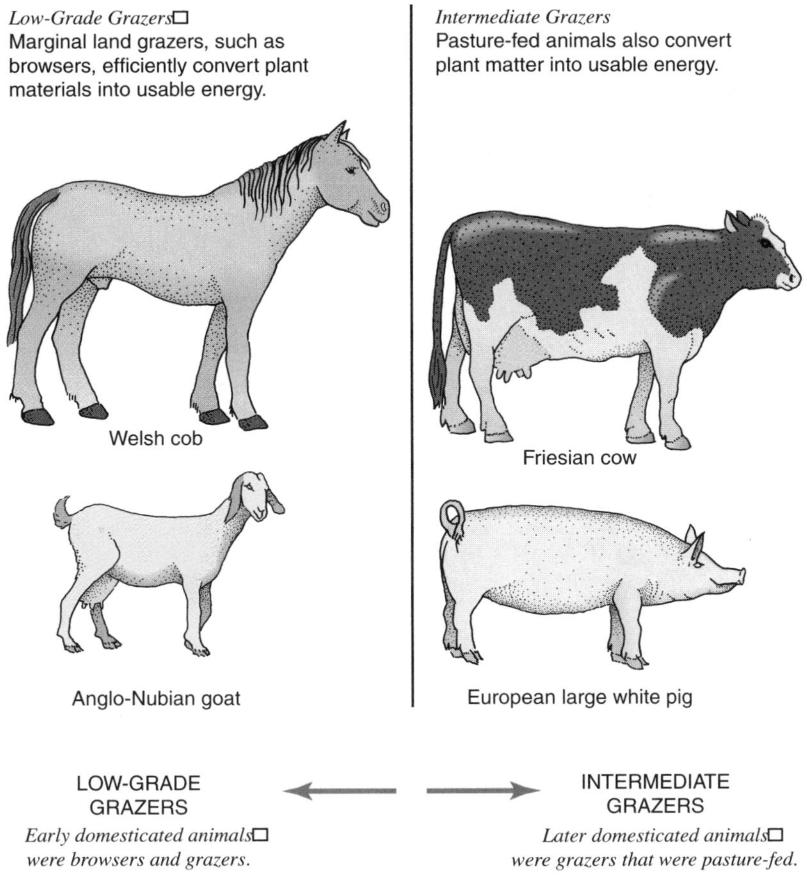

Low-Grade Grazers
Marginal land grazers, such as browsers, efficiently convert plant materials into usable energy.

Welsh cob

Anglo-Nubian goat

Intermediate Grazers
Pasture-fed animals also convert plant matter into usable energy.

Friesian cow

European large white pig

LOW-GRADE GRAZERS
Early domesticated animals were browsers and grazers.

INTERMEDIATE GRAZERS
Later domesticated animals were grazers that were pasture-fed.

Figure 23-6 Several animals domesticated by humans are illustrated here.

proved agricultural practices reduced the number of people needed to grow food. Many people left the farms and settled in villages and cities, where they entered into various crafts and businesses. The cities grew and eventually became centers for commerce, trade, government, and religion.

23.3 Section Review

1. Why is the plow considered to be the turning point for modern technology?
2. What were the advantages of agriculture to primitive social groups?
3. Explain the impact of domestication of plants and animals on cultural evolution.

LABORATORY INVESTIGATION 23

Pairing Wild and Domestic Animals and Plants

PROBLEM: *To identify the wild species from which a domesticated animal or plant descended*

SKILLS: *Observing, classifying*

MATERIALS: *A set of pictures of a variety of domestic plants and animals, a corresponding set of pictures of wild plants and animals*

PROCEDURE

1. From the pictures your teacher gives you, select a domestic animal and match it to the wild animal from which it descended.
2. Select a domesticated plant and match it to its wild ancestor.

OBSERVATIONS AND ANALYSES

1. Compare the appearance of the domestic animal with that of its wild ancestor. Determine the traits that were selectively bred into the domestic animal you chose.
2. Compare the appearance of the domestic plant with that of its wild ancestor. Determine the traits that were selectively bred into the domestic plant you chose.
3. In what ways do domesticated animals and plants differ from their wild ancestors?
4. Why are domesticated species more at risk of extinction through disease than are wild species?
5. Of what importance is the maintenance of stocks of wild representatives of domesticated species?

Chapter 23 Review

Answer these questions on a separate sheet of paper.

Vocabulary

The following list contains all the boldfaced terms in this chapter.

anthropologists, archaeologists, artifacts, bipedal gait, culture, domestication, hominids, hunter-gatherers, middens, nomads, paleontologists, plow, pottery, slash-and-burn agriculture, Stone Age, technology, tool

Fill In

Use one of the vocabulary terms listed above to complete each sentence.

1. Hominids that walk on two feet have a ______.
2. Scientists who study the fossil record to learn about past life-forms are ______.
3. ______ are scientists who study the remains and artifacts of early civilizations.
4. ______ are objects made by humans or through human activity.
5. The technique used to clear rain forest vegetation to prepare land for planting is called ______.

Multiple Choice

Choose the response that best completes the sentence or answers the question.

6. The spread of technology is accomplished through *a.* culture. *b.* language. *c.* adaptation. *d.* problem solving.
7. An example of an artifact is *a.* a Cro-Magnon tool. *b.* a fossil hominid bone. *c.* a dinosaur footprint. *d.* pollen grains.
8. Hominids first appeared in *a.* Asia. *b.* Europe. *c.* Africa. *d.* North and South America.
9. The hominid that most closely resembles modern humans is *a.* Neanderthal. *b.* Cro-Magnon. *c.* *Homo erectus.* *d.* *Australopithecus.*
10. What material was added to clay to make pottery less brittle after firing? *a.* obsidian *b.* ground antler *c.* straw *d.* water

Short Answer (Constructed Response)

Use the information you learned in this chapter to respond to the following items.

11. Describe three characteristics of hominids
12. Give two examples of hominid artifacts.
13. What are middens?
14. What animals were probably the first to be domesticated by humans?
15. What are tools?
16. Where do scientists think hominids originated?

Essay (Extended Response)

Use the information in the chapter to respond to these items.

17. Describe how the development of a bipedal gait led to changes in hominid society.
18. What were the characteristics that made seed grasses attractive for domestication?
19. Why do rain forest soils fail to produce crops after several plantings?
20. Language and the use of tools are considered important parts of the social behavior that define humans. However, researchers have shown that apes such as chimpanzees use tools and can learn to communicate through symbols. What implication does the use of tools by chimpanzees have for the study of primates?

Research Projects

- Explore the use of tools by other mammals and birds.
- Trace the historical development of the domestication of an animal or plant.

CHAPTER 24
Technological Society

When you have completed this chapter, you should be able to:

Describe the factors that led to the growth of cities.

Explain how the Industrial Revolution affected the environment.

Discuss the population explosion.

Define and give examples of urban ecosystems.

Since the founding of the first cities about 6000 years ago, the growth of cities has followed the development of civilization. Our cities reflect the best and worst that technology has bestowed on our societies. In our cities, resources are collected and distributed to the population, information and culture are shared, and technology is dispersed. By the year 2000, Earth's 10 largest cities each held more than 10 million people, and were classified as megacities. In addition to people, plants and animals also live in cities. In this chapter, you will learn about the growth and development of cities.

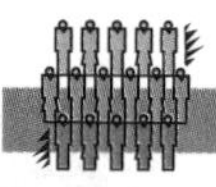

24.1 LEAVING THE STONE AGE

In the first agricultural regions, populations were protected against those factors that had limited the growth of previous societies. Food was grown in quantity and stored for use during times of drought. Permanent homes were built to protect the population from severe changes in weather or climate. In areas that were secure from environmental hazards, villages were constructed with defenses against invasion from outsiders. A **village** is a collection of rural households linked by culture, family ties, or occupational relationships. In a **rural area**, most residents of villages depended on agriculture for their livelihood.

The Evolution of Cities

The transition from an **agrarian**, or farming, society to an industrial one was accompanied by urbanization. Small villages with a farming population grew to become cities. Historically, a **city** is defined as a community that has abundant resources and a population large enough to allow its people to specialize in a wide variety of professions.

For the first time in history, large numbers of people lived close to one another in cities. Communicable diseases passed easily from one person to another. Lack of knowledge of sanitary practices led to contaminated water supplies and the rapid spread of deadly diseases such as typhus, cholera, and dysentery. The water-filled trenches used for irrigation became stagnant breeding grounds for disease-carrying mosquitoes. Epidemics of viral diseases, such as measles, smallpox, and influenza, spread quickly.

The growth of civilizations led to increased travel between countries and, as a result, spread diseases throughout the world. Epidemics and plagues were often responsible for the deaths of a quarter to a third of the population of a region. The rate of death due to city living was much greater than it had been among individuals who followed a nomadic lifestyle. Yet many people were drawn to cities because of the varied economic opportunities that were available.

In 1800, only 3 percent of the world's population lived in cities. As the cities grew in size, they formed large urban areas. In an **urban area**, the members of the population are not dependent on an agricultural or natural resource-based economy to make their living. By 1950, the number of people in cities increased to 30 percent of the population. Shortly after the year 2000, the world became predominantly urban; more than 50 percent of the population resides in cities. The United States Census Bureau considers any incorporated community to be a city and defines any city with a population of more than 250,000 as urban. Furthermore, cities are reaching exceptional sizes. By the beginning of the twenty-first century, there were 23 megacities in the world. A **megacity** is a city with a population in excess of 10 million people. (See Table 24-1.)

Cities are artificial, structural habitats for human use, which cause changes to the natural environment and consume many of the biosphere's resources. They are dominated by bricks, concrete, asphalt, and steel and have few natural areas or parkland. As cities grow, they cover areas that were previously used for agriculture, and they require more and more water, energy, and food for their populations. Wastes produced through everyday living are disposed of in the surrounding environment, which pollutes large areas of the biosphere. The high concentration of people in urban areas exposes them to the dangers posed by pollutants, the risk of disease epidemics, and the stresses associated with living in a densely populated area.

TABLE 24-1 THE 10 LARGEST URBAN AREAS IN THE WORLD, 2000 AND 2015

Largest Urban Areas, 2000 *(population in millions)*		Largest Urban Areas, 2015 *(projected population in millions)*	
1. Tokyo, Japan	26.44	1. Tokyo, Japan	27.19
2. Mexico City, Mexico	18.07	2. Dhaka, Bangladesh	22.77
3. São Paulo, Brazil	17.96	3. Bombay, India	22.58
4. New York City, United States	16.73	4. São Paulo, Brazil	21.23
5. Bombay, India	16.09	5. Delhi, India	20.88
6. Los Angeles, United States	13.21	6. Mexico City, Mexico	20.43
7. Calcutta, India	13.06	7. New York City, United States	17.94
8. Shanghai, China	12.89	8. Jakarta, Indonesia	17.27
9. Dhaka, Bangladesh	12.52	9. Calcutta, India	16.75
10. Delhi, India	12.44	10. Karachi, Pakistan	16.20

24.1 Section Review

1. How is a city different from a village?
2. Explain how the growth of cities affects the natural environment.
3. Use reference materials from your local library or historical society to prepare a report on the history of the village or city in which you live.

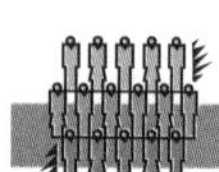

24.2 THE INDUSTRIAL REVOLUTION

A drastic change in the relationship between the human population and the biosphere came about with the Industrial Revolution. The **Industrial Revolution** was the change that occurred in manufacturing, from small-scale production by hand to large-scale production by machine. It began in England during the 1700s and from there gradually spread to other parts of the world. It started in the United States in the 1800s. The economic transformation brought about by the Industrial Revolution is still progressing. This transformation has had a major impact on all the world's cultures.

Industrialization is the process of transforming the economy of a nation or region through the development and application of technology. The modernization that follows requires the widespread use of energy sources to fuel the technological processes that are developed. Industrialization is the main reason that the developed nations of the world have become urban centered, energy consuming, broadly educated, and centrally governed.

The Growing Demand for Energy

The Industrial Revolution resulted in an increased demand for energy, which led to the destruction of forests and widespread timber shortages. As wood became costly, the British turned to coal, a fossil fuel. The emerging technologies accelerated the change to an industrial society and at the same time put some new stresses on the environment.

With the use of coal-powered machines, manufacturing became more energy intensive and economy driven and less labor intensive.

Figure 24-1 **Modern tractors have helped farmers increase crop production.**

The move away from hand labor meant that more importance was placed on machines and production and less importance was placed on people. Increased production led to the development of new means of transportation to move goods between cities. The steam engine was used to power trains and ships. As new industries sprang up within cities, the need for fuel, raw materials, and factory workers rose sharply. With this came increased demands for food and shelter for the workforce.

Mechanization also affected agricultural practices. Technological advances, such as the cast-iron plow and the reaper, increased agricultural production. The internal combustion engine made horse-drawn and cattle-drawn farming implements obsolete. (See Figure 24-1.) Fewer farmworkers were needed, which added to the migration of rural populations to cities.

As local supplies were depleted, industrialized nations looked for new sources of needed resources. Increased production stimulated industrial nations to seek new markets for their products. During the early part of the Industrial Revolution, these factors led to the spread of colonialism by industrialized nations. Unsound exploitation of colonies by industrialized nations often led to widespread destruction of the environment.

Environmental Effects

As we entered the twenty-first century, industrialization's effects on the environment become clearer. Toxic wastes contaminate the land

and groundwater that surround many former or present industrial sites. In developing countries, the rush to increase production often adds to this problem by repeating the ecological mistakes of the past. The use of fossil fuel has strained Earth's limited supply and led to pollution of the atmosphere and hydrosphere. Oil spills have killed wildlife and degraded beaches. Ozone depletion threatens plant and animal life due to increased exposure to ultraviolet radiation. The humans face an increased risk of skin cancer, and mutations are occurring in amphibian populations. Greenhouse gases threaten Earth with global warming or cooling.

Industrialization and new technologies have greatly broadened our knowledge of the world, improved public health, led to a better-educated society, and provided an abundance of material goods to improve our lives. At the same time, the environmental costs of industrialization have been excessive in terms of pollution of the land, waters, and atmosphere; deterioration of the landscape; and the potential for catastrophic effects on the biosphere in the future.

24.2 Section Review

1. Explain the role that coal played in the Industrial Revolution.
2. What changes did the Industrial Revolution make in agricultural practices?
3. Describe the factors that led to the spread of colonialism by industrialized nations.

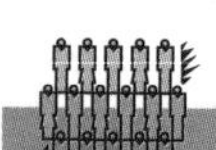

24.3 THE POPULATION EXPLOSION

Fossil evidence suggests that early humans spread over Earth very slowly. For the first few million years of their existence on the African continent, the total human population is estimated to have been no more than 1 million people. Migrations to other continents allowed the human population to increase. By the time agriculture was started about 10,000 years ago, the population had climbed to about 5 million people. By 2000 years ago, the human population had reached 150 million. In 1600, the population was approaching

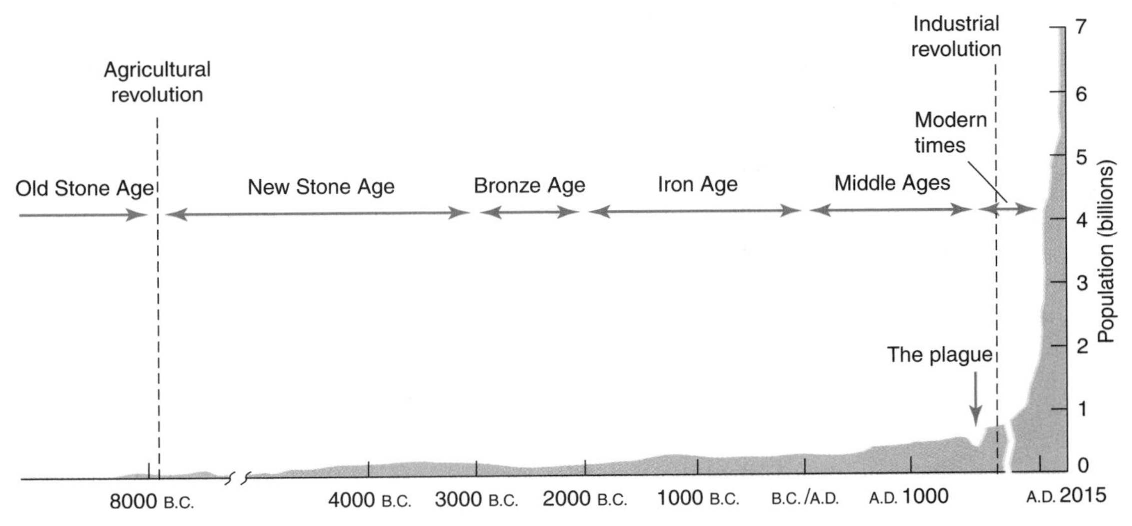

Figure 24-2 The world's population growth is exponential.

500 million. It doubled to 1 billion by 1850. The next doubling, to 2 billion in 1930, took only 80 years. By 1976, the population had reached 4 billion. It had doubled in only 46 years. By the year 2000, the population had exceeded 6 billion. By 2050, the world's population is expected to be more than 9 billion people. Every day, more than 200,000 people are added to the world's population. (See Figure 24-2.)

The **population growth rate** is the annual surplus of births over deaths. To determine this, the **mortality rate**, or annual number of deaths, is subtracted from the annual birth rate, the number of births in one year. In addition, you must subtract the number of emigrants, people who move out of the country, and add the number of immigrants, people who move into a country. A stabilized population is one that has achieved **zero growth rate**, or no change in population. Only a handful of countries, all of them in Europe, have reached a zero growth rate.

The Birth Rate Question

The recent, great increase in the human population is called a **population explosion**. Some scientists theorize that the human population is increasing so rapidly that widespread famine is inevitable in the near future. Our food supply is one step ahead of the rising demand. This has led to a debate over methods of slowing the human population explosion. But there are no simple answers to the population problems that will face our world in the near future.

Developing countries that have high population growth rates usually have higher birth rates than industrialized countries. Is population growth a result of higher birth rates? By studying the historical record, scientists have determined that high birth rates in underdeveloped countries are typical of humans through time. The lower birth rates in developed countries are not typical. Population growth is therefore thought to be a result of advances in modern medicine and public health, which have brought about a steady decrease in mortality rates so that more children live to become adults.

In the industrialized countries, improved agricultural practices have increased the food supply and prevented famine from devastating whole populations. Childhood diseases have been controlled, shifting the population toward a higher percentage of younger, childbearing individuals.

Yet in poorer countries, food is in short supply. Production cannot meet the demands of the growing population. The expanding urban areas in these countries often suffer from poor water quality, lack of proper sanitation, inefficient waste collection, and minimal health services. These factors promote the spread of many diseases.

A population controversy exists that sets the rights of individuals against the needs of society. Some people have strong feelings against government interference in an individual's right to have children. They question placing limits on population growth. Others maintain that a continued high birth rate strains the limited resources of society and will lead to starvation, death, and misery for many. There are a multitude of moral and social questions involved in the controversy. Can the information obtained by using basic ecological principles aid the world in planning a future in which the population can live on the available resources?

Earth's Carrying Capacity

The carrying capacity of an ecosystem refers to the population that the available resources can support. When the carrying capacity is exceeded, populations decrease. Will this apply to the human population as well? Historical evidence indicates that, unlike other organisms, humans have increased the carrying capacity of the environment through innovation and technology. Therefore, less land can support more people. Some scientists view Earth as having limited resources. They say that the planet has an ultimate carrying capacity, which places limits on the human population. Others think that advances in technology will be able to increase Earth's carrying capacity.

The human population is not distributed evenly over Earth's land areas. The greatest numbers of people are gathered where favorable climate and the presence of resources make a higher population density possible. Population growth drives environmental decline and is an important factor in poverty, forced migration, and social conflicts. The higher population growth rate in developing countries creates greater stress on the environment than in industrialized countries. But industrialization itself causes stress to the environment.

The human population has made a considerable impact on the biosphere. Earth's tropical rain forests once covered almost 15.5 million square kilometers. More than one-third of the rain forests have been cut down, and the rate of loss is increasing. Since 1945, more than 11 percent of Earth's land area has suffered moderate to severe damage. This land has lost its capacity to grow crops or even support the region's natural vegetation. Due to pressures from agriculture and urban use, water resources are scarce in many regions. The growth in size and number of urban areas has increased air pollution and caused many respiratory diseases in the urban population. Increases in the two most important greenhouse gases, carbon dioxide and methane, threaten to change the global climate. The accelerating loss of natural habitat has led to the extinction of countless animal and plant species.

24.3 Section Review

1. Calculate the approximate year the world population will reach 10 billion people.
2. What are some effects the population explosion has had on the natural environment of developing nations?
3. Choose one side of the population controversy and defend it in a report or a class debate.

24.4 URBAN ECOSYSTEMS

The growth of the human population has led to more and more cities and towns spreading out across Earth's surface. This, in turn, has caused the destruction of large parts of the biomes that once covered Earth and the extermination of many species. Yet some species have been able to make a better living in the new environments that were created. These artificially created environments in cities have become a new ecosystem, the **urban biome**.

Various species have adapted to an urban lifestyle. New opportunities opened up for some species that did not exist in a rural environment. These organisms have been able to make use of a wide range of resources related to food, shelter, and survival, and each has adapted to life in the city in its own particular fashion.

Urban Wildlife

Starling

Figure 24-3 **These birds were introduced into the United States from England. They have displaced many native birds.**

A variety of urban habitats attract wildlife. These include ponds, lawns, parks, vacant lots, landfills, and even the roofs of apartment buildings. Wherever they are found, these environments tend to attract the same kinds of plants and animals. Starlings and English sparrows are now abundant in almost every city in the United States, but they were introduced to North America only 100 years ago. (See Figure 24-3.) These birds have adapted to and flourished in this new urban environment. Due to the availability of nesting spaces on buildings and the lack of predators, pigeons are year-round inhabitants of most cities, too.

Wild animals often migrate to cities, where the pickings are easy. Coyotes, rats, foxes, and raccoons prowl through our garbage; squirrels search for food in city parks; peregrine falcons hunt pigeons in the canyons between city skyscrapers; and escaped pets, such as canaries, finches, and parakeets, feed on the seeds of grasses and weeds in gardens and vacant lots.

Swifts, swallows, and bats, which usually nest in hollow trees or caves, also nest on buildings. Since cities offer many more buildings to nest on than there are hollow trees in a forest, the populations of these species often increase in an urban environment. Bats find food around the bright streetlights that attract many insects at night.

Our homes provide a wide range of habitats for organisms. The warm, cozy attics of many houses attract various creatures. Wood-boring beetles infest beams, bats and martins nest under the eaves, and spiders spin their webs to catch some of the insects attracted to the attic warmth. Beneath the roofs of houses, moths, centipedes, ticks, daddy longlegs, fleas, flies, and crickets seek shelter during the winter months. Mounds that indicate the nests of ants dot many lawns.

Homes offer many hiding places, a moderate climate, and an abundant food supply. The creation of stable indoor environments allows colonization by tropical plants, animals, and insects that would otherwise never have been able to exist in temperate regions. The cockroach is a tropical insect that has adapted to the cozy confines of human habitations worldwide. The housefly is an insect pest that feeds on any organic material and often spreads bacterial contamination. The cockroach and fly have probably lived in association with humans since the time of early hunter-gatherers. (See Figure 24-4.)

Field mice and house mice live in the walls of our homes. They come out to feed on the scraps and crumbs left from meals. House mice have been associated with humans for so long that they have become dependent on us for survival. The brown rat lives in basements or sewers, where it finds plentiful food. The habits of these animals often lead to the spread of diseases.

Many organisms live on things we don't usually think of as food.

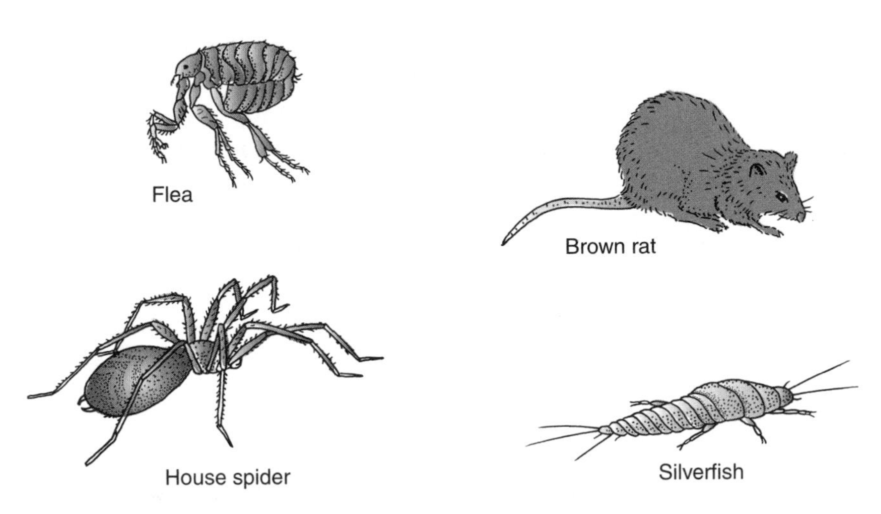

Figure 24-4 Some examples of urban wildlife.

CONSERVATION

Protecting Piping Plovers

The coastal town of Cape May, New Jersey, is a popular tourist resort that becomes very crowded in the summer. It also is a refuge for the largest and most varied group of migrating birds of any spot in North America. However, the birds' nesting and feeding territory is being threatened by development in the area. The New Jersey Department of Environmental Protection (DEP), the federal government, and conservationists are working with local officials and developers to protect the birds' nesting sites. In fact, the birds contribute to the local economy by attracting bird-watchers and people who enjoy nature.

Among the birds that visit Cape May are the piping plovers. Piping plovers nest on beaches that have little vegetation. Because plovers are not social birds, they nest about 100 meters apart from each other. The state has set up protected areas for these birds. Conservationists rely on programs that teach people that these birds need space and quiet to survive. By keeping out people and predators, the state protects vulnerable young birds while they mature.

In 1999, a 490-acre stretch of Two Mile Beach was transferred from the Coast Guard to the Cape May National Wildlife Refuge. The beach is closed annually to provide an undisturbed nesting habitat for the piping plover, which is on the federal list of threatened species. During the spring and fall, the beach also
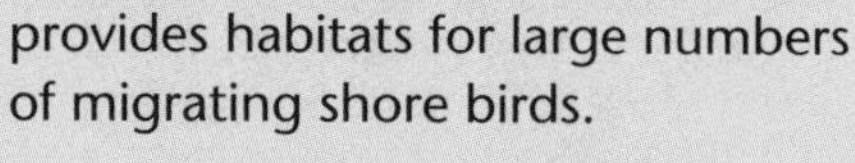
provides habitats for large numbers of migrating shore birds.

Piping Plover

The policy has had some success. Recently, the plover population in New Jersey has increased. To be successful, efforts to restore birds to their natural habitats must reduce the harmful direct and indirect effects of people.

The larvae of the clothes moth and several species of beetles feed on woolen clothing. The introduction of artificial fibers in recent years has caused these organisms to all but disappear from our homes. Silverfish eat paper and often destroy books. Slugs and lice eat mold that forms on damp paper products. The furniture beetle is a wood-boring beetle that feeds on indoor wood. Earthworms are abundant in rich garden soils. They tunnel under the surface, taking in soil and extracting the organic materials it contains. The tunnels excavated by earthworms aerate the soil and provide drainage for water.

Few suburbs are without backyard bird feeders. The high-energy food in the feeders attracts a large bird population. Many migratory birds remain in the city, attracted by the steady supply of food. This has a considerable impact on their survival. When these birds are faced with harsh winter conditions, and the food supply is cut off, many birds perish.

The Urban Jungle

Imported tropical plants are used widely for decorative purposes. (See Figure 24-5.) These rain forest plants are adapted to tolerate low-light conditions and are well suited for growth indoors. Shopping malls, houses and apartments, business offices, and stores thus become home to these plants and to the alien insect pests that feed on them. The insects, which often arrive in or on the plants as eggs, thrive in the artificial environments they find ideal. The alien insects migrate outdoors and attack native plants. Since the alien insects have few natural enemies in this new habitat, they soon become unwanted pests.

Figure 24-5 African violets are houseplants that come from the rain forest.

Figure 24-6 The ailanthus, originated in Asia, can now be found growing in many places in the United States.

Weeds are plants that grow in areas where they are not wanted. When humans transport alien plants to new habitats, these plants often grow unchecked and become weeds. The American prickly pear cactus has become a weed in the Mediterranean region and Australia. The water hyacinth is a tropical plant that was introduced to Florida, where it has become a pest by blocking waterways with mats of vegetation. The ailanthus is a weed tree imported from China during the early part of this century that has rapidly spread throughout the northeastern United States. The fast-growing ailanthus has a rough bark and long tapering leaves and looks like a tree in size and shape. It shoots up so quickly that it soon outgrows the rest of the flora around. (See Figure 24-6.)

The climate of most cities is warmer than that of the surrounding region. Concrete, stone, and asphalt used in the construction of buildings and roads absorb heat and then slowly radiate it as the air temperature drops. Thus the city cools more slowly than the area around it. Because of the warmth, some plant species flower earlier and store more food during summer. Animal species lay their eggs earlier. This gives these species a biological advantage over competitors and allows these creatures to extend their range and displace other species.

For plants and animals, there are many disadvantages to city living, including exposure to human activity, lack of natural vegetation, lack of exposed soil, and pollution. However, cities do offer advantages in the form of readily available food sources; a lack of predatory species; an abundance of places to safely live and breed; buildings and other structures to roost upon; underground systems like subways, sewers, and cable lines to breed in and move through; and waters high in nutrients that aid plant growth.

24.4 Section Review

1. Discuss how the urban environment opens new areas for species to colonize.
2. Discuss the advantages and disadvantages that city living presents to an organism.
3. Select an organism that has adapted to an urban lifestyle and report on its interactions with the human population.

LABORATORY INVESTIGATION 24

Uninvited Guests in Buildings and Gardens

PROBLEM: *What living things may be found in buildings and gardens or lawns?*

SKILLS: *Observing, listing, classifying*

MATERIALS: *Paper, pen, magnifying lens, trowel*

PROCEDURE

1. Examine the interior of your home or school. Make a list of all the organisms that you find.
2. Examine the exterior of your home or school. Make a list of all the animals and plants that you discover.
3. Examine the lawn and/or garden of your home or school.
4. Make an inventory of the plant community.
5. List the animals found living underground, on the soil, and on the plants.
6. Be on the lookout for signs of unseen visitors:
   a. animal droppings (are they from insects or other animals?)
   b. exit holes made by burrowing insects on wooden surfaces, leaves, stems, flowers, or fruit of plants
   c. signs of animals eating the plants
   d. animal or bird tracks

OBSERVATIONS AND ANALYSES

1. Classify the organisms you found inside the school or home as plants, fungi, or animals.
2. Classify the organisms you found outside the school or home as plants, fungi, or animals.
3. Describe some of the organisms that are native to the region in which you live.
4. Determine the interrelationships among the organisms you found.
5. Describe some of the organisms that are alien invaders of your ecosystem. Where did these organisms originally come from?
6. What are some of the environmental conditions that may have attracted these invaders?

Chapter 24

Review

Answer these questions on a separate sheet of paper.

Vocabulary

The following list contains all the boldfaced terms in this chapter.

agrarian, city, Industrial Revolution, industrialization, megacity, mortality rate, population explosion, population growth rate, rural area, urban area, urban biome, village, zero growth rate

Fill In

Use one of the vocabulary terms listed above to complete each sentence.

1. In a(an) ________, the members of the population do not depend on agriculture to make a living.
2. The process of transforming an economy through technology is called ________.
3. The ________ began with the first use of machines for large-scale production.
4. The rapid growth of the human population due to the development of agriculture is called the ________.
5. New environments that have been created within cities are ecosystems known as ________.

Multiple Choice

Choose the response that best completes the sentence or answers the question.

6. Which of the following is a weed that was imported from China, which grows like a tree in the United States? *a.* ailanthus *b.* prickly pear *c.* hyacinth *d.* dandelion
7. About how many years ago did cities begin to develop? *a.* 1000 *b.* 6000 *c.* 10,000 *d.* 20,000
8. In 1800, the percent of the world's population that lived in cities was *a.* 3. *b.* 10. *c.* 29. *d.* 50.
9. Which of the following is an undesirable mammal that has become dependent on humans for survival? *a.* dog *b.* rat *c.* horse *d.* bat
10. An insect that often destroys books is the *a.* cockroach. *b.* silverfish. *c.* termite. *d.* house fly.
11. Compared with the surrounding area, the climate of most cities is usually *a.* cooler. *b.* warmer. *c.* more humid. *d.* less humid.
12. People who move out of a country are called *a.* immigrants. *b.* travelers. *c.* emigrants. *d.* tourists.

13. One advantage to animals in adapting to an urban environment is the *a.* lack of human activities. *b.* lack of natural predators. *c.* lack of natural vegetation. *d.* presence of pollutants.
14. A population that has achieved zero growth rate is said to be a *a.* population explosion. *b.* megacity. *c.* stabilized population. *d.* city.
15. People who move into a country are called *a.* immigrants. *b.* travelers. *c.* emigrants. *d.* tourists.

Short Answer (Constructed Response)

Use the information you learned in this chapter to respond to the following items.

16. What is meant by the term *agrarian society*?
17. What is a village?
18. In early human history, what type of occupational relationship held the residents of small villages together?
19. In what ways did the Industrial Revolution transform human society?
20. List the factors that determine a nation's population growth rate.
21. What is meant by a zero growth rate?
22. In the year 2000, which city had the largest urban area?

Essay (Extended Response)

Use the information in the chapter to respond to these items.

23. As cities grew, why was mortality greater in cities than in rural areas?
24. Explain the impact cities have on their surrounding areas.
25. Describe some wildlife habitats that are found in a city.

Research Projects

- Trace the historical growth of a megacity.
- Research and report about some of the environmental problems that are unique to magacities.

CHAPTER 25
Human Society and the Biosphere

When you have completed this chapter, you should be able to:

Discuss the ways cities get rid of their wastes.

Define pollutant and pollution and give an example of each.

Explain the success and impact of alien species.

A city is like a living organism. It needs energy to carry out its activities. A city takes in raw materials that are used to supply the needs of its population, and it produces waste materials that must be eliminated. Like a living organism, a city has its own metabolism. As cities grow, the quantity of waste materials produced strains the city's ability to dispose of them. The processes that the city uses to eliminate wastes often create pollution problems elsewhere in the biosphere.

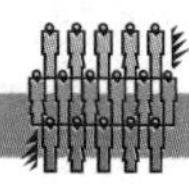

25.1 URBAN WASTES

Most of us are familiar with the unsightly conditions often found in and around urban areas. There are litter-lined streets and roads, trash-filled parks and beaches, alleys and lots covered with rotting garbage. In addition, you find open dumps filled with piles of junked cars, rusting machines, and worn tires. Most of this waste consists of consumer products. We can send astronauts to the moon and machines to explore the outer reaches of the solar system, but we have not been able to find efficient ways of safely disposing of all the wastes that clutter our environment.

Garbage and Solid Wastes

The average American throws away almost 2 kilograms (about 5 pounds) of trash and garbage each day. **Garbage** includes food waste from domestic or commercial sources. It is **biodegradable**; that is, decomposers in the environment can break it down. **Trash** is non-food waste, such as glass, aluminum, some plastics, and tin cans. Some of the trash, such as wood, cardboard, and paper, is biodegradable. The combination of household and commercial garbage and biodegradable trash produced in the United States each year is more than 300 million metric tons.

The quantity of solid wastes generated within the United States is staggering. Every year we produce 15 million metric tons of paper products, 2 million metric tons of plastics, 30 billion bottles, 60 billion cans, 80 million lead auto batteries, 250 million tires, 9 million automobiles, and millions of major appliances. (See Figure 25-1.) Much of these materials is dumped into landfills or abandoned on streets. It is estimated that as many as 2 billion discarded tires have already accumulated in dumps.

The **waste stream** describes the flow of waste materials into the environment. Many of the materials in the waste stream would be valuable resources if they were not mixed with other materials. Since mixing makes separation both difficult and expensive, the materials wind up as part of the trash in the waste stream.

Figure 25-1 Leaving junked cars to rust and not recycling them wastes resources.

Landfills

When the quantity of waste materials produced by cities was small, it was common practice to bury trash and garbage in areas called **landfills.** (See Figure 25-2.) Bacteria in the soil of the landfill would break down most of the wastes, converting the biodegradable portion into carbon dioxide, water, and methane. However, the huge quantities

Figure 25-2 This landfill, in the Staten Island section of New York City, received much of that city's solid wastes. It was closed in 2001.

of wastes that the population now generates are often contaminated by toxic substances. These wastes also attract insects, rodents, and birds such as ravens and gulls. It is estimated that 75 to 85 percent of the waste generated in the United States ends up in landfills.

Because open landfills create risks to both public health and environmental quality, they are closely regulated and controlled. To decrease insect and rodent populations and to reduce unpleasant odors, landfills are required to compact and cover the wastes with a layer of soil every day. Seepage of rainwater can spread hazardous and toxic substances into adjoining areas. **Toxic substances** can cause damage to living tissues through contact or absorption. Impermeable linings of clay or plastic and drainage systems are now used to control seepage. To prevent contamination of groundwater, newer sites are located in areas away from rivers, lakes, and aquifers.

Incineration

Incineration, the burning of solid wastes, is another method used extensively by cities. (See Figure 25-3.) Since combustion is never complete, organic compounds, ashes, soot, and heavy metals often enter the atmosphere. Unburnable residues, which represent 10 to

Figure 25-3 Burning municipal wastes in incinerators reduces the volume they occupy in landfills.

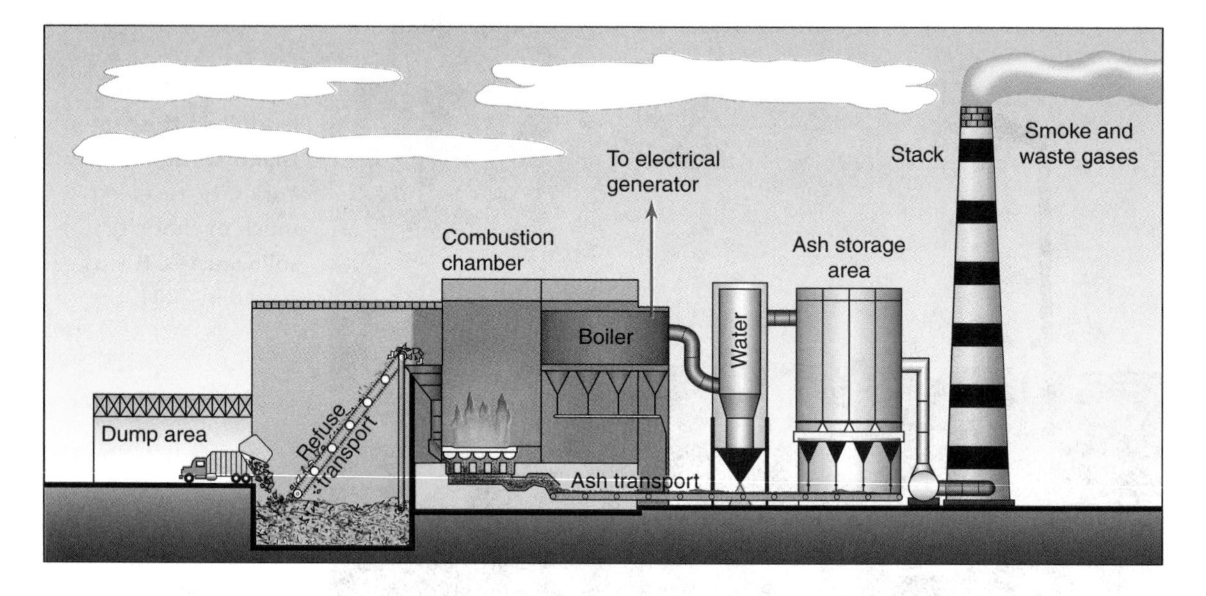

20 percent of the original volume of the waste, are usually disposed of in landfills. Disposal is easier since burning significantly reduces the volume of garbage. But the ash produced often contains toxic materials that can create environmental hazards if not disposed of properly. Incineration degrades air quality, reduces visibility, causes a variety of health problems, and damages the environment. Incineration often replaces problems of land contamination with problems of air and water pollution.

Sewage

Human and animal wastes cause the most serious health-related water pollution problems. More than 500 different types of pathogens, that is, disease-causing organisms, can be spread through human or animal excrement. This includes the bacteria that cause typhoid, cholera, and dysentery; viruses that cause hepatitis; and parasitic worms, flukes, and protozoa.

Early in human history, nomadic tribes simply left their wastes behind and moved on to new areas. But when humans abandoned the nomadic lifestyle and established cities, these habits could no longer be tolerated. However, the development of methods to deal effectively with the disposal of wastes took time.

The first city sanitation systems were built about 5000 years ago in cities of the Indus River valley. Sewers were built in Rome, Italy, more than 2000 years ago and are still in use today. These systems usually discharged wastes into nearby rivers, lakes, or seas. The sewers made cities more hospitable, but the problems of pollution were merely being transferred to other areas.

Even today in poorer countries of the world, people go out into the fields or forests to relieve themselves as humans did throughout history. Where the population density is low, natural processes can quickly and easily decompose these wastes. But the high population densities in cities and some rural areas make this impractical.

The problems of sanitation faced by large cities grew more and more complex and were the cause of public concern. This led to the development of the flush toilet, modern sewer systems, and wastewater treatment facilities. By controlling the disposal of wastes, cities became better able to protect public and environmental health. Cities utilize two different types of sewer systems. Sanitary sewers carry domestic and industrial waste. Storm sewers carry runoff from city streets. Since runoff often contains many undesirable materials, dumping it into the environment leads to pollution problems. Therefore, cities usually connect both systems and route all sewer water to a treatment facility.

Wastewater treatment facilities use a three-step sewage-treatment process. (See Figure 25-4.) **Sewage** is wastewater from homes, businesses, or industry that contains cooking, cleaning, or bathroom wastes. Primary treatment involves straining large solids from the waste stream. The remaining liquid, called **effluent**, is pumped into a sedimentation tank and allowed to settle. Organic materials collect on the bottom of the tank as a layer of sludge. **Sludge** is treated solid sewage or organic matter produced by sewage-treatment plants, paper mills, and refineries.

After settling, the liquid effluent is then pumped into a tank for secondary treatment, which consists of the biological decomposi-

Figure 25-4 Modern sewage treatment includes primary, secondary, and tertiary stages.

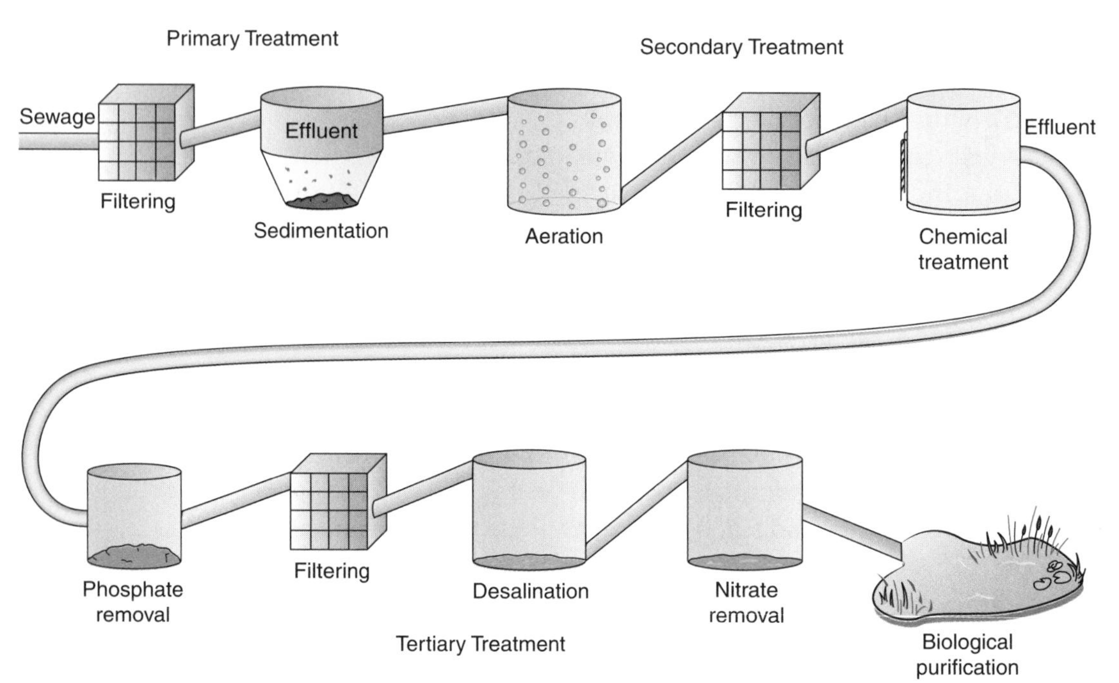

tion of suspended organic materials, filtration, and aeration. More sludge settles to the bottom and is separated from the remaining liquids. The sludge that is removed may be incinerated, composted, dumped in landfills, dumped into oceans, or dried and used as fertilizer. The wastewater effluent that remains is often treated with chlorine to kill any remaining pathogens.

Tertiary treatment refers to advanced procedures for treating the effluent from secondary treatment. It involves the removal of nitrates, phosphates, and organic materials. Rather than adding expensive chemicals, an alternate method of tertiary treatment is biological purification. The effluent percolates through large artificial lagoons or marshes. Here, algae and water plants, such as water hyacinth and duckweed, remove chemicals from the standing water. The green plants can detoxify many chemicals, using only the energy from sunlight. Following tertiary treatment, the water can be safely used for irrigation.

In 1948, only 33 percent of Americans were served by municipal sewage systems. By 1980, 70 percent of the urban population lived in areas that treated sewage. However, only 26 percent of Americans had water that got more than primary treatment. The Clean Water Act of 1972 required that by 1988 all cities supply at least secondary treatment to sewage wastes. By 1996, as a result of this act, only 10 percent of the United States population was not served by secondary treatment facilities. No major cities in the United States discharge raw sewage into rivers or lakes. Compare these conditions in the United States with those in Latin America, where only 2 percent of urban sewage receives any treatment.

Ocean Dumping

For many years, the oceans have been the dumping ground for solid wastes, untreated sewage, and sludge. The deep ocean basins were thought to be the most remote and therefore the safest places to dump wastes. However, we know too little about these areas to understand just how dumping affects them, or to determine the environmental changes that dumping creates.

Until recently, both municipal and industrial wastes were disposed of through ocean dumping. These materials often contained toxic substances and hazardous materials that then were dispersed through the environment.

Every year, thousands of metric tons of consumer packaging, including bottles, cans, and plastic containers, are dumped at sea. Toxic substances are being detected on beaches and in ocean trenches as far from civilization as Antarctica. The New York Bight is heavily polluted from years of waste discharge and ocean dumping from surrounding cities. Up until 1988, many cities in the United States dumped their municipal waste into the oceans. Federal legislation now forbids this practice. In 1992, New York City became the last municipality to stop offshore disposal of sewage sludge.

25.1 Section Review

1. Briefly describe the three steps involved in sewage treatment.
2. What are the drawbacks to incineration of solid wastes?
3. Contact your municipal sanitation department to determine the waste disposal practices used in your locality.

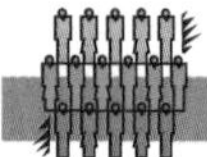

25.2 URBAN POLLUTION

Pollution is unwanted environmental change caused by human activity. A **pollutant** is a chemical, such as DDT and nuclear wastes, or a physical agent, such as light or heat, that when added to the environment threatens people, wildlife, plants, or the normal functioning of an ecosystem. Pollutants may be naturally occurring substances that have increased in concentration due to human activity, or substances that result from modern technology. A substance considered a pollutant under certain conditions may be a natural part of the environment under others. Smoke from a factory is a pollutant, especially if it contains toxic substances. However, smoke from wildfires is a normal part of the environment.

As cities grow, they change the natural environment by using water, food, and energy in greater amounts. Wastes are emptied into the ecosystem. The burning of home heating fuel puts tons of gases and smoke particles into the atmosphere. Gasoline, burned to run public transportation as well as automobiles, spills gaseous pollutants into the air. Power plants burn fossil fuels and contaminate the air with their emissions. Agricultural regions, which supply food to the city, produce chemical runoff that pollutes the environment.

The waste products of the city's metabolism impact directly and indirectly on the whole biosphere. Not only does the city pour chemical pollutants into the atmosphere and hydrosphere, but also high levels of noise, heat, and light all adversely affect the environment. It has become increasingly clear that drastic strategies are required to reduce the quantity of pollutants we pour into the atmosphere and to remove those already there. To preserve Earth for future generations, many people think that we must *reduce, reuse,* and *recycle.*

Thermal Pollution

Waste heat is generated during the burning of fossil fuels and in nuclear reactors that produce electricity for cities. Almost two-thirds of the heat content in fuels is released into the immediate environment. **Thermal pollution** is waste heat released into the environment. When waste heat is released into rivers or other large bodies of water, it raises the temperature of the water, changing the ecosystem. The heat decreases the amount of oxygen dissolved in the water, which can kill some aquatic organisms. Higher-than-normal water temperatures can also interfere with species' migrations, spawning, the hatching of eggs, and the development of young. Thermal pollution is by far the most common environmental pollutant in urban areas.

Due to thermal pollution, the local climate of a large city is often very different from that of surrounding areas. The most obvious difference occurs in nighttime temperatures. During the daytime, the concrete and asphalt pavements absorb more heat than areas covered by vegetation. (See Figure 25-5 on page 442.) At night, the heat is released from the concrete and asphalt, which warms the air. Where large city buildings reduce surface winds, the warm air is kept from circulating and being blown away. Pollution particles in city air trap the heat and add to the atmospheric warming.

Figure 25-5 Why is the climate in a large city different from that of surrounding areas?

Light and Noise Pollution

Light pollution refers to light from cities that interferes with observations of the night sky. In some cities, half the light produced is wasted on the sky. Many modern city residents have never seen the full glory of a starry sky. Instead of being able to see thousands of stars, only several hundred may be visible from a big city. Light pollution around cities has greatly reduced the effectiveness of optical telescopes in many observatories.

Noise pollution is an increasing problem in industrialized society. **Noise** is annoying or undesirable sounds. It disrupts our activities, disturbs our sleep, and interferes with our concentration. Noise can cause stress, raise blood pressure, and damage hearing.

Noise levels are measured in units of sound called **decibels**. A quiet, undisturbed room might have a decibel level between 0 and 20. A library reading room is about 30 decibels. Normal conversation is about 60 decibels. A truck or motorcycle engine is in the 90-decibel range. A heavy metal rock concert is more than 100 decibels. A jet plane taking off is between 120 and 150 decibels. A rocket engine is 180 decibels. Prolonged exposure to sounds louder than 90 decibels can cause hearing loss by permanently damaging the middle ear.

25.2 Section Review

1. Describe some of the factors that make the climate in cities different from that of surrounding areas.
2. Explain the condition under which smoke would not be an atmospheric pollutant.
3. The yellow-flowered forsythia is one of the first shrubs to bloom in spring. The flowering is triggered by temperature. Chart the dates for the flowering of forsythia, or other early spring flower, in 10-kilometer zones that extend out from your city.

25.3 ALIEN INVADERS

As humans have spread out over Earth, they purposely and accidentally brought other organisms along with them. These organisms can be thought of as biological pollutants. Many of these creatures have caused irreversible damage to the natural environment. Freed from predators and competitors that would normally keep their numbers in check, these immigrant organisms rapidly increased their populations and crowded out native species. **Alien organisms**, also called exotic species, are plants and animals that are introduced into a new environment from an outside source. **Native organisms** are the original animal and plant inhabitants of an ecosystem.

Trains, trucks, cars, boats, and planes accidentally carry alien organisms as stowaways into new areas. Species arrive constantly, and some find the right conditions in the new environment to become firmly established. The pineapple weed that came from northeast Asia is now a common roadside weed in North America. Its seeds cling to mud and thus to car tires. Cars have spread the plant quickly and efficiently to all areas reached by our modern highway system.

Plant and animal species accompany people wherever they go. Some organisms may be brought along purposely as pets, such as dogs and cats, while others are brought along unintentionally, such as fleas, roaches, mice, and rats. Some insects come along with our food. The doglike dingo was transported to Australia more than

TABLE 25-1 THE IMPACT OF ALIEN SPECIES IN THE UNITED STATES

Species	Origin	Impact
African "killer" bees	Africa	Displaced native honeybees and are responsible for human deaths
Argentine fire ant	Argentina	Crop damage
Dutch elm disease	Europe	Killed American elms
European starling	Europe	Competition with native songbirds
European gypsy moth	Europe	Destroys crops and vegetation
Kudzu	Japan	Crowded out native plants
Nutria	Argentina	Crop destruction; habitat destruction
Sea lamprey	North Atlantic Ocean	Destruction of Great Lakes trout and whitefish
Water hyacinth	South America	Clogs waterways

3000 years ago by aboriginal colonizers. It was, in some ways, responsible for the extinction of the native thylacine "wolf." Today, it is a pest that preys on native kangaroos and wallabies, as well as on domesticated sheep and calves.

Many of the new species in an area are hardly noticed, while others become severe problems. (See Table 25-1.) The Japanese beetle accidentally brought here from Asia has become an agricultural pest. These insects eat leaves, flowers, and fruit. The young insects, called grubs, burrow into the soil where they eat the roots of plants. They are responsible for the destruction of many valuable flowers and fruits. (See Figure 25-6.)

The mongoose is a small mammal that was introduced into Puerto Rico in 1877 to control the rat population that had been particularly destructive to the sugarcane crop. For a time, the rat population was reduced. But the rats learned to avoid predation by mongooses by climbing into trees. The mongoose turned to other sources of food, attacking poultry, small birds, and lizards. The reduction in the lizard population led to an explosion in the population of white grubs, the larvae of the June beetle. The white grubs soon replaced the rats as pests by damaging many crops.

The nutria, a fur-bearing rodent, was introduced into the United States in the early part of the twentieth century. People hoped to breed them for their fur. When it was discovered that nutrias do not breed well in captivity, they were released into the wild. Since their release, they have established themselves over large areas of marshy

Figure 25-6 The harmful Japanese beetle eats plant leaves, flowers, and fruit.

land in Texas, Louisiana, and Mississippi. Here the nutrias compete with the native wildlife for food and habitat. Nutrias also are agricultural pests that feed on and damage root crops.

The Argentine fire ant arrived in the 1890s from Brazil in coffee shipments. With no natural enemies here, they have spread through much of the southern United States, causing the displacement of native ants, the destruction of crops, and the death of ground-nesting birds. Scientists had predicted that fire ants could not tolerate cold winters. However, a new type of fire ant has been discovered that burrows deep underground to escape the cold. This may enable these ants to extend their range. (See Figure 25-7.)

Kudzu is a fast-growing Asian vine that was imported and planted in the United States to control erosion. This plant can grow a foot a day and often chokes other vegetation. Kudzu's rapid growth is responsible for more than 50 million dollars in annual losses to the farm and timber industries.

Figure 25-7 The alien invaders nutria and fire ant damage crops.

25.3 Section Review

1. What are some of the factors that contribute to the rapid population growth of an alien species?
2. Describe some of the methods by which alien organisms enter new territories.
3. Investigate an alien organism in your locale. Determine where it originated and how it got to your area. What are some of the effects it has had on your local environment?

LABORATORY INVESTIGATION 25

Atmospheric Particulates

PROBLEM: *To observe particulate matter in the atmosphere*

SKILLS: *Observing, classifying*

MATERIALS: *Microscope, 2 coverslips, petroleum jelly, 2 microscope slides, hand lens, glass-marking pencil*

PROCEDURE

1. Label one microscope slide A and the other B. Apply a thin coating of petroleum jelly to one side of each slide.
2. With the coated surface facing upward, place slide A on an outside window ledge that is sheltered from the rain. Place slide B on an indoor surface. Leave them uncovered and exposed to the air for 24 hours.
3. Retrieve the slides and examine the petroleum-jelly-coated surface of each with a hand lens.
4. Place a coverslip on each slide and examine them with the low-power objective of the microscope.
5. For each slide, categorize the types of particles you find on the petroleum jelly. Try to distinguish among soot, ash, dust, spores, and pollen.
6. For each slide, count the number of particles of each type within the field of view. This is the particulate count.
7. Using the steps above, take the particulate count each day for one week. Note the weather for each day during this period. Prepare a table that indicates the atmospheric particulate count observed over the one-week period.
8. Compare the weekly particulate count with local weather for the same period.

OBSERVATIONS AND ANALYSES

1. Describe what you observed on the slides you examined with the hand lens.
2. Where did this material come from?
3. Were the particulate counts higher indoors or outdoors?
4. How was the particulate count affected by local weather?

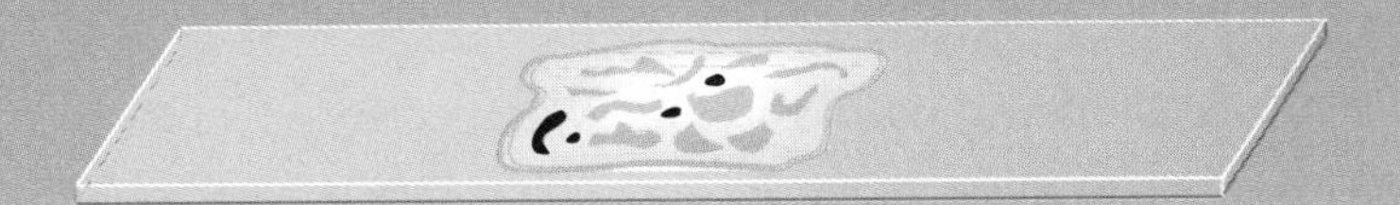

Chapter 25 Review

Answer these questions on a separate sheet of paper.

Vocabulary

The following list contains all the boldfaced terms in this chapter.

alien organisms, biodegradable, decibels, effluent, garbage, incineration, landfills, light pollution, native organisms, noise, pollutant, pollution, sewage, sludge, thermal pollution, toxic substances, trash, waste stream

Fill In

Use one of the vocabulary terms listed above to complete each of the following sentences.

1. ________ materials are broken down naturally by decomposer organisms.
2. Living tissues can be damaged by ________ in the environment.
3. Many cities bury trash and garbage in areas called ________.
4. ________ is undesirable environmental change due to human activity.
5. Species that are introduced into a new environment are called ________.

Multiple Choice

Choose the response that best completes the sentence or answers the question.

6. The amount of trash and garbage thrown away each day by the average American is *a.* 2 kilograms. *b.* 20 kilograms. *c.* 2 metric tons. *d.* 20 metric tons.
7. An example of a material that is considered to be garbage is *a.* wire hangers. *b.* plastic grocery bags. *c.* steak bones. *d.* Styrofoam packing.
8. An example of a material considered to be trash is *a.* potato skins. *b.* aluminum pie pans. *c.* eggshells. *d.* coffee grinds.
9. Where were the first city sanitation systems built about 5000 years ago? *a.* Great Britain *b.* Crete *c.* Troy *d.* Indus Valley
10. What types of wastes are removed in the primary treatment of sewage? *a.* effluents *b.* organic chemicals *c.* solids *d.* nitrates and phosphates

11. The solid part of treated sewage is called *a.* refuse. *b.* sludge. *c.* toxins. *d.* hazardous waste.
12. What percent of the total waste generated in the United States winds up in landfills? *a.* 10 to 20 percent *b.* 30 to 40 percent *c.* 55 to 65 percent *d.* 75 to 85 percent
13. An example of an alien species that has established itself in the United States is the *a.* water hyacinth. *b.* pineapple. *c. Aedes* mosquito. *d.* whitefish.
14. The flow of waste materials is called *a.* the waste stream. *b.* pollution. *c.* toxins. *d.* hazardous materials.
15. What type of pollution is caused by waste heat? *a.* light *b.* noise *c.* thermal *d.* organic

Short Answer (Constructed Response)

Use the information you learned in this chapter to respond to the following items.

16. How is sludge different from effluent?
17. How is garbage different from trash?
18. Name the types of pollution.
19. How does light pollution affect human activities?
20. What are three dangers of noise pollution?

Essay (Extended Response)

Use the information you learned in the chapter to respond to the following items.

21. How do large cities cause changes in the climate of their region?
22. What are some of the factors that led to legislation against ocean dumping?
23. Why do alien organisms often win the battle for survival against competing native organisms?
24. Write an essay or story that describes what your local environment will be like in 50 years if present practices in regard to the environment are not changed.

UNIT SEVEN

EARTH'S RESOURCES

All living things depend on Earth's natural resources. For a long time, modern societies acted as if they could squander Earth's natural resources because there would always be more where they came from.

As resources are used, the biosphere is changed, often for the worse, as illustrated by the false-color image of Indonesia, which shows the smoke from forest fires near cultivated fields. Environmental problems arise as large quantities of resources are consumed. The consumption of resources produces wastes, some of which cannot be decomposed or recycled.

If we want to preserve our environment, we must change the ways we use our resources. In addition, we must learn to produce wastes in a form and at a rate that the biosphere can absorb effectively.

CHAPTER 26
Earth's Riches

When you have completed this chapter, you should be able to:

Discuss the intelligent use of nonrenewable resources.

Explain why renewable resources must be conserved.

Describe the ways the environment has been degraded.

Discuss the ways we can preserve our natural resources.

Earth's riches divided into two categories—renewable and nonrenewable resources. Nonrenewable resources, such as coal or mineral ores, cannot be replaced once they have been removed from Earth's crust and consumed. On the other hand, renewable resources, such as agricultural crops, forests, and marine fisheries, have the potential to be restored after they have been harvested. Earth's resources affect the size of the human population that the biosphere can support. Human societies depend on natural resources to support their population and economic growth. In this chapter, you will learn about Earth's resources.

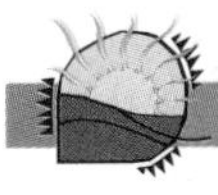

26.1 NONRENEWABLE RESOURCES

A **natural resource** is a useful material that living things get from the environment. Earth's crust is a vast storehouse of natural resources, some renewable and others nonrenewable. The **nonrenewable resources**, which are depleted with use, include fossil fuels, minerals, and fertile soils. The extraction and use of these resources reduce the amount available and lead to scarcities. Since these resources are limited, continual use will eventually cause them to run out.

Though vast amounts of resources may be available, they are useful only if they can be recovered. Recovery involves locating deposits of a resource, removing the material that contains the resource, and separating that material to obtain the resource. For example, many tons of ore have to be mined, transported to mills, and then refined to extract small quantities of precious metals such as gold or silver.

Minerals

A mineral is a naturally occurring element or inorganic compound obtained from Earth's crust. (See Table 26-1.) Rocks that contain useful minerals are called ores. An ore must be refined, or separated, to

TABLE 26-1 METALS OF ECONOMIC VALUE

Metal	Ores	Uses
Aluminum	Bauxite	Food and beverage packaging, electronic devices, vehicles, cookware
Chromium	Chromite	Production of steel alloys, plating
Copper	Chalcopyrite, malachite	Electrical and electronic industries, cookware
Iron	Hematite, magnetite, taconite	Manufacture of steel, production of machinery, automotive industry
Lead	Galena	Batteries, paint
Mercury	Cinnabar	Electronics, thermometers, batteries
Nickel	Garnierite	Production of steel alloys, chemical industry processes, batteries

obtain the desired mineral. Copper and iron are metallic minerals, sulfur and phosphorus are nonmetallic minerals, and salt and calcium carbonate are mineral compounds.

Strategic minerals are those minerals a country uses but does not produce itself. A shortage of strategic minerals can cripple the economy of a nation. For this reason, wealthy industrial nations stockpile strategic minerals in times when prices are low and supplies are abundant. The United States is dependent on foreign sources for such strategic minerals as aluminum, tin, chromium, and manganese. Developing nations depend heavily on the export of strategic minerals for foreign exchange.

Fossil Fuels

Fossil fuels are the hydrocarbon remains of plants or animals that have been changed by natural processes. These substances include coal, crude oil or petroleum, natural gas, and peat. Hydrocarbons are composed mainly of the elements hydrogen and carbon. However, because fossil fuels were derived from things that were once alive, they also contain nitrogen and sulfur, which are found in proteins. Fossil fuels can be burned to release energy, and this is a major source of the world's energy and a primary cause of environmental pollution. The combustion of all hydrocarbon fuels releases carbon dioxide gas into the atmosphere, which is believed to contribute to global warming through the greenhouse effect. Combustion of fossil fuels also produces oxides of sulfur and nitrogen.

Fossil fuels are at the center of many worldwide environmental and political concerns. Petroleum and natural gas are not distributed evenly throughout the world; many of the deposits are located in the Middle East. The dependence of the United States and Europe on these reserves is a continuing source of diplomatic and military conflict. The energy crisis of the 1970s and the Persian Gulf War were motivated by reliance on fossil fuels as a source of energy.

Peat is the residue of partly decomposed plant material from ancient swamps, bogs, and marshes. (Refer to Chapter 20.) Peat is a low-grade fuel, which produces little heat. For centuries, people in northern countries such as Scotland and the Netherlands have been using peat as a fuel, because it is more abundant in these areas than are coal and oil.

Coal is a brownish-black, combustible, sedimentary rock. It is derived from plants that lived about 300 million years ago. Over time, geologic processes exerted pressure on the organic materials and changed them. (See Figure 26-1.) Most coal contains small amounts of sulfur, nitrogen, and other trace elements. Although coal is widely distributed around the world, the United States, China, and the former Soviet Union account for more than half the recoverable reserves.

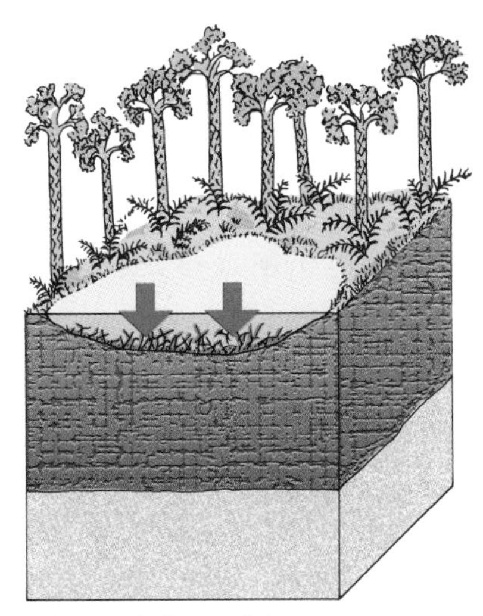

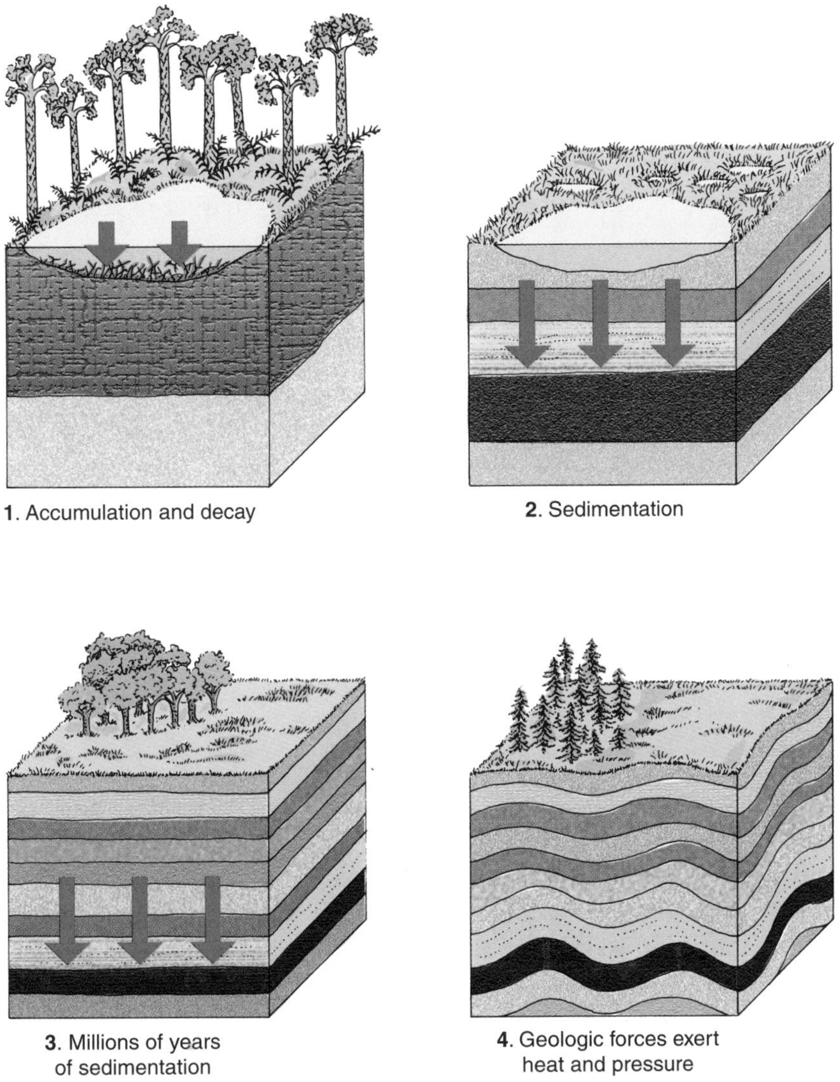

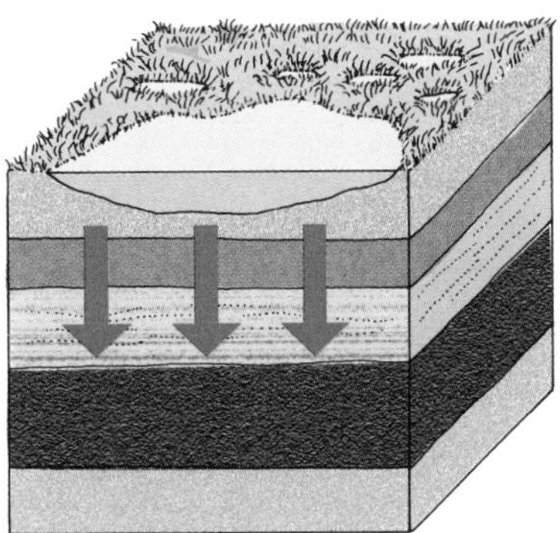

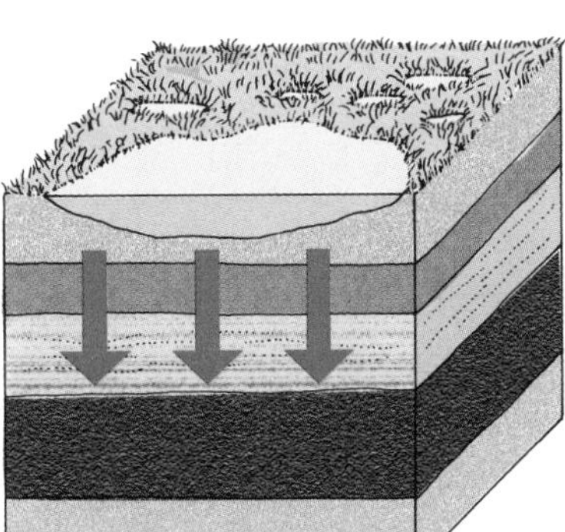

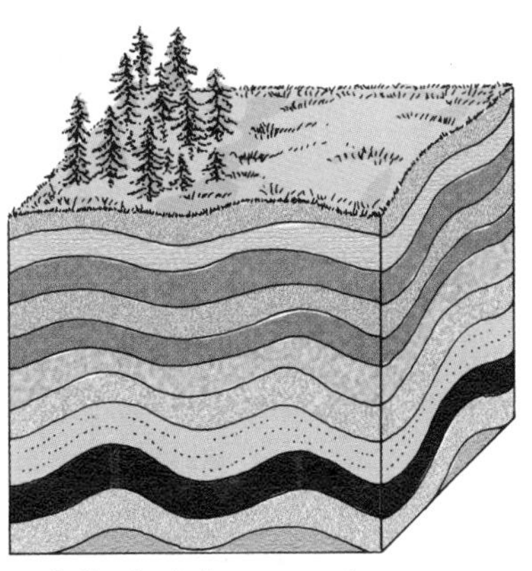

Figure 26-1 **Coal is formed over time from the remains of plants.**

Petroleum is a naturally occurring, combustible, liquid hydrocarbon. Petroleum deposits formed from organic debris that sank and collected at the bottom of ancient seabeds many millions of years ago. High pressure and temperature caused chemical reactions that led to the formation of both petroleum and natural gas. Petroleum deposits have been found on every continent as well as beneath the continental shelf. Drilling a well into the deposit allows the compressed natural gas to force the petroleum to the surface, where it can be collected. Modern oil-drilling techniques began in Pennsylvania in 1859.

The known reserves of petroleum in the world are estimated at 3.5 trillion barrels. The current reserves, those that are known and can be recovered, are about 1 trillion barrels. At our current consumption rate of 20 billion barrels per year, these reserves will be used up in 50 years. It is estimated that another 800 billion barrels may become economically feasible to bring to market in the future when prices rise. Some reserves may still lie undiscovered.

Natural gas is a mixture of hydrocarbon gases; the main component is methane. It is a convenient, inexpensive, and relatively clean fuel compared with coal and oil. When burned, natural gas produces few pollutants and far less carbon dioxide than other hydrocarbon fuels. It is extracted from wells and transported over land through pipelines. Natural gas cannot be readily transported over oceans in its gaseous form and therefore must first be liquefied, which adds to its cost. A worldwide change to natural gas as a fuel could reduce global warming. But the world is approaching the limits of easily accessible reserves.

Topsoil

Soil is a complex, layered mixture of weathered rocks and minerals, decomposed organic material, and living organisms. The uppermost soil layer, called topsoil, ranges in thickness from about a meter in grasslands to being just about absent in some deserts. Topsoil builds up in a slow, natural process. The development of most terrestrial ecosystems relies on the materials found in topsoil. Fertile, tillable soil for growing crops is a significant agricultural resource and, with care, can be sustained indefinitely. But erosion, degradation, and

Figure 26-2 Erosion forms gullies on land stripped of vegetation.

the depletion of nutrients can affect the ability of soil to sustain agriculture. (See Figure 26-2.)

New topsoil usually forms to replace the soil eroded away. In those areas where the people have changed the natural vegetation for agricultural purposes, erosion rates have increased dramatically. This usually is due to a decrease in ground-cover plants, which exposes large surface areas to wind and rain. When the rate of erosion is greater than the rate at which new soil forms, the productivity of the soil decreases.

A decrease in soil productivity increases the cost of food production since chemical fertilizers must be added to ensure the growth of crops. Irrigation has to be intensified because of increased runoff from the degraded soil. Herbicides have to be applied, since the exposed, nutrient-poor soil is the perfect habitat for weeds. Soil that is washed away may collect in irrigation systems, rivers, and reservoirs, thereby interfering with their use.

26.1 Section Review

1. What steps are necessary to recover a mineral resource?
2. Describe the role strategic minerals play in the world economy.
3. List the different metals found in your home and describe their uses.

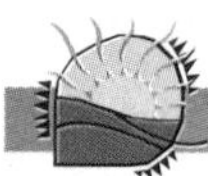

26.2 RENEWABLE RESOURCES

After harvesting, **renewable resources** can be replaced. Timber, marine fisheries, and agricultural crops are examples of renewable resources. (See Figure 26-3.) These biological resources, powered by sunlight, are basically self-renewing.

Since many renewable resources have critical population sizes and habitat preferences, overharvesting and mismanagement can lead to diminished stocks and eventual extinction. Overharvesting caused the American bison, beaver, and sea otter populations to become endangered at one time. The extinctions of the Steller's sea cow, great auk, and passenger pigeon were caused by overharvesting. Many marine fisheries have been overharvested in recent years, causing their populations to reach dangerously low levels.

Renewable resources can become scarce when the demand exceeds their rate of replenishment. Many forests disappeared as the demand for firewood, timber, grazing, and farmlands increased. Now some forests are selectively logged or replanted to prevent this.

Agriculture

Crops supply food to people and livestock. Though considered a renewable resource, agricultural lands can be depleted of nutrients and become unfit for farming. To meet their economical needs, farmers often must employ practices that deplete the resources

Figure 26-3 **Tree farms produce wood, a renewable resource.**

within the soil. This leads to a spiraling need for chemical fertilizers, pesticides, and irrigation to produce crops. The escalating costs of these materials can lead farmers into debt. Thus they must raise food prices and attempt to increase the yields from yearly harvests.

In this hungry world, more and more land is being converted to agricultural use. But this often only worsens the crisis, as the quest for economic gain depletes the resources of the ecosystem. To sustain agricultural productivity, we must preserve our soil resources through prudent and practical farming practices. The growth of crops for human consumption is actually more efficient and productive than growing crops as feed for livestock, which then are used as food by people. Energy is lost when humans act as secondary consumers rather than primary consumers.

Fisheries

Fishing for food is probably as old as humanity. The use of fishing gear and fishhooks is thought to go back more than 10,000 years. On cave walls, pictographs more than 5000 years old show humans fishing. For many years, we have considered our fish stocks as being indestructible. But in many parts of the world, fish stocks have been seriously depleted due to overharvesting. Technological developments in the fishing industry have produced huge, high-strength nets that have made possible larger catches than in the past. Modern fishing boats are larger, can remain at sea for longer periods, and can clean, cook, store, and freeze their catch so that it can reach markets without spoiling.

Our global fisheries are a resource that must be shared by all nations, because neither the fish nor the ocean ecosystems recognize national boundaries. The 200-nautical-mile economic zone enjoyed by nations that border the oceans and seas does not serve as a barrier to migrating schools of fish. The protection of fish stocks therefore requires international cooperation.

26.2 Section Review

1. Why is wood considered to be a renewable resource?
2. Compare today's fishing boats with those in use 100 years ago.

3. Collect five different types of paper products used in your home. Describe the differences in texture, strength, and color among the various types.

26.3 DEGRADATION OF THE ENVIRONMENT

On every continent, human activities have changed productive land into deserts. This process, called **desertification**, is caused by overgrazing; deforestation for agriculture, timber, and firewood; poor irrigation techniques; soil depletion through unsound agricultural practices; and global warming. The creation of deserts has led to famines that kill millions of people each year. (See Figure 26-4.)

Land Degradation and Desertification

Forestlands are lost when trees and other woody plants are chopped down for timber or firewood, or to clear land for agriculture. When the vegetation is removed, the soil no longer absorbs water from rain. As the water runs off the land, it carries away the topsoil because there are no root systems left to bind the soil.

Overgrazing by cattle, sheep, and goats can also lead to desertification. When these animals graze on dry grasslands, they eat more vegetation than the land can replace. Their hooves compact the soil so that water cannot penetrate it. This prevents plants from receiving nourishment and leads to the loss of covering vegetation. With no roots to hold the soil, erosion proceeds rapidly.

The development of agriculture probably has done more to alter the environment than any other human activity. As soon as people began to plant for harvest, they changed their ecosystem. To plant crops, native grasses had to be destroyed and trees and shrubs cut.

Faulty agricultural practices often lead to degradation of the environment and desertification. Overfarming, or planting the same crop in the same place year after year, causes soil depletion. When

Figure 26-4 These sand dunes in the Sahara were photographed from the Space Shuttle *Columbia.*

crops are grown in soil that is low in nutrient content, the land needs time between plantings to restore lost nutrients. Continued cultivation leads to erosion of topsoil by wind and rain. Runoff from cultivated fields, which may have inadequate ground cover, hastens erosion and washes away nutrients. Ground cover reduces soil erosion. Leaves protect the soil from rain while roots hold the soil particles in place, keeping them from being washed away. The use of chemical fertilizers, herbicides, and pesticides also affects soil productivity.

Environmental Contamination

Toxic substances in the environment cause many human diseases and disorders. Some of these poisonous substances are pollutants that enter the environment through industrial processes, the combustion of fuel, urban wastes, and agricultural practices. When consumed in large quantities, almost any substance can be toxic, even common food items such as sugar and salt.

INSECTICIDES AND PESTICIDES

In addition, some poisons are purposely introduced into the environment as insecticides and pesticides. These substances are supposed to kill pests while leaving people and other desirable species unharmed. However, many pesticides and insecticides have been found to be harmful to a variety of organisms. Here is an example.

DDT is an insecticide that was used extensively after World War II. Later, scientists discovered that DDT interfered with reproduction in many bird species because it caused thinning of their eggshells. It was also linked to breast cancer in humans. DDT, dioxin, PCBs, and other chlorinated hydrocarbons are persistent chemicals. A **persistent chemical** breaks down slowly, so it remains in the environment for a long time. Chlorinated hydrocarbons are insoluble in water but highly soluble in fat. Once ingested, they cannot be excreted easily from the body. Accumulation of these chemicals in living tissues can have long-term toxic effects. As shown in Figure 26-5, the higher an organism is on the food chain, the greater the quantity of these persistent chemicals that it accumulates in its tissues.

Hazardous substances pose a threat to human health or to the environment. Hazardous substances include strong acids or bases; solvents such as toluene, benzene, and acetone; heavy metals such

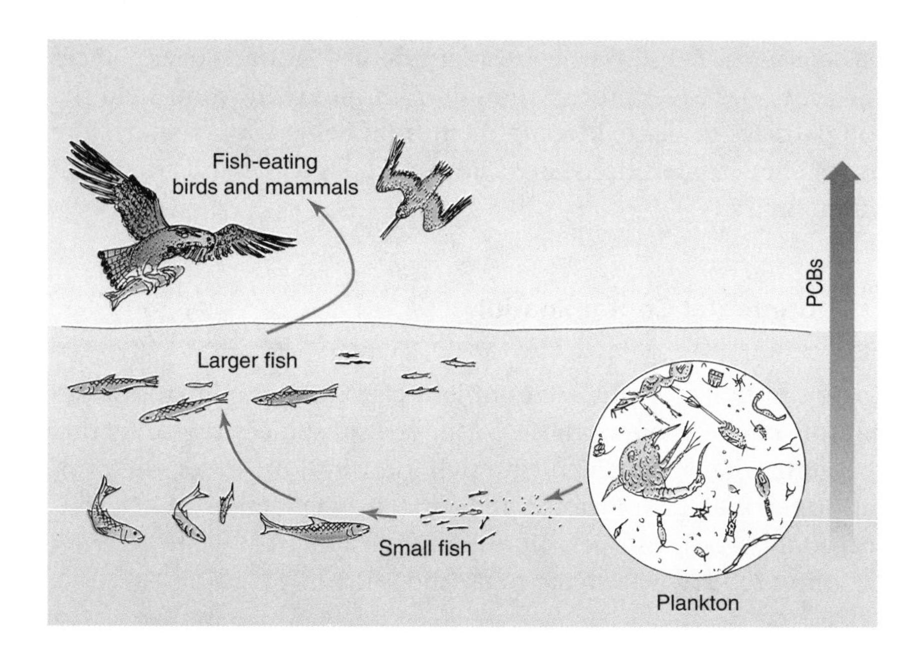

Figure 26-5 PCB concentration in organisms increases as you go up a food chain.

TABLE 26-2 HAZARDOUS INDUSTRIAL AND BUSINESS WASTES

Industry or Business	Wastes Generated
Chemical	Acids, bases, solvents, reactive materials
Construction	Paint wastes, solvents, inorganic dust
Dry cleaning	Chemical solvents
Electronics	PCBs, heavy metals
Medical	Contaminated dressings, biologicals, used syringes and needles, blood and blood products
Paper	Paint wastes with heavy metals, flammable solvents, acids and bases
Printing	Solutions of heavy metals, ink, dyes, solvents

as mercury, arsenic, lead, cadmium, and antimony; and medical wastes.

Hazardous wastes are those wastes that are toxic, corrosive, flammable, reactive, or cause disease. Toxic wastes are poisons that are harmful when ingested or absorbed. Corrosive wastes eat through metals and can harm or irritate living tissues. Flammable wastes readily catch fire. Reactive chemicals are unstable and when mixed with water can explode or release toxic fumes. (See Table 26-2.)

In the past, hazardous waste materials were often buried in pits, pumped into rivers and streams, or stored in large metal drums. Over time, many of the drums corroded, allowing the wastes to leak into the environment and poison large areas. Poisonous chemicals built up in our waterways, reaching dangerous levels. The cleanup of these areas is an ongoing environmental problem.

The famous Love Canal area near Niagara Falls in New York State was a hazardous waste landfill for industrial chemicals. After the landfill was closed, houses and schools were built on the site. Over time, the chemicals seeped into the basements of nearby homes and created health problems, such as cancers, for residents. The population had to be relocated to protect them from the contaminants. It is estimated that 80 percent of the population of the United States lives somewhere near a hazardous waste site.

Household hazardous waste is a significant cause of environmental pollution. These wastes include paint, pesticides, cleaning chemicals, batteries, pharmaceuticals, and automotive oil. The amount of used oil discarded each year is equal to almost 30 *Exxon Valdez* oil spills. Household batteries are a source of heavy metals such as nickel, zinc, mercury, and cadmium. When batteries are disposed of improperly, they leak these dangerous metals into the environment. Consumer electronics are also sources of heavy metal contaminants.

MINING

Mining has always had an impact on the environment. More material is mined from Earth's crust each year than all the world's rivers erode from the landscape. In recent years, growing public concerns about the environment have led to strict government regulation of the mining industry. Mining can create hazards to both the public and the environment. Networks of abandoned mine tunnels are often near urban areas. In addition to containing hazardous substances, these tunnels can collapse, or subside, producing environmental disasters. Unsightly piles of mining waste, called tailings, can litter the landscape. (See Figure 26-6.) Tailings can contain toxic heavy-metal contaminants such as arsenic, mercury, and cadmium. Precipitation can cause these toxic materials to leach from the tailings and leak into the ecosystem. The toxic materials are carried away in runoff and contaminate streams and watersheds.

Coal-mining operations create many environmental problems. Miners are often subjected to health hazards from dust that causes such illnesses as black lung disease and emphysema. Surface min-

Figure 26-6 **Tailings around an abandoned mine leach pollutants into the environment.**

ing activities scar the landscape and produce runoff that pollutes streams. The combustion of coal pollutes the atmosphere, and the high sulfur emissions contribute to the problems of acid rain. **Acid rain** is precipitation that contains sulfuric and nitric acids, which form when sulfur and nitrogen compounds combine with moisture in the air.

In open pit and strip mining, the rock that covers mineral deposits is removed. The ore is then mined and transported away. The dug up area may be replaced with fill, forming long ridges called spoil banks. These banks are susceptible to erosion because they have no natural covering of topsoil and are slow to regenerate. Natural succession is limited, and the original biotic community does not return. These areas often mar otherwise beautiful landscapes. Today, federal and state laws require that after mining is complete, mining companies restore the land to its original vegetation and appearance, a process called **reclamation**.

The processing of mineral ores to produce metals can release toxic materials that are more hazardous to the ecosystem than the mining activities. Smelting and roasting are processes that heat ores to release metals. The roasting process is a major cause of air pollution since it produces sulfur emissions. These emissions can form acids, which can destroy ecosystems thousands of miles away from the processing plant.

FOSSIL FUELS

The operations of the petrochemical industry offer many opportunities for pollutants to escape into the environment. Even small amounts of some hydrocarbons can build up over time and degrade both human health and the environment. Combustion of petrochemical fuels releases atmospheric pollutants such as carbon monoxide, sulfur dioxide, nitrogen oxides, and hydrocarbons. Drilling operations have been responsible for fatal fires and explosions, as well as air, water, and ground pollution.

Oil spills are caused by natural seepage from offshore deposits, well blowouts, breaks in pipelines, and tanker spills. Petroleum is transported across oceans in large quantities by huge oil tankers. In 2002, the *Prestige* snapped in two and sank off the coast of Spain, spilling nearly 10,000 tons of its 77,000-ton oil cargo. Leaking may continue into 2006. The spill endangered fishing grounds and beaches. Though marine oil spills are not frequent, they release almost 4 million metric tons of oil into the world's oceans each year. They cause the most visible of all environmental disasters and are responsible for the most outspoken public reactions.

The 1989 *Exxon Valdez* oil spill in Prince William Sound off the coast of Alaska took many years to clean up and cost billions of dollars. Many groups concerned about the long-term effects of drilling operations and spills oppose proposals for more exploratory drilling in Alaska.

The extraction of natural gas from deep wells can cause subsidence, a sinking of Earth's surface. In the Long Beach Harbor in Los Angeles, California, the ground over well sites has dropped almost 10 meters. The draining of bogs for peat production also damages local ecosystems, watersheds, and wildlife.

26.3 Section Review

1. Why are persistent chemicals most dangerous to the top predators in food chains?
2. How can tailings contaminate streams and watersheds?
3. Monitor your household hazardous wastes for one week. List any potentially hazardous materials that have been discarded. Contact your municipal sanitation department to determine the proper methods of disposal for these materials.

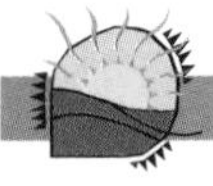

26.4 PRESERVING OUR NATURAL RESOURCES

The production of goods requires the use of natural resources. The continued overuse of nonrenewable resources denies them to future generations. In a world where price is based on the quantities available and user demand, the cost of nonrenewable resources rises as they are depleted. The cost of renewable resources rises solely with demand, since they are replenished as they are used.

Conservation is the wise and careful use, protection, and preservation of our resources to prevent their exploitation, destruction, or neglect. Resource management is the purposeful action taken to reduce the effects of overuse of resources. In primitive societies, resource management was not essential, since there were natural controls on overuse. In technological societies that have high population densities, management is essential since limited resources are available.

Agricultural and Forest Management

Managing the soil through prudent agricultural practices will insure that future generations will not go hungry. (See Figure 26-7.) Some efficient soil management practices include:

- **Terracing** is the cutting of flat areas, called terraces, into steep hillsides that are susceptible to erosion. This diminishes the effects of runoff and protects against the loss of topsoil.
- Rotation of crops by varying the crops planted and introduction of crops such as peas and alfalfa, which naturally restore nutrients to the topsoil.
- The use of crop residues and mulch (sawdust, paper, compost) as organic fertilizers, which reduces the reliance on expensive chemical fertilizers and protects productivity of the soil.
- Irrigation practices that protect against soil erosion can increase yields without the danger of topsoil loss.
- Reduction of livestock herds prevents overgrazing and compacting of the topsoil.

The United States Forest Service administers the 200 million hectares of public land that make up the 154 national forests. Within the national forests, land is available for growing commercial timber, mining, and recreation. Management decisions are

Figure 26-7 Terracing (left) and strip cropping (right) are soil conservation practices.

under constant criticism from a wide variety of groups. Conservationists feel that too much timber is taken from these lands by the lumber industry and that it is overgrazed by ranchers. Others feel that the United States Forest Service is insensitive to the economic needs of local communities that wish to increase the use of the forests. As forest use increases due to exploitation for recreational purposes, the character of the ecosystem changes. The once-pristine lands may become disfigured and marred with wastes.

Environmental Cleanup

Cleaning up after an oil spill requires many types of tools. Floating barriers called booms are used to collect the oil. Boats skim spilled oil from the water's surface. Large sponges soak up oil. Chemical dispersants break down oil into its components. Oil can be burned as it floats on the water's surface. Special vacuums can remove oil from the water or the beach. High-pressure hoses are used to wash the beaches of oil. Heavy equipment is used to move oil-saturated sand to areas where it can be washed.

Society is being overwhelmed by the problem of disposing of almost a billion metric tons of garbage and solid waste that are produced each year. The economic and environmental costs of getting rid of these wastes are huge. Landfills are becoming saturated. There are fewer and fewer places to dispose of waste. Even when disposed of in landfills, garbage can create serious environmental problems. Runoff from landfills can poison surrounding water supplies and soil. The formation of methane gas creates the danger of fire and explosion.

Recycling is a process of collecting and reprocessing waste materials for reuse. Paper, cardboard, glass, aluminum and tin cans, plastics, yard wastes, and even whole automobiles can be recycled and used over again. One example is the tree-free paper products made from recycled paper. Recycling can greatly help our communities if we all strive to collect recyclables and bring them to our municipal recycling centers. Reusing materials is one way of conserving resources, reducing garbage and waste, and making landfills safer for the environment. Recycling is the final step in the three-step process of reuse, reduce, and recycle. (See Figure 26-8.)

Figure 26-8 At a recycling center, materials are collected so they can be processed for recycling.

Environmental Legislation

Cooperation among federal and state governments led to legislation to protect forestlands. Tax incentives along with technical and financial assistance encouraged the productive management of forests. Laws were enacted to protect forests from fires, insect infestations, and disease. Public lands were acquired to expand the forests under protection. These actions restored the forests so that they could once again serve as an important natural resource.

The Environmental Protection Agency (EPA), established in 1970, formulates and enforces regulations governing activities that may adversely affect the environment or public health. The programs run by the EPA are based on laws passed by Congress. A summary of some of these laws is given in Table 26-3 on page 470.

26.4 Section Review

1. Describe some of the uses of United States Forest Service lands.
2. How do landfills pollute the environment?
3. Research and report on the ways your city or town promotes recycling and what materials are recycled.

TABLE 26-3 SUMMARY OF ENVIRONMENTAL LEGISLATION

Law	Action
Freedom of Information Act of 1966 (FOIA)	Ensured that government information would be available to everyone
National Environmental Policy Act of 1969 (NEPA)	Requires developers, loggers, and others to describe the effects their project will have on the environment and develop plans to minimize those effects
Occupational Safety and Health Act of 1970 (OSHA)	Provides for and ensures worker and workplace health and safety
Federal Insecticide, Fungicide, and Rodenticide Act of 1972 (FIFRA)	Provides for federal control of pesticide distribution, sale, and use
Endangered Species Act of 1973 (ESA)	Provides programs for the conservation of threatened and endangered species of plants and animals and the habitats in which they are found
Safe Drinking Water Act of 1974 (SDWA)	Protects the quality of drinking water in the United States
Resource Conservation and Recovery Act of 1976 (RCRA)	Gives the EPA authority to control the generation, transport, treatment, storage, and disposal of hazardous wastes
Toxic Substances Control Act of 1976 (TSCA)	Gives the EPA the ability to track industrial chemicals produced or imported into the United States
Federal Food, Drug, and Cosmetic Act amended in 1976 (FFDCA)	Regulates the manufacture and interstate commerce of foods, drugs, devices, and cosmetics to protect against alteration
Clean Air Act of 1977 (CAA)	Regulates emissions into the air from all sources and authorizes the EPA to establish National Ambient Air Quality Standards (NAAQS)
Clean Water Act of 1977 (CWA)	Sets standards regulating the discharge of pollutants into United States waters and gives the EPA authority to implement pollution control standards for industry
Comprehensive Environmental Response, Compensation, and Liability Act of 1980 (CERCLA or Superfund)	Creates a tax on the chemical and petroleum industries to pay for the cleanup of hazardous waste sites
Superfund Amendments and Reauthorization Act of 1986 (SARA)	Amends CERCLA by stressing the importance of finding permanent remedies for hazardous waste cleanup, providing for new enforcement tools, and increasing state and local involvement
Emergency Planning and Community Right-To-Know Act of 1986 (EPCRA)	Helps communities protect public health, safety, and the environment from chemical hazards
Oil Pollution Act of 1990 (OPA)	Streamlines and strengthens the EPA's ability to prevent and respond to catastrophic oil spills
Pollution Prevention Act of 1990 (PPA)	Calls for a reduction in pollution through cost-effective changes in production, operation, and raw materials used by industry
Chemical Safety Information, Site Security, and Fuels Regulatory Relief Act of 1999 (Amends the Clean Air Act)	Provides for reporting and disseminating information under the CAA, as it pertains to flammable fuels and worst-case scenario data

LABORATORY INVESTIGATION 26

Making Recycled Paper

PURPOSE: *To recycle waste newspaper*

SKILLS: *Manipulating, observing*

MATERIALS: *Old newspapers, water, wire mesh screen (approximately 20 cm × 20 cm), several sheets of wax paper larger than the screen, 5- to 10-L bucket, plastic tray (approximately 25 cm × 30 cm), potato masher or kitchen spatula, plastic sheet (approximately 20 cm × 20 cm) or plastic bag, heavy weight such as books, hand lens*

PROCEDURE

1. Overnight, soak six sheets of old newspaper in the bucket of water.
2. Drain some of the excess water, then mash the wet paper into a pulp. Examine the pulp mixture with the hand lens.
3. Half fill the plastic tray with the pulp mixture.
4. Slide the wire mesh screen into the mixture and lift it out so that the mesh is covered with pulp. Allow some of the water to drain from the pulp-covered wire mesh.
5. Place a sheet of wax paper on a flat surface and then place the mesh on top of it, pulp side down. Remove the mesh so that a sheet of pulp remains on the wax paper. Place another sheet of wax paper on top of the pulp and press it down firmly.
6. You may repeat steps 4 and 5 several times until all the pulp has been used, alternating the sheets of wax paper and pulp to form a stack.
7. Place several books on top the stack to help squeeze out the moisture. You can protect the books from moisture by placing them on a sheet of plastic or by putting them in a plastic bag.

8. After several hours or overnight, when most of the moisture has drained out, gently peel the layer of pulp from the sheets of wax paper. Place the pulp layers on a sheet of newspaper until they have dried completely.

OBSERVATIONS AND ANALYSES

1. Describe the appearance of the water after the paper soaked overnight.
2. What impurities or wastes does the water contain?
3. Describe the composition of the pulp mixture you examined with the hand lens.
4. Why is bleaching an important process in recycling paper?

Chapter 26 Review

Answer these questions on a separate sheet of paper.

Vocabulary

The following list contains all the boldfaced terms in this chapter.

acid rain, conservation, desertification, fossil fuels, hazardous substances, natural resource, nonrenewable resources, persistent chemical, reclamation, recycling, renewable resources, strategic minerals, terracing

Fill In

Use one of the vocabulary terms listed above to complete each of the following sentences.

1. Resources that can be restored after harvesting are ______.
2. A chemical that breaks down slowly and remains in the environment for long periods is called a(an) ______.
3. Substances that pose a threat to human health or the environment are ______.
4. ______ is the wise and careful use and protection of resources.
5. ______ contains sulfuric and nitric acids, which form when sulfur and nitrogen compounds combine with moisture in the atmosphere.

Multiple Choice

Choose the response that best completes the sentence or answers the question.

6. Ground cover acts as an important check on soil erosion by *a.* compacting the soil. *b.* aiding runoff. *c.* adding nutrients to the soil. *d.* absorbing the impact of rainfall.
7. Examples of a persistent environmental pollutant are *a.* PCBs. *b.* biodegradable materials. *c.* flammable solvents. *d.* contaminated medical dressings.
8. Which of the following is a heavy metal pollutant? *a.* iron *b.* copper *c.* silver *d.* mercury
9. The Love Canal region of New York State was a former *a.* coal mine. *b.* petroleum well. *c.* cattle ranch. *d.* toxic waste dump.
10. Naturally occurring elements or compounds that are found in Earth's crust are called *a.* minerals. *b.* renewable resources. *c.* pulp. *d.* tailings.
11. Unsightly and often toxic piles of mine waste are called *a.* tailings. *b.* ores. *c.* ground cover. *d.* strategic minerals.

12. Reclamation is *a.* recycling paper. *b.* restoring vegetation at open pit and strip mine sites. *c.* heating metal ores. *d.* compacting the soil.
13. In how many years do experts estimate that the known reserves of petroleum will be used up? *a.* 10 *b.* 50 *c.* 100 *d.* 500
14. Which group contains only renewable resources? *a.* oil, gold, silver *b.* trees, fish, mercury *c.* coal, oil, natural gas *d.* trees, fish, cattle
15. Chlorinated hydrocarbons are *a.* insoluble in water. *b.* soluble in fat. *c.* persistent chemicals that concentrate at the top of food chains. *d.* all of the above.

Short Answer (Constructed Response)

Use the information you learned in this chapter to respond to the following items.

16. List three examples of nonrenewable resources.
17. What is an ore?
18. What are strategic minerals?
19. Explain the term *recycling.*
20. List two environmental problems caused by landfills.

Essay (Extended Response)

Use the information you learned in the chapter to respond to the following items.

21. Why is a decrease in soil productivity accompanied by increased economic costs?
22. How can increased use of natural gas as a fuel slow global warming?
23. Explain why recycling is an important part of the process of conserving our natural resources.
24. In every section of the country, there are endangered plants and animals. Contact local conservation and environmental groups to discover what species in your area are threatened.

Research Project

Investigate the changes brought about in your community by the Clean Air Act or the Clean Water Act.

CHAPTER 27
Freshwater

When you have completed this chapter, you should be able to:

Discuss the importance of freshwater to human society.

Describe the causes and effects of cultural eutrophication on a body of water.

Relate the forms of water pollution to their harmful effects.

"Water, water, every where, Nor any drop to drink." Unfortunately, these familiar lines from "The Rime of the Ancient Mariner" by Samuel Taylor Coleridge describe the water situation in many parts of the world. All living things need water to survive. Water shortages threaten the survival of people as well as other animals and plants. Some water shortages are caused by changes in rainfall patterns, others by waste of water resources. In addition, pollution has made the water unfit to drink in many areas. In this chapter, you will learn about Earth's freshwater resources.

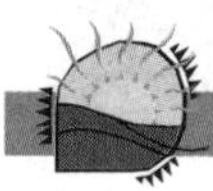

27.1 POTABLE WATER

Living things need water. A person needs approximately 2 liters of water per day. However, the use of water by the human population goes far beyond that figure. In the United States, water use exceeds 5000 liters per person per day. Much of this water is used to irrigate crops, maintain livestock, and manufacture goods. In developing nations, water use may be only 50 liters per person per day.

More than 1.5 billion people lack an adequate supply of fresh water. Contaminated water, due to poor sanitation, is a major cause of disease in many of the world's developing nations. Many areas have shortages of drinking water, often due to changing rainfall patterns caused by global warming, overgrazing, improper agricultural practices, the depletion of natural supplies, and population shifts.

Water fit for human consumption is known as potable water. Potable water is free of all contaminants and contains no discernible pollutants. Water is a renewable resource that is continually regenerated and purified as it circulates through the hydrologic cycle. (See Chapter 11.) Human threats to the water cycle include global climate changes due to greenhouse warming, deforestation, atmospheric and water pollution, and the population explosion.

The Impact of Water

Freshwater is essential for all living things and plays an important role in most human activities.

- Water is the primary component of all plant and animal cells. The human body is approximately 65 percent water by weight.
- Water has the ability to dissolve so many other substances that it is called the universal solvent. This property of water is crucial to all life, since minerals must be dissolved in water before plants can make use of them, and oxygen must be dissolved in water before animals can use it in respiration.
- Many ecosystems exist solely within water environments.
- Water plays a key role in photosynthesis by supplying hydrogen for the synthesis of carbohydrates.

- The distribution of plant, animal, and human populations is determined primarily by the accessibility of water.
- Water plays an important role in regulating Earth's surface temperatures and thus controls climate.

The presence and availability of water have always determined the location of human settlements and affected their economic growth. Water has shaped the history of the world and will surely play an ever-increasing role in the future. Nations that do not have a renewable supply of water are at an extreme economic disadvantage unless they have resources they can offer in trade.

Since the development of cities, human societies have used wells, reservoirs, and delivery systems to make efficient use of water supplies. Wells are drilled into the crust and down to the water table to tap groundwater reserves. Groundwater, obtained through wells, supplies almost 40 percent of the water used for agricultural and domestic purposes in the United States. Groundwater is often withdrawn from aquifers faster than it can be replaced. This lowers the water table and often causes shallower wells in surrounding areas to go dry. (See Figure 27-1.) **Reservoirs** are large areas on the surface of the land that are used for water storage. They may be lakes or rivers and streams that are dammed to allow the water to build in depth. Delivery systems, such as aqueducts and pipelines, transport water from reservoirs to urban areas.

An aquifer is a porous, water-bearing layer of sand, gravel, or rock.

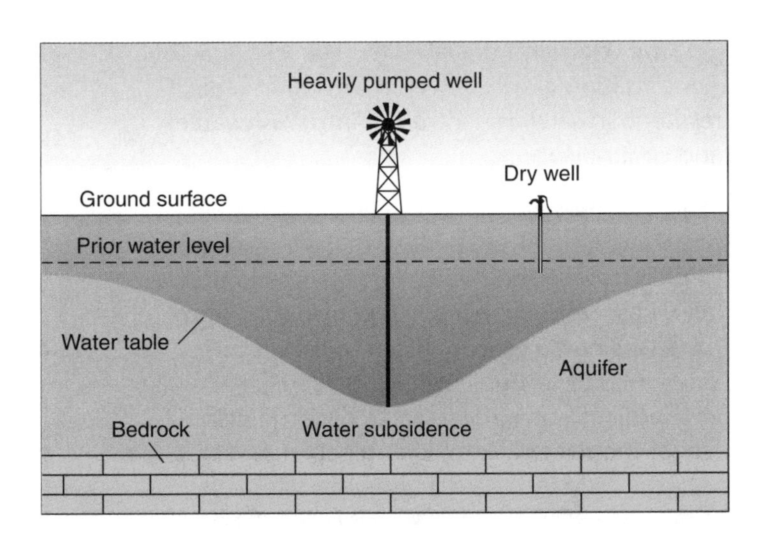

Figure 27-1 **A deep well drilled into an aquifer may lower the water table.**

INTERACTIONS

Restoring the Kissimmee River

The Kissimmee River in Florida flows into Lake Okeechobee. The waters of Lake Okeechobee in turn flow south, creating the Everglades. The Everglades, called the "River of Grass" because most of its 610 thousand hectares are underwater, is the largest subtropical wilderness in the United States. This water-covered area is made up of brackish wetlands and the shallow estuarine waters of Florida Bay. The Everglades is a patchwork of sawgrass marshes, mangrove forests, and junglelike pine and hardwood hammocks. Its freshwater prairies are home to many rare and endangered species.

The Kissimmee River once meandered 166 kilometers through a productive floodplain, before it was channeled into an 84-kilometer-long canal in 1961. The canal became known as "The Ditch." Dams were built to straighten the river as a method of flood control and to aid the growth of agriculture. In doing this, the U.S. Army Corps of Engineers reduced the river's natural flow and destroyed more than 12,000 hectares of wetlands. What was once a rich wildlife habitat became a sterile canal blocked by concrete dams and flanked by earthen levees. Agriculture and livestock activities sprang up along the river's banks.

Within two years, the project affected the waters of Lake Okeechobee. The lake's ecosystem was dying. The wastes from cattle ranches, pesticides and fertilizer from farms, and sediment from the river were poisoning Okeechobee. In the past, the marshlands formed by the meandering waters of the Kissimmee River had purified the water of waste before it reached the lake. The marshes also held back the silt, keeping it from clogging the lake. The reduced flow of water, due to the construction of irrigation canals, lowered the water table, resulting in saltwater intrusion into freshwater and groundwater.

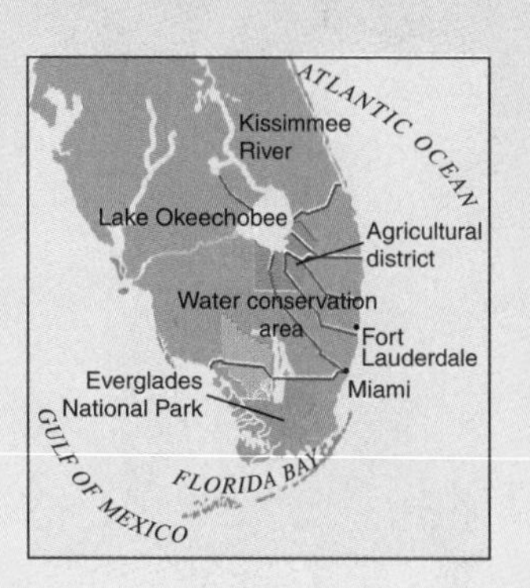

In 1983, steps were taken to reverse the damage. New dams were built to divert the river into its old channel and flood the marshes along the river. Reservoirs were built to control flooding. Land was purchased and used to return the ecosystem to its original state. In 2001, a ten-year project was completed to divert 69 kilometers of the river back to its original channel.

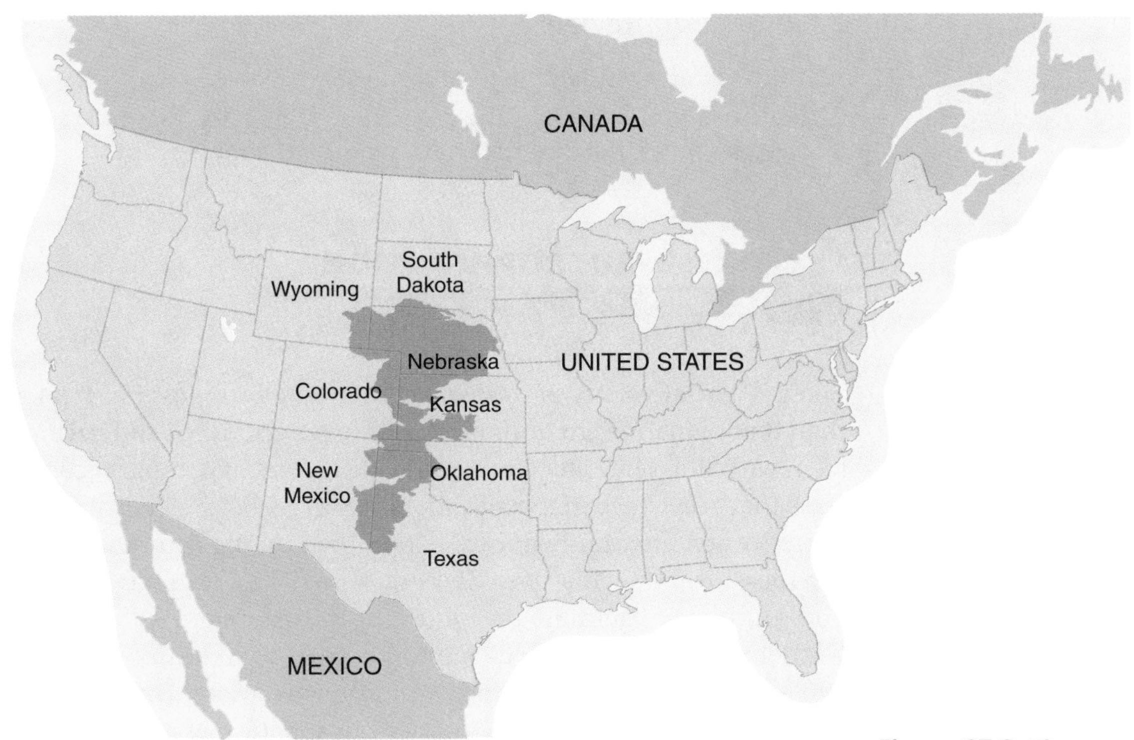

Figure 27-2 The Ogallala Aquifer extends from South Dakota to Texas.

The Ogallala Aquifer, located in the Great Plains region of the United States, is the largest body of freshwater in the world. (See Figure 27-2.) It ranges in thickness from 10 to 400 meters. The aquifer supplies water for the agricultural and livestock industries of the eight states that it lies under. Because of the large quantities of water that are withdrawn yearly from the aquifer, it is being depleted faster than it can be refilled naturally. This has caused the water table to drop as much as 50 meters in some areas and reduced the ability of some wells to supply water for agriculture.

The development of safe, potable water has led to a major, worldwide decrease in waterborne diseases such as typhoid, cholera, and dysentery. Yet the populations of many nations still do not have safe drinking water. In some developing nations, a daily 2-kilometer walk to obtain water is not unusual. Unsanitary conditions, due to the dense populations that surround available water, often contaminate the water, causing disease epidemics. The ponds and rivers, which are often the main sources for drinking water, are also the sites for toilets.

27.1 Section Review

1. Why is water vulnerable to pollution?
2. Prepare a table that lists the countries lacking adequate water supplies and countries that have abundant supplies.

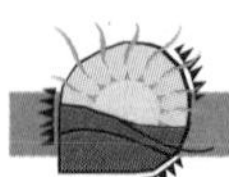

27.2 EUTROPHICATION

There is a great variety of freshwater environments, possibly greater than that of marine environments. Lakes, ponds, rivers, and streams teem with life. They all comprise a wide variety of complex ecosystems that recycle nutrients within the water. The minerals present in freshwater environments are determined by the composition of the rocks over which a river flows or on which a lake has formed. These minerals determine the biota that the ecosystem will support. Nutrient-rich water that will support a large community of plants and animals is called eutrophic. Nonproductive water is called oligotrophic. Phosphorus, in the form of phosphates, serves as the primary regulator of the trophic state of a body of freshwater.

Eutrophication is the natural process that enriches water with the dissolved minerals needed to support plant and animal life. It is part of the natural process of succession in a lake or river. But the meaning of eutrophication has changed over time to mean also the unnatural overnourishment of a body of water. This is commonly called cultural eutrophication to differentiate it from natural eutrophication. (See Figure 27-3.) Cultural eutrophication is caused by the addition of excess nutrients from agricultural runoff, industrial discharge, and sewage. These nutrients accelerate the natural eutrophication process. Aquatic plants and algae take over the area and block the sunlight. When the plants die, their decomposition depletes the water of oxygen. This causes fish in the water to die. Cultural eutrophication associated with pollution occurs at a faster rate than natural eutrophication. Increased water temperatures due to the discharge of warm water from power plants, and increased sunlight due to the removal of nearby trees and shrubs, can also lead to cultural eutrophication.

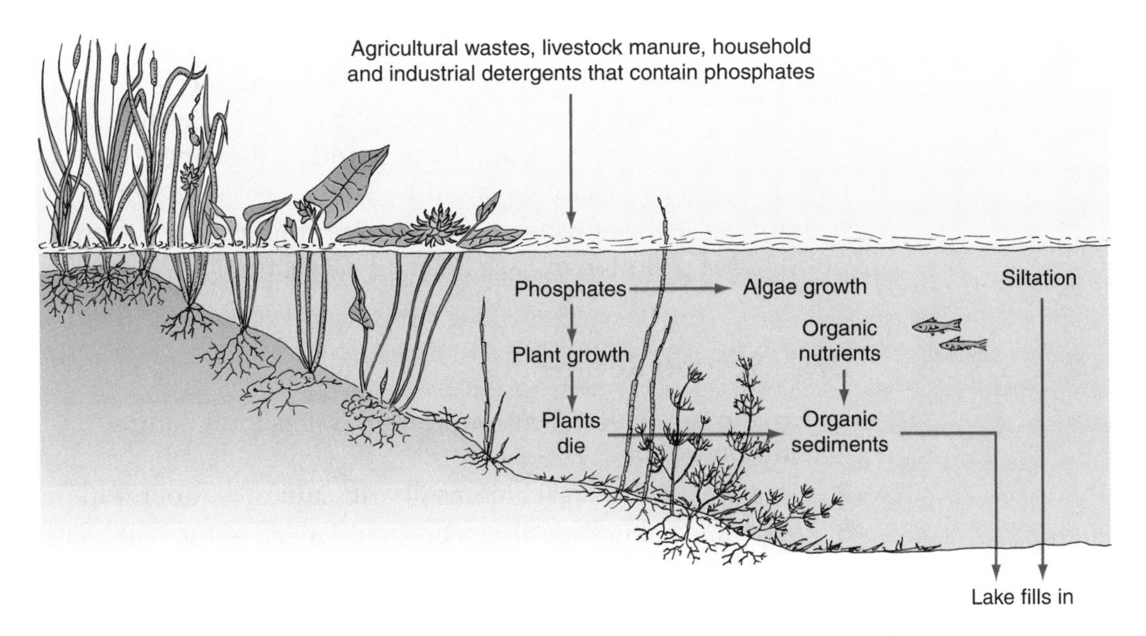

Figure 27-3 Phosphate pollutants speed up plant growth.

Lake Ecosystems

Lake ecosystems are endangered in many ways. Acid rain and pollution can make waters inhospitable to life. Sewage discharge and fertilizers and manure in agricultural runoff cause cultural eutrophication. These substances add nutrients, such as nitrates and phosphates, to bodies of water, which contaminate lakes and streams and stimulate the growth of bacteria and algae. The rapid growth of algae, known as algal blooms, soon chokes the ecosystem by covering the water's surface. The algae die, and their decomposition uses up the available oxygen, causing aquatic plants and animals to die. Then the lake dies. The water becomes cloudy, and the decomposing plants and animals produce unpleasant tastes and odors. The water may become toxic and no longer potable. Organic sediments (produced by the decomposers) and silt build up on the bottom and eventually fill in the lake.

Lakes near large population centers have been affected most by cultural eutrophication. In less than 100 years, the Great Lakes became severely polluted. Lakes do return to their natural trophic states. When pollution is stopped, the water cleanses itself naturally as it becomes oxygenated and pollutants in bottom sediments are prevented from circulating to surface waters. Biological agents in sediments can then decompose the pollutants into harmless substances.

Many detergents once contained phosphates. The millions of metric tons used in the United States and flushed into sewer systems from homes, factories, laundries, and offices were a main contributor to eutrophication. They were the primary cause of phosphorus pollution of Lake Erie and Lake Ontario.

27.2 Section Review

1. Explain how cultural eutrophication differs from natural eutrophication.
2. Describe the role algal blooms play in cultural eutrophication.
3. Why can chemical fertilizers be good for agriculture but bad for aquatic resources?

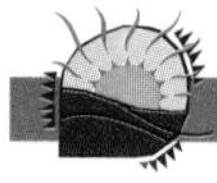

27.3 WATER POLLUTION

Water pollution is any physical, chemical, or biological change in water quality that adversely affects living things or the environment. This is an environmental problem that is easily recognized and should be of concern to everyone because it can cause water to become unsuitable for human uses.

Water is considered contaminated when it contains a pollutant that is toxic and causes death or disease. Dirty water is an ideal habitat for **pathogenic**, or disease-causing, organisms. Because many sources of water cross international boundaries, pollutants produced in one country often end up in the water of another country. The actions of one nation can have serious consequences on a global scale. For example, the dumping of wastes in rivers, accidental industrial spills, and uncontrolled pollution can affect commercial fisheries worldwide. Water pollution has increased considerably since the Industrial Revolution, due to a variety of factors that include:

- The human population's rapid growth
- The increased and often extravagant use of water for personal hygiene, manufacturing, cooling, and agriculture

- The use of water sources for the disposal of human wastes, animal wastes, and garbage

Society often tries to solve pollution problems by dilution. Unfortunately, this often saturates the receiving water within a short period of time and leads to the death of whole ecosystems. Some pollutants are biodegradable and can be broken down by living organisms such as bacteria or fungi. Sunlight and high temperatures can also contribute to the decomposition of some pollutants. Pollutants that are **nonbiodegradable** cannot be broken down by living organisms. Thus, these compounds can persist in the environment for a long time.

Pathogens

Of all water pollutants, pathogens pose the most serious threat to human health and are the main source of waterborne diseases. These organisms can be bacteria, viruses, or parasites. Waterborne pathogens originate in untreated human wastes, animal wastes carried in runoff from feedlots and fields, and food-processing plants with inadequate waste treatment facilities. Sewage treatment plants and pollution control measures have greatly reduced or eliminated most of these infectious agents from our nation's water supplies. But they still pose a serious health threat in developing nations. Worldwide, almost two-thirds of the deaths of children under five years of age are associated with waterborne diseases.

Detecting specific pathogenic organisms in water supplies is a costly, time-consuming process. Therefore, water is usually analyzed for the presence of **coliform bacteria**. These are the bacteria that live in the intestines of humans and other animals. It is assumed that if coliform bacteria are present in a water sample, pathogens are also present. A high level of coliform bacteria in a sample indicates recent contamination by untreated human wastes. Contaminated waters are usually closed to human activity until the number of coliform bacteria they contain decreases to a safe level.

A **pollution indicator organism** is a species of animal, plant, or microorganism that is not normally present in an aquatic environment unless the water is polluted. For example, *Escherichia coli* bacteria are not normally found in unpolluted water; their presence

Organisms Found in
Nonpolluted Waters

Water shrimp

Water flea

Freshwater clam

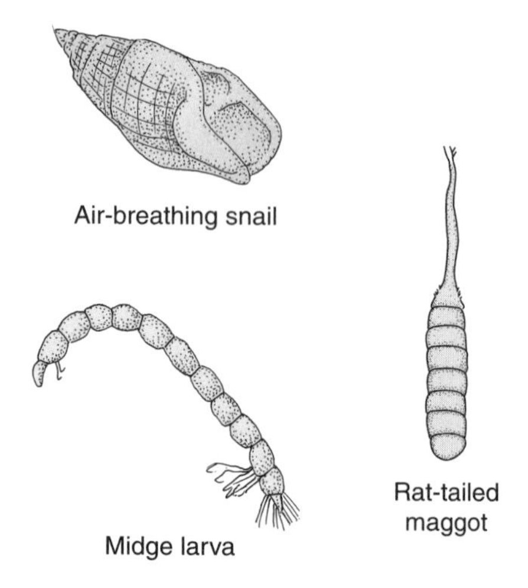

Figure 27-4 These organisms serve as a biological index to monitor water quality.

indicates that the water is contaminated with **fecal material**, solid waste of human or animal origin.

Small aquatic invertebrates such as freshwater clams, water fleas, mayfly nymphs, and freshwater shrimp are sensitive to water pollutants. These creatures serve as a **biological index** for monitoring water pollution from industrial waste and municipal sewage. Ecologists can determine whether water is polluted by collecting water samples and counting the number of these invertebrates present. The presence of large numbers of any of these organisms in a sample indicates that the water is relatively free of pollution. Rat-tailed maggots, midge larvae, and air-breathing snails indicate poor water quality. These organisms tolerate the low level of oxygen found in polluted water. (See Figure 27-4.)

Toxic Inorganic Chemicals

Industrial activities can often release toxic chemicals into water supplies. Salts, acids, nitrates, and chlorides are inorganic materials that can adversely affect water quality. Of most serious concern are the **heavy metals** such as cadmium, lead, mercury, nickel, arsenic, and selenium, and their compounds. Although heavy metals are a natural part of the environment, these substances become a problem when they enter the food chain. Heavy metals are persistent and accumulate in organisms at the top of food chains. Heavy metal poisoning is responsible for a wide variety of human health problems: nickel damages the lungs, which leads to respiratory problems; cadmium can damage the kidneys and cause kidney failure. Nickel and cadmium are found in nicad batteries.

Mercury is a by-product of smelting and mining activities, as well as being part of industrial discharges. Often, mercury is released in large quantities into rivers and streams, where it builds up in sedi-

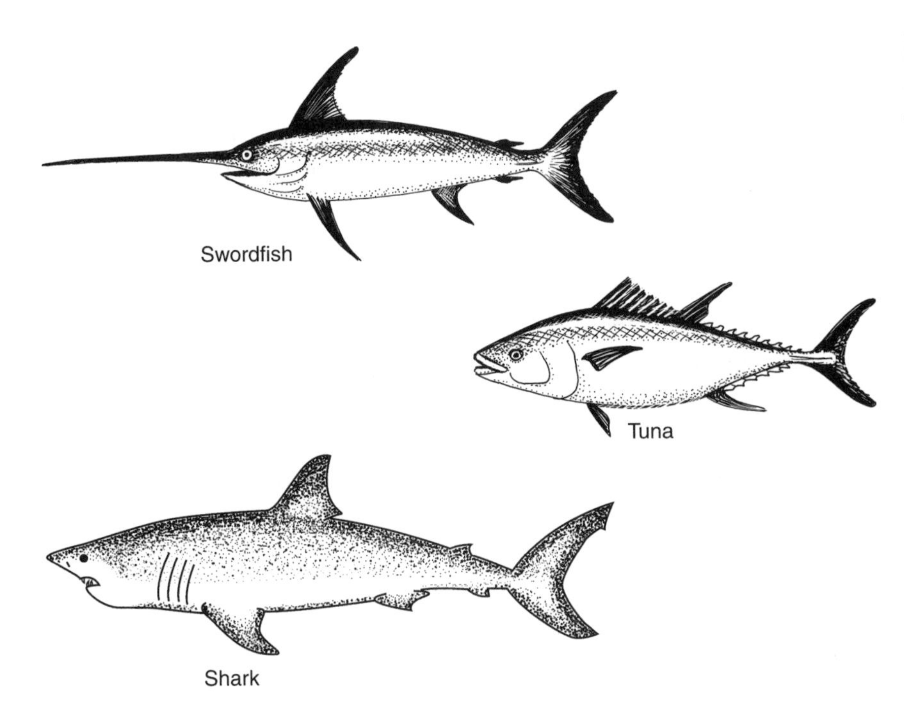

Figure 27-5 These organisms are at the top of their food chain.

ments and enters the food chain. Once in the food chain, mercury's concentration increases at each trophic level because the large organisms each eat many smaller organisms. The highest concentration is found in those organisms at the top of the food chain. This process is called **bioaccumulation**, or biomagnification. Mercury has been found in high concentrations in the tissues of marine fish such as tuna and swordfish that are at the top of the food chain. (See Figure 27-5.) This poses serious consequences for populations that rely on fish as their main source of protein. Mercury attacks the central nervous system and can cause mental retardation in children. Because the human body cannot quickly get rid of mercury, it accumulates in tissues.

In the 1950s, an outbreak of mercury poisoning in the Minamata Bay region of Japan raised public awareness of the dangers of mercury. Seafood there had become contaminated with methyl mercury, which was discharged into the bay by an industrial facility. This caused several hundred poisonings, the results of which ranged from neurological disorders to paralysis and death. Since that time, the bay has been reclaimed. In July 1997, the governor of Kyushu reported that fish and shellfish taken from Minamata Bay now have mercury concentrations below the level considered to be dangerous by the Japanese government.

Lead pipes, which for many years have been used to carry drinking water, are a serious source of pollution. Water that is even slightly acidic leaches lead from the pipes and carries it into the water supply. Lead-based paints are another source of this metal in the environment. Overexposure to lead can severely damage the circulatory, digestive, and central nervous systems. Lead poisoning can lead to mental retardation in children, particularly those younger than six years old. In recent years, many laws have been enacted to reduce lead levels in the environment. In the United States, for example, only unleaded gasoline and lead-free paints are sold.

Salts

Depletion of aquifers in coastal regions can cause salt water from the ocean to contaminate freshwater. (See Figure 27-6.) As freshwater is pumped out of an aquifer, salt water can intrude upon and mix with the freshwater. This occurs in Florida and on Long Island, New York. Freshwater into which salt water has intruded may still

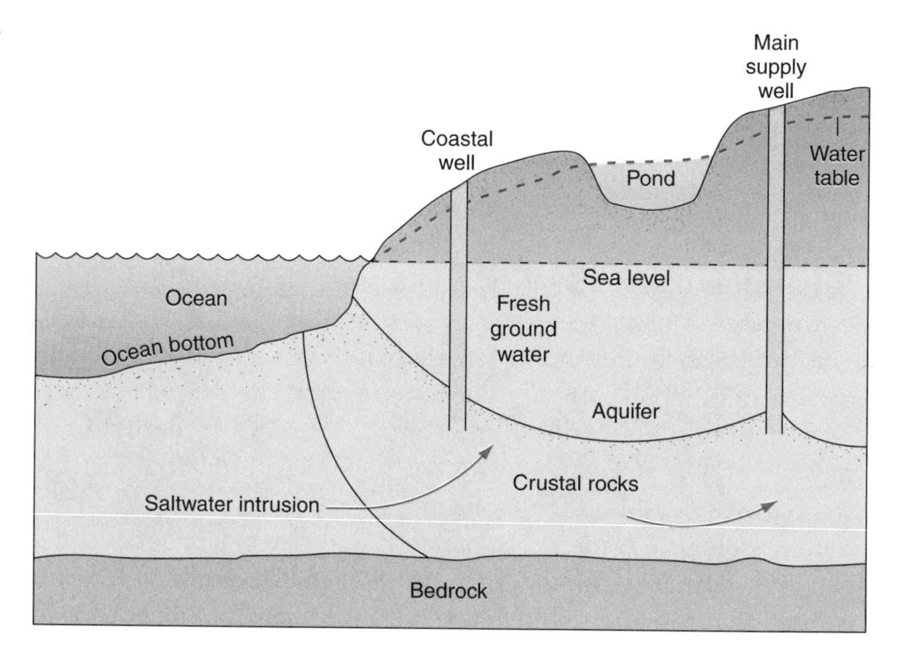

Figure 27-6 As freshwater is removed, salt water intrudes into the aquifer.

be used for irrigation, although it cannot be used as drinking water for livestock or humans.

In areas that have winter snows, salts are applied to roads to melt the ice and snow. These salts are carried into streams and groundwater by runoff when the ice melts. This can contaminate water supplies and kill species that cannot tolerate salt.

Acid Rain

Acid rain describes precipitation that is acidic. Rainfall is naturally slightly acidic because carbon dioxide in the atmosphere readily dissolves in water to form carbonic acid. This weak acid poses little or no environmental problems. Sulfur dioxide and various oxides of nitrogen are produced during the combustion of fossil fuels. These oxides dissolve in water to produce sulfuric acid and nitric acid, two strong acids. The presence of these acids in precipitation can have serious environmental consequences.

The acidity (pH) of most water is about 7.0, which is neutral. Figure 27-7 illustrates the pH scale. Water that contains dissolved carbon dioxide can have an acidity of 5.6, which is slightly acidic. The acidity of acid rain ranges between 5.5 and 2.1, which is highly acidic. Exposure to acid rain can deteriorate limestone and marble buildings and statues.

Whole forests in eastern Europe and Canada have been destroyed by acid rain. Acid rain is damaging to the ecosystems of lakes; it is deadly to many young and small fish and water plants. Acid rain changes the quality of water over large areas that are downwind from the source of the pollution. Because the pollutants that produce acid rain can be carried great distances by prevailing winds, this problem is of international concern.

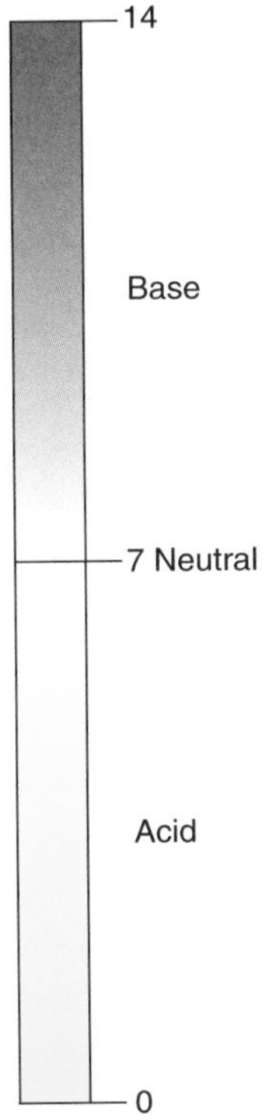

14
Base
7 Neutral
Acid
0

Figure 27-7 The pH scale ranges from 0 to 14.

Organic Pollutants

Thousands of different organic compounds are used in the production of plastics, pesticides, pharmaceuticals, paint pigments, and synthetic fibers. Many of these chemicals are toxic and can cause birth defects, genetic mutations, and allergic responses. Others are

carcinogens, compounds that can cause cancer. Many of these organic compounds are nonbiodegradable. They persist in the environment for extended periods and through bioaccumulation build up to dangerously high levels. The improper disposal of household and industrial wastes and runoff from agricultural lands and urban streets are the main sources of these pollutants. Dangerous levels of pesticides, such as DDT, and other chemicals, such as PCBs and dioxin, have built up over time to contaminate much of our waters.

DDT, *d*ichloro-*d*iphenyl-*t*richloroethane, was used for many years in the United States as an agricultural insecticide. It entered rivers and lakes in runoff from farmlands and became a serious environmental pollutant. Once in the rivers, it entered the food chain, where it was readily stored in the fatty tissues of small fish.

DDT is not soluble in water and so is not readily excreted. It builds in concentration up the food chain through bioaccumulation, as small fish are eaten by larger fish, which in turn are eaten by birds, mammals, and humans. DDT is especially harmful to fish-eating birds. This insecticide interferes with the birds' use of calcium. The birds lay eggs with thinner than normal shells, which can break easily. This lowers their reproductive success. Even though DDT was banned in the United States in 1972, it is still manufactured here in large quantities for export to developing nations.

PCBs, *p*oly*c*hlorinated *b*iphenyls, are synthetic organic compounds that are carcinogenic to humans and animals. PCBs are highly persistent, and once they enter the food chain, bioaccumulation occurs. Even though the concentration of PCBs may be small in a body of water, it builds up to high levels in the tissues of animals. Because PCBs have high electrical resistance and are fire resistant, they are produced for the manufacture of transistors, transformers, and capacitors. They were also used in paint, plastics, fire-retarding materials, dyes, lubricants, and hydraulic fluid. In 1979, the Environmental Protection Agency prohibited the manufacture of PCBs and called for a five-year phaseout of their use. (See Figure 27-8.)

Dioxins are a family of compounds known as polychlorinated dibenzo-para-dioxins. They are not manufactured directly but are formed as impurities during the production of **herbicides**, compounds that are used to kill weeds. Health effects related to exposure to dioxins include skin rashes, liver and kidney failure, and the formation of tumors. On July 10, 1976, dioxin was accidentally

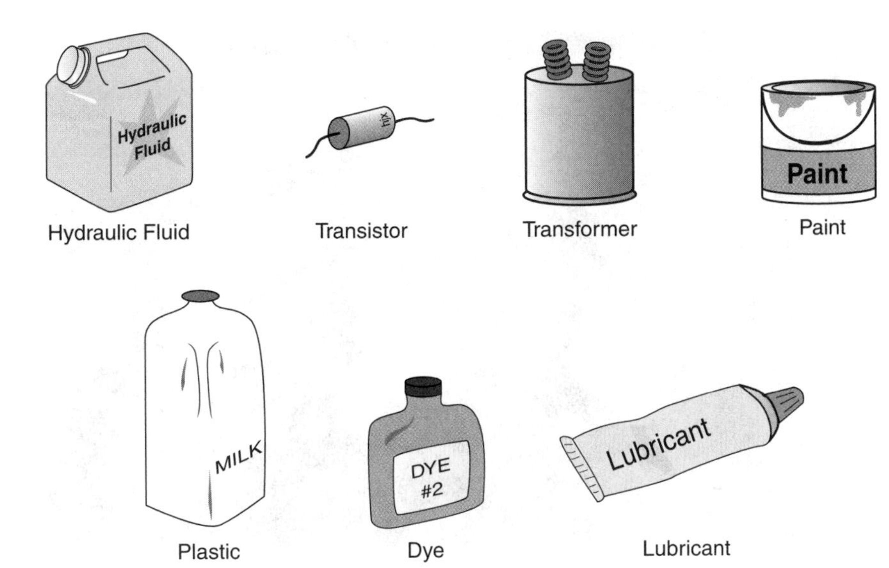

Figure 27-8 Manufactured items that contribute to PCBs in the environment.

released from a chemical plant in Seveso, Italy. The cloud of vapor spread over the surrounding area, which contained 40,000 people. The town was evacuated when children began to develop rashes and domestic and wild animals began to die.

During the Vietnam War, a combination of herbicides was used to destroy crops and eliminate enemy cover. This herbicide combination, called Agent Orange, was contaminated with dioxins. Soldiers and civilians exposed to Agent Orange have suffered many health problems. There also have been environmental problems caused by its use.

Agricultural Pollutants

In addition to causing cultural eutrophication, agricultural practices affect water supplies in a number of ways. Lakes, reservoirs, and streams can become filled with silt, due to the erosion of soil from farmlands. Manure from domestic livestock can wash into waterways, contaminating the water supply with bacteria. (See Figure 27-9 on page 490.) Pesticides applied to crops also can wash into rivers and lakes. These chemicals work their way up the aquatic food chains, often poisoning a variety of organisms.

A feed lot is an area where large numbers of cattle or pigs are fat-

Figure 27-9 Manure can contaminate local water supplies.

tened for market. In feed lots, the diet of the animals can be carefully regulated so that they gain weight faster than they would by grazing. However, having all these animals concentrated in a relatively small area creates a large amount of manure. Farmers store the manure in ponds and spread it over their fields as fertilizer. However, the manure eventually washes into waterways, polluting the water with bacteria, viruses, sediment, and nutrients, which cause algal blooms. In the United States, the number of feed lots has decreased while the number of animals has increased. The EPA has recently held meetings in Arkansas, Oklahoma, and Texas to study the problem.

27.3 Section Review

1. Describe how the pollution of one nation's water can cause environmental problems for other nations.
2. Why do nonbiodegradable pollutants often pose the most serious threat to public health?
3. Find out how the water quality of your town compares with that of some other cities nationwide.

LABORATORY INVESTIGATION 27

Carbonic Acid

PROBLEM: *How does the addition of carbon dioxide affect water?*

SKILLS: *Manipulating, observing*

MATERIALS: *100-mL bottles (2), water, limewater [$Ca(OH)_2$], marking pencil, drinking straw, litmus paper (or pH paper)*

PROCEDURE

1. Half fill two 100-mL bottles with water. Label one A and the other B. Test the acidity of the water in each bottle with litmus paper or pH paper. The pH paper will give a more accurate measurement than the litmus paper.
2. Insert the straw into the water in bottle A. For about 30 seconds, exhale through the straw into the water. The air you exhale contains carbon dioxide. Test the acidity (pH) of the water.
3. Add 5 mL of limewater to bottle A. Observe. (Limewater turns milky in the presence of carbon dioxide.) Test the acidity (pH).
4. Add 5 mL of limewater to bottle B. Observe. Test the acidity (pH).

OBSERVATIONS AND ANALYSES

1. What was the original acidity (pH) of the water in bottle A?
2. After you exhaled into bottle A, what was the acidity (pH) of the water?
3. What did you observe when limewater was added to bottle A? To bottle B?
4. Why did you add the limewater to both bottles?
5. What was the acidity (pH) of the water in bottle A after the addition of limewater? In bottle B?

Chapter 27 Review

Answer these questions on a separate sheet of paper.

Vocabulary

The following list contains all the boldfaced terms in this chapter.

bioaccumulation, biological index, carcinogens, coliform bacteria, fecal material, heavy metals, herbicides, nonbiodegradable, pathogenic, pollution indicator organism, reservoirs, water pollution

Fill In

Use one of the vocabulary terms listed above to complete each sentence.

1. Areas used for water storage are called ________.
2. Pollutants that cannot be broken down by living organisms are ________.
3. ________ are compounds used to kill weeds.
4. Compounds that can cause cancer are called ________.
5. In water, the presence of ________, which live in the intestines of humans and animals, indicate that pathogens are also present.

Multiple Choice

Choose the response that best completes the sentence or answers the question.

6. Compounds of which element are the primary regulators of the trophic state of water? *a.* sulfur *b.* phosphorus *c.* nitrogen *d.* mercury
7. The main cause of plant and animal death due to cultural eutrophication of lakes and ponds is *a.* toxic chemicals. *b.* oxygen depletion. *c.* bacterial diseases. *d.* phosphate poisoning.
8. The presence of a high level of coliform bacteria in a water sample indicates *a.* heavy metal contamination. *b.* contamination with human wastes. *c.* cultural eutrophication. *d.* no pollution is evident.
9. The largest body of freshwater in the world is *a.* Lake Superior. *b.* the Mediterranean Sea. *c.* the Ogallala Aquifer. *d.* the Mississippi River.
10. Wastes that can be broken down by bacteria and fungi are *a.* contaminants. *b.* pathogens. *c.* blooms. *d.* biodegradable.
11. Bioaccumulation is characteristic of *a.* typhoid bacteria. *b.* PCBs. *c.* acid rain. *d.* sulfur dioxide.

12. Which oxides are responsible for the formation of acid rain? *a.* iron and sulfur *b.* iron and nitrogen *c.* nitrogen and sulfur *d.* carbon and nitrogen
13. Nonproductive freshwater is said to be *a.* eutrophic. *b.* oligotrophic. *c.* trophic. *d.* toxic.
14. The addition of excess nutrients to a body of water causes *a.* cultural eutrophication. *b.* cultural oligotrophism. *c.* oxidation. *d.* bioaccumulation.

Short Answer (Constructed Response)

Use the information you learned in this chapter to respond to the following items.

15. Name three examples of dangerous, persistent metals that accumulate in food chains.
16. Give one example of pathogenic bacteria.
17. What are pollution indicator organisms?
18. Why are coliform bacteria indicators of serious pollution?
19. What is potable water?

Essay (Extended Response)

Use the information in the chapter to respond to these items.

20. Why do some areas that previously had adequate supplies now have water shortages?
21. How do dense human populations often contribute to contaminated water supplies?
22. Why has water pollution increased dramatically since the Industrial Revolution?
23. Contact your local water company. Find out where your water comes from. Are there any potential pollution problems associated with your water supply? Have there been any pollution problems in the past? What was done to correct past problems and to keep them from recurring? Report on your findings.

Research Project

Answer the following questions about your local water supply.

- What are its sources?
- How is the water supplied?
- How is it purified?
- Does it contain any additives?
- How does it compare with water from other communities?

CHAPTER 28
Protecting Our Water Resources

When you have completed this chapter, you should be able to:

Compare the effectiveness of methods used to improve water quality.

Describe ways you can conserve freshwater ecosystems.

List three steps governments have taken to reduce marine pollution.

Most people agree that our freshwater resources must be protected from pollution. After all, we need safe water for drinking, cleaning, and agriculture. However, we also use our rivers and streams as a convenient place to dump our wastes. Our uses for water resources conflict with each other. In this chapter you will learn about the measures we take to protect our freshwater resources.

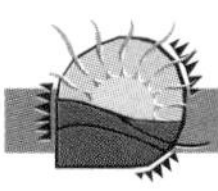

28.1 MAINTAINING WATER QUALITY

While most health departments claim that our drinking water is safe, no one knows what "safe" really means. Drinking water standards vary from place to place and cover only a fraction of the harmful pollutants that may be found in the water. Our tap water, or drinking water, is generally free of all microorganisms and pollutants that can adversely affect human health. But tap water may contain unknown pollutants that over time can accumulate in the body and cause health problems. Very small amounts of carcinogenic chemicals can be found in almost every water supply.

Water Quality Criteria are standards that define the maximum concentration level (MCL) of contaminants permissible in drinking water. In the United States, the Environmental Protection Agency sets water quality standards. The EPA requires that water suppliers test for dozens of chemical and bacterial pollutants and advise customers of the results. In addition to setting standards for water, the Safe Drinking Water Act banned the use of lead in public water systems. Of all the contaminants regulated by the EPA, only bacteria and nitrates pose an immediate threat to public health if unsafe levels are reached.

Water quality standards depend on the specific use of water, such as a public water supply, an industrial supply, water used for recreational purposes, and water for aquatic habitats. MCLs are set for concentrations of certain organic and inorganic chemicals, coliform bacteria, certain radioactive substances, and turbidity. **Turbidity** is a measure of the cloudiness of water. Water that does not meet those standards set by the EPA has to be treated before it can be used. **Water treatment** involves processing well water or surface water before distribution to the public. Water treatment does not completely purify water, but it does produce potable water for cities and towns.

Water quality can be adversely affected by local conditions. Septic waste systems may leak pollutants into nearby wells. In homes, old plumbing systems that contain lead pipes or joints can leach lead into drinking water. Hazardous materials that contaminate water supplies often originate in runoff from waste dumps, agricultural areas, feedlots, industrial areas, military bases, or underground petroleum storage sites.

Water Treatment

Examination and regulation of the quality of public water supplies began in the late 1800s. Initially, drinking water was treated with chlorine to control waterborne diseases. Modern water treatment practices were instituted in the early 1900s. Water was first filtered and then treated with chlorine to insure that it met standards for purity. From 1899 through 1970, major water pollution legislation included the prohibition of dumping in navigable waters without a permit, water quality control, and aid to municipalities for the construction of plants to treat drinking water and wastewater. The Public Health Service monitored water pollution until 1970, when these functions were turned over to the EPA.

Disinfection is a process that destroys pathogens, or harmful microorganisms, in water. Standards for the disinfection of drinking water vary from place to place. Cities that use groundwater from protected, deep aquifers are not required to disinfect their water at all. In cities that obtain drinking water from rivers and streams, there is a danger of contamination by microorganisms, industrial chemicals, and surface runoff. These waters require extensive treatment, and water costs are high. The potential contaminants in drinking water obtained from lakes and reservoirs are similar to those of rivers.

Water passes through several processes as it is made fit to drink. **Sedimentation** is a process in which water is allowed to stand undisturbed, so particles such as sand and dirt can settle to the bottom and be removed. (See Figure 28-1.) **Coagulation** is a chemical process used to remove dirt and debris particles from water. Alum is mixed with the water to cause the particles to coagulate, or clump together, forming larger particles. The larger particles settle out of the water and are removed by filtration. **Filtration** removes solid particles from water. The smaller the holes in the filter, the smaller the particles that will be screened out of the water. This method does not remove viruses and small bacteria. **Chlorination** is a form of chemical disinfection that uses chlorine to kill bacteria and remove objectionable tastes and odors. The use of chlorine has been questioned, since it reacts with other chemicals in the water to form carcinogens. **Irradiation**, treatment with ultraviolet light, is another method of disinfection. In the process of **aeration**, water is sprayed

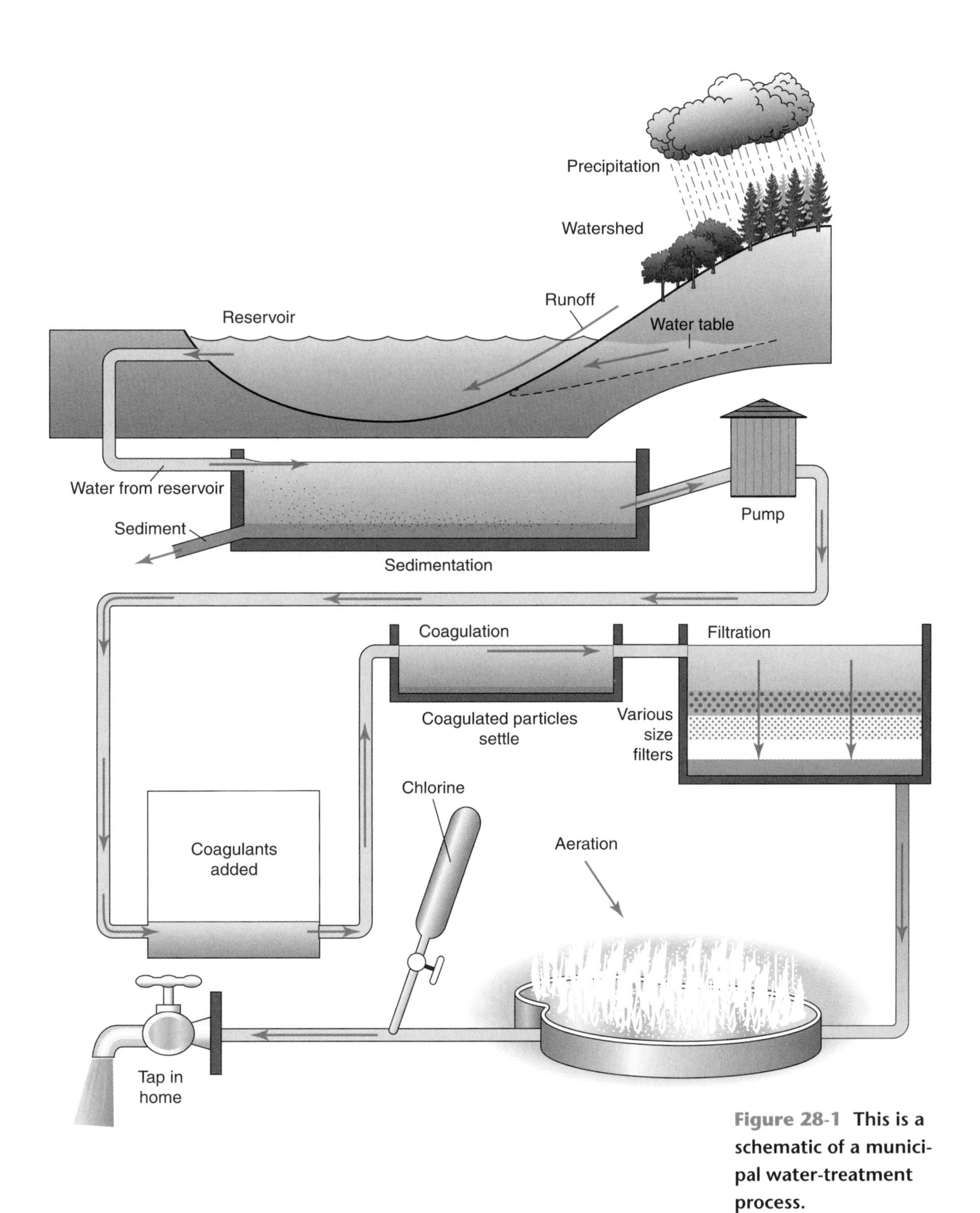

Figure 28-1 This is a schematic of a municipal water-treatment process.

into the air. This allows oxygen from the air to dissolve in the water, which kills anaerobes and improves water quality.

Despite water treatment, drinking water is sometimes contaminated with disease-causing organisms. During 1993, almost 400,000 residents of Milwaukee, Wisconsin, developed diarrhea as a result of the presence of the parasite *cryptosporidium* in their drinking water. For more than a week, the population had to boil all water before drinking it or using it to wash food. In the recent past, Carrollton, Georgia; Medford, Oregon; and Braun Station, Texas, also had outbreaks of this parasite in their drinking water. All of these outbreaks were traced to breakdowns in the operations of water treatment facilities.

No water treatment system can produce pure water that is free of any and all chemicals. Most tap water contains some mineral pollutants that may cause changes in the color, taste, or odor of the water. While these chemicals do not necessarily mean that water is unfit to drink, they may indicate a decline of water quality. Well water may contain high concentrations of calcium and magnesium ions that have dissolved from the surrounding rocks. This is called hard water. The minerals that are present in most supplies of drinking water are not potentially harmful and may in fact be beneficial.

Since the 1960s, fluoride has been added to the drinking water in many cities. Fluoride strengthens tooth enamel against the effects of decay-causing acids. Almost two-thirds of the population in the United States has a fluoridated water supply. This practice has greatly reduced the incidence of cavities in the permanent teeth of children. However, excessive fluoride levels can lead to the mottling, or spotting, of children's teeth. Therefore, the government sets the standards for fluoride levels allowed in drinking water.

Water Management

Most of the water taken from streams and aquifers never returns to its source. For example, 80 percent of the water used for agriculture evaporates. The loss of water causes rivers to flow at a fraction of their natural rates and may dry up aquifers. This often leads localities to search for new water supplies instead of managing their resources more responsibly.

Water management is a method of increasing both the quantity and quality of the world's water supplies. It requires investment in dams and storage areas, as well as controlling processes involved in the hydrologic cycle. Constructing dams on rivers not only aids water storage but also controls flooding, generates hydroelectric power, provides irrigation water for agriculture, and opens areas for recreational activities. (See Figure 28-2.)

There are ways to improve the quality of water in the home. Aeration devices remove foul smells, volatile pollutants, and radioactive gases such as radon. Carbon-based filtering systems can control most taste- and odor-causing chemicals. They remove radon, chlorine, organic chemicals (pesticides and industrial waste), and some toxic minerals. But they are not effective against bacteria or heavy metals and often become the breeding grounds for disease-causing organisms.

Figure 28-2
This photograph shows Hoover Dam and Lake Mead behind it.

Physical filters such as wire screens can remove particles of sediment, rust, or dirt but cannot remove disease-causing organisms. Ultraviolet light disinfection units kill bacteria and viruses but do not remove chemical pollutants and are ineffective against spores.

Distillation removes all minerals, heavy metals, and most organic chemicals and also kills all bacteria and viruses. However, distillation requires a large amount of heat energy to boil the water. Water obtained from distillation, called distilled water, is pure and drinkable but bland tasting.

Water used to cool machinery in industrial plants is returned to the environment. Because of the water's high temperature, it devastates ecosystems. This is the cause of thermal pollution. A nonpolluting alternative involves using the heated water to boost productivity in greenhouses or warm fishponds.

Increasing Water Supplies

Cloud seeding is a process used to make rain in times of drought or to lessen the intensity of storms. To initiate precipitation, high-flying aircraft sprinkle dry ice or potassium iodide crystals on saturated clouds. These solids act as condensation nuclei around which raindrops form. However, cloud seeding in one area can prevent rain from falling elsewhere.

The idea has been proposed to supply freshwater to arid regions by towing icebergs from the Arctic or Antarctic. Icebergs are composed of freshwater, and it is theorized that only a very small fraction would be lost through melting. At this time, the energy costs in moving an iceberg prohibit tapping this potential resource. In some areas, the high cost of increasing the water supply makes it impractical.

Throughout the world, several processes are in use for **desalinization**, converting salt water into freshwater. (See Figure 28-3.) Distillation is a process that uses heat energy to evaporate seawater and condensation of the vapor to produce pure water. Those forms of distillation that use sunlight as the source of energy for evaporation are the most cost-effective but are very slow. Using various fuels as an energy source is more expensive but produces freshwater at a faster rate. Reverse osmosis is a desalinization process that uses pressure to force water through pores in a semipermeable membrane.

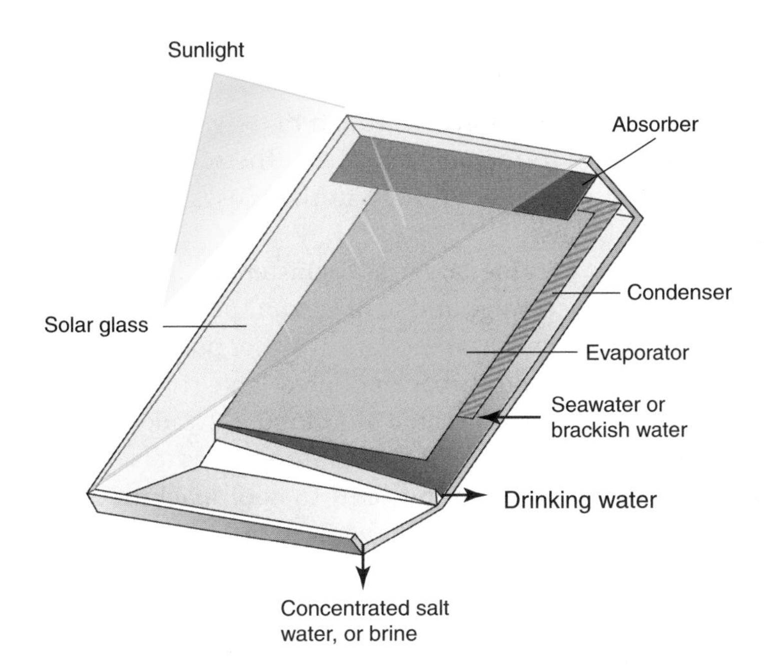

Figure 28-3 **A solar-powered evaporation desalinization device converts salt water to freshwater.**

The freshwater passes through the pores, while salts and minerals are excluded.

28.1 Section Review

1. Explain why most of the water taken from streams and aquifers never returns to its source.
2. What local factors can adversely affect water quality?
3. Describe three processes for removing solid impurities from water.

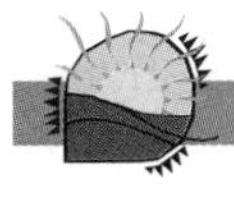

28.2 CONSERVING FRESHWATER ECOSYSTEMS

Throughout history, people have disposed of sewage and garbage by dumping it into rivers, lakes, and seas. In the past, rivers' running water carried away the wastes while the deep waters of lakes and seas buried them in bottom sediments. There was usually little evidence that dumping had occurred. Some communities that used streams as sewage dumps might have noticed pollution in the area of the dumping. However, several miles downstream the water would again appear to be clean and uncontaminated. Sewage dumped in lakes and seas spread through the large volume of water and eventually sank to the bottom. There, the sewage was broken down by decomposers and returned to the ecosystem. There were no pollution problems.

As communities became larger and clustered together to form cities, the amount of sewage and garbage dumped became so great that the water could not accommodate it. Water pollution became a problem. Raw sewage and garbage polluted large stretches of lakes and rivers, and these waters became unfit for use. To add to the problem, factories dumped chemical wastes into the waters, poisoning many plants and animals. Water used to cool machinery caused thermal pollution when it was poured back into the environment. The increased temperatures made the water inhospitable to the original biota. Agricultural runoff added nutrients to the water, which accelerated the natural aging process of rivers and lakes.

Eventually the problems reached dangerous proportions. In large lakes, the commercial fishing industry came to a halt because many fish died. Algal scums and detergent froths covered shorelines. Beaches were closed to swimming. Bacteria contaminated the waters. Drinking water had to be treated. And often, after treatment, the water was discolored and had an unpleasant odor. Water intake pipes were clogged with sludge. The impact pollution made on the natural balance of ecosystems and on the human communities surrounding the polluted waters required immediate action.

The Clean Water Act of 1972 focused on the most visible pollution sources, such as sewage treatment outlets and industrial waste discharges. These are known as **point sources**, because pollution from them can be traced to its point of origin. But the act overlooked **nonpoint sources**, pollution sources that cannot be traced to a single point of origin. Nonpoint sources include runoff from urban streets and lawns, faulty septic systems, erosion from disturbed hillsides, and pollution caused by agricultural fertilizers, pesticides, and manure. (See Figure 28-4.) These types of pollution are difficult to control since they originate from scattered sources. In 1987, an amendment to the Clean Water Act attacked the problem of nonpoint source pollution management. The EPA's job is to identify and control violators.

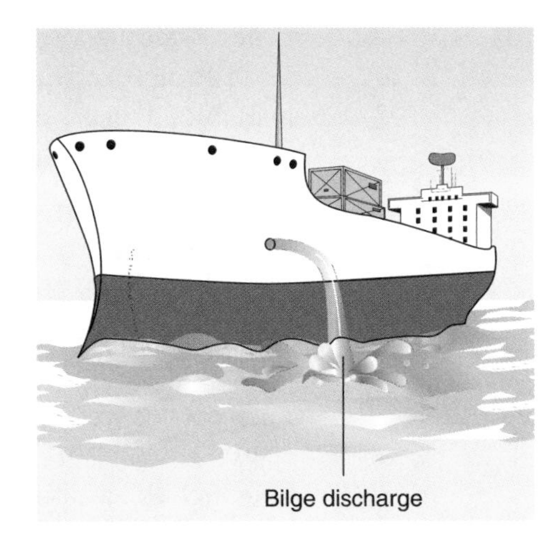

Figure 28-4 **An animal feedlot is a nonpoint source of pollution. A ship is a point source of pollution.**

In a 1995 report, the Environmental Protection Agency stated that despite environmental controls, nearly 4 out of every 10 lakes, rivers, and estuaries remained polluted. Fishing, swimming, and other recreational uses had to be discontinued in these areas year-round. The EPA report also indicated that the most common pollutants were sewage, disease-causing bacteria, toxic metals, fertilizers, and petroleum products. Due to pollution, about 37 percent of the nation's lakes and estuaries and 36 percent of its rivers are unsafe for fishing and swimming.

Water Conservation Practices

We could save as much as 50 percent of the water we now use without great changes in our lifestyle. Simple conservation measures by agriculture, industry, and individuals could go a long way toward preventing projected shortages. Almost half the water used in the United States is used for cooling electrical power plants and manufacturing facilities. This use could be reduced through the installation of dry cooling systems. Further reductions could be made through multiple use of discharged water for heating, irrigation, or **aquaculture**, the cultivation of fish or shellfish in artificial ponds. Improved farming techniques could dramatically reduce the amount of water used for agriculture. These techniques include reducing evaporation with mulches, repairing leaks in irrigation canals, trickle irrigation, and reducing water lost through runoff.

Many people are not aware of the volume of water they use each day. People use water to excess because it seems to be so readily available. Before water becomes a scarce commodity, we must rethink its use. Here are some water-saving tips:

- A shower is better than a bath. Each shower can save between 50 and 100 liters of water.
- Repair leaking faucets to save up 100 liters of water per day.
- Do not let the water run when washing dishes or brushing your teeth.
- Keep cool water for drinking in the refrigerator instead of running the tap until the water gets cold.
- Reduce the amount of water wasted each time you flush by placing a brick in the reservoir tank of your toilet or changing to a low-volume tank.
- Reduce lawn areas by substituting gravel or mulch. Plant a natural lawn using native plants. Construct a rock garden.
- Limit watering of lawns to times of need. Do not run sprinklers on hot, windy days when water loss through evaporation is high.
- Limit the water used to wash your car. Wet the car, turn off the water, and wash it with a pail of soap and water, then rinse off the car. Or bring the car to a car wash that recycles its wash water.

28.2 Section Review

1. Describe some of the factors that endanger the health of rivers.
2. Why are point sources of pollution easier to control than nonpoint sources?
3. Why is it not wise to water lawns on hot, windy days?

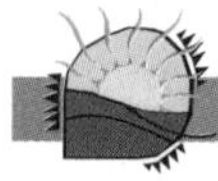

28.3 CONSERVING MARINE ECOSYSTEMS

More than 80 percent of the pollution in marine ecosystems comes from land-based activities such as ocean dumping of sewage and industrial waste. For many years, ocean dumping was considered an inexpensive and convenient method of waste disposal. Solid wastes that do not end up in or on the land or in the air end up in the water. The worst conditions are found in enclosed areas such as the

Figure 28-5 The Mediterranean Sea has a narrow outlet, the Strait of Gibraltar.

Mediterranean Sea and the North Sea. (See Figure 28-5.) In these areas, pollution levels are so high that wildlife and human health are threatened. The major sources of pollution are sewage from the towns and cities that fringe the coasts and islands, wastes from industrial activities, and agricultural runoff carried to the sea by rivers and streams. Diseases such as cholera, typhoid, dysentery, and hepatitis are common in the enclosed Mediterranean region. These diseases are associated with bathing in contaminated water or eating contaminated seafood.

Ocean dumping of sewage, industrial wastes, municipal wastes, and agricultural and urban runoff releases many dangerous pollutants into the marine ecosystem. Throughout the United States, concerns regarding the effects of dumping on the environment have increased steadily. They reached a peak in 1988, when illegally dumped sewage, medical waste, and garbage washed up along thousands of miles of the coastline from Maine to Texas. Many beaches were closed, much marine life died, and the tourism industry lost millions of dollars.

Although the oceans are vast, signs of human activity can be observed in the most remote places. From the deepest regions of the seafloor to the beaches of newly emerging volcanic islands, wastes associated with human society litter the landscape and pollute the biosphere. Plastic containers, glass bottles, discarded fishing nets, petroleum slicks, aluminum cans, and Styrofoam packaging are found from the ocean surface to the seafloor sediments and beach sands. Plastic makes up a large proportion of all the solid litter. Industrial wastes that contain toxic chemicals such as mercury and dioxin contaminate the waters. Coastal areas, which have the highest concentrations of marine life, are most severely affected.

The rings of a plastic six-pack holder can be deadly to marine life. Mammals and birds often become entangled in them and are choked to death. Their bodies, with the rings still around them, wash up on beaches. As with all items you bring to a beach or aboard a boat, what you take in, you should take out. Cut apart six-pack holders before disposing of them properly in a trash receptacle. (See Figure 28-6.)

Trash composed of natural materials such as wood, paper, and cardboard is broken down by decomposer organisms within the environment. Those materials are biodegradable. Other materials, including some types of glass, certain metals, and some ceramic materials, break down when exposed to sunlight, moisture, or atmospheric gases. These materials do not cause long-term pollution problems. However, nonbiodegradable substances persist in the environment for a long time. Many plastics are nonbiodegradable.

Phytoplankton are the primary producers in marine food chains. They form huge pastures for marine animals and play an important role in recycling oxygen within the biosphere. In addition, plankton produce sulfide gases. These gases rise in the atmosphere where they are the source of nuclei for the condensation of water droplets that form clouds. Chemical and thermal pollution harm plankton. Oil slicks, which block sunlight from penetrating the surface water, pre-

Figure 28-6 This American coot was killed by a six-pack holder.

vent phytoplankton from carrying out photosynthesis. These factors can lead to plankton die-offs, which affect the health of the entire marine ecosystem as well as change global weather patterns.

A steady decline in coastal fisheries, or fishing industries, has been noted for many years. In part, this decline has been traced to ocean dumping of sludge from municipal sewage treatment plants. The nutrients in sludge encourage the rapid growth of phytoplankton, or algal, blooms in coastal waters. The dense growth of plankton, called tides, can cause the water to turn red, brown, or green. One of the most dangerous types is the toxic red tide, which can kill fish, marine mammals, and birds. People can become ill and even die from eating shellfish contaminated by the red tide organisms. Dense blooms of brown tide phytoplankton can prevent sunlight from penetrating beneath the water's surface. These blooms lead to oxygen depletion, and fish in these areas often suffocate. Toxic materials in sludge also harm marine organisms. The contamination of fish and shellfish has led to the closing of many commercial and recreational fisheries to protect human health.

The decline of many marine species can be traced to pollution and its effects on the environment. Pollution and human encroachment on wetlands have destroyed spawning and breeding grounds. Environmental changes caused by human activity, such as shifting ocean currents and ocean warming, have also reshaped marine ecosystems. Overfishing has reduced many fish stocks and affected organisms at the top of marine food chains. Recognizing the global dangers posed by marine pollution, nations have banded together to formulate international treaties with laws designed to protect the environment.

Marine Laws

In 1958, an international agreement, the Convention on the High Seas, was set up to protect the marine environment and its resources. The London Convention of 1972 is a treaty that governs the disposal of wastes in the oceans. It resolved to phase out all ocean dumping by 1995. In particular, it banned the dumping of all radioactive wastes at sea. The 1972 Marine Protection and Sanctuaries Act required that a permit be obtained for any ocean

CONSERVATION

Fish Protection Treaty

In December 2001 when Malta became the 30th nation to ratify it, the United Nations Fish Stocks Agreement went into effect. As of July 2004, 52 countries, including the United States, Canada, most of the European Community, India, South Africa, and the United Kingdom among others had signed the agreement. Japan, a nation with a large fishing industry had not ratified the treaty. The World Wide fund for Nature has warned that the treaty to protect marine fish stocks is endanger of being ineffective, since it has not yet been ratified by all the important nations involved in marine fisheries. It is hoped that eventually, the treaty will promote greater international cooperation in managing marine fish stocks.

The treaty covers both large migratory fish such as tuna, sharks, and billfish, as well as smaller fish whose range straddles the 300-kilometer exclusive economic zone of at least one country. Coastal nations usually manage fishing within their own zone but have no control of fishing on the high seas. This new treaty is the first to place precautionary principles into fishery management.

Overfishing has depleted the seas of many desired fish stocks such as cod, tuna, shark, swordfish, and haddock. As the stocks of these species fall, other species are harvested to replace them. Thus, species such as Chilean sea bass, monkfish, and orange roughy, became economically important. These species are often not native to United States waters and must be imported. Some stocks, such as orange roughy will crash since the species reproduces slowly and the population does not rebound quickly. The wasteful practice of removing shark fins and throwing the rest of the carcass back in the sea has depleted the populations of many shark species. Because most sharks are slow-growing fish that produce few offspring, intensive fishing has been devastating. In addition, shark-fishing techniques endanger marine turtles and mammals.

With more people adding fish to their diet, the industry is in crisis. Environmental groups now make consumer recommendations as to those species that are struggling and those that are in good shape. These recommendations prompted many restaurants to remove swordfish from their menus until the stocks recovered.

dumping. It also prohibited the dumping of any sludge after December 31, 1981.

The International Convention for the Prevention of Pollution from Ships, formulated in 1975, set limits on the amounts of sewage, garbage, petroleum, and hazardous materials that can be discharged by ships at sea. In some areas, oil discharge was prohibited altogether. The convention also called for "fingerprinting" oil cargoes. In fingerprinting a cargo, chemicals are added to the cargo to identify the ship that carried the oil. This makes it easier to identify polluters.

In 1976, the countries that border the Mediterranean Sea put aside national rivalries in an attempt to clean up the waters. They joined in the Barcelona Convention for the Protection of the Mediterranean Sea Against Pollution to preserve this huge ecosystem. Marine laboratories now carry out long-term monitoring of land-based pollution. Controls have been placed on the dumping of industrial waste, sewage, and agricultural pesticides. Sewage treatment plants are projected for all large cities. Protection has been given to endangered marine species such as the Mediterranean monk seal and the Mediterranean sea turtle.

In 1982, 117 nations signed the United Nations Convention on the Law of the Sea. This treaty deals with pollution, fishing rights, and conservation of marine resources. The United States and several other industrialized nations refused to sign this treaty because they disagree with the proposals regarding deep-sea mining rights.

The United States passed the Ocean Dumping Ban in 1988. This act prohibits the dumping in coastal waters of industrial wastes, sludge, and contaminated materials.

The Decline of Marine Fisheries

For many maritime nations, fish are the most important source of animal protein. Worldwide, fish supply almost 25 percent of this necessary nutrient. Large commercial fisheries play a crucial role in the economy of these countries. For example, Japan relies on the oceans for 60 percent of its animal protein. Japan has the world's third largest fishing fleet. The growing population in such nations has increased the demand for fish. Animal feed supplements (fishmeal and fish oil) and processed fish flour also place heavy demands on fish stocks. These factors have spurred the development of many new technologies in the fishing industry, which have resulted in improved methods of locating, catching, and processing fish.

Today, factory ships accompany large commercial fishing fleets. The factory ships process and store the catch and have greatly increased the number of fish caught. Overharvesting removes such a large portion of a population that breeding stocks can no longer maintain enough fish for future harvests. The fishing industry has put enormous stress on certain fish populations. This often leads to the collapse of once-productive fisheries. The herring fishery in the North Sea, the Atlantic cod fishery, and the anchovy fishery off the coast of Peru have all collapsed in recent times. In 1992, a coastal ban on all cod fishing was imposed. By the end of 2002, the stock of cod had not rebounded; the population in the Gulf of Maine was the lowest in 30 years. Restrictions were extended, closing a huge area to fishing, and greatly reducing the daily catch. Economic losses were estimated at more than 20 million dollars.

Destructive modern fishing techniques played a major role in these catastrophes. Giant fishing trawlers that use heavy nets and equipment churn up the spawning beds and feeding banks, destroying the fragile ecosystem. Everything is dragged from the bottom, including young fish, invertebrates, and noneconomic species. Purse seining uses large, fine nets to scoop fish from the sea. Only the economically desirable fish are kept. The rest of the catch, which is either dead or dying, is thrown back. (See Figure 28-7.)

Huge drift nets, some 50 kilometers long, have caused severe damage to pelagic, or open ocean, species such as tuna, dolphin, sea turtles, sharks, marine mammals, and birds. Drift nets are dropped into the ocean and allowed to drift with the currents. When fish try to swim through the nets, their heads get stuck in the openings. As the fish struggle, they become more entangled and die. A United Nations ban on drift-net fishing went into effect in 1992. Before the ban, approximately 700 vessels set drift nets each year. Between 1996 and 1998, some illegal drift-net fishing was detected in the North Pacific, and one vessel was seized. The catch from these vessels was most likely too small to have had much of an impact on fish stocks.

The world's fisheries are dominated by a few nations that have large, highly efficient fishing fleets. These nations include the former

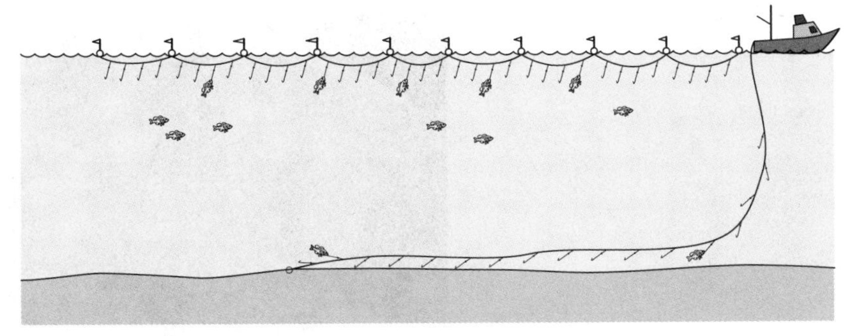
Hooks and long-line fishing

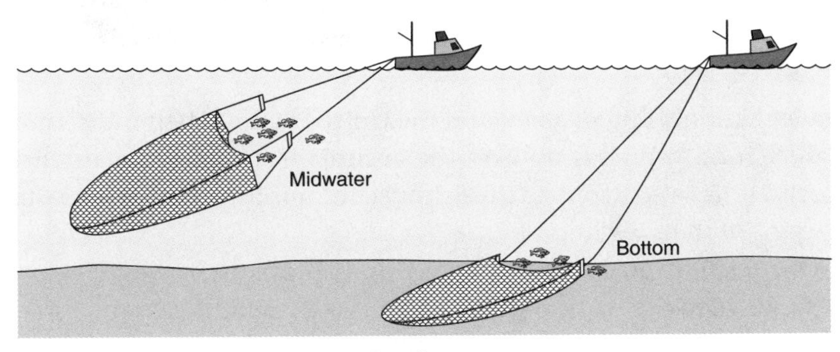

Trawling nets

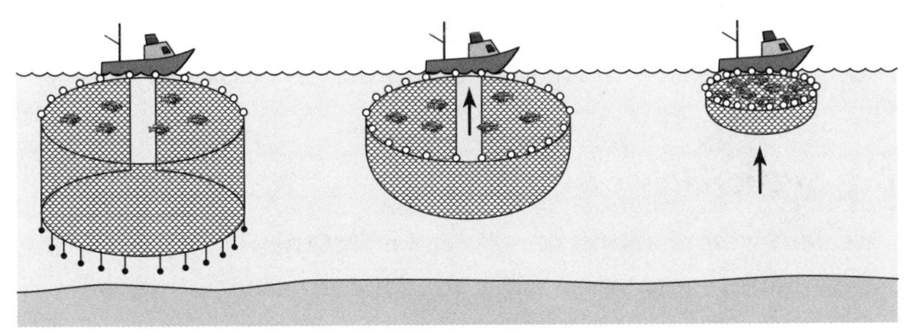
Purse seine net

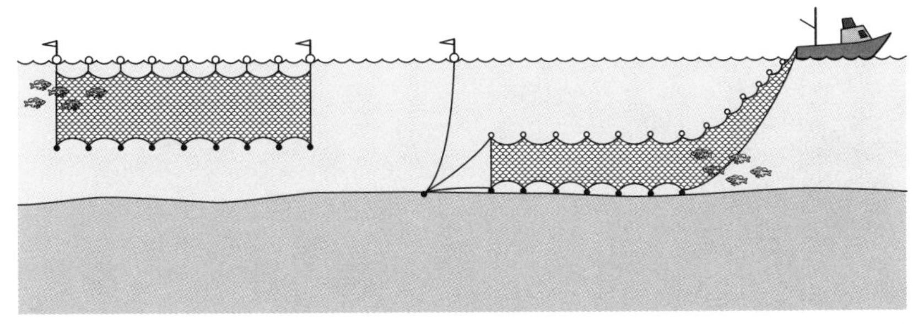
Gill nets

Figure 28-7 This illustration shows four common commercial fishing methods.

Figure 28-8 **Hybrid striped bass are farmed in California.**

Soviet Union, China, Japan, and the United States. Their indiscriminate fishing, which is now on the decline, has depleted many fish stocks. It has also deprived the fishing communities in poorer coastal nations of their livelihood.

Our present methods of harvesting fish are often wasteful and destructive to the marine environment. As a result, the young fish of many species as well as commercially undesirable fish species are all under pressure. An alternative is aquaculture, raising and harvesting fish or shellfish under controlled conditions within ponds or inside enclosed areas in lagoons and estuaries. (See Figure 28-8.)

28.3 Section Review

1. Explain why plankton die-offs can affect the entire biosphere.
2. How have governments acted to reduce marine pollution?
3. Describe some of the adverse economic effects caused by marine pollution.

LABORATORY INVESTIGATION 28

Removing a Salt From Its Solution

PROBLEM: *How can copper sulfate be removed from a solution?*

SKILLS: *Estimating, manipulating*

MATERIALS: *Ring stand, ring, funnel, 250-mL beaker, filter paper, wire mesh, copper sulfate, 15-cm watch glass, 2 test tubes, 50 mL of water, test tube holder, heat source (Bunsen burner), tongs or clamp*

PROCEDURE

1. In a beaker, dissolve a small amount of copper sulfate in 50 mL of water. Make sure to add enough copper sulfate to turn the water blue.

2. Fold a piece of filter paper, dampen it, and place it in a funnel. Place the funnel in a ring on a ring stand. Hold a test tube under the funnel.

3. To examine the process of filtration, pour about 20 mL of the copper sulfate solution into the filter paper in the funnel. Collect the filtrate in a test tube. Examine the filtrate. Remove the funnel from the ring.

4. Put the wire mesh on the ring on the ring stand. To examine the process of distillation, place the beaker with the remaining copper sulfate solution on the mesh. Use the heat source to bring the copper sulfate solution to a gentle boil. (CAUTION: Do not let the water boil violently.)

5. Attach the watch glass to a clamp. When the solution begins boiling, hold the watch glass about 5 centimeters above the top of the beaker, so that water begins to condense on its lower surface. By tilting the watch glass slightly, the water on its lower surface can be made to drip from the edge. Collect this water in a test tube. (CAUTION: Use a test tube holder since the water is still hot.)

6. Compare the color of the original copper sulfate solution you prepared with the filtrate collected from the filter paper and the condensate collected from boiling.

OBSERVATIONS AND ANALYSES

1. Describe any differences in color you found between the original solution, the filtrate, and the condensate. Explain reasons for the color differences.
2. In what ways can this be used to demonstrate the relative effectiveness of filtration and distillation as water purification tools?
3. What happens to dissolved solids during the process of distillation?

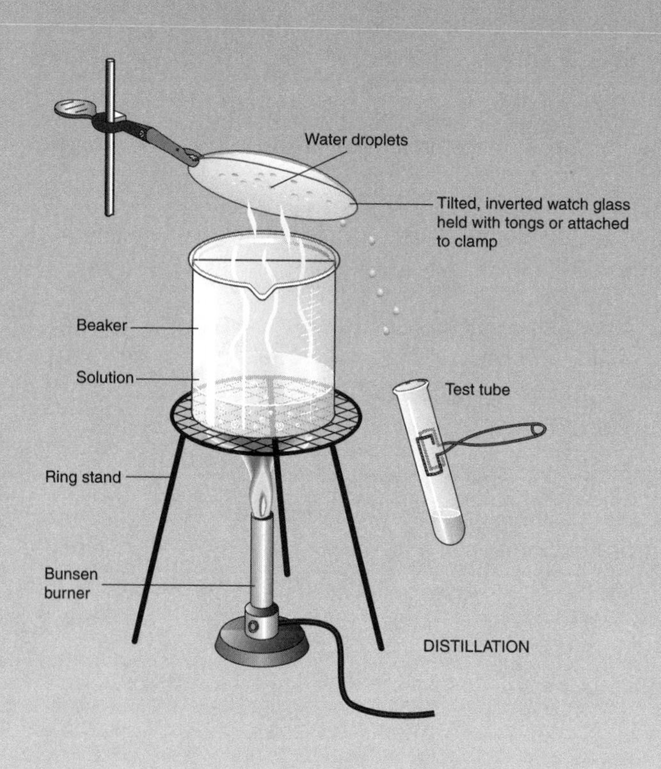

Chapter 28 Review

Answer these questions on a separate sheet of paper.

Vocabulary

The following list contains all the boldfaced terms in this chapter.

aeration, aquaculture, chlorination, cloud seeding, coagulation, desalinization, disinfection, filtration, irradiation, nonpoint sources, point sources, sedimentation, turbidity, Water Quality Criteria, water treatment

Fill In

Use one of the vocabulary terms listed above to complete each of the following sentences.

1. _______ is a measure of the cloudiness of water.
2. The process of converting seawater to freshwater is _______.
3. Raising and harvesting fish or shellfish under controlled conditions within ponds is called _______.
4. During times of drought, the process of _______ with dry ice crystals may be used to cause rain.
5. _______ involves the processing of well water or surface water before distribution in public water systems.

Multiple Choice

Choose the response that best completes the sentence or answers the question.

6. The process used to kill bacteria and remove odors and tastes in water is *a.* filtration. *b.* sedimentation. *c.* aeration. *d.* chlorination.
7. An example of a nonpoint source of pollution is *a.* agricultural runoff. *b.* sewage plant outlets. *c.* factory waste pipes. *d.* power plant coolant water.
8. Most of the solid litter that pollutes the oceans is composed of *a.* dioxin. *b.* paper. *c.* glass. *d.* plastic.
9. Which treatment method increases the amount of oxygen in water? *a.* distillation *b.* chlorination *c.* aeration *d.* sedimentation
10. The amount of marine pollution that originates in land-based activities is about *a.* 20 percent. *b.* 40 percent. *c.* 60 percent. *d.* 80 percent.
11. Organisms that cause disease are called *a.* carcinogens. *b.* heavy metals. *c.* pathogens. *d.* herbicides.

12. Which is *not* related to water management? *a.* It is used to increase the quantity of water. *b.* It controls processes involved in the hydrologic cycle. *c.* Dams are built on rivers. *d.* Sewage treatment plants are built.
13. Which is *not* a water conservation technique? *a.* using warm water discharged from power plants in aquaculture *b.* reducing evaporation in irrigation *c.* washing the car and leaving the water running *d.* taking a shower rather than a bath
14. Which is *not* a biodegradable substance? *a.* plastic *b.* wood *c.* paper *d.* cardboard

Short Answer (Constructed Response)

Use the information you learned in this chapter to respond to the following items.

15. In the United States, which group sets water quality standards?
16. List two methods used to disinfect water supplies.
17. Why do some communities add fluoride to drinking water?
18. What does "fingerprinting" oil cargoes mean?
19. Explain how drift nets damage marine ecosystems.

Essay (Extended Response)

Use the information you learned in the chapter to respond to the following items.

20. Why is filtration not an effective method for removing viruses and bacteria from water?
21. How does overharvesting of marine species affect the economy of a region?
22. Describe some of the ways you can practice water conservation around your home.

Research Projects

- Investigate methods used by your community to promote water conservation.
- Explore methods of increasing water supplies for your community.

CHAPTER 29
Fresh Air

When you have completed this chapter, you should be able to:

Identify different forms of air pollution.

Explain the causes of global warming and global cooling.

Discuss the ways air pollution affects human health.

In addition to polluting our freshwater resources, humans are also polluting our atmosphere. Air pollution affects all living things. It causes serious health problems as well as damage to trees, stone buildings, and statues. Another effect of this pollution is that the atmosphere is warming at an alarming rate. We have seen reports in the newspapers and on television that a hole developed in the ice covering the North Pole. At the South Pole, some glaciers are melting. In the future, this may raise sea levels and flood coastal areas. Farmland may turn to desert, reducing food production and causing worldwide famine. In this chapter, you will learn about the forms and effects of air pollution.

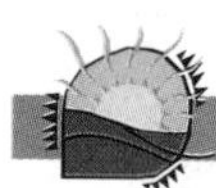

29.1 WHAT IS POLLUTION?

Air is a mixture of gases, whose concentrations are regulated by activities within the biosphere. (See Figure 29-1.) Some pollutants make their way into the atmosphere naturally through wind erosion, forest fires, and volcanic eruptions. Other pollutants enter as a result of human activity, for example, automotive exhaust gases, smoke from burning coal and oil, and discharge of industrial gases. Pollutants can be solid particles, such as dust and spores; liquid droplets, such as acid rain and hydrocarbons; or gases, such as carbon monoxide and sulfur dioxide.

Some air pollutants are substances that do not occur naturally in the atmosphere. Others are natural atmospheric components that have been raised to dangerous concentrations. The incomplete combustion of fossil fuels adds numerous pollutants to the air. In addition to carbon, fossil fuels contain nitrogen and sulfur from proteins. As a result, pollutants such as sulfur oxides, carbon monoxide, soot, and hydrocarbons are introduced. Levels of carbon dioxide and water vapor, natural components of the atmosphere, are also increased.

Major Air Pollutants

Some pollutants, such as hydrocarbons that are released directly into the air, are called **primary pollutants. Secondary pollutants**

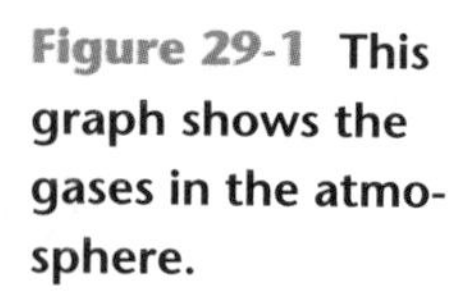

Figure 29-1 **This graph shows the gases in the atmosphere.**

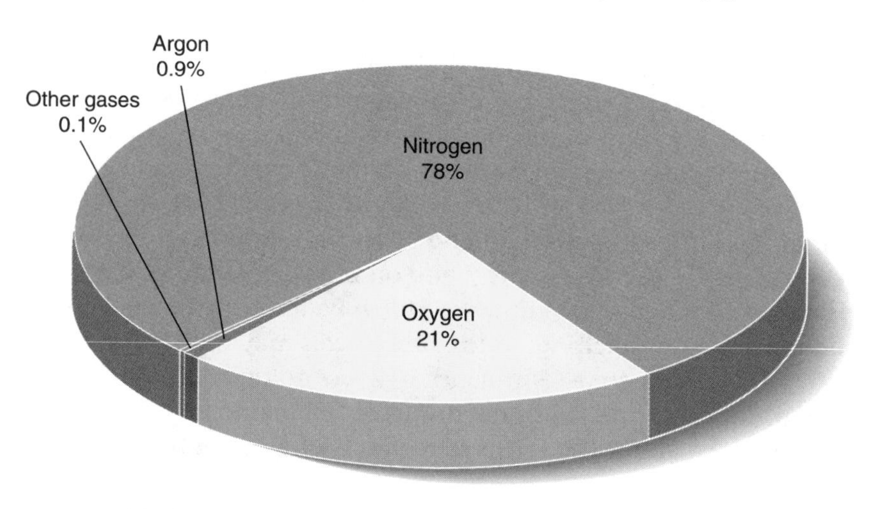

TABLE 29-1 MAJOR AIR POLLUTANTS

Pollutant	Description	Main Sources
Sulfur dioxide	SO_2 gas	Burning coal and oil that contain high levels of sulfur
Carbon oxides	CO and CO_2 gases	Burning fuel in internal combustion engines
Nitrogen oxides	NO and NO_2 gases	Fuel combustion and the generation of electricity
Hydrocarbons	Benzene, toluene, phenol, formaldehyde	Released by power plants, chemical manufacturers, petroleum refineries, and vehicles
Photochemical oxidants (smog)	Ozone, peroxyacetyl nitrate	Atmospheric products formed through the action of solar radiation
Lead and lead oxides	Pb and PbO	Burning leaded gasoline, smelting ores, waste disposal
Particulates	Ash, soot, smoke, dust, aerosols	Industrial processes, urban activities, agricultural processes

are substances that enter the air and are made hazardous by reactions with natural atmospheric components. Photochemical smog and acid rain are examples of secondary pollutants. The Clean Air Act of 1970 set maximum levels for the seven pollutants that pose the most serious threats to human health. (See Table 29-1.)

SULFUR OXIDES

The combustion of sulfur-containing fuels, especially low-grade coal, releases large amounts of sulfur dioxide into the atmosphere. The extraction of metals from sulfide ores also contributes sulfur pollutants to the atmosphere. Sulfur dioxide is a colorless gas that smells like rotten eggs. It is corrosive to both plant and animal cells. Sulfur dioxide can be extremely damaging to lung tissue. In the air, some sulfur dioxide reacts with oxygen to form sulfur trioxide. Dissolved in water, sulfur trioxide forms sulfuric acid, thus contributing to problems of acid rain. In the air, sulfur compounds produce a haze that reduces visibility.

CARBON OXIDES

The most common oxide of carbon in the air is carbon dioxide. It is colorless, odorless, and nontoxic. Carbon dioxide is a natural component of air, but the increase in atmospheric concentration is

TABLE 29-2 CAUSES OF COMMON POLLUTANTS

Fossil Fuel Component	Process	Pollutants
Hydrocarbons	Incomplete combustion	Carbon monoxide, unburned hydrocarbons (soot)
Hydrocarbons	Complete combustion	Carbon dioxide
Sulfur	Combustion	Sulfur dioxide, sulfur trioxide
Minerals	Combustion	Particulates (for example, lead, mercury)

affecting global climate. Use of fossil fuels is the greatest contributor to atmospheric carbon dioxide pollution. The amount of carbon dioxide released by plants and animals during respiration is naturally balanced by the amount green plants use in photosynthesis.

Carbon monoxide (CO) is also a colorless, odorless oxide of carbon. But carbon monoxide is toxic. In the blood, carbon monoxide binds to hemoglobin, displacing oxygen. Thus it limits the blood's ability to carry oxygen to the cells. Carbon monoxide is produced during the incomplete combustion of fuels, the anaerobic decomposition of organic material, and the incineration of solid wastes. Most atmospheric carbon monoxide is produced by internal combustion engines, such as those in cars and trucks. (See Table 29-2.) Ozone-producing reactions within the atmosphere remove much of the carbon monoxide.

NITROGEN OXIDES

Nitrogen oxides are gases formed when nitrogen in the air is heated to high temperatures in the presence of oxygen and by the decomposition of organic materials by soil bacteria. Nitric oxide (NO) oxidizes in the atmosphere to form nitrogen dioxide (NO_2), a reddish-brown gas that produces photochemical smog. Nitrogen oxides combine with atmospheric water to produce nitric acid (HNO_3), a major component of acid rain. Most of the polluting nitrogen oxide gases come from fuel combustion in transportation and power generation and from the decay of organic fertilizers. (See Figure 29-2.) These gases contribute to global warming.

HYDROCARBONS

The most common hydrocarbon pollutant is methane. This gas is produced by bacteria that live in the gut of ruminants (cattle, sheep,

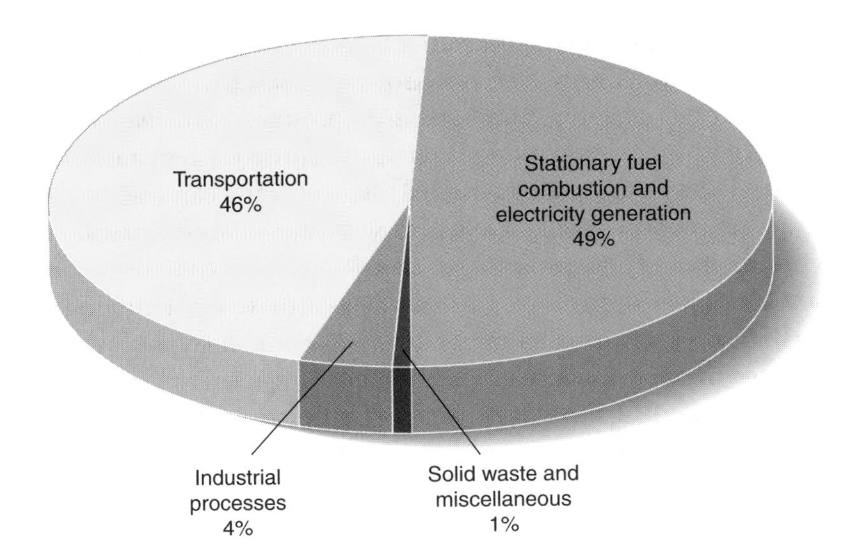

Figure 29-2 The percentage of nitrogen oxide pollutants in the air from various sources is shown.

deer) and termites. The bacteria ferment the food the animals eat. Methane is also called swamp gas because it is produced by the bacteria that anaerobically decompose organic materials in wetlands and rice paddies. Rising methane levels in the atmosphere may contribute to global warming. Hydrocarbons are responsible for the formation of photochemical oxidants, and many are potentially carcinogenic.

The production of synthetic organic compounds is a major source of pollution. Burning of organic materials and fossil fuels in power plants, petroleum refineries, chemical plants, and motor vehicles releases substances into the atmosphere. These pollutants include benzene, phenol, and toluene, as well as chlorinated hydrocarbons such as chloroform, chloromethane, formaldehyde, trichloroethylene, and vinyl chloride.

PHOTOCHEMICAL OXIDANTS

Smog is an atmospheric condition in which visibility is reduced due to air pollution that contains high levels of particulates or photochemical oxidants. The word *smog* is derived from *sm*oke and f*og*. There are two types of smog. Particulate smog is formed when smoke mixes with atmospheric gases. Particulate smog forms in a layer of polluted air that hangs close to Earth's surface. Recently instituted air pollution controls have reduced this form of smog. **Photochemical smog** is most often caused by sunlight acting on motor vehicle exhaust emissions that form ozone. Car-choked cities, such as Houston, Los Angeles, New York, and San Diego, suffer most from this type of smog. But long hot summers can lead to the buildup of photochemical smog across the entire United States. The high ozone levels associated with this type of smog can lead to crop destruction, health problems, loss of visibility in scenic areas, and environmental degradation.

Smog is most dangerous to human health when temperature inversions occur. During a temperature inversion, a layer of warmer air moves over a cooler layer of smog and stalls there, preventing the smog from dissipating. With the smog trapped in the same location for many days, concentrations of pollutants build up to unhealthful levels. (See Figure 29-3.) The Los Angeles area is prone to these inversions because nearby mountains allow warm air to flow in and trap smog. From May through October, the smog is noticeable as a brownish-orange haze on the horizon.

Ozone (O_3) is a photochemical oxidant that is produced by ultraviolet radiation in the stratosphere. In the stratosphere, ozone shields the biosphere from ultraviolet solar radiation. However, ozone at ground level is a pollutant. It causes oxidation that can damage vegetation; degrade paint, rubber, and plastic; and irritate the eyes and respiratory system. Ozone has a distinctive odor and is often a component of photochemical smog.

LEAD

Lead is a toxic metal present in air. Elemental lead as well as lead compounds can harm the digestive, circulatory, and central nervous systems. Young children are most vulnerable to lead's effects.

For many years, lead was added to gasoline to increase engine performance. The lead entered the atmosphere with the engines' exhaust gases. Pollution control standards led to the removal of leaded gasoline from the market in the United States. Atmospheric levels have dropped sharply since the use of leaded gasoline was phased out by many industrialized nations. However, leaded gasoline often is the only type of fuel available in developing countries. Lead also enters the atmosphere as a result of smelting ores, waste incineration, and electrical and electronics manufacturing.

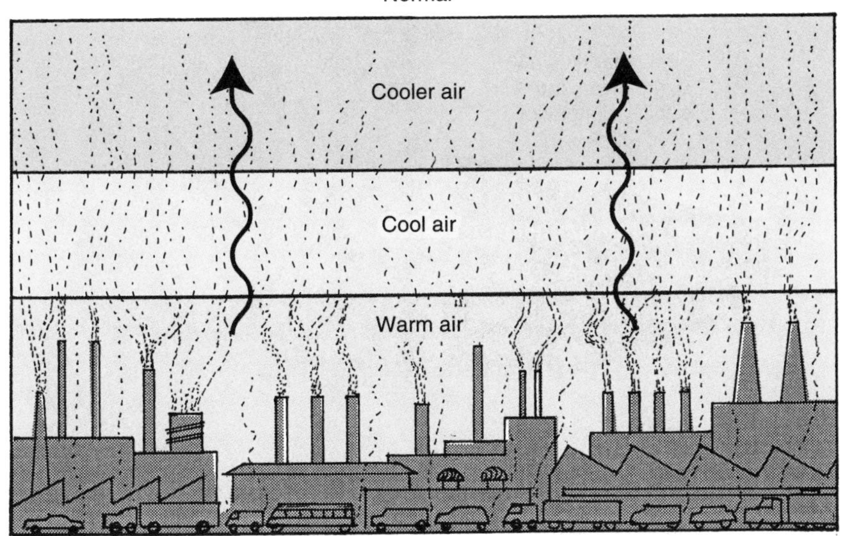

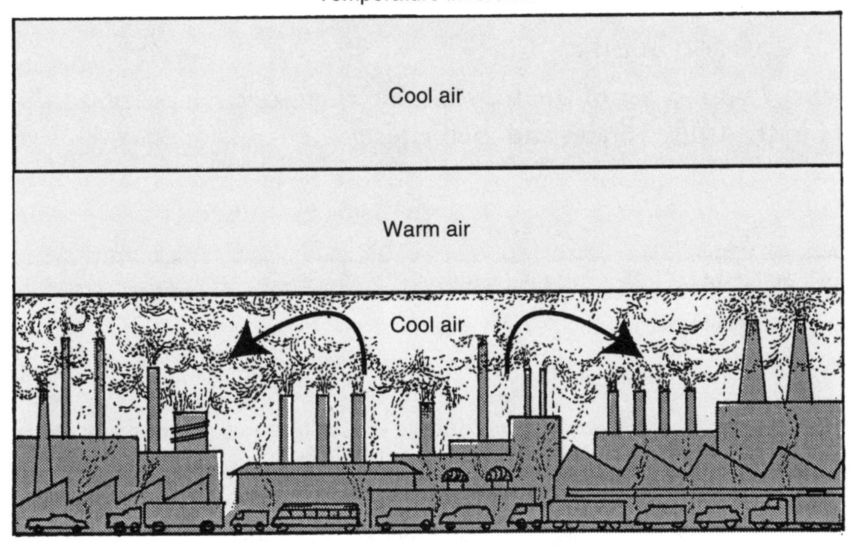

Figure 29-3 A temperature inversion traps pollutants in the layer of air near the ground.

PARTICULATES

Particulates are solid materials or tiny droplets of liquid that are suspended in air. These include fly ash, soot, dust, bacteria, spores, pollen, asbestos fibers, cigarette smoke, and aerosols. (See Figure 29-4 on page 524.) Aerosols consist of particles 1 micrometer or less in diameter that are dispersed in air. Particulates are the most obvious form of air pollution since they reduce visibility and leave dirty

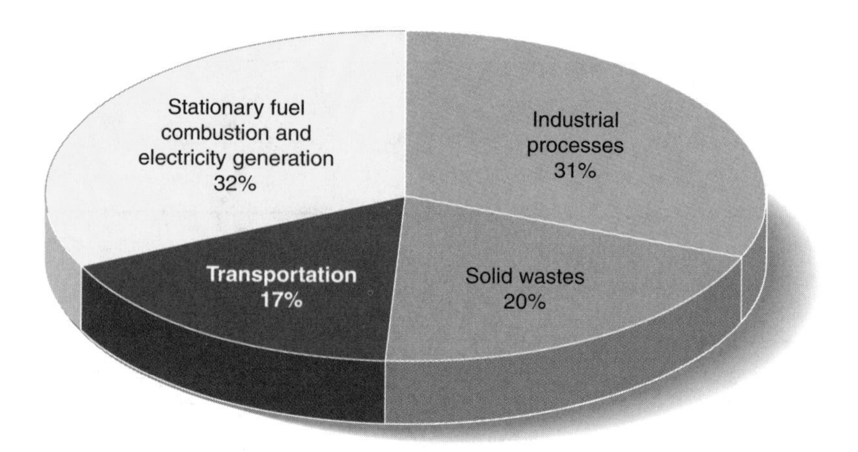

Figure 29-4 The percentage of particulate pollution from various sources is shown in this graph.

deposits on surfaces. The smallest particles are damaging to respiratory tissues. Cigarette smoke and asbestos fibers are potentially the most dangerous because they are carcinogens.

TOXIC AIR POLLUTANTS

Table 29-3 is a list of the top 10 toxic chemicals released into the air in the United States and their effects.

TABLE 29-3 THE EFFECTS OF TOXIC POLLUTANTS

Toxic Pollutant	Effects
1. Methanol (methyl alcohol)	Blindness; liver damage
2. Toluene	Irritates eyes, nose, and throat; causes dizziness and death
3. Ammonia	Burns skin and eyes; irritates respiratory organs; causes death
4. Acetone	Damages kidneys, liver, and nervous system
5. Methyl chloroform	Damages kidneys and liver; irritates skin and eyes; affects heart; causes death
6. Xylene	Damages bone marrow, kidneys, and liver; causes headaches, nausea, and death
7. Methyl ethyl ketone	Irritates skin, eyes, and nose; damages nervous system
8. Carbon disulfide	Causes changes in personality; damages eyes, nose, and throat; contributes to cardiovascular disease
9. Hydrochloric acid	Irritates skin and respiratory organs
10. Dichloromethane	Damages liver and brain; causes cancers

Old man's beard

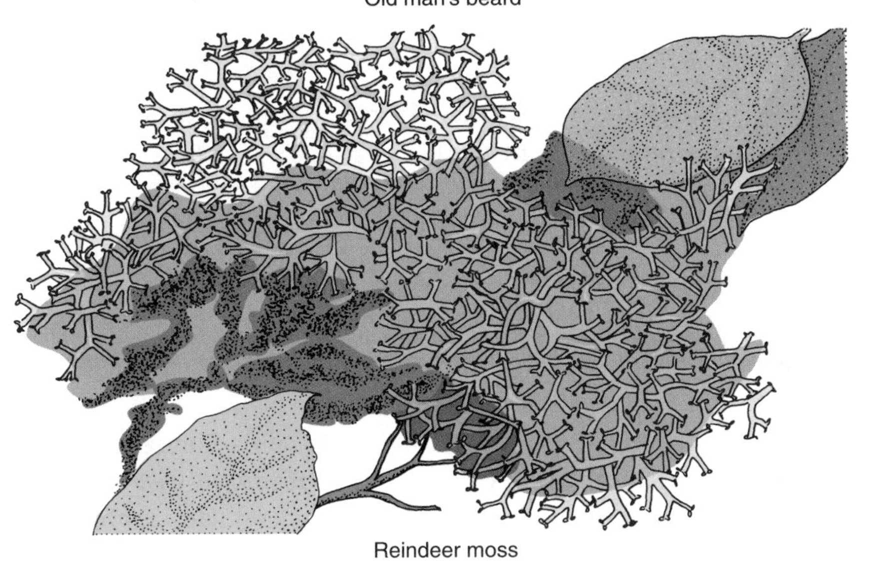

Reindeer moss

Figure 29-5 Types of lichens are illustrated.

Air Pollution Indicators

Lichens are unique organisms formed by a partnership between an alga and a fungus. (See Figure 29-5.) Lichens are one of the few organisms able to grow on bare rock. They often form thick, spongy mats that cover rocky outcrops. The lichens' scaly leaflike surfaces insulate against the cold. On the tundra, temperatures within the lichen mats are often 11°C higher than outside temperatures. Many insects live beneath the lichen mats, protected from the arctic weather. Caribou, musk oxen, and lemmings eat lichens during the long winters.

Ecologists are able to use lichens to measure the impact of human activity on the atmosphere. Lichens are extremely sensitive to atmospheric pollutants and thus are indicators of air quality. Due to high concentrations of pollutants in the air, few lichens grow in or around cities or industrial areas. However, the farther away you go from these areas, the more numerous and diverse the number of lichen species become. The presence of lichens and the variety exhibited can be used to determine the air quality of a region. One species, which exhibits a treelike form, always indicates excellent air quality.

29.1 Section Review

1. How is particulate smog different from photochemical smog?
2. How is ozone harmful to human health?

29.2 GLOBAL WARMING VERSUS GLOBAL COOLING

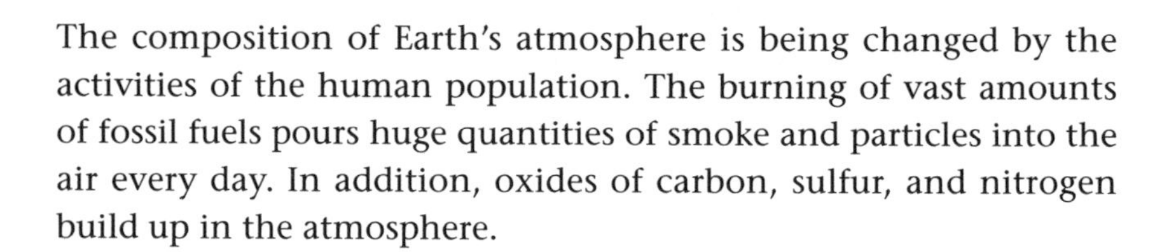

The composition of Earth's atmosphere is being changed by the activities of the human population. The burning of vast amounts of fossil fuels pours huge quantities of smoke and particles into the air every day. In addition, oxides of carbon, sulfur, and nitrogen build up in the atmosphere.

The Greenhouse Effect

The atmosphere warms naturally as carbon dioxide, methane, and water vapor in the air absorb infrared radiation emitted from Earth's surface. This natural greenhouse effect keeps our planet warm. However, "greenhouse effect" most often refers to the excessive warming that results from the increase of greenhouse gases such as carbon dioxide and methane. These gases trap heat within the atmosphere in much the same way glass traps heat within a greenhouse.

Over the past 100 years, atmospheric concentrations of methane have increased. Methane is produced naturally by the action of microorganisms in the digestive system of various species of animals and in swamps. It is discharged into the atmosphere by the host organisms. Because rice paddies generate methane, the increase in world rice production is another factor in the increase of methane in the atmosphere.

The burning of fossil fuels pours vast amounts of smoke and carbon dioxide into the atmosphere. Fossil fuels are composed of carbon compounds, which produce carbon dioxide and particulates when burned. This has led to a steady increase in the amount of carbon dioxide and other pollutants in the air. Over the past 100 years, the increase may be as much as 25 percent by volume. (See Figure 29-6.) Since 1900, the average temperature throughout the world has risen by more than 0.5°C. It is expected to rise another 1° by the year 2025.

Some scientists have projected that late in the twenty-first century world temperatures will have risen by as much as 5°C. Almost half of this predicted warming would be due to increased concentrations of CO_2 from the burning of fossil fuels. The increase in temperature could cause the polar ice caps to melt. The resulting rise in

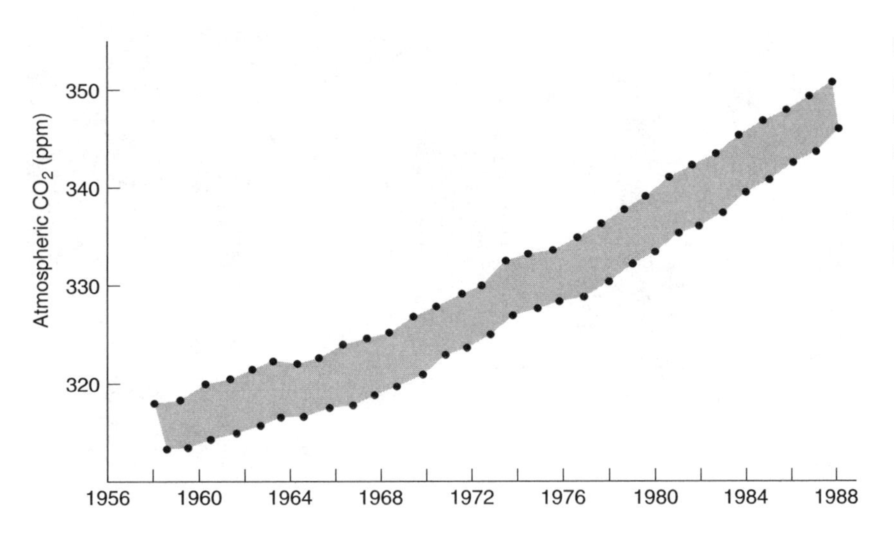

Figure 29-6 The amount of carbon dioxide in the atmosphere has been increasing. What are some reasons for this increase?

sea levels would cause flooding of many coastal areas, damage wetlands, and destroy the agricultural capability of many regions. It also could cause worldwide changes in climate.

Global warming is the increase in Earth's average temperature caused by the greenhouse effect. The problem of global warming has stimulated a great deal of international activity, as government leaders try to find solutions and establish environmental controls. The United States, Europe, Russia, and Japan account for about two-thirds of all pollution that causes global warming.

Global Cooling

Volcanic eruptions or collisions with comets or meteors could blanket the planet with a cloud of dust. This would block out much of the sunlight and lower Earth's average temperature. Loss of sunlight would interfere with photosynthesis, and many green plants would die, thus reducing the concentration of oxygen in the air and raising the levels of carbon dioxide. As shown in Figure 29-7, there is evidence that objects from space have hit Earth in the past. In fact, some scientists theorize that this led to the extinction of the dinosaurs, as well as other mass extinctions that appear in the fossil record.

In 1991 during the Persian Gulf War, Iraq set fire to many Kuwaiti oil wells. Great quantities of soot and fumes rose into the atmo-

Figure 29-7 Meteor Crater in Arizona was caused by a meteorite that struck Earth about 50,000 years ago. The crater is about 174 meters deep and 1265 meters across.

sphere. Wind patterns spread them around the globe. These materials persisted in the atmosphere for several years, affecting the health of Kuwait's population and the native biota. Due to the pollution, local weather patterns were altered during this time. The 1991 eruption of Mount Pinatubo in the Philippines also released huge quantities of particles and aerosols into the atmosphere. These pollutants persisted in the atmosphere for several years. During this time, global temperatures dropped on average about 1°C.

Our industrialized society pours huge amounts of particulates into the atmosphere. This continued pollution can absorb radiation from the sun and lead to cooling. This would shift world climate zones toward the equator. Agriculture in temperate zone nations would be devastated. A drop in the average daily temperature of only 2 or 3°C would cut wheat production by half in Canada and Russia. This would be felt throughout the world as food supplies diminished. Reduced rainfall and temperatures in tropical and subtropical regions would wipe out thousands of species. Northern cities could become icebound year-round.

29.2 Section Review

1. Explain how particulate emissions can alter world climate.
2. Describe the potential effects of global warming on your region's agriculture.

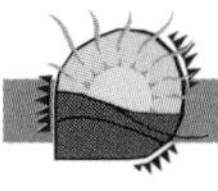

29.3 POLLUTION AND HEALTH

The government of the United States has spent a great deal of money and invested much effort to control air pollution. But we are often exposed to greater health risks from airborne pollutants inside our homes and offices than from air pollution outdoors. Since houses and workplaces are often closed tightly during air-conditioning and heating seasons, toxic air pollutants can build up to dangerous levels. Because we spend more time inside than outside, higher levels of these contaminants are inhaled.

Tobacco

Smoking and the effects of environmental tobacco smoke (ETS) are serious health threats. **Environmental tobacco smoke**, or **second-hand smoke**, is smoke that is released from burning cigarettes, is exhaled by smokers, and inhaled by persons nearby. Effects from tobacco smoke include lung cancer, emphysema, circulatory diseases, and stroke. The Surgeon General's Office estimates that more than 480,000 people in the United States and 2.5 million worldwide die each year from smoking-related illnesses. Smoking can even affect the unborn. The average birth weight of children born to women who smoke is lower than that of children born to nonsmokers.

The active drug in cigarettes, **nicotine**, is addictive. Nicotine is a poisonous alkaloid produced by the tobacco plant. Its mood-enhancing properties cause users to develop a habit that is extremely hard to break. In addition to nicotine, tobacco and tobacco smoke contain carcinogenic compounds, including aromatic hydrocarbons and particulates. Tobacco products include cigarettes, cigars, chewing tobacco, and snuff. All these products contain nicotine and carcinogens in varying concentrations.

Due to an intensive antismoking campaign, cigarette production and consumption in the United States has fallen dramatically. Cigarette production in most industrial countries is sluggish or falling, as more people reject smoking. It is estimated that more than 70 percent of the world's 1.2 billion smokers live in developing countries.

Indoor Pollution

In addition to tobacco smoke, hydrocarbons, such as formaldehyde from building materials, benzene from consumer products, and vinyl chloride from plastics, impact public health. Formaldehyde and vinyl chloride are carcinogenic. Asbestos, once used extensively in floor and ceiling tiles, insulation, soundproofing, and plaster, also has been identified as a carcinogen. (See Figure 29-8.) People who live in older homes are often exposed to high levels of these toxic substances.

Lead pollution is still a problem in many older homes. Lead-based paints often flake from walls and ceilings. Young children sometimes

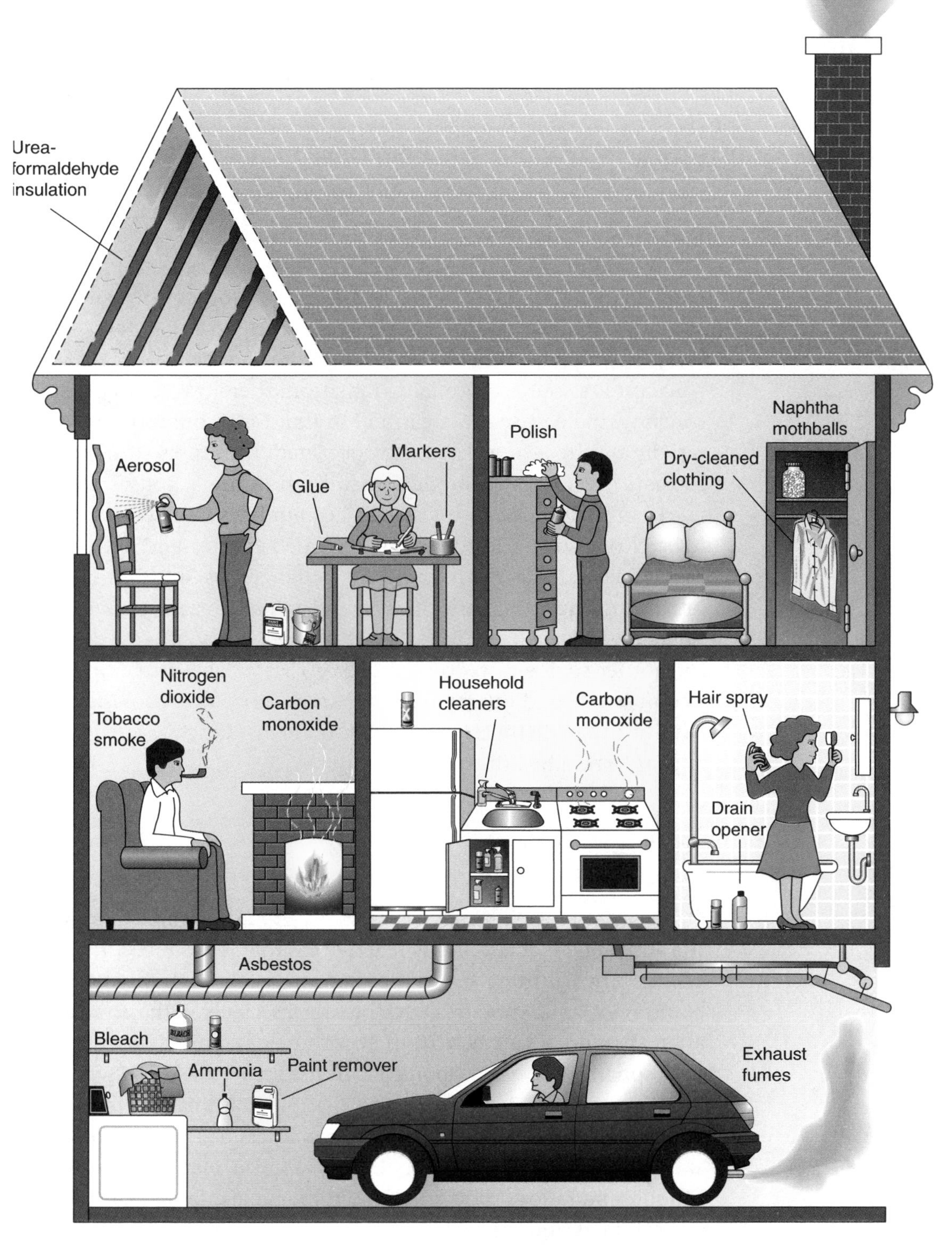

Figure 29-8 There are many sources of chemical pollutants in a home.

eat these paint flakes, which can cause lead poisoning, a condition that can impair neurological development.

Radon

Radon is an inert gas that is colorless, odorless, and radioactive. Ores of radium and uranium, which are widely distributed through Earth's crust, decay to produce radon gas. Radon-emitting minerals are found in rock that underlies much of the northeastern United States, the upper Midwest, and the northern Rockies.

Radon is a source of ionizing radiation; therefore, it is a potential carcinogen. Because it is a gas, radon that forms outdoors is diluted rapidly by the atmosphere. However, radon diffuses through rock formations, filters through the soil, and enters houses through cracks in their foundations. Radon accumulates in basements or the lower floors of buildings. Tightly sealed homes and offices trap radon gas and allow it to build up to dangerous levels. Well-ventilated buildings dissipate the gas and maintain safe radon levels. There are kits available to test your home for radon.

Another source of radon exposure is water obtained from wells. Radon is released during bathing, showering, and cooking and is readily inhaled. Ingestion of radon in drinking water does not appear to be a health threat.

The Ozone Hole

Ozone, triatomic oxygen (O_3), is usually present in small amounts in the atmosphere. It is a highly reactive gas that is produced in the troposphere by lightning and in the stratosphere through the absorption of ultraviolet radiation by O_2. Ozone forms a layer in the atmosphere at an altitude of between 20 to 30 kilometers. This layer affects the atmospheric processes responsible for regulating the world's climates.

The ozone layer protects Earth's plant and animal life from ultraviolet radiation. Excess ultraviolet radiation is a **mutagen**, a substance that causes mutations that can harm plants and animals. Through mutation, ultraviolet radiation can cause skin cancers and eye cataracts and can disrupt the immune systems of humans and other animals. Ultraviolet radiation reduces the growth rate of plants. The atmosphere's ozone layer shields the entire biosphere.

Over the last 55 years, scientists have observed a decline in the concentration of atmospheric ozone and the appearance of holes in the ozone layer. The **ozone holes** are large areas over the polar regions that show a seasonal drop in ozone concentration.

Chlorofluorocarbons (CFCs) and halons are compounds used in aerosol sprays, air-conditioning and refrigeration units, and the manufacture of plastic and plastic foam. CFCs and halons migrate to the stratosphere, where ultraviolet radiation causes them to break down and release chlorine and bromine atoms. These atoms react with and destroy ozone. Since chlorine destroys atmospheric ozone without itself being consumed, a little pollution can cause a great deal of destruction. Methylchloroform and carbon tetrachloride are other gases that can contribute to depletion of the ozone layer.

The formation of the ozone holes is linked to ice clouds in the stratosphere. During the winter season, ice clouds help release chlorine atoms from CFCs and bromine atoms from halons. On the surface of ice cloud particles, the reaction between chlorine and ozone is speeded up. In the Antarctic, cloud particles form mainly in the ozone layer at the low temperatures encountered during the winter. For this reason, destruction of the ozone layer by pollution is most apparent in what has become known as the Antarctic ozone hole. This hole, discovered during the 1980s, contains 50 percent less ozone than normal and has reached the size of the continental United States. Recent data indicate that a similar ozone hole forms each year above the Arctic. The Arctic ozone hole is more dangerous because it indicates a drop in the ozone level over many of the world's northernmost cities.

Acid Rain

The pH scale, which ranges from 0 to 14, measures acidity and basicity. A neutral substance has a pH of 7. Acids have a pH less than 7 while bases have a pH between 7 and 14. Rainfall is slightly acidic naturally. Carbon dioxide in the atmosphere dissolves in the water to form weak carbonic acid. Precipitation with a pH less than 5 is called acid rain. The burning of fossil fuels produces oxides of sulfur and nitrogen, which react with water to form sulfuric and nitric acids. These strong acids greatly increase the acidity of precipitation. Electric generators, industrial plants, and internal combustion engines that burn fossil fuels are the primary sources of the pollutants that form acid rain.

Once acid-causing pollutants enter the atmosphere, they may be carried long distances before they fall as acid rain. Global winds and the mixing of the atmosphere allow air pollution produced by one country to affect all countries. This extends the environmental problems caused by acid rain. Therefore, the causes of acid rain are an international problem. Due to the easterly circulation of air over the United States, rain and snow that fall in eastern states have a lower pH and are therefore more acidic than precipitation that falls in other parts of the country.

Much environmental harm is caused by acid rain. It contributes to the corrosion of iron, steel, zinc, and painted surfaces. Automobile exteriors are damaged by acid rain. It dissolves the stone used in some buildings and statues. Although exposure to acid rain is not a direct threat, the particles in the air that cause acid rain do pose a risk to human health. The gases that cause acid rain interact in the atmosphere forming sulfate and nitrate particles that when inhaled can lead to disorders such as asthma, bronchitis, and emphysema.

Oxides of sulfur and nitrogen released by industrial plants in the Midwest are responsible for the acid rain that falls on the forests in New England and New York State. (See Figure 29-9 on page 532.) This acid rain has acidified many small lakes in these regions, destroying their aquatic ecosystems. Organisms in many lakes and streams are sensitive to changes in pH. Acidification can decrease the ability of these aquatic organisms to reproduce. This is especially true for many species of fish and amphibians that reproduce in early spring. Runoff from melting snow and ice will significantly lower the pH of lakes and streams causing "acid shock" that destroys the eggs of these organisms. Many species of amphibians cannot reproduce in waters that have a pH below 5.

The coniferous forests have been particularly hard hit, since their acidic soil cannot neutralize the effects of acid rain. Runoff flowing into lakes and streams destroys their delicate acid balance. The young of many aquatic organisms cannot survive in acidic waters. The Adirondack Mountains of New York State contain many lakes that have been "killed" by acid rain.

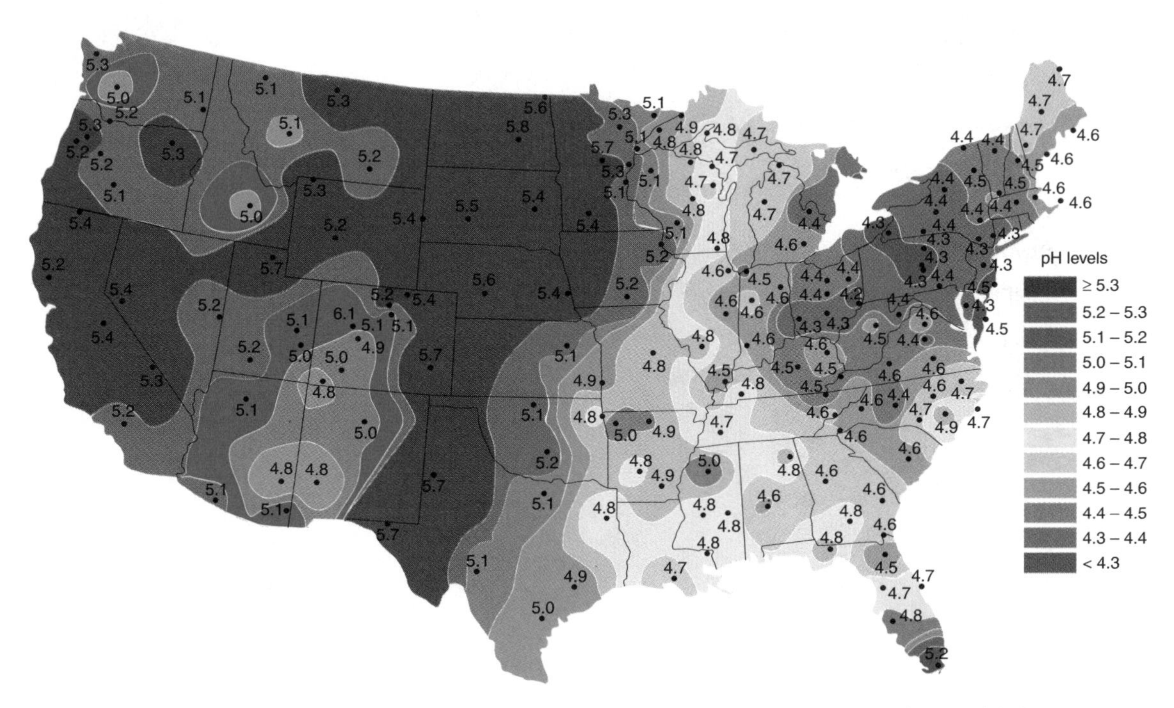

Figure 29-9 This map shows the 2001 acid rain concentrations in the United States.

Some California forests, located far from cities, are dying due to acid rain. Here the major source of pollution is the exhaust gases of the many automobiles that clog the freeways around the large coastal cities. The costs of damage to the environment due to acid rain must be balanced against the potential cost in dollars of controlling pollutants. Debate rages over who will pay the costs. Will it be industry or the public?

You can easily demonstrate the effects of acid rain on building materials, such as marble and limestone. Chalk is made of calcium carbonate, the same material found in marble and limestone. Place one piece of chalk in a jar filled with vinegar. Place another piece, exactly the same size as the first piece, in a jar filled with water. Let the jars stand overnight. Compare the two pieces of chalk. Describe differences you see. What does the vinegar represent? What is the purpose of the jar filled with water?

29.3 Section Review

1. What are some indoor pollutants that affect public health?
2. Explain how CFCs and halons affect Earth's ozone layer.

LABORATORY INVESTIGATION 29

Atmospheric Pollutants

PROBLEM: *How do atmospheric pollutants affect your health?*

SKILLS: *Observing, interviewing, collecting data*

MATERIALS: *Air Pollution Index or Air Quality Report from your local newspaper, health questionnaire*

PROCEDURE

1. During each day of the study, save the Pollution Index or Air Quality Report from your local newspaper.
2. At the end of each day during the week of your study, distribute a health questionnaire to the members of your homeroom class.

Health Questionnaire

Date ______________

1. How did you feel today? (Check one.)
   Excellent ❑ Good ❑ Fair ❑ Poor ❑ Sick ❑
2. Were your eyes red, watery, or dry? If yes, explain.
3. Were your breathing passages constricted or clogged?
4. Did you experience any itching of the skin?
5. Did you feel drowsy?
6. Do you suffer from asthma?
7. Do you suffer from hay fever?
8. Do you suffer from any other allergies? If yes, to what are you allergic?
9. How many hours did you spend in an air-conditioned environment?

3. At the end of your study, make a table that shows the air quality for each day.

	Mon.	Tues.	Wed.	Thurs.	Fri.	Sat.	Sun.
Pollution Level							

4. For each day, tabulate the results of the questionnaires. Compare the results with the Pollution Index or air quality for that day.

OBSERVATIONS AND ANALYSES

1. What correlation did you find between students' health and air quality?
2. What specific pollutants were responsible for the greatest number of health problems?
3. Do your results indicate that being in an air-conditioned environment eases pollution-related problems?
4. What recommendations would you make to students whose health was affected by air pollutants?

Chapter 29 Review

Answer these questions on a separate sheet of paper.

Vocabulary

The following list contains all the boldfaced terms in this chapter.

environmental tobacco smoke, global warming, mutagen, nicotine, ozone holes, particulates, photochemical smog, primary pollutants, radon, secondary pollutants, secondhand smoke, smog

Fill In

Use one of the vocabulary terms listed above to complete each sentence.

1. Exposure to ultraviolet radiation can cause genetic damage. It is therefore considered a ________.
2. ________ is the increase in Earth's average temperature caused by the greenhouse effect.
3. The activities of humans have created ________ over the North and South Poles.
4. ________ is a colorless, odorless radioactive gas that can accumulate in basements.
5. ________ is an atmospheric condition in which visibility is reduced due to air pollution that contains high levels of particulate or photochemical oxidants.

Multiple Choice

Choose the response that best completes the sentence or answers the question.

6. Which of the following is an example of a primary pollutant? *a.* photochemical smog *b.* hydrocarbons *c.* acid rain *d.* sulfur dioxide
7. Which of the following is an example of a secondary pollutant? *a.* acid rain *b.* an oil spill *c.* carbon monoxide *d.* cigarette smoke
8. A poisonous product of the incomplete combustion of fossil fuels is *a.* carbon dioxide. *b.* carbon monoxide. *c.* nitrogen dioxide. *d.* methane.
9. A dangerous metal released into the atmosphere during smelting of some metal ores is *a.* lead. *b.* nitric oxide. *c.* asbestos. *d.* radon.
10. A pollutant given off during the burning of low-grade coal is *a.* PbO. *b.* SO_2. *c.* benzene. *d.* phenol.

11. The processes of respiration and photosynthesis balance the atmospheric concentration of *a.* sulfur dioxide. *b.* ozone. *c.* nitric oxide. *d.* carbon dioxide.
12. The reddish-brown gas often responsible for photochemical smog is *a.* ozone. *b.* carbon monoxide. *c.* nitrogen dioxide. *d.* methane.
13. Which organism is composed of a microscopic green alga and a thick-walled fungus? *a.* lichen *b.* ruminant *c.* termite *d.* conifer
14. Which gas is produced by microorganisms in the digestive system of sheep and cattle? *a.* carbon dioxide *b.* carbon monoxide *c.* ozone *d.* methane
15. This inert gas is colorless, odorless, and radioactive. It is produced by ores of radium and uranium. *a.* halon *b.* CFCs *c.* ozone *d.* radon

Short Answer (Constructed Response)

Use the information you learned in this chapter to respond to the following items.

16. Why is carbon monoxide especially dangerous to people?
17. Give three examples of particulate pollutants.
18. Describe two effects of secondhand smoke on human health.

Essay (Extended Response)

Use the information in the chapter to respond to these items.

19. How does photochemical smog develop?
20. How do temperature inversions create health hazards?
21. Why are coniferous forests most severely affected by acid rain?
22. Explain how lichens can be used to determine air quality.
23. Explain how ozone protects the biosphere.
24. Explain the effects of ultraviolet radiation on the biosphere.
25. List and explain the pros and cons of subsidizing public transportation to relieve air pollution.

Research Projects

- Look around your community and report on any sites that may produce harmful air pollution.
- Research and report on the methods used by your town or city to reduce particulate emissions.

CHAPTER 30
Protecting Our Atmosphere

When you have completed this chapter, you should be able to:

Describe the different ways air pollution is measured.

Compare the effectiveness of different conservation practices.

Discuss the effectiveness of air quality legislation.

Our health is threatened by the vast amounts of pollutants that we pour into the atmosphere. In addition, these air pollutants damage the environment and may alter Earth's climate. The atmosphere has been able to renew itself through its interactions with the hydrosphere and biosphere. Now, this ability is threatened. Burning of fossil fuels, industrial production, and the increasing use of motor vehicles pour more pollutants into the atmosphere than it may be able to handle. On the positive side, over the last decade, air pollution has been reduced in the United States. In this chapter, you will learn about the measures that are being taken to protect our atmosphere.

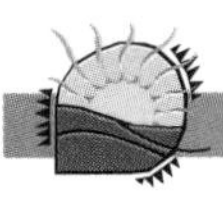

30.1 AIR POLLUTION STANDARDS

Under the provisions of the Clean Air Act, the Environmental Protection Agency sets standards for **maximum concentration levels** (MCLs) of dangerous pollutants in the air. State agencies are then responsible for monitoring regional air quality and maintaining safe levels of pollutants. The EPA develops strategies to improve air quality and imposes penalties when the standards are not met. **National Ambient Air Quality Standards** are the specific pollution levels set by the EPA. **Emission standards** limit the level of pollutants that may be discharged by specific sources, such as automobile exhaust-power plants, and factories.

Air Pollution Index

In 1976, the Council on Environmental Quality standardized the nation's air pollution indexes. Maximum concentration levels were set for the five most hazardous pollutants: oxidants (ozone), carbon monoxide, sulfur dioxide, nitrogen dioxide, and particulates.

When the quality of the air is unhealthful, air pollution alerts may be issued to warn people to stay indoors and avoid strenuous activities. Restricting physical activity lessens exposure of respiratory tissues and organs to potentially damaging pollutants. During these alerts, air-conditioned homes and offices offer the least polluted environments.

In recent years, the **Air Quality Index** was created. This index measures the concentration of a number of dangerous pollutants, including ozone, sulfur dioxide, carbon monoxide, nitrogen dioxide, and particulates. Throughout the United States, the Air Quality Index is part of the daily weather report. Index readings between 0 and 30 are good, those between 30 and 300 range from moderate to unhealthful, while those over 300 are hazardous to human health.

30.1 Section Review

1. Why do air pollution alerts advise people to stay indoors?
2. What does an Air Quality Index of 30 indicate?

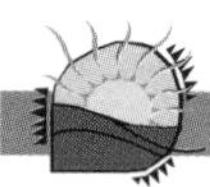

30.2 CONSERVATION PRACTICES

Improving energy efficiency in transportation, industry, business, and the home is the fastest, most cost-effective method of reducing atmospheric pollutants. Greater use of mass transit would reduce the number of cars on the roads and improve air quality. Reducing energy consumption by using more efficient lighting, heating, and appliances can also lead to improved air quality.

The automobile is responsible for up to 75 percent of the air pollution that affects most American cities. It is estimated that half of this pollution is caused by only 10 percent of the cars on the roads, older cars and cars that have had their emission controls tampered with. The exhaust fumes from all internal combustion engines are dangerous to our health. However, diesel engines emit exhaust fumes that are more carcinogenic than other automotive pollutants.

California has led the way in legislation to reduce atmospheric pollution from motor vehicles. State regulations require the phasing in of cleaner-burning fuels and low-pollution-emitting vehicles. The Clean Air Act provides incentives for other states to adopt California's high standards. Table 30-1 shows how emission levels have changed since 1970.

California passed legislation requiring by 2003 that at least 10 percent of all new vehicles sold be electric vehicles (EVs). A switch to electric cars will improve air quality, because electricity produced by utilities creates less pollution than do cars powered by internal combustion engines. Figure 30-1 illustrates the batteries used in an electric vehicle.

For the time being, electric cars are not as practical as internal

TABLE 30-1 EMISSIONS OF PRINCIPAL AIR POLLUTANTS 1970–2000 (MILLIONS OF TONS)

Emission	1970	1980	1990	2000
Carbon monoxide	129	117	99.1	109
Nitrogen oxides	20.9	24.3	24.1	24.4
Hydrocarbons	30.9	26.3	21.1	20.3
Particulates	12.3	6.25	3.34	2.93
Sulfur oxides	31.1	25.9	23.6	18.1

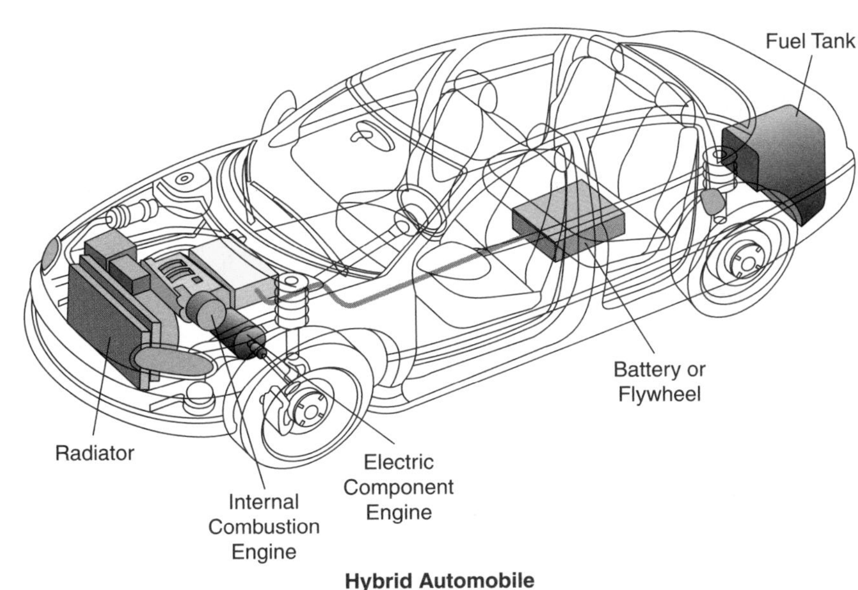

Figure 30-1 **Large storage batteries in this vehicle are recharged between uses.**

combustion cars. Battery technology limits the distance these cars can travel between charges. Each type of battery that has been proposed has drawbacks. Zinc-chloride batteries can store a large charge but have short lives. Sodium-sulfur batteries can also store large charges but must operate at temperatures over 300°C, 200 degrees hotter than boiling water. Alkaline batteries are expensive to manufacture. Lithium batteries store a large charge and have a long life but are the most expensive to produce. Mercury- and cadmium-based batteries also offer excellent potential but contain toxic environmental pollutants. These problems will have to be solved before the electric car replaces the internal combustion car. Though the vehicles are considered to be pollution-free, the generation of the electricity needed to supply their energy is not pollution-free. Therefore, EVs indirectly impact on the environment.

Hybrid Vehicles

Hybrid vehicles increase fuel economy by combining the best features of internal combustion engines and electric motors. In hybrid cars, the gasoline engine and the electric motor are connected to the wheels through the same transmission. The engine can be smaller and lighter because the motor also powers the car. A computer decides whether to use the motor or engine and when to store electricity in batteries for future use. The electric motor is used for low-speed cruising or as an aid in acceleration. When braking or decelerating, the electric motor acts as a generator to produce and store electricity. Unlike vehicles that are completely electric, hybrid vehicles do not have to be plugged in to recharge the batteries.

Federal tax laws allow a tax deduction for vehicles propelled by a clean-burning fuel. Since hybrid vehicles obtain greater fuel efficiency and produce fewer emissions, buyers are eligible for a deduction of up to $2000 in the year the vehicle was purchased.

Other Alternative Fuel Vehicles

Flex-fuel vehicles have a single fuel tank, fuel system, and engine. These vehicles are designed to run on a mixture of regular unleaded gasoline and alcohol (either ethanol or methanol). The fuel mixture reduces the amount of pollutants in the emissions, since burning alcohol produces carbon dioxide and water.

Bi-fuel vehicles have two separate fuel systems and are capable of switching from one to the other easily. To assure a ready source of fuel, one system is either gasoline or diesel. The second system is usually designed to run on compressed natural gas (CNG) or liquefied propane gas (LPG). CNG and LPG fuels are nonpolluting. In addition, gasoline engines can be modified to use compressed natural gas or liquefied propane.

The Catalytic Converter

Emissions from motor vehicle internal combustion engines have been greatly reduced by the adoption of new technologies, such as the catalytic converter and fuel injection. As shown in Figure 30-2, the **catalytic converter** is an air pollution control device that is part of an automobile's exhaust system. It reduces the levels of hydrocarbons and carbon monoxide in the exhaust gases, converting them to carbon dioxide and water vapor. Some catalytic converters also remove nitrogen oxides from exhaust gases.

Fuel injection systems modify carburation in auto engines. **Car-**

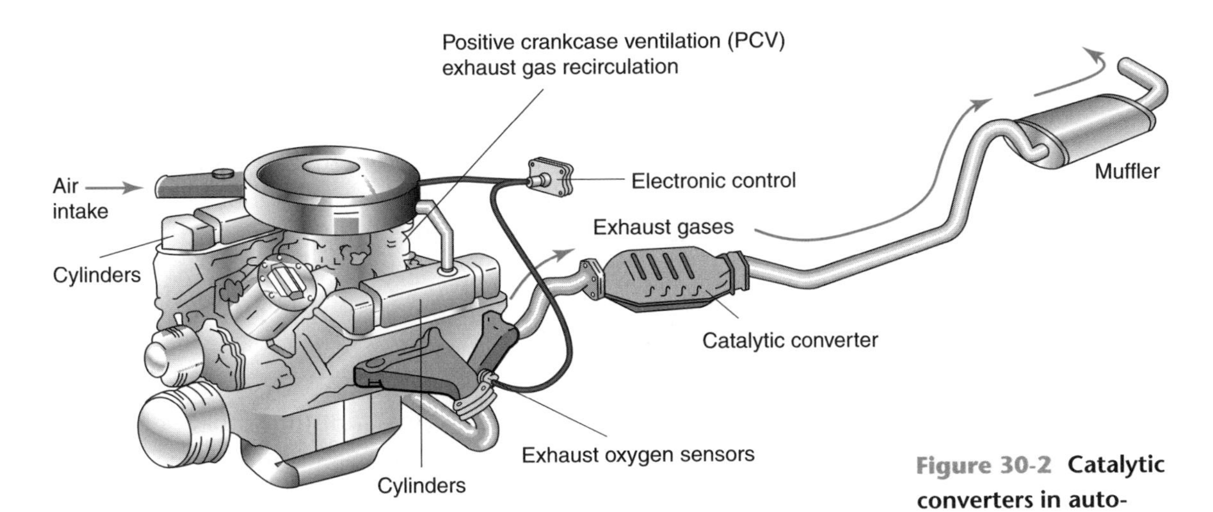

Figure 30-2 Catalytic converters in automobiles change waste gases to water and carbon dioxide.

buration is the process by which fuel and air (oxygen) are mixed before the fuel is burned. Fuel injection prevents the evaporation of gasoline and improves engine efficiency.

Oil that escapes from around the pistons and unburned fuel are channeled back to the engine to be burned. This system, called **positive crankcase ventilation**, prevents evaporation of hydrocarbons and volatile organic compounds, thus reducing pollution.

Reformulated Fuels

Clean Air Act amendments required that in nine cities with severe ozone problems, cleaner-burning, reformulated gasoline be sold starting in 1995. These nine cities are Baltimore, Chicago, Hartford, Houston, Los Angeles, Milwaukee, New York, Philadelphia, and San Diego. Altering the refining process, which changes the chemical composition of the fuel, produces **reformulated gasoline**. The reformulated gasoline reduces emission of ozone-producing organic compounds.

Beginning with the winter of 1992–1993, oil companies were required to boost the oxygen content of their winter gasoline blends sold in 39 major United States cities. This step was taken to reduce carbon monoxide emissions. Adding ethanol (ethyl alcohol) to the fuel is the main method used to boost oxygen content.

Gasohol is an oxygen-rich fuel composed of 90 percent gasoline and 10 percent ethanol. The ethanol is often obtained by fermentation of corn and other plant products. Using gasohol reduces our dependence on fossil fuels. Controversy surrounds gasohol use, since it increases the cost of gasoline and reduces fuel economy.

Conversion to other cleaner-burning fuels can also cut carbon emissions from automobiles. In some areas, buses use compressed natural gas for fuel. This reduces pollution in crowded cities.

Fuel Efficiency Standards

Fuel efficiency is a measure of the consumption of fuel per unit of distance (kilometer or mile) traveled. In 1965, the average new car had a fuel efficiency of 5.9 kilometers per liter (14 miles per gallon) of gasoline. Fuel efficiency standards became law in 1975. By 1985, average fuel efficiency was supposed to increase to 11.7 km/L (27.5 mpg). In 2004, it was 11.7 km/L (27.5 mpg). The average fuel efficiency is based on a manufacturer's complete line. Therefore, some models could have fuel efficiency greater than 11.7 kilometers per liter and others less, as long as the line averaged 11.7. Some methods used to improve fuel economy were reductions in vehicle size and weight and improved engine design. The fuel efficiency for sport-utility vehicles and other light trucks in 2004 was only 8.8 km/L (20.7 mpg). In general, sport-utility vehicles are not as fuel efficient as cars.

Another idea that might encourage government, industry, and consumers to invest in energy efficiency is taxing fuels based on their carbon content. The increased price of energy from fossil fuels would make them more competitive with more costly renewable energy sources. In addition, imposing more taxes on gasoline should encourage people to purchase energy-efficient vehicles and reduce vehicle use overall.

In March 1993, the Chicago Board of Trade added a new tradable commodity, "permits to pollute the atmosphere." Buyers paid millions of dollars for the right to emit a total of 150,000 tons of sulfur dioxide into the air each year. This represents less than 1 percent of the average annual emission of this pollutant. The permits are a new step in pollution control, giving industry greater flexibility in meeting pollution standards. Polluters can buy a permit from another polluter, which may have reduced its own emissions more than required. Polluters that invest in cleaner technologies can now pay for their investment by selling their emission permits.

Pollution Control Devices

There are two general methods of reducing air pollution: removing pollutants from emissions and converting harmful pollutants to harmless substances. Filtration is often used to control particulate emissions in smoke or stack gases discharged from factories, refineries, and power plants. As shown in Figure 30-3, **air filters** are made of a mesh of interwoven fibers composed of cotton, spun glass, or cellulose. While air freely passes through the filters, the mesh traps solids. Filters must be changed regularly to function efficiently. In many plants, smoke is passed through a series of cloth bags that can remove up to 99 percent of the particulates in smoke.

Electrostatic precipitators also are used to control particulate emissions from power plants. As shown in Figure 30-4, precipitators are composed of charged plates (electrodes) set up inside chimneys. When smoke particles pass between the plates, they pick up an electrostatic charge from one electrode. The particles migrate to and are deposited on the oppositely charged electrode, the collecting plate. When the precipitator is turned off, the particulates fall to the

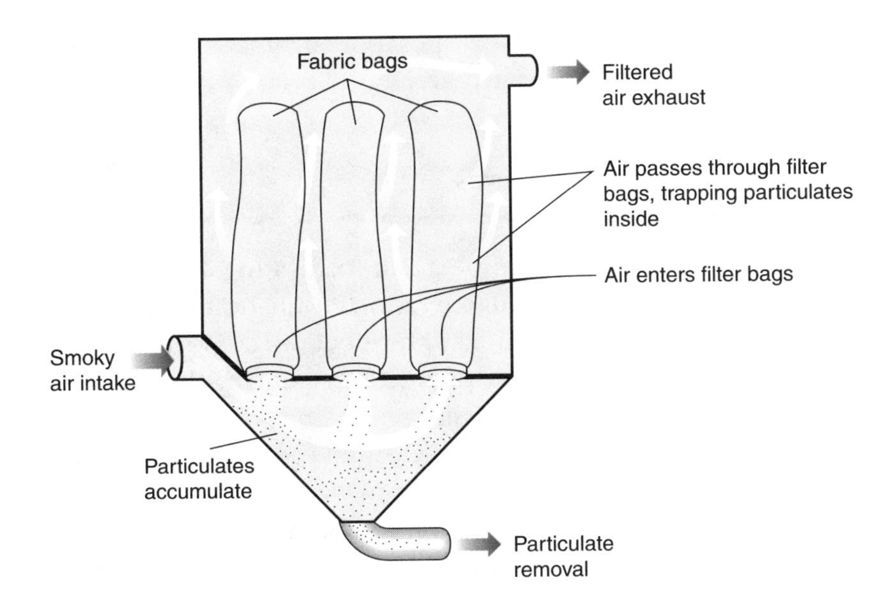

Figure 30-3 Smokestack filters reduce the pollutants that reach the air.

Figure 30-4
An electrostatic precipitator reduces particulate pollution.

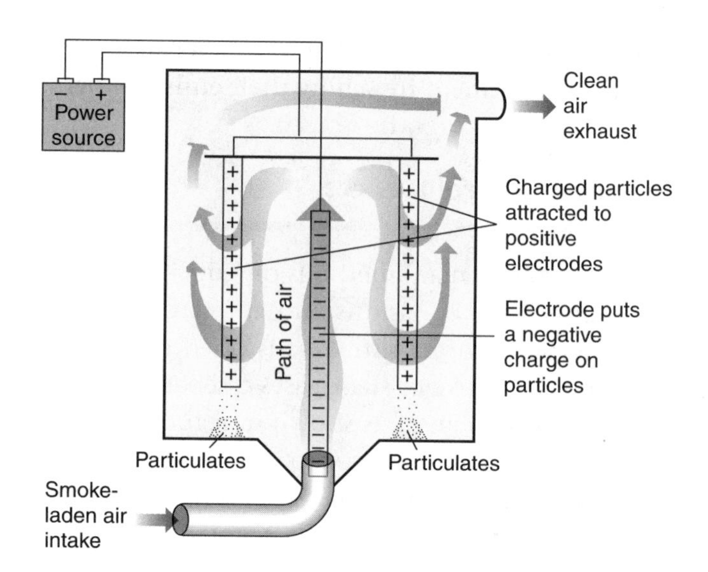

bottom of the stack and are removed. Precipitators use a great deal of energy but are very efficient at removing particulates.

Scrubbers remove both particulates and gases, such as sulfur dioxide, from exhaust emissions. The stack gases are passed through a fine mist of water and lime, which traps particulates and sulfur dioxide gas. The gas is changed into insoluble sulfur compounds, which can be trapped. The cylone device is effective only for large particulates.

Removing pollutants from stack gases helps clean the air. However, it does have drawbacks. The particulates and gases collected contain hazardous materials. Disposal of this material can create environmental problems elsewhere.

Other Conservation Measures

The simultaneous production of steam for heating and electricity for light and power is called **cogeneration**. Waste heat from the production of electricity can be used to heat buildings. Waste heat also can be recycled to boost energy production, reducing the use of fossil fuels and cutting carbon emissions.

While growing, the trees absorb CO_2 from the atmosphere. When these trees are burned for firewood or to clear land for agriculture, they return CO_2 to the atmosphere. Excess levels of atmospheric CO_2 can lead to global warming. By curbing deforestation and encouraging replanting of trees, people can slow the global buildup of atmospheric CO_2. Reversing deforestation is especially important in the tropics, where the rate of rain forest destruction is great. Not only will the environment benefit from these actions, but the inhabitants of these regions can profit economically by harvesting and selling rain forest products. (See Figure 30-5.)

Figure 30-5 **Bananas are an important rain forest crop.**

30.2 Section Review

1. What effect does reformulated gasoline have on the environment?
2. Why will the use of electric cars improve air quality?
3. What actions can you take to improve air quality in your community, school, and home?

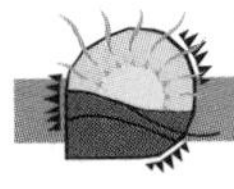

30.3 AIR QUALITY LEGISLATION

To insure that air pollution would be reduced, individual states and the United States government have enacted air quality legislation. On a global scale, international agreements were signed to protect the atmosphere. Following are some examples of air quality legislation.

United States—The Clean Air Act

The Clean Air Act of 1963 was the first national legislation in the United States aimed at controlling air pollution. Funds were provided for states to set and enforce air quality regulations. In 1965, amendments were added that set standards for allowable levels of carbon monoxide and hydrocarbons in automobile exhaust emissions.

The Clean Air Act of 1970 designated seven pollutants as the major cause of degradation of the air and the most serious threats to human health: sulfur dioxide, carbon monoxide, nitrogen oxides, hydrocarbons, photochemical oxidants, lead compounds, and particulates. The act resulted in new emission standards for automobiles and new industries and air quality standards aimed at protecting human health and the environment.

The Environmental Protection Agency was established in 1970 under provisions of the Clean Air Act. The EPA was given responsibility for implementing the provisions of the Clean Air Act as well as regulatory and enforcement powers. Though responsibilities are delegated to the states, the EPA is ultimately responsible for enforcing all regulations and prosecuting polluters.

The Clean Air Act was amended in 1977 and 1990 to limit industrial growth in areas where pollution was exceeding air quality standards. Provisions in the 1977 amendments:

- Allowed revised plans by states that had not met standards.
- Encouraged state automobile inspection stations.
- Required states to balance new pollution-worsening developments with offsetting reductions elsewhere.
- Extended the antitampering prohibition for automobile emission control devices.
- Required the EPA to set new standards for NO_2 emissions.

Fuel efficiency standards are important for combating air pollution and the environmental problems it causes. The 1990 Clean Air Act introduced new ways of reducing motor-vehicle-related air pollution. For the first time, fuel was considered a potential source of pollution. The act mandated that improved gas formulations be sold in selected, highly polluted cities to reduce carbon monoxide and hydrocarbon emissions. Reduced vehicle emission standards were introduced to stimulate the production of cleaner vehicles and fuels.

In April 2003, the National Research Council of the National Academy of Sciences reported that carbon monoxide levels have dropped dramatically. In 1971, 90 percent of the carbon monoxide monitors around the country registered violations. Now, only a few places exceed the standards on one day or two days a year.

California

In southern California, counties in and around Los Angeles have developed a comprehensive program to control air quality. The South Coast Air Quality Management District has instituted a

20-year plan aimed at meeting regional air quality standards. Provisions of the plan include:

- Having employers offer employees incentives to join car pools, take public transportation, or ride bikes to get to work.
- Constructing car-pool lanes on freeways.
- Banning trucks from freeways during rush hours.
- Restricting areas where future businesses and housing developments may be located.
- Enforcing the use of vapor-recovery nozzles on gasoline pumps. (See Figure 30-6.)
- Curbing vehicles that emit high levels of pollutants.
- Developing a "clean-fuels" program for buses, rental cars, and fleet vehicles.

Figure 30-6 This vapor-recovery nozzle prevents gasoline vapors from entering the atmosphere.

International

The first international meeting to promote human development without damaging the environment was held in Stockholm, Sweden, in 1972. Following this meeting, the wealthy nations started to clean up their own air and water sources but continued to use forest and mineral resources elsewhere.

The Montreal Protocol on Substances That Deplete the Ozone Layer is an international agreement that was signed in 1987. It established a schedule for reducing the production of halons and chlorofluorocarbons in participating countries. The use of these ozone-destroying chemicals in aerosol spray cans has decreased sharply, and most nations have ceased production of these gases. (See Figure 30-7 on page 552.) In addition, air conditioners in cars no longer use freon, a chlorofluorocarbon, as a coolant.

At the 1992 Earth Summit, held in Rio de Janeiro, Brazil, world leaders signed a nonbinding accord to protect the environment while strengthening the economies of the poorer nations. Yet, the world's population has continued to explode, poverty has deepened in nations throughout Africa and Asia, marine fisheries have crashed, and levels of greenhouse gases have increased.

The 1997 Kyoto Protocol called on industrialized nations to reduce carbon dioxide emissions, crucial to reversing global warming, to 1990 levels by 2012. The protocol went into effect on February

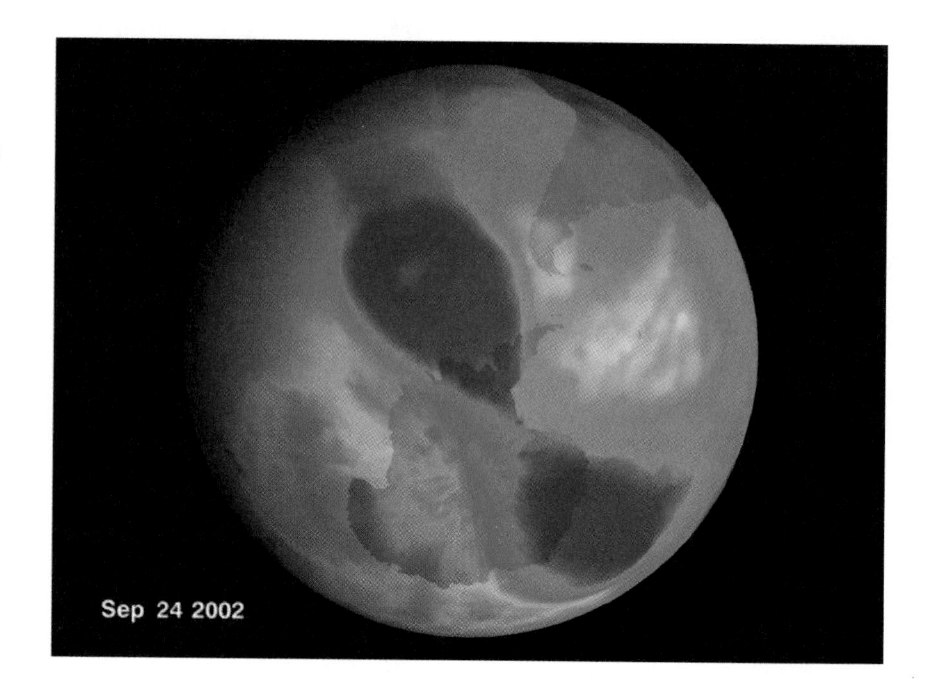

Figure 30-7 For several years, a hole has formed in the ozone layer over the South Pole.

16, 2005 after being ratified in November of 2004 by Russia. A controversy evolved between the developing nations and the wealthy, developed nations. The developing nations say that the wealthy nations are the largest polluters and should clean up first. The wealthy nations, especially the United States, feel all nations should be bound by the protocol.

A United Nations World Summit on Sustainable Development was held in Johannesburg, South Africa, in 2002. World leaders developed a plan to protect Earth's atmosphere, lakes, oceans, and wildlife while focusing on the links between poverty and environmental degradation. It was agreed that reducing poverty was a central element to the plan. Yet questions remain on how to do this and still protect Earth's natural resources.

30.3 Section Review

1. How does car pooling reduce atmospheric pollution?
2. What agency is responsible for implementing the provisions of the Clean Air Act?
3. Should cigarette smoking be banned in public places? Who does, or should have, more rights, the smoker or the nonsmoker?

LABORATORY INVESTIGATION 30

Effects of Acid Rain

PROBLEM: *What effect does acid have on different materials?*

SKILLS: *Observing, measuring, weighing, comparing*

MATERIALS: *5 test tubes with stoppers; water or rainwater; vinegar (dilute acetic acid) or acid rain; chips of marble, limestone, slate, and concrete; 2 identical pieces of cotton fabric; triple-beam balance*

PROCEDURE

1. Determine the mass of each chip and record your findings.
2. Label the test tubes A, B, C, D and E. Place a piece of marble in A, a piece of limestone in B, a piece of slate in C, a piece of concrete in D, and one piece of cotton fabric in E.
3. To each test tube, add 5 mL of acetic acid and enough water to completely cover the sample. If you are using acid rain, do not add water; use enough rainwater to cover the sample.
4. Carefully examine the samples and record your observations.
5. Let the samples stand for five days. Record your observations daily.
6. As directed by your teacher, empty the liquid from each test tube. Rinse the samples in freshwater. Let the samples dry overnight.
7. Examine each sample. Note any change in its appearance.
8. Determine the mass of the samples and record your findings.
9. Compare the masses obtained in step 7 with those from step 1.
10. Compare the cotton fabric soaked in acetic acid with the untreated fabric.

OBSERVATIONS AND ANALYSES

1. Prepare a table that shows the mass of each sample before exposure to acid and after exposure to acid.

Sample	Mass Before Soaking	Mass After Soaking	Change in Mass

2. Describe any changes in appearance that occurred in the samples after exposure to acetic acid.
3. Suggest possible causes for the changes you observed.
4. What effect did the acid have on the cotton fabric?
5. Which of the samples would be affected by acid rain?

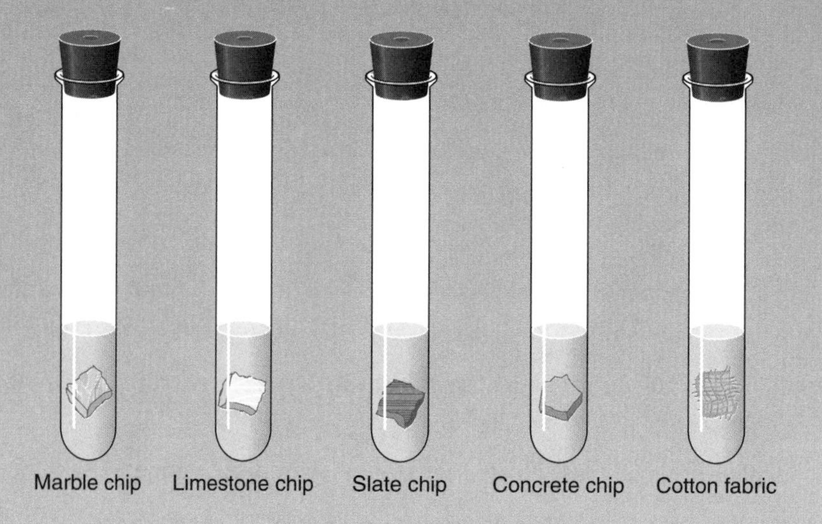

Chapter 30 Review

Answer these questions on a separate sheet of paper.

Vocabulary

The following list contains all the boldfaced terms in this chapter.

air filters, Air Quality Index, carburation, catalytic converter, cogeneration, electrostatic precipitators, emission standards, fuel efficiency, gasohol, maximum concentration levels, National Ambient Air Quality Standards, positive crankcase ventilation, reformulated gasoline

Fill in

Use one of the vocabulary terms listed above to complete each sentence.

1. The air pollution control device that is part of a car's exhaust system is the ______.
2. Devices composed of charged plates that are used to remove particulates from smoke are ______.
3. The simultaneous production of steam for both heating and to produce electricity is known as ______.
4. ______ is a measure of the consumption of fuel per unit of distance.
5. By returning oil and unburned fuel to the engine for combustion, ______ reduces air pollution caused by automobiles.

Multiple Choice

Choose the response that best completes the sentence or answers the question.

6. Which number is an example of a hazardous Air Quality Index value? *a.* 20 *b.* 80 *c.* 100 *d.* 400
7. A mesh of fibers is used in *a.* gasohol. *b.* precipitators. *c.* air filters. *d.* cogenerators.
8. What device in automobiles makes gasoline use more efficient? *a.* carburetor *b.* catalytic converter *c.* scrubber *d.* fuel injector
9. Gasohol is composed of *a.* 10 percent gasoline and 90 percent ethanol. *b.* 90 percent gasoline and 10 percent ethanol. *c.* 10 percent gasoline and 90 percent methanol. *d.* 90 percent gasoline and 10 percent methanol.
10. A device used to remove particulates and sulfur dioxide from smoke stack gases is called *a.* a scrubber. *b.* an electrostatic precipitator. *c.* a cylone. *d.* a filter.

11. Which of the following does *not* increase atmospheric levels of CO_2? *a.* deforestation *b.* burning fossil fuels *c.* hydroelectric dams *d.* internal combustion engines
12. Which of the following was *not* designated a major pollutant by the Clean Air Act of 1970? *a.* carbon dioxide *b.* particulates *c.* nitrogen oxides *d.* photochemical oxidants
13. The fumes from diesel engines are more dangerous than those from other types of engines because they *a.* contain more CO. *b.* are more carcinogenic. *c.* contain SO_2. *d.* contain more O_2.
14. One drawback to using electrostatic precipitators is they *a.* remove oxygen. *b.* use a great deal of energy. *c.* add carbon dioxide. *d.* remove nitrogen.
15. Legislation passed in California states that by the year 2003 *a.* 10 percent of new cars must be electric. *b.* all cars must be electric. *c.* no one may drive cars. *d.* all new cars must be solar powered.

Short Answer (Constructed Response)

Use the information you learned in this chapter to respond to the following items.

16. Which atmospheric pollutants does the Air Quality Index measure?
17. What does an Air Quality Index value of 20 indicate?
18. What is gasohol?
19. What percent of the air pollution affecting most cities is thought to be caused by automobiles?
20. What seven pollutants were designated as the most serious threats to human health in the Clean Air Act of 1970?

Essay (Extended Response)

Use the information in the chapter to respond to these items.

21. Explain the controversy involved in the use of gasohol as a fuel for automobiles.
22. Describe the main features of the hybrid vehicles.
23. Explain how professional interests might lead manufacturers either to minimize or exaggerate the potential environmental dangers of a new product or process.
24. Based on the information in Table 30-1, has the Clean Air Act had any effect on air pollution? What trends do you see?

UNIT EIGHT

FILLING SOCIETY'S NEED FOR ENERGY

What consequences would we face if the United States were to run out of fossil fuels? There would be no gasoline for cars, diesel fuel for trucks and buses, or aviation fuel for planes. Transportation would grind to a halt. Without coal, power plants would not be able to generate electricity. Our cities would be dark. We could not enjoy our computers, television, or air conditioning. Speaking of comfort, no oil or natural gas would mean no heat in the winter. These prospects are alarming.

We couldn't really run out of energy—could we? In this unit, you will learn about how we use and abuse our energy resources. You will also learn how important it is to conserve existing resources while developing new sources of energy.

CHAPTER 31
Nonrenewable Energy Sources

When you have completed this chapter, you should be able to:

Discuss the use of nonrenewable fuels.

Describe the environmental costs of using nonrenewable fuels.

Identify alternative sources of fuel.

When we review the history of energy use, we find that our choice of fuel has shifted from wood to coal to petroleum and natural gas. Except for wood, these fuels are nonrenewable fossil fuels. Most industrialized nations use fossil fuels to fulfill their needs for energy. However, our wasteful use of these resources is causing them to run out faster than need be. As supplies decrease, costs increase, not only in terms of money but also in terms of environmental damage. As resources run out, new technologies must be developed to meet our need for energy. In this chapter, you will learn why the Age of Fossil Fuels may be but a brief chapter in human history.

31.1 NONRENEWABLE FUELS

Fossil fuels contain solar energy that has been stored in the form of hydrocarbons, that is, compounds made of carbon and hydrogen. These fuels, which include coal, petroleum, natural gas, and peat, are formed from the remains of plants and animals that lived in Earth's distant past. Fossil fuels are still being formed today, but the rate is so slow that our supply will be gone before a new supply forms. Therefore, fossil fuels are a nonrenewable resource: once burned, they are lost forever. Today, nonrenewable fossil fuels supply almost 90 percent of the world's commercial energy demands. The widespread use of this resource is also the leading cause of many environmental problems. The gases released as these fuels are burned contribute to atmospheric pollution, are harmful to human health, and may cause global warming.

Wood, a renewable fuel, was the first fuel. The use of fossil fuels dates back to ancient civilizations in the Near East. Shallow deposits of bitumen, or asphalt, were mined and exported to Egypt where chemicals used in the preservation of mummies were extracted. Bitumen was mixed with sand and fibers to make bricks. It was used to caulk ships, as a bonding material for tools and weapons, and as a road-building material. About 300 years ago, coal came into general use as a fuel. But it was not until the eighteenth century and James Watt's coal-burning steam engine that coal use became widespread. The Industrial Revolution created a demand for coal, and this demand rapidly increased with industrial growth until coal became the world's major commercial energy source. For much of the world, coal now has been replaced by petroleum. The use of petroleum and natural gas, which are the world's most recently developed fossil fuel sources, began little more than 100 years ago.

While more than 90 percent of fossil fuels are used as fuels, virtually every product is made from substances derived from fossil fuels or fossil fuels are used in its manufacture or transport. These products include synthetic fibers, such as nylon, rayon, acrylics, and polyesters. Paints, detergents, cleaning solutions, cosmetics, shampoos, food additives, and a wide range of plastics also rely on fossil fuels. In addition, products are packaged in materials derived from

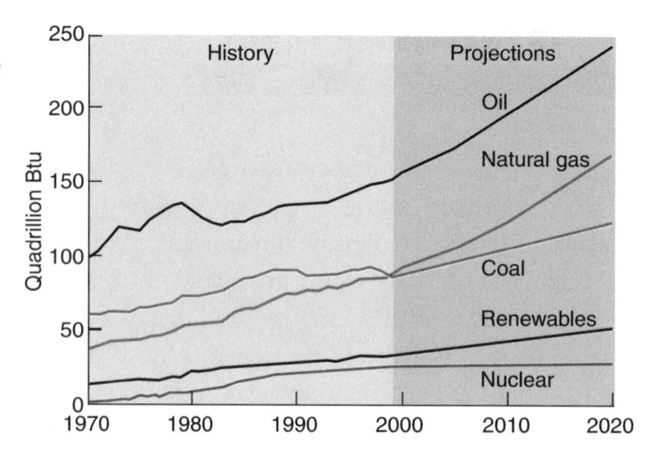

Figure 31-1 World energy consumption by fuel type, 1970–2020.

fossil fuels. These activities also have negative effects on the environment and human health.

During the past 300 years, the consumption of fossil fuels has increased at nearly four times the rate of the world's population. In recent times, consumption has fluctuated. From 1960 to 1973, fossil fuels were relatively inexpensive and consumption grew very rapidly. In 1973, shortages and rising prices forced consumers to conserve and use fuels efficiently. By 1990, supply had again exceeded demand and fuel use increased again. The 1991 Persian Gulf War created shortages and cut consumption.

In 2000, worldwide consumption of fossil fuels declined by 0.2 percent for the second straight year. While consumption of coal fell for the fourth year in a row, natural gas and oil consumption rose. World oil consumption increased at more than 2 percent per year. At that rate, oil reserves will be consumed in less than 40 years. This allows very little time to develop technology that is not based on oil. The world's wealthiest countries, with only 20 percent of the world's population, consume 80 percent of all energy, including fuel oil. Unless these nations reduce their consumption, increased energy consumption in developing nations will deplete energy reserves at an even faster rate. (See Figure 31-1.)

Coal

Coal is the world's most abundant fossil fuel. It is composed of fossilized, decomposed plant material that was buried in sediments and compacted. The formation of most coal began in lush, tropical swamps during the Carboniferous Period between 280 million and 345 million years ago, when Earth's climate was much warmer and wetter than it is now. That is why coal is nonrenewable: it takes millions of years to form.

Worldwide deposits of coal are huge, almost 10 times that of petroleum and natural gas combined. North America, China, and the former Soviet Union contain more than two-thirds of the world's known reserves of coal. China still derives three-fourths of its energy from coal. However, in other parts of the world, the environmental costs of mining and burning coal have caused its decline as a fuel source.

Petroleum

Like coal, petroleum is a fossil fuel. However, petroleum is derived from the remains of marine plants and animals that were broken down by microscopic organisms. Between 30 million and 180 million years ago, vast numbers of plankton sank to the bottom of the oceans and were buried in sediments. Over millions of years, the partially decomposed organic matter was subjected to heat and pressure, causing it to change chemically to petroleum. Folding and deformation of rock strata create pockets of porous rocks that are saturated with petroleum. Petroleum and natural gas usually accumulate under layers of impermeable rocks, such as shale. As shown in Figure 31-2, the petroleum is not found in a vast chamber, but fills the spaces in porous rock, like water in a sponge. A layer of natural gas often caps these reservoirs.

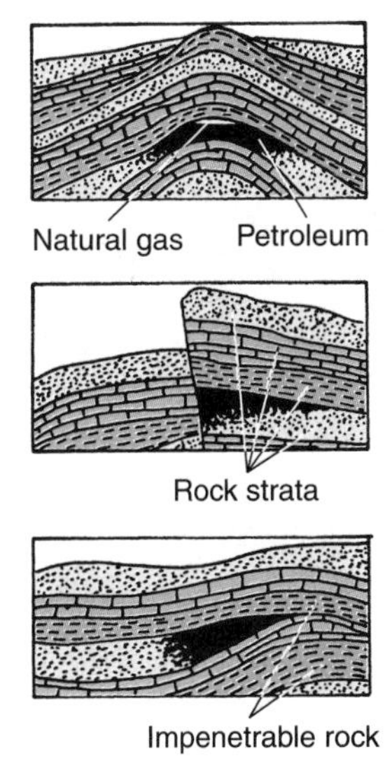

Figure 31-2 Petroleum can be trapped between impermeable strata.

The nations of the Persian Gulf region possess almost two-thirds of the world's petroleum resources. Saudi Arabia contains almost 25 percent of the proven reserves, while tiny Kuwait contains 10 percent. The accessibility of petroleum reserves affects their cost. The cost of producing a barrel of oil in Alaska or the North Sea can be up to 10 times greater than it is in the Persian Gulf. This explains why the bulk of the world's petroleum wealth is concentrated in five of the eleven members of the Organization of Petroleum Exporting Countries (OPEC): Saudi Arabia, Kuwait, the United Arab Emirates, Iran, and Iraq.

DISCOVERY

Life in a Pressure Cooker

Until recently, scientists thought all life was divided into just two lines of descent, the prokaryotes and the eukaryotes. The major difference between these two lines is that the prokaryotes, the bacteria, do not have a nuclear membrane surrounding their genes. In the eukaryotes, the genes are located in a nucleus enclosed within a nuclear membrane. All other unicellular and multicellular life-forms, including humans, are eukaryotes. Or so they thought.

A microorganism collected 3 kilometers down on the Pacific seafloor has been confirmed as a new type of life-form. *Methanococcus janaschii* (a) lives around volcanic vents (b), where temperatures reach 85°C and pressures are greater than 252 atmospheres. This unusual life-form is unlike bacteria or eukaryotic organisms. These microorganisms, which belong to the Archaea, have no nucleus, but their genes behave like those of the "higher" organisms. While some of *M. janaschii*'s genes resemble bacterial genes and others resemble eukaryotic genes, 60 percent of its genes are unlike any yet observed.

The Archaea live in some of the most extreme environments: near deep-sea volcanic vents, hot springs, extremely acidic waters, extremely alkaline waters, and extremely salty waters. They also live in the digestive systems of cows and other ruminants, termites, and marine organisms. Archea are found in muddy marshes, on the ocean bottoms, and in petroleum deposits within Earth's crust, thriving in an environment lacking oxygen. Since they are the only organisms that can survive in these extreme environments, they are often called extremophiles.

Also of interest to people is the fact that *M. janaschii* produces methane gas as a by-product of its metabolism. They may even lead to new, inexpensive sources of energy.

(a)

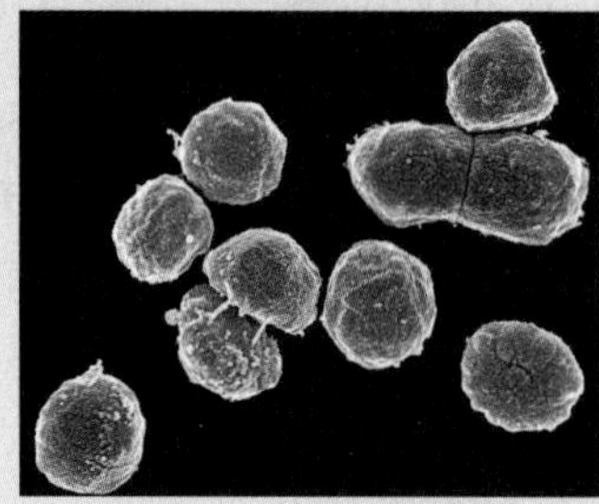

(b)

Natural Gas

Natural gas is composed primarily of the hydrocarbon methane. It is the world's third largest commercial fuel energy source. The use of natural gas has many advantages over coal and petroleum since it is clean burning and relatively inexpensive. Methane burns to form water and carbon dioxide, producing only half as much carbon dioxide as an equivalent amount of coal or petroleum. Natural gas is also free of the contaminants found in other fossil fuels. The main disadvantage of natural gas as a fuel is that, in its gaseous state, it is difficult to ship or store. North Americans are fortunate to have an abundant and readily available supply of natural gas located within the continental crust. An extensive pipeline network has been constructed to deliver it to homes, schools, and factories. The bulk of the known reserves of natural gas are located in the former Soviet Union and Iran.

31.1 Section Review

1. Explain how the energy in fossil fuels is actually derived from the sun.
2. Describe how the invention of the steam engine affected the worldwide use of fossil fuels.
3. Determine the types of fuels used in the homes in your neighborhood.

31.2 ENVIRONMENTAL COSTS

Both the extraction and use of fossil fuels create environmental problems, affect the quality of life, and hinder the economic development of many nations. Each fuel creates its own set of problems. People should keep in mind these environmental costs of using fossil fuels.

Coal

The mining and use of coal have an extensive impact on the environment and human health. Contaminated and highly acidic water from mines often flows into rivers and streams, poisoning wildlife and destroying water quality. Coal mines create mountains of tailings, or waste rock. Underground coal-mining activities have been subject to cave-ins, fires, explosions, poisonous gas leaks, and contamination of water supplies. Gases may leak from mine areas, causing health problems in miners and nearby residents. In addition, miners may develop black lung, a disease caused by inhaling coal dust.

The land is polluted by surface mine operations as topsoil is scoured away. Strip-mining and surface-mining operations are safer for workers than are underground mines. But these mining activities scar the landscape and destroy scenic areas. Surface operations use huge machines to remove surface layers of dirt and rock, called **overburden.** Once the overburden is removed, the seams and beds of coal are exposed. As shown in Figure 31-3, erosion of the steep slopes formed from piles of overburden may cause the land to slide downhill, burying forests, streams, and farmland.

Restoration of the environment following surface coal-mining activities is required by the Federal Surface Mine Reclamation Act. Topsoil is set aside and used to reshape the land after mining operations end. The land is then replanted with native vegetation. Even when government-recommended reclamation practices are followed, huge, ugly scars are left on the land. Erosion, loss of cover for

Figure 31-3 Erosion caused by surface-mining activities.

wildlife, and degradation of the natural landscape make strip-mining and other surface-mining operations undesirable environmental practices.

Air pollution results directly from coal-mining activity and indirectly from vehicles used to transport coal. Coal-fired electrical power plants pollute the atmosphere with tons of particulates and gases. The polluting gases include nitrogen oxides and sulfur oxides, which contribute to acid rain.

Natural Gas and Oil

Large quantities of natural gas are produced in conjunction with oil from wells and refineries. Most countries cannot afford to construct pipelines to deliver to consumers the natural gas found in association with coal mines and petroleum wells. Therefore, this gas is treated as waste and burned, which squanders a potentially valuable resource.

Natural gas extraction can cause **subsidence**, a sinking of Earth's surface in the area around the wells. The Long Beach Harbor region of Los Angeles, California, has dropped more than 9 meters in elevation due to subsidence around gas wells.

Trans-Alaska Pipeline

In 1967, a huge oil field was discovered on the North Slope near Prudhoe Bay, in Alaska. Prudhoe Bay is located north of the Arctic Circle and is subject to subfreezing temperatures for most of the year. Since the area is inaccessible to ships during most of the year, other methods had to be devised to transport the oil to refineries. The 1300-kilometer-long Trans-Alaska pipeline, extending from Prudhoe Bay oil fields to the port of Valdez in southern Alaska, was proposed to solve the problem. The 1.2-meter-diameter pipeline was designed to carry petroleum overland to the warmer waters off Valdez. There it is loaded on tankers and shipped to the West Coast of the United States.

The pipeline proposal sparked immediate debate between oil interests and conservationists. The environmentalists argued that the pipeline would have a negative impact on wildlife, the warm fuel oil would melt the permafrost, and the increased tanker traffic would raise the chances of a major oil spill along the Alaskan coast. Native Americans claimed that they held rights to the land and had the authority to grant permission for construction. The oil interests held that supplies would stimulate the economy of Alaska and the United States. The petroleum would reduce the nation's dependence on imported fuel, and the operations would create many new jobs throughout the state.

To protect the permafrost, the pipeline was placed on refrigerated supports. In addition, the supports raised the pipeline high enough above the ground so that it did not interfere with animal (mainly caribou) migrations.

In 1989, the worst possible scenario for an environmental disaster occurred. The supertanker *Exxon Valdez,* carrying oil from Prudhoe Bay, ran aground in Prince William Sound. About 40 million liters of petroleum were spilled into the water. The spill caused extensive damage to the ecosystem. The oil killed thousands of birds, fish, and sea otters. The beaches were covered with oil. Many people participated in the cleanup, washing birds and rocks to remove the oil. The ecosystem is slowly returning to its natural state.

31.2 Section Review

1. What are some ways coal mines and oil wells indirectly pollute the environment?
2. How does the accumulation of mine gases threaten the environment?

31.3 CONSERVATION OF ENERGY RESERVES

For most of human history, energy has been taken for granted and used wastefully. The energy obtained from fossil fuels forms the lifeblood of society. Fossil fuels supply about 90 percent of the world's commercial energy. Yet it is carelessly consumed and treated as an inexhaustible commodity. However, there have been times when supplies were dangerously low. During these fuel shortages, industry grinds to a halt, agricultural production declines, vehicles cannot move, and millions of people are put out of work.

There is a great difference in the amount of fossil fuel used by the world's developing nations compared with that used by highly industrialized nations. For example, the United States consumes one quarter of the world's energy. We waste more fossil fuel than is used by the world's developing nations put together. On the other hand, more than 2 billion people who live in developing nations still rely on wood as a source of energy. Most of these nations face growing shortages of wood, which further increases the worldwide demand for fuel oil. As countries become more and more dependent on fuel oil to run their economy, their fuel purchases become a major source of national debt. Since the effects of energy use have an impact on the entire biosphere, they are of global concern.

Alternatives

Estimates of the world's total petroleum supply do not take into account unconventional oil resources, such as shale oil and tar sands. **Shale oil** is a type of petroleum recovered by processing shale, a fine-grained sedimentary rock. Some shale contains a material called kerogen. This shale is broken into fine pieces and heated to high temperatures to vaporize the kerogen. The heavy, oily liquid that is recovered by condensation of the vapors is the shale oil. Shale oil processing is illustrated in Figure 31-4 on page 570. The largest shale oil reserves are the Green River deposits in Utah, Colorado, and Wyoming.

There are environmental costs in oil shale production. The shale is strip-mined, which degrades large tracts of land. Processing the shale produces huge amounts of solid waste, called spent shale. The spent shale creates the same type of environmental problems as do mine tailings.

Tar sands are sandy surface deposits that contain bitumen, a petroleumlike material with a high sulfur content. These sands are surface-mined. Tar sands are processed with heated water to collect the bitumen. The bitumen is then refined to produce crude oil.

Figure 31-4 Oil can be removed from shale and processed.

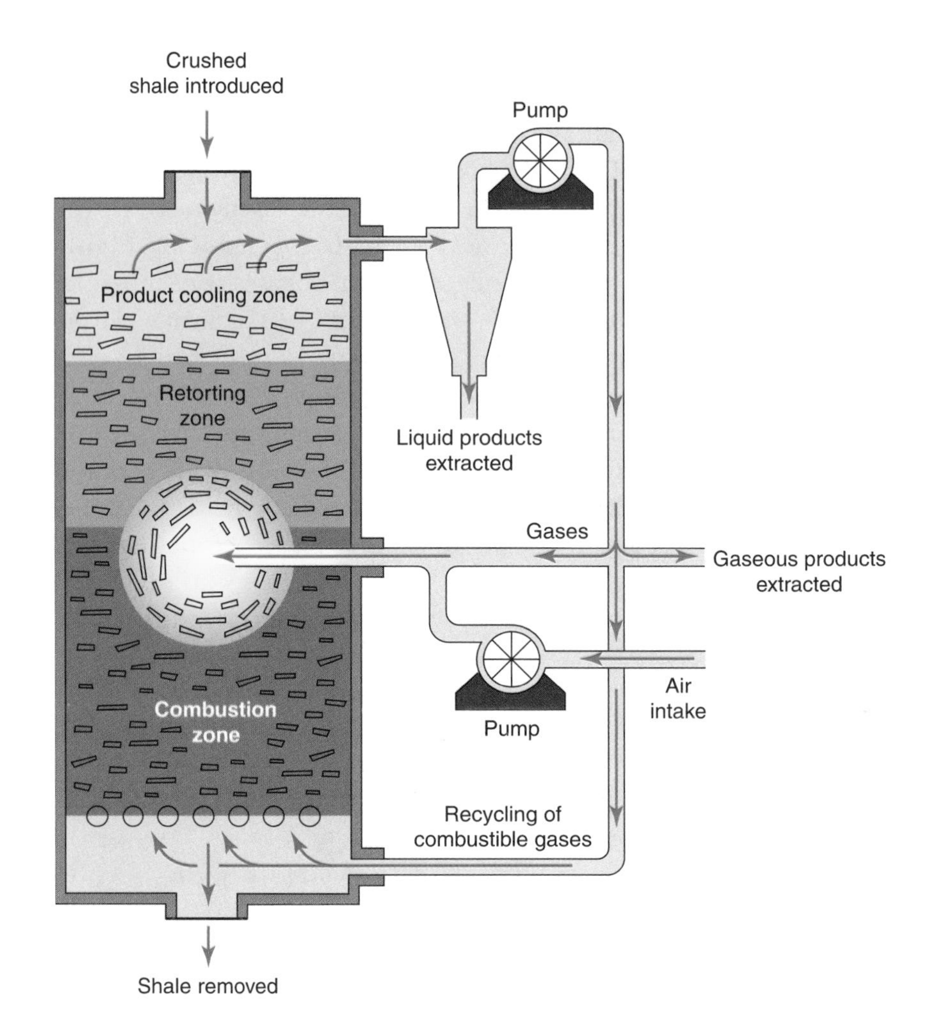

There are large tar sand deposits in Canada (in Alberta), Venezuela, and the former Soviet Union. Smaller deposits are found in the United States, mainly in Utah.

Even though the Alberta tar sand deposits represent the world's largest known petroleum field, environmental factors restrict their development as a fuel source. Tar sand–processing plants create huge quantities of toxic sludge and consume and contaminate billions of liters of water a year. (See Figure 31-5.)

The abundance of coal has created interest in the development of new technologies. **Coal gasification** produces combustible gases. **Coal liquefaction** produces an oil that can be refined the same way as petroleum. These technologies react hydrogen with coal to form

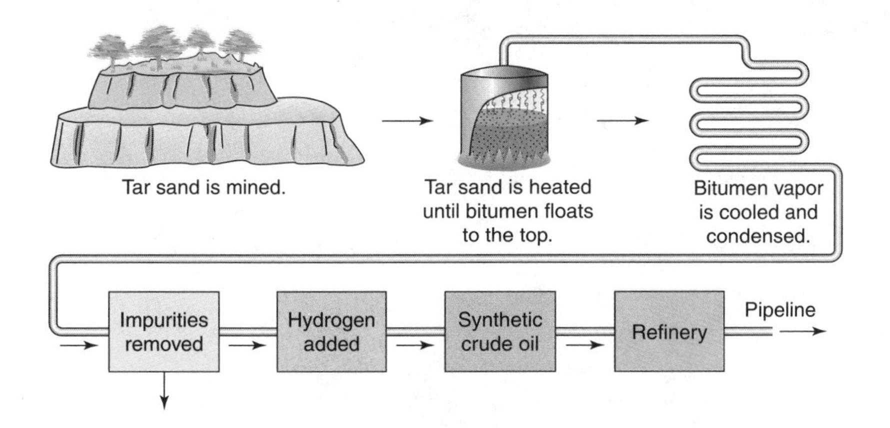

Figure 31-5 The removal of synthetic crude oil from tar sand.

a synthetic fuel, or **synfuel.** At present, these processes produce significant amounts of air and water pollutants and are too costly to compete with natural fuels.

Most natural gas is obtained as a by-product of petroleum drilling. But there are unconventional sources of natural gas that remain untapped. Methane pockets often form within coal seams and other rock formations. The gas is also produced as a by-product of sewage treatment as well as by microbial action in landfills. Some cities collect and use this methane. **Animal waste digesters**, or **biomass digesters**, are used in some areas of the United States and in China to produce methane. Waste digesters utilize decomposer organisms to convert biomass into methane gas. The biomass may consist of animal wastes or crop residues. The gas produced can be used for heating, lighting, and cooking.

Figure 31-6 Propane tanks provide the fuel used in barbecue grills.

In areas not served by the natural gas pipeline, propane is widely used as a fuel. Propane is a colorless gas found in association with petroleum and natural gas. It is also used as a cooking fuel for barbecues and for heating recreational vehicles. (See Figure 31-6.) It is stored and distributed in pressurized cylinders, as liquefied propane gas.

31.3 Section Review

1. Describe how shale oil recovery can have a negative impact on the environment.
2. Discuss some unconventional sources of methane gas.
3. List some ways you could help to conserve fossil fuels.

LABORATORY INVESTIGATION 31

Generation of Coal Gas

PROBLEM: *How can coal or wood be used to generate another fuel?*

SKILLS: *Manipulating, observing, inferring*

MATERIALS: *Test tube, stand, clamp, 3 pea-size pieces of coal, 3 wood splints, one-hole rubber stopper, 7.5-cm length of straight glass tubing (fire polished at both ends), heat source, hand lens*

PROCEDURE

1. Place several pieces of coal into the test tube. Carefully insert the fire-polished glass tubing into the stopper. Stopper the test tube.
2. Set up apparatus as shown in laboratory diagram.
3. Heat the test tube. When gases start coming off the coal, place a lighted splint at the tip of the glass tubing extending from the stopper.
4. After the material has cooled, examine it with a hand lens.
5. Repeat the investigation using the pieces of wood splint.

CAUTION: The gases produced by this investigation are combustible and have a foul odor. It is recommended that the experiment be carried out in a well-ventilated hood.

OBSERVATIONS AND ANALYSES

1. What substances were produced inside the test tube?
2. What happened when the lighted splint was placed near the glass tubing?
3. Describe what happened to the wood splints that were heated in the test tube.
4. How is the process that occurred in the test tube different from burning?
5. Describe the substances remaining in the test tube after it cooled.
6. How do you think charcoal is produced?

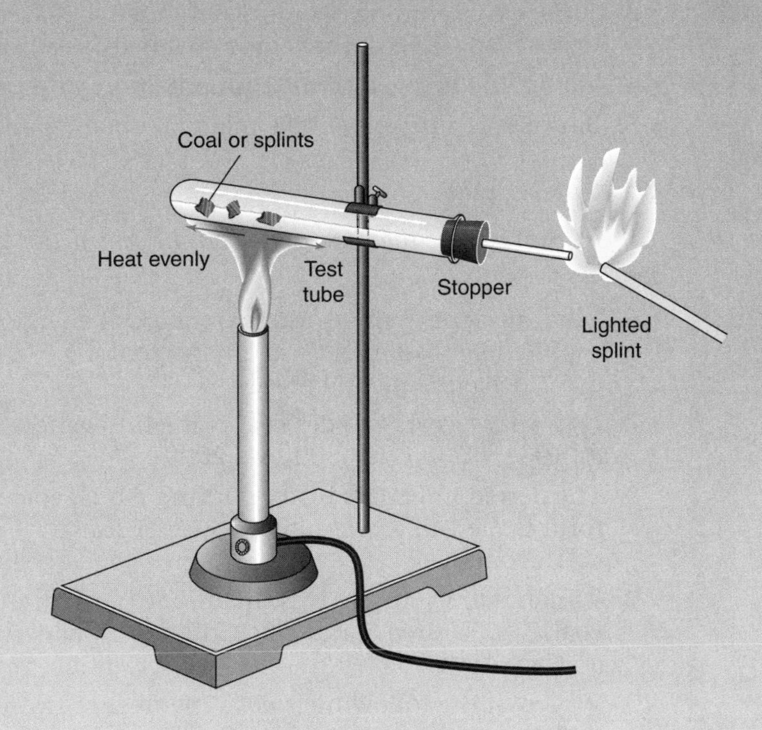

Chapter 31 Review

Answer these questions on a separate sheet of paper.

Vocabulary

The following list contains all the boldfaced terms in this chapter.

animal waste digesters, biomass digesters, coal gasification, coal liquefaction, overburden, shale oil, subsidence, synfuel, tar sands

Fill In

Use one of the vocabulary terms listed above to complete each of the following sentences.

1. Coal can be reacted with hydrogen to produce ______.
2. The sinking of Earth's surface around well areas is known as ______.
3. Surface coal-mining operations remove the layers of dirt and rock called ______ that cover coal beds.
4. Vaporizing kerogen from sedimentary rocks produces ______.
5. Bitumen is a petroleumlike material obtained from ______.

Multiple Choice

Choose the response that best completes the sentence or answers the question.

6. What percent of the world's commercial energy demands are supplied by fossil fuels? *a.* 50 percent *b.* 75 percent *c.* 90 percent *d.* 100 percent
7. How many years ago did coal come into general use as a fuel? *a.* 100 *b.* 300 *c.* 500 *d.* 2000
8. The period in Earth's history during which coal started to form is the *a.* Carboniferous. *b.* Permian. *c.* Silurian. *d.* Devonian.
9. Which nation still derives three-fourths of its energy from coal? *a.* United States *b.* China *c.* Saudi Arabia *d.* Canada
10. Which of the following is *not* a member of the Organization of Petroleum Exporting Countries? *a.* Kuwait *b.* Canada *c.* Iran *d.* Saudi Arabia
11. The first fuel used by humans was *a.* coal. *b.* peat. *c.* methane. *d.* wood.
12. The main disadvantage to the use of natural gas for energy is *a.* limited supply. *b.* no demand. *c.* difficulty with shipping and storage. *d.* difficult to use.

Short Answer (Constructed Response)

Use the information you learned in this chapter to respond to the following items.

13. List four uses of fossil fuels, other than as fuel.
14. Fossil fuels are composed of which two elements?
15. Why is wood considered a renewable fuel?
16. List two synthetic fibers derived from fossil fuels.
17. What hydrocarbon is natural gas composed of?
18. What is shale oil?
19. What is bitumen?
20. What two technologies are used to produce synfuels.

Essay (Extended Response)

Use the information in the chapter to respond to these items.

21. Describe how coal forms.
22. Describe how petroleum forms.
23. What are the advantages of using natural gas as a fuel?
24. How does the use of wood for energy impact on the environment?
25. How does energy use in the United States compare with its use in other countries?
26. Imagine a world without automobiles and trucks. What impact would that have on the economics of the nation and the world? What changes would you see in the environment?
27. How has air conditioning impacted our society? How has it affected our energy demands, our leisure activities, the workplace, and the environment?

Research Project

United States policy has been to make energy as freely available and inexpensive as possible. However, European nations charge their consumers many times more for fuel. Research in the library or on the Internet and report on how these policies have affected lifestyles here and in Europe.

CHAPTER 32
Renewable Energy Sources

When you have completed this chapter, you should be able to:

Define renewable energy resources and give some examples.

Explain how kinetic energy is related to renewable sources of energy.

Describe the advantages and disadvantages of using the various renewable energy resources.

When the United States can satisfy its energy needs using only renewable resources, we can become energy independent. This independence would mean that we would no longer have to rely on foreign sources of fossil fuels. However, we must also be responsible to future generations by producing this energy without damaging the environment.

At this time, many of the renewable energy resources are more expensive to use than fossil fuels. As fossil fuel reserves are exhausted, energy from renewable resources will become more competitively priced. In this chapter, you will learn about the advantages and disadvantages of renewable energy sources.

32.1 RENEWABLE ENERGY

A **renewable energy** resource is one that is readily regenerated by natural processes at a rate that exceeds its rate of use by people. At present, the sun is our major source of renewable energy. Each day, Earth receives more than enough **solar energy**, or sunlight, to meet all of our foreseeable energy needs. We use sunlight as a source of light and heat. We can use as much of this renewable resource as we want, since nuclear processes within the sun continually produce more energy. Energy from sunlight is stored by plants during photosynthesis in the form of chemical energy, or biomass energy. Heat from the sun drives the winds, creating wind energy.

Every inhabited region of the world has a potential form of renewable energy. The map in Figure 32-1 on page 578 indicates places in the Unites States that have potential renewable energy resources. The tropics have abundant thermal energy stored in ocean waters, long daily periods of sunlight, and environmental conditions best suited for the use of biomass energy. The subtropical deserts are exposed to long periods of intense sunlight, which is advantageous in using solar energy. The temperate zones contain differing regions; some may have sunlight for the growth of biomass crops. In other regions, there may be wind or fast-moving rivers and streams. The environmental conditions in each locality dictate the renewable sources that have the best potential for use. The higher latitudes have hydroelectric and wind energy resources.

In 2004, about 9.2 percent of the world's energy was supplied by renewable sources. In the United States, about 6.6 percent of our energy was renewable. Renewable energy sources usually are used to produce electricity. The potential for expanding our use of renewable energy sources is very high, but the extent to which this potential becomes a reality depends on many factors.

Energy Efficiency

The ratio of useful energy produced to energy used for production is known as **energy efficiency**. In all cases, energy efficiency is less

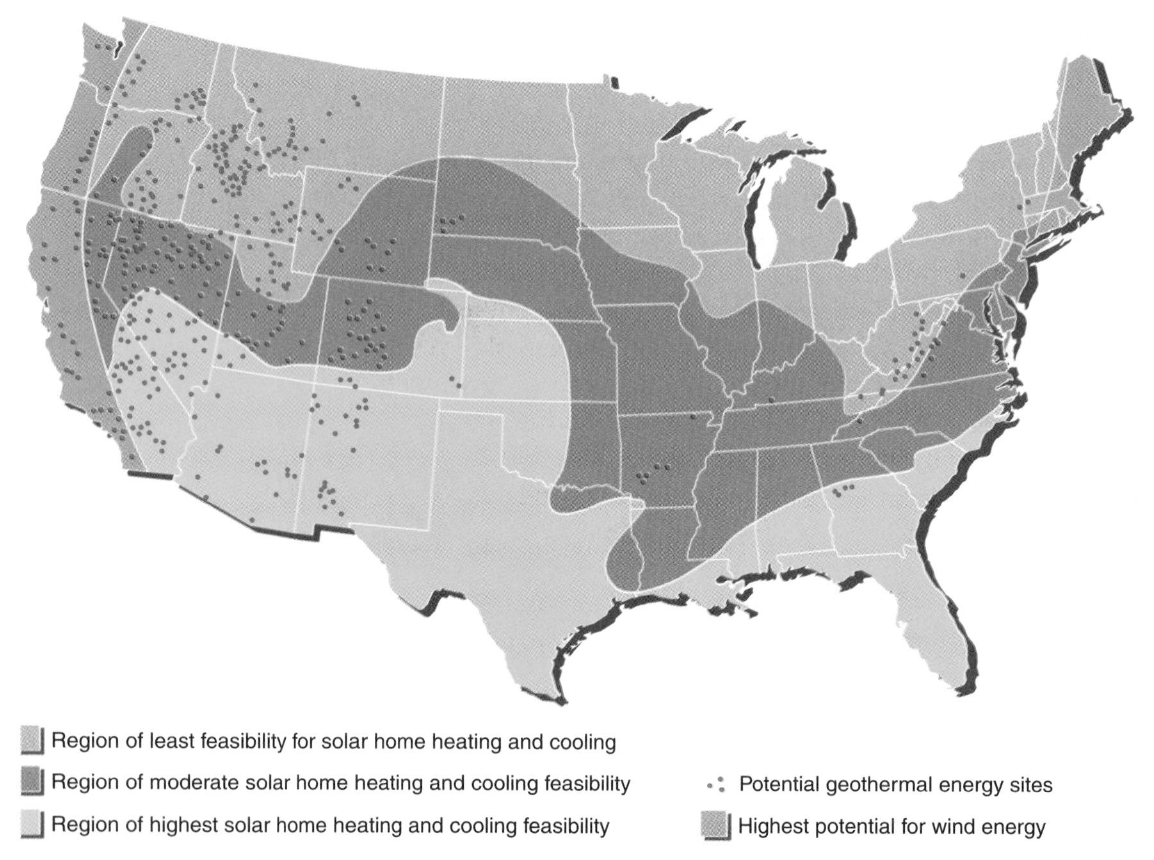

Figure 32-1 Alternate energy forms can be used economically in some parts of the United States.

than 100 percent, because useful energy is lost in the form of waste heat. Electric generators, electric motors, and batteries are energy efficient, producing little waste heat. On the other hand, internal combustion engines, jet engines, and coal-fueled and oil-fueled steam engines are inefficient. They produce large amounts of waste heat. Improving the efficiency of our energy production can reduce energy costs and extend our resources. Improving the efficiency of energy generation also reduces its impact on the environment.

The following two points are of primary concern when considering the introduction of alternative methods of energy production from renewable sources.

- The production of energy must be carried out with a high degree of efficiency:

$$\text{energy efficiency} = \frac{\text{energy output}}{\text{energy input}}$$

- Energy production must have little effect on the environment.

32.1 Section Review

1. Why is sunlight our most important source of renewable energy?
2. What are some examples of the most energy-efficient devices?
3. Survey your home and list ways you can make it more energy efficient.

32.2 WATER POWER

Kinetic energy is energy possessed by moving objects. The energy of a moving object depends on its the mass and velocity. More massive, or heavier, objects and faster-moving objects possess greater kinetic energy than lighter or slower-moving ones.

Hydroelectric power uses the kinetic energy of moving water to spin turbines. Hydroelectric power derives its energy from the solar heat that drives the hydrologic cycle and from Earth's gravity. In the production of hydroelectric power, **thermal energy**, or heat, from the sun and kinetic energy in moving water are converted to electrical energy. The spinning turbines turn generators that produce electricity. (See Figure 32-2 on page 580.) Hydroelectric power is the largest contributor to the nation's renewable energy sources. During 2000, hydroelectric plants generated about 2.88 percent of the total energy produced in the United States. Hydroelectric generators are efficient producers of energy. They convert almost 85 percent of the total energy stored in moving water into electricity.

To make use of the energy stored in moving water, dams are built across rivers to block the flow of water. A tall dam causes the water to back up and rise to a sizable depth. The water then flows through the dam and picks up kinetic energy before it reaches the turbines at the base of the dam. As shown in Figure 32-3 on page 580, the Grand Coulee Dam on the Columbia River in the state of Washington is

Figure 32-2 The energy of moving water can produce electricity.

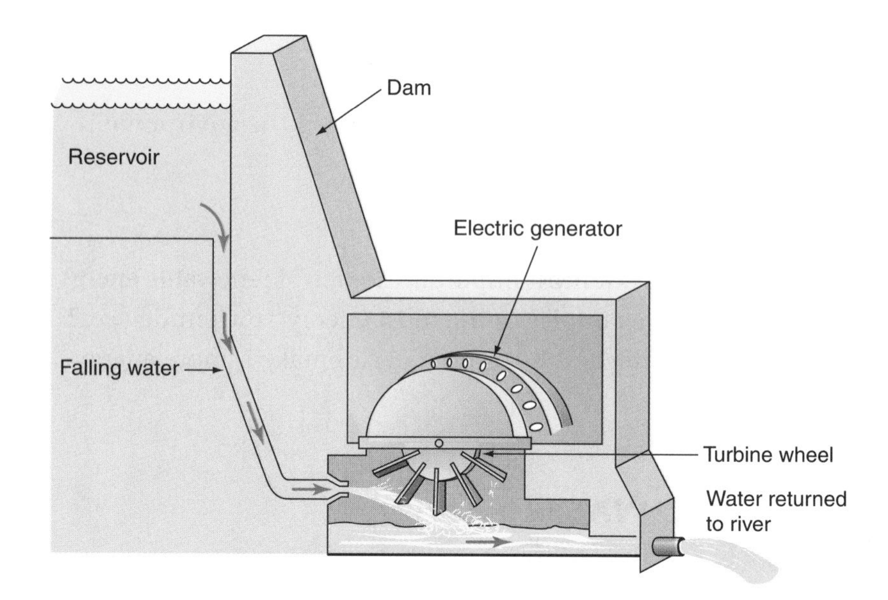

1600 meters long and 170 meters high. Roosevelt Lake, formed by the dam, extends 240 kilometers upriver. This giant lake serves to power the electric generators at the dam and as a recreational facility for the region.

China is constructing the Three Gorges Project on the Yangtze River. It will have the world's largest hydroelectric power dam,

Figure 32-3 The water held behind this dam can be used to generate electricity.

which will be 183 meters high and create a lake 610 kilometers long. The lake will flood much farmland, cover many cities and towns, and displace 1.3 million people.

Advantages and Disadvantages

Hydroelectric power has many advantages over other means of generating electricity. It is less expensive, does not produce air pollution, and is a renewable resource. As with all forms of energy production, there are disadvantages to hydroelectric power. Huge reservoirs lose enormous amounts of freshwater due to surface evaporation. In the United States, this energy resource has been extensively utilized, and many of the accessible sites for construction of hydroelectric plants already have been used. Those sites that remain often are protected by environmental organizations that desire to keep the rivers in their wild and natural state as a legacy for future generations.

The construction of hydroelectric plants can damage the environment. Building dams across wild and scenic rivers converts them to lakes and slow-moving streams, thus destroying the natural ecosystem. Salmon and other fish that need access to the upper portions of streams to reproduce disappear from these rivers. To help salmon reach their spawning grounds, fish ladders are built at dams. As shown in Figure 32-4, these ladders allow fish to make their way around the dam. The flooding of rivers destroys commercial forests, waterfowl nesting areas, and shoreline habitats.

Figure 32-4 Fish ladders help fish reach their spawning grounds.

Hydroelectric projects also can impact negatively on human lives. Flooding may destroy communities behind the dam. Fishing industries downriver from the dam often are wiped out. Because water flow is slowed, pollutants can accumulate. Fish may become contaminated with heavy metals such as mercury, which makes them unfit for human consumption. The banks of rivers that are diverted may suffer from extensive erosion. In addition, silt that had previously fertilized farmlands builds up behind dams, which limits the length of time the dams are useful.

As a result of the advantages and disadvantages of hydroelectric power, conflicts arise among citizens. Some groups prefer to have rivers remain in their natural state to preserve their beauty or for conservation reasons. Others, living on land that will be flooded if a dam is constructed, may argue that their rights as property owners are being violated. Residents elsewhere in the region may find economic rewards in the inexpensive, nonpolluting energy that dammed rivers provide. These issues often result in long delays for government permits to construct dams. Frequently, the courts must settle the conflicts between environmentalists who advocate free-running rivers and advocates of hydroelectric power.

32.2 Section Review

1. What types of energy are used to produce hydroelectric power?
2. What are some problems that might arise regarding a proposal to build a new hydroelectric plant within a national park?

32.3 WIND POWER

About 2 percent of the sunlight that strikes Earth's surface is converted into the thermal energy that produces winds. Wind is caused by the unequal heating of the atmosphere. Air at or near the equator receives more solar heat energy than air at the poles. The warmer air rises and is replaced by cooler air from the upper atmosphere. These air movements within the atmosphere are called vertical wind currents. Vertical wind currents create regions of varying air pressure. Air moves horizontally along Earth's surface from regions of higher pressure to regions of lower pressure. These horizontal currents give rise to steady wind patterns called global winds.

Prevailing winds blow regularly and usually from one direction. But prevailing winds are not always a reliable source of power. There are often periods, called calms, when there is no wind. Since most energy use requires a steady flow, extended periods of calm can affect energy efficiency. For example, energy use in cities fluctuates. During daylight hours, the operation of factories consumes large amounts of energy while residential use is low. As factories shut for the evening, the consumption of electricity for homes, street lighting, and advertising increases and then tapers off. A steady flow of electricity is therefore required 24 hours a day.

Harnessing the Wind

The wind is one of our oldest sources of renewable energy. In the Netherlands, individual windmills have been used for centuries to pump water, process grain into flour and meal, and generate small amounts of electrical energy. Electrical energy can be generated on a larger scale by a group of windmills called a **wind farm**. (See Figure 32-5.) The windmills turn electric generators that produce clean

Figure 32-5 These giant windmills are able to generate electricity.

energy from a renewable source. Wind farms require large amounts of land. However, this land may be used at the same time for agriculture or livestock grazing. Wind energy is free and readily available. The construction of wind farms can provide jobs and safe, nonpolluting energy for developing nations.

Locations with steady winds offer the best possibility for large-scale development of wind power. An example of this is the island of Oahu, Hawaii. Oahu lies in the path of the trade winds, the steady winds that carried early explorers across the oceans. The state of Hawaii plans that, eventually, wind farms will supply 10 percent of its energy. California has already found several excellent locations for wind farms, including the Tehachapi Mountain Wind Resource Area and San Gorgonia Wind Resource Area. (See Figure 32-6.) The amount of energy generated by wind power in California is equal to the production of one nuclear power plant. As the need for clean and renewable energy sources becomes more pressing, more wind farms will certainly be built across the country. In 2000, only 0.05 percent of the total energy produced in the United States came from the wind.

A method of increasing the availability of wind-derived energy is to use large storage batteries. A storage battery is a device used to store electrical energy. These batteries are charged while the wind is

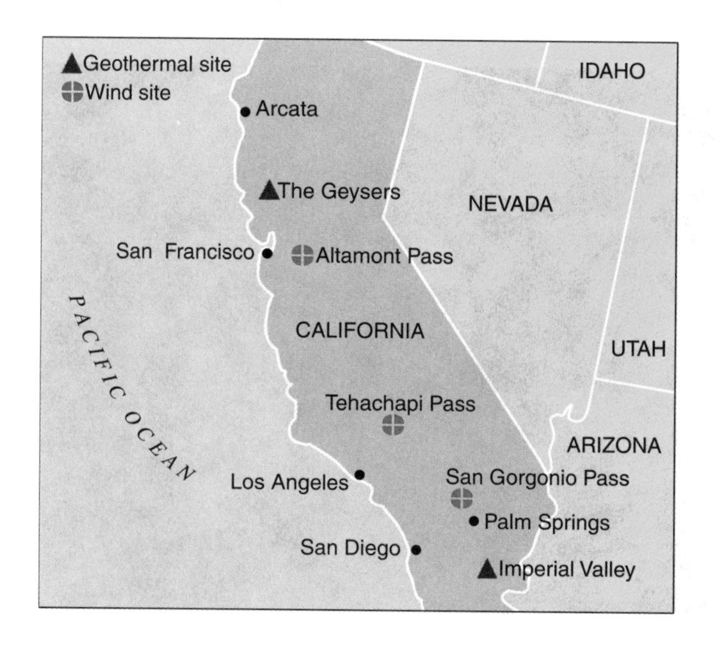

Figure 32-6 **Wind and geothermal energy sites in California.**

blowing, thus storing some of the electricity produced. The stored energy can be used during calm periods.

Another solution to the problem of unreliable winds is the use of wind farm energy to pump water into reservoirs. During periods of calm, the water in the reservoir is discharged through turbines and used to produce hydroelectric power.

A third solution is to construct windmills with large, heavy, low-friction flywheels. A **flywheel** is a massive wheel that spins like a top. Like a gyroscope, once a flywheel has started spinning it continues to do so for extended periods of time. The flywheel can store large amounts of kinetic energy. Flywheels can be spun by wind power. When the winds slow or stop, the energy stored in the spinning flywheel can be used to generate electricity. Finally, wind farms need not be a primary power supply. By supplying energy during windy periods, they may be used to supplement nonrenewable energy sources.

Disadvantages

Often, people say that wind farms are not attractive and do not want one near their home. Windmills make noise as they turn, which some people find disturbing. Birds often fly into the spinning blades and are killed. Currently, electricity produced on wind farms costs about 40 percent more than electricity produced by oil-powered generators. This is due to the high cost of manufacturing windmills and their low efficiency. This problem will be overcome as technological advances result in more efficient windmill designs. As more windmills are manufactured, cost per unit should be reduced. In considering the future of windmills, it must be remembered that energy from fossil fuels is nonrenewable. As fossil fuel supplies run low, their costs will increase, and this should make wind energy more attractive.

32.3 Section Review

1. Why has wind power not been more fully utilized as an energy source?
2. How can a flywheel increase the efficiency of wind energy production?

32.4 ENERGY IN THE EARTH

The sun warms Earth's surface, but the interior is heated by the hot, molten rock that forms the core. Heat caused by tremendous pressures and radioactive processes maintains Earth's internal temperature. Earth's internal heat is called **geothermal energy**. Water from rain and snow enters Earth's crust through cracks and crevices in it. As you descend into Earth's crust, the temperature increases by about 2°C for each 100 meters of descent. Several kilometers below the surface, temperatures reach that of boiling water. At these depths, water in rocks changes to steam. Through cracks in the crust, steam forces its way up to the surface, where it forms geysers and hot springs. The next rain or snow replenishes the water. Under the proper conditions, this heated water can be used as a source of energy. (See Figure 32-7.) About 0.32 percent of the energy produced in 2000 by the United States was geothermal energy.

Advantages and Disadvantages

Most geothermal energy can be tapped without the energy potential being diminished. But some geothermal sources are isolated deposits and are in actuality nonrenewable. These underground deposits of hot water or steam can be depleted when they are used at rates that exceed their rate of replenishment. The impact created by their use often is not recognized until the source is depleted.

There are geothermal energy sources in northern California, the Philippines, New Zealand, Iceland, and Italy. The geothermal energy stored beneath the United States is equal to the energy of trillions of tons of coal. But use of this energy is practical only in those areas where it is at or near the surface. The area in and around Yellowstone National Park, in Wyoming, is the largest geothermal field in the United States. There has been a great deal of controversy over the impact that the construction of a geothermal plant just outside of Yellowstone Park would have on the geology of the region. Environmentalists argue that waters removed from the crust would pollute the Yellowstone River and that the withdrawal of water could

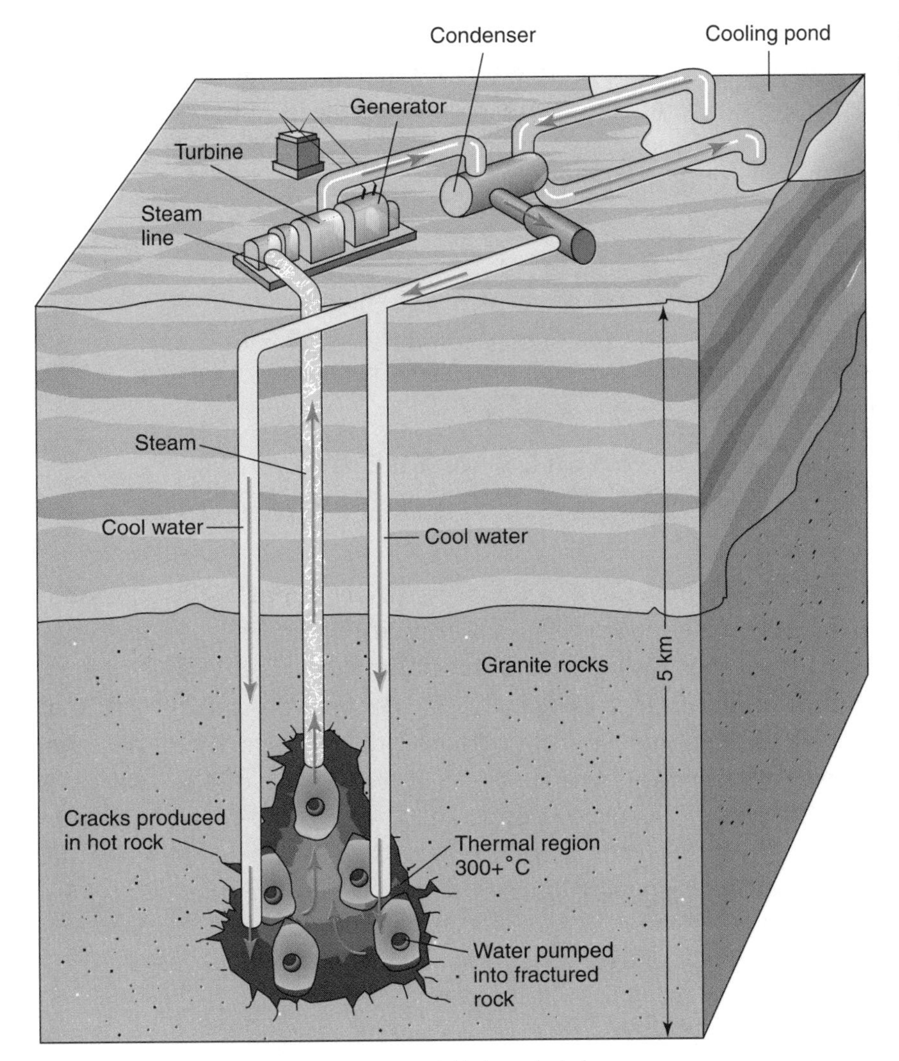

Figure 32-7 Geothermal energy can be used to generate electricity.

lead to subsidence of the region. In New Zealand, plans for geothermal plants were cut back when it appeared that the plants would endanger a major tourist attraction.

One example of the use of geothermal energy is the power plant at The Geysers, in northern California. At this site in 1960, the Pacific Gas and Electric Company began using steam from Earth's crust to power electric generators. In most areas, the steam issuing from Earth's crust contains dissolved salts that corrode generating equipment. The steam at The Geysers is free from dissolved salts. It can efficiently and safely drive turbines to generate electricity.

Unfortunately, areas where clean steam is available are rare; thus the cost of geothermal energy is higher than that of energy from fossil fuels. Advances in technology should lead to the development of geothermal energy as a major resource in the near future.

32.4 Section Review

1. Why is geothermal energy usually renewable?
2. What are some of the drawbacks in the use of geothermal energy?

32.5 ENERGY IN THE OCEANS

The constant movement of the oceans gives a hint of the enormous amount of kinetic energy stored in their waters. The kinetic energy present in ocean waves is a potential source of renewable energy. But technology has not been able to overcome the punishment that the sea gives to such energy-producing devices. The thermal energy stored when sunlight warms the upper layers of the oceans is another potential energy source. The energy can be most efficiently harvested in the tropics, where there is a substantial temperature difference between warm surface waters and deep colder layers. (See Figure 32-8.)

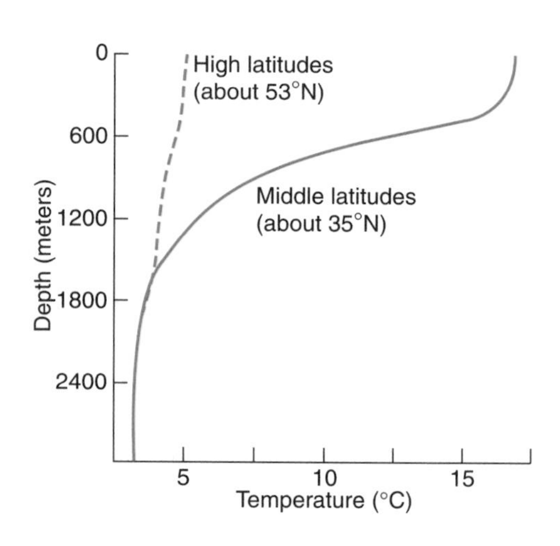

Figure 32-8 Thermal energy is available because there is a temperature gradient in the oceans.

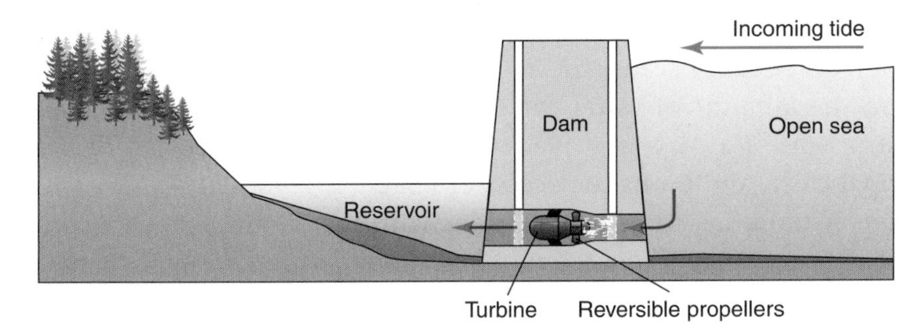

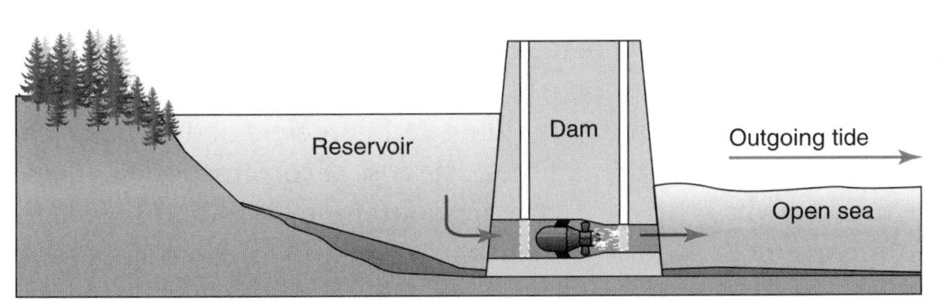

Figure 32-9 **The movement of tides can be used to generate electricity.**

The gravitational attraction among Earth and the moon and sun causes the tides. Because the tides rise and fall, they have kinetic energy. This kinetic energy, called **tidal power**, can be used to generate electricity. To construct a tidal power plant, the entrance to a tidal basin or estuary must be closed by a dam. (See Figure 32-9.) As the incoming tide enters the basin, it passes through dam openings that contain turbines. The water turns the turbines, which produce electricity. After the tide comes in, the dam is closed, and water is stored in the tidal basin. As the tide recedes, the dam is opened to slowly release the stored water. Once again, the water passes through the turbines, generating additional electricity. Locks are constructed in the dam to permit ships to enter and leave the tidal basin.

In 1966, the first large tidal power plant was built in France. It is at the mouth of the Rance River on the Atlantic Ocean. The plant generates 240 megawatts of electricity, about one-quarter the capacity of a modern coal-burning plant. This is a considerable amount of energy to be produced by a renewable source.

Disadvantages

There are a limited number of locations suitable for a tidal power plant. The basic requirements are an ample difference in height between high and low tides and a large tidal basin. A large tidal basin is needed because the amount of electricity generated depends on the amount of water that passes over the turbines. The construction of a dam at a river entrance changes the ecology of the tidal basin and limits the access of boats to the ocean. It is easy to understand that these factors, along with the high cost of dam construction, have limited the number of tidal power plants.

The United States once considered constructing a huge tidal energy-generating project at Passamaquoddy Bay, Maine. The projected output of the generator was to have been 1000 megawatts, equivalent to a modern coal-burning generating plant. Construction of this plant required building 11 kilometers of dams and 160 water gates to house the turbines. The cost of construction has made it unlikely that this project will be attempted in the foreseeable future. As more is learned about the short- and long-term effects of energy production, difficult choices will have to be made between low-cost energy and conservation of the environment.

32.5 Section Review

1. What are some drawbacks associated with tapping the oceans for energy?
2. Examine a map of the east or west coast of the United States. What are some sites that might be suitable for the generation of tidal power? What problems would you foresee in the construction of tidal power plants in these areas?

32.6 BIOMASS ENERGY

Biomass is the total amount of living material in an ecosystem. It includes all Earth's biota: plants, animals, and microorganisms. **Biomass energy** comes from solar energy stored by plants during photosynthesis. Although they were formed from material that was once living, fossil fuels are not considered part of the biomass. Biomass also differs from fossil fuel in that it is a renewable resource. In the United States in 2000, 3.28 percent of the energy produced came from wood, wastes, and alcohol blended into gasoline.

Biomass supplies about 20 percent of the world's energy needs. Most of this energy comes from two sources: forest products and agricultural crops. **Fuelwood**, wood used directly as a fuel, is the main source of biomass energy. Animal manure and animal fats and oils are other products used as biomass fuels. Biomass can be burned directly as fuel, or converted into methane or alcohol. In China, dried animal manure is an important fuel used for both heating and cooking.

Wood as Fuel

Biomass has been used as a source of energy since earliest times. People obtain energy directly from biomass by burning fuelwood. Fuelwood is the primary source of energy for more than 80 percent of the people in the world's developing nations. Consumption of large amounts of wood, much of it used for heating and cooking, has resulted in the destruction of many of the world's forests. (See Figure 32-10.) This in turn has led to the loss of habitat for plants

Figure 32-10 This area was stripped of trees and shrubs due to their overuse for fuelwood.

and animals. Loss of habitat is a critical factor in the extinction of many species.

Another destructive consequence of the misuse of forestlands is soil erosion. Without tree roots to stabilize soil and absorb runoff, rainwater washes topsoil into rivers and from there into the seas. By making the land unfit for agriculture, misuse of forestland can have a negative impact on a nation's economy. This often creates further destruction of forestlands and reduces supplies of needed fuelwood.

The paper and lumber industries consume large amounts of forest wood. They have been able to conserve energy by using waste wood from their manufacturing processes as an energy source. Wood scraps are burned to heat water and drive steam turbines that produce electricity. In some cases, more than 40 percent of the energy needed to run a lumber mill or paper plant can be derived from waste produced during the manufacturing process. Paper and lumber companies find it a wise practice to replant trees in the areas they cut. However, the new forests usually contain only one species of tree. These monoculture, or one species, forests are subject to the rapid spread of disease and attack by insect pests.

Fuelwood currently is used as a minor heating material in the United States. When burned, wood yields about the same amount of energy per kilogram as coal. In the United States, 1 in every 10 homes uses wood as either a primary or alternate heating source. As you travel through rural areas, you see many homes with large woodpiles, which indicate a reliance on wood for fuel.

The United States Forest Service, a division of the Department of Agriculture, is responsible for the management of more than 73 million hectares of national forests. Its function is to protect our forest reserves so that our nation always has supplies of lumber and pulpwood. Forests are also maintained as a recreational resource. The various uses of our forests often lead to conflicts between lumber interests and environmental groups.

Although wood is a readily available and renewable energy source, burning it creates pollution from carbon monoxide and particulates, solid particles such as smoke and dust. Wood stoves produce 10 times more particulates and carbon monoxide than do gas or oil furnaces.

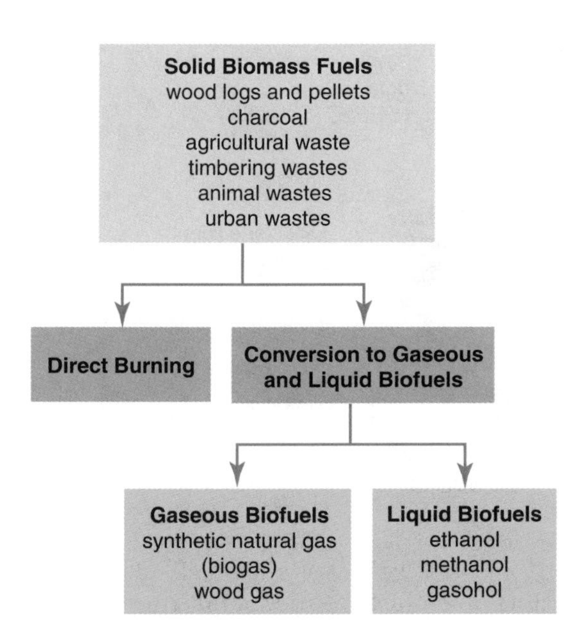

Figure 32-11 Principal biomass fuels and derivatives.

Energy From Waste

Biomass from crop residues (such as sugarcane leaves), manure, and urban waste is plentiful and easily obtained. Instead of presenting a disposal problem, these wastes can be converted into synthetic fuels, or synfuels. (See Figure 32-11.) The most practical way to produce synfuels is bioconversion. In bioconversion, decomposers such as bacteria or fungi carry out chemical changes. Some bacteria decay organic matter in the absence of oxygen, a process called anaerobic fermentation. Anaerobic fermentation converts organic matter into biogas, which is a mixture of methane and carbon dioxide. Biogas can be burned directly to produce energy for cooking and heating or used to produce electricity. Sludge, a waste product of the fermentation process, is rich in nutrients and can be used as a fertilizer for crops.

To manufacture biogas and sludge by anaerobic fermentation, an airtight container called an anaerobic digester, or biogas generator, is used. As shown in Figure 32-12 on page 594, the biogas generator is filled with organic waste and sealed. As biogas is produced, it is collected from the generator. When production of biogas slows, the

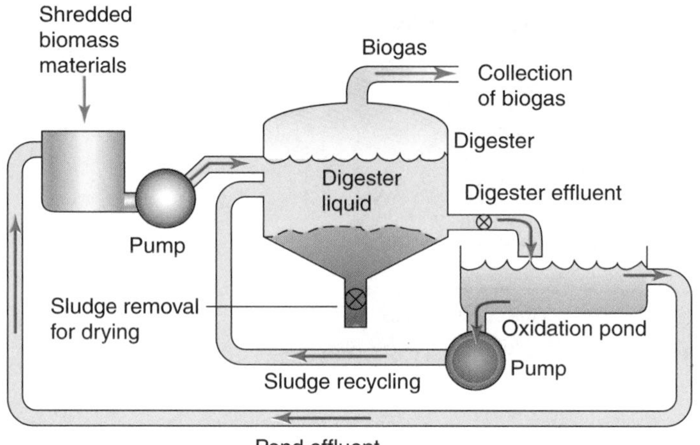

Figure 32-12 **A biogas generator converts organic wastes into fuel that can be used for heating or cooking.**

digester is emptied, and the nutrient-rich sludge is spread in the fields as fertilizer. Digesters are then refilled and the process repeated. This technology is widely used in China, and it is well suited for use in developing countries with limited technology. In more developed nations, such as the United States, the same biological processes can be used to retrieve biomass energy from wastes in landfills.

Biomass can be used to produce a variety of synfuels. These include ethanol and methanol. The process of fermentation has been used for thousands of years to produce alcoholic beverages and breads. Using the same process, alcohol can be produced from any biomass material containing carbohydrate that can be converted to a sugar. The sugar is fermented by yeast to produce alcohol:

$$\underset{\text{GLUCOSE}}{C_6H_{12}O_6} \xrightarrow{\text{YEAST}} \underset{\text{ETHANOL}}{2\ CH_3CH_2OH} + \underset{\text{CARBON DIOXIDE}}{2\ CO_2}$$

Commercial ethanol production uses grains (such as corn or wheat), sugarcane, or sugar beets. The fermentation process breaks down sucrose, a sugar, to produce alcohol and carbon dioxide. The carbon dioxide gas is released as a waste product of this process.

Because it is often produced from grains, ethanol is called grain alcohol. Ethanol can be mixed with gasoline to produce gasohol or used as an additive to enhance the octane rating of gasoline. Although the energy content of ethanol is less than that of gasoline, it has a very high octane rating. Methanol (CH_3OH), which used to be produced from cellulose, the carbohydrate found in wood, is known as wood alcohol. Methanol is extremely toxic; it can cause blindness, liver and kidney damage, and death. The main benefit of alcohols is that because they contain oxygen, when alcohols burn, they produce less carbon monoxide than nonoxygenated fuels.

Although burning biomass fuels releases carbon dioxide into the atmosphere, the amount released is equal to the amount extracted during the growth of the biomass organisms. Thus these fuels are nonpolluting, so long as they are not burned any faster than they are grown.

Advantages and Disadvantages

Very few of our renewable energy resources are economically competitive with nonrenewable sources. Those that are economically competitive are frequently destructive to the environment. Wood can be cut from our remaining forests at low cost. However, the destructive effects of extensive deforestation far outweigh the value of the cheap energy harvested.

Large-scale production of fuel from biomass energy crops is becoming cost-competitive with conventional fuel sources. Grains, sugar crops, rapid-growing species of trees, and algae all show potential as biomass materials that, by fermentation, can be converted readily into alcohol fuel.

32.6 Section Review

1. Describe the problems associated with the use of fuelwood.
2. Describe the process that is used to produce biogas.

LABORATORY INVESTIGATION 32

Energy From the Wind

PROBLEM: *To construct a wind indicator and gather information on the suitability of wind power in your area*

SKILLS: *Measuring, observing*

MATERIALS: *Protractor, fine monofilament line, Ping-Pong ball, needle, stand, clamp, tape, wind speed conversion chart*

PROCEDURE

1. Tape the protractor to the stand as shown in Figure 32-13.
2. With the needle, pierce the ball through the middle. Insert the monofilament line through the ball. Tie a knot in one end of the line so that the ball cannot slip off the line. Tie the other end of the line to the clamp and attach the clamp to the stand as shown in Figure 32-13.
3. Set the indicator up in an open site so that the line holding the ball is perpendicular to the ground. Point the curved edge of the protractor into the wind. Record the angle that the line makes with the perpendicular.
4. Take measurements at four different times during one day or once a day on four different days. Determine the average angle of the line for these four measurements. Use Table 32-1 to find the average wind speed for the site. A wind speed between 16.0 and 31.5 kilometers per hour is most effective for wind power generation.

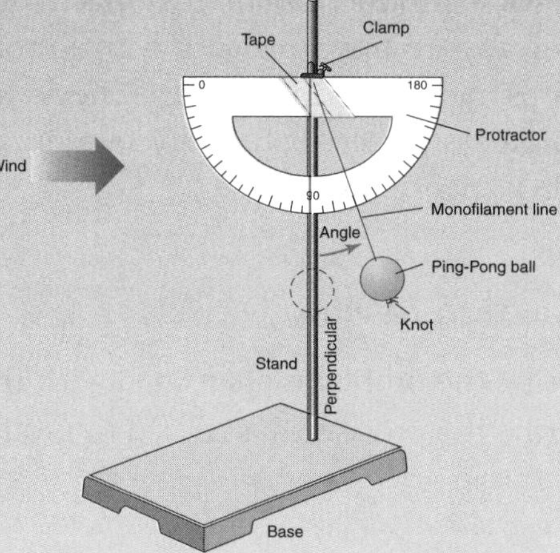

Figure 32-13

TABLE 32–1 WIND SPEED CONVERSION CHART

Angle of Monofilament Line	Wind Speed (k/hr)
0°	0
5°	7.0
10°	13.0
15°	16.0
20°	19.0
25°	21.5
30°	24.0
35°	26.5
40°	29.0
45°	31.5
50°	34.0
55°	37.5
60°	40.0
65°	42.5

OBSERVATIONS AND ANALYSES

1. Measurements of angle: ________, ________, ________, ________
2. Average angle during period of measurement: ________
3. Average wind speed: ________
4. Would your area benefit from wind power? Explain your answer.

EXTENSION

Collect data for at least one month. For your data to be most useful in determining the suitability of wind power for your area, the average annual wind speed is needed.

Chapter 32 Review

Answer these questions on a separate sheet of paper.

Vocabulary

The following list contains all the boldfaced terms in this chapter.

biomass energy, energy efficiency, flywheel, fuelwood, geothermal energy, hydroelectric power, kinetic energy, renewable energy, solar energy, thermal energy, tidal power, wind farm

Fill In

Use one of the vocabulary terms listed above to complete each of the following sentences.

1. Energy that is readily regenerated by natural processes is ________.
2. The ratio of energy delivered to energy supplied is the ________.
3. Energy from moving water is used to spin generators and produce ________.
4. During photosynthesis, plants store solar energy as ________.
5. ________ is heat produced within Earth.

Multiple Choice

Choose the response that best completes the sentence or answers the question.

6. Energy efficiency is *a.* energy output plus energy input. *b.* energy input multiplied by energy output. *c.* energy output divided by energy input. *d.* energy input divided by energy output.
7. Thermal energy is stored in winds as *a.* geothermal energy. *b.* kinetic energy. *c.* electromagnetic energy. *d.* atomic energy.
8. Tidal energy is harnessed in a way most like *a.* solar energy. *b.* wind energy. *c.* hydroelectric energy. *d.* geothermal energy.
9. What part of the world's energy is supplied by biomass? *a.* 20 percent *b.* 40 percent *c.* 60 percent *d.* 80 percent
10. Biogas production in anaerobic digesters requires *a.* biomass and bacteria. *b.* biomass and oxygen. *c.* biomass and heat. *d.* biomass and carbon dioxide.
11. What percent of the solar energy reaching Earth is converted into winds? *a.* 2 percent *b.* 10 percent *c.* 20 percent *d.* 50 percent

12. For every 100 meters of descent below Earth's surface, temperature increases by *a.* 2°C. *b.* 5°C. *c.* 10°C. *d.* 20°C.
13. Why was a tidal power plant *not* built on Passamaquoddy Bay? *a.* Hydroelectric power generates too much pollution. *b.* Hydroelectric power is unreliable. *c.* Construction would be too costly. *d.* Environmental groups blocked its construction.

Short Answer (Constructed Response)

Use the information you learned in this chapter to respond to the following items.

14. What is the source of geothermal energy?
15. Describe a wind farm.
16. What is kinetic energy?
17. Define renewable energy.
18. What is gasohol?

Essay (Extended Response)

Use the information in the chapter to respond to these items.

19. Describe some environmental problems associated with the construction of hydroelectric power plants.
20. Is the construction of wind farms in wilderness areas an environmentally sound procedure? Explain your answer.
21. How is moving water used to supply energy?
22. Explain how tides are used to supply energy.
23. What are some advantages of using biomass fuels?
24. Look around your school to determine whether energy is wasted. Develop a list of energy conservation measures for your school. Meet with school administration and engineering staff to determine the feasibility of instituting your proposals.
25. Based on the information presented in this chapter, construct a bar graph showing the percent of energy the United States produces from the renewable resources mentioned.

Research Project

Use yeast and various grains, vegetable material, fruits, etc., to compare the rate of alcohol production. In your report, describe your hypothesis, procedure, and conclusion.

CHAPTER 33
Solar Energy

When you have completed this chapter, you should be able to:

Distinguish between passive and active solar heating.

Explain why solar energy does not supply all our energy needs.

Sunlight is Earth's most important form of renewable energy. The sun's massive nuclear furnace supplies Earth with virtually unlimited energy. We get light and heat from the sun; it drives the weather; and it provides the energy plants need for photosynthesis. Each year, the sun bathes Earth's surface with thousands of times more energy than Earth's population could use. In this chapter, you will learn how, by harnessing solar energy, it might be possible to meet all our foreseeable energy needs inexpensively and at the same time have a minimal impact on the environment.

33.1 PASSIVE SOLAR HEATING

Solar energy comes from the sun in the form of electromagnetic radiation. Vast amounts of solar energy enter Earth's upper atmosphere each day. As shown in Figure 33-1, the atmosphere absorbs and scatters some of this radiation. Depending on the atmospheric conditions, between 45 and 89 percent reaches Earth's surface. The average amount of radiation falling daily on 1 hectare (a square 100 meters on each side) of the continental United States is equivalent to the energy in 50 barrels of fuel oil. The sun's heat warms Earth's surface, driving the planet's wind systems; warms the oceans, forming currents; and moves moisture into the atmosphere, maintaining the hydrologic cycle. Biomass energy is derived from sunlight by photosynthetic organisms that convert light into chemical energy.

Sunlight is the basic physical ingredient that maintains all life in

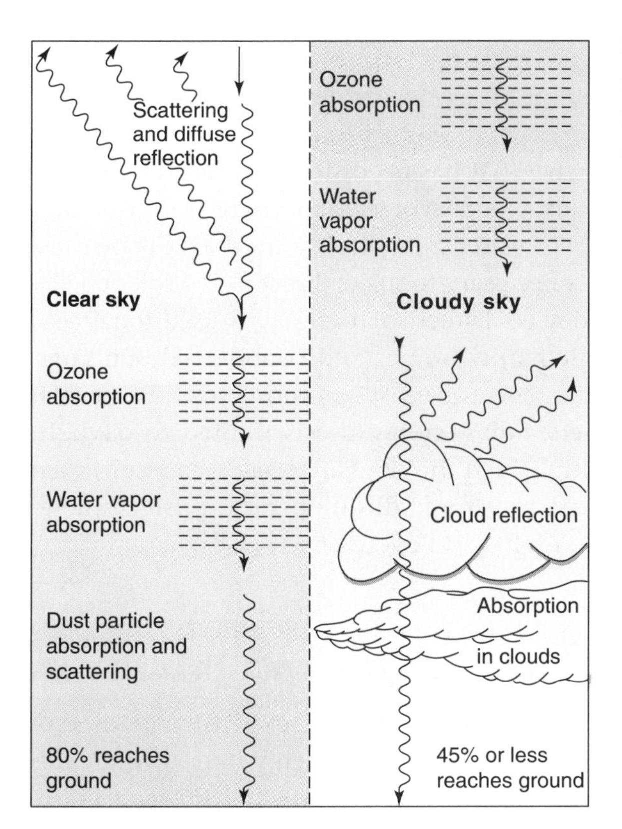

Figure 33-1 Solar radiation can be reflected or absorbed.

Figure 33-2 A radiometer and a light meter convert sunlight directly into other forms of energy.

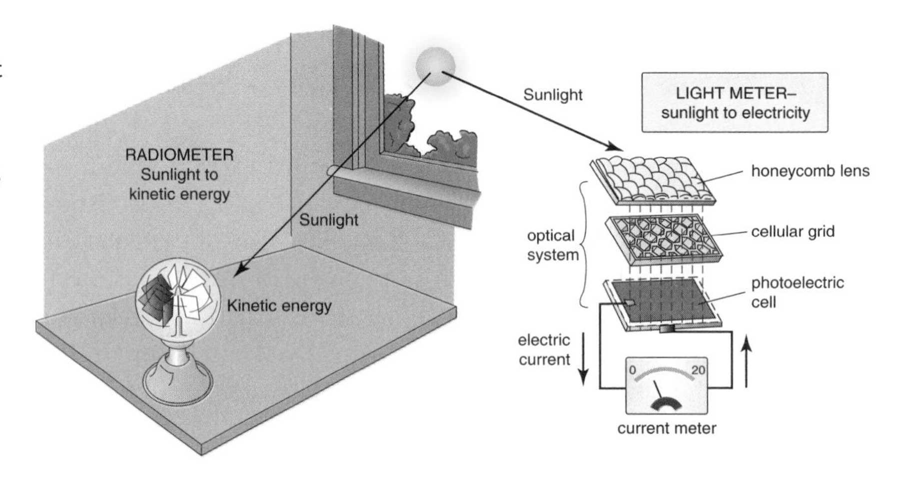

the biosphere. However, the quantity of radiation received at Earth's surface varies from location to location. Solar energy use can be classified as either direct or indirect. Direct use of solar energy occurs where it is absorbed to produce thermal, electrical, or other forms of energy. For example, a **radiometer** is a device that converts sunlight into kinetic energy. The kinetic energy spins the blades of the radiometer. A photographic light meter is an electronic device that converts light into electric energy. (See Figure 33-2.) Indirect use of solar energy occurs when we use energy that has been stored by Earth's environment in the form of winds, biomass, or thermal energy in the oceans.

The main advantages to using solar energy are that it is free and nonpolluting. All that is necessary to make direct use of solar energy is a simple collection device. Direct solar energy is used today as a source of heat for single-family homes and commercial buildings. Using sunlight to heat water is the most common direct use of solar energy by home owners. Solar energy use is limited to daylight hours and periods during which the skies are clear. As a result, solar energy must be collected and stored during periods of sunshine to have it available during nights and overcast days.

Using the Sun to Heat a Home

Passive solar heating is the most direct and inexpensive method of making use of solar energy, since the heat energy is used where it is collected. **Passive solar heating** systems use sunlight to heat air, floors, or walls in a house or to heat water. Energy from the sun is absorbed directly and stored in **collectors**. The collectors slowly radiate the heat at a later time to provide warmth. Thermal insulation maintains interior temperatures. Collectors are constructed of dark-colored materials, which are most efficient at converting light into heat. This process is familiar to anyone who has observed the heating of an asphalt road on a sunny day. To allow for the storage of large quantities of heat energy, efficient collectors tend to be massive. Stone walls and floors, and tanks of water are examples of efficient heat collectors.

A common passive solar heating system is the greenhouse. The glass in a greenhouse is transparent to sunlight but not readily transparent to heat. Sunlight passing through the glass is absorbed by the soil and other materials in the greenhouse (collectors) and converted into heat. The collectors warm up and in turn radiate some of the heat energy absorbed. Trapped inside the greenhouse by the glass, the radiated heat energy in turn warms the air. As long as sunlight is available, warmer temperatures than those outside can be maintained inside the greenhouse.

Well-insulated homes, which are designed using the principle of the greenhouse, can save up to 90 percent of the cost of home heating. As shown in Figure 33-3, these solar-energy homes are

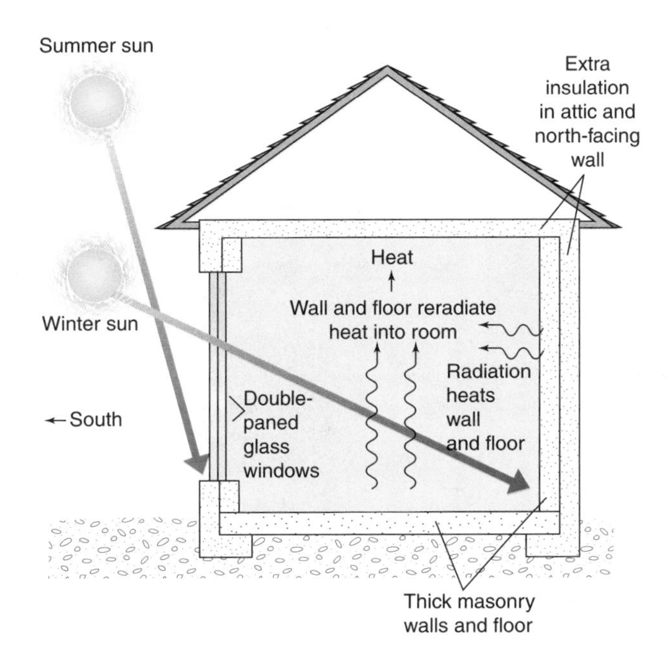

Figure 33-3 **Passive solar heating uses energy absorbed from the sun for heat.**

constructed with south-facing glass walls exposed to a maximum amount of sunlight daily. Collectors, such as walls made of thick masonry or brick, absorb sunlight during daylight hours. The air in the sun-filled room is thus warmed and spreads throughout the house. In the evening, shutters are drawn over the glass walls, trapping the heat in the house and maintaining the temperature during the night. Most homes designed to use passive solar heating have an auxiliary source of heat for use on cloudy days.

Cooking With Solar Energy

The **solar stove**, or solar cooker, is a passive-heating device that has been introduced in developing countries as a means of reducing the reliance on fuelwood and reducing air pollution. One type of solar stove uses parabolic mirrors to concentrate the sun's energy on a cooking area. A **parabolic mirror** is curved to reflect light so that it can be focused at a specific point. This produces high temperatures and bright light, both of which can be dangerous. As shown in Figure 33-4, a simpler, less expensive solar cooker is composed of an insulated box with a dark interior and a transparent glass top. Pots

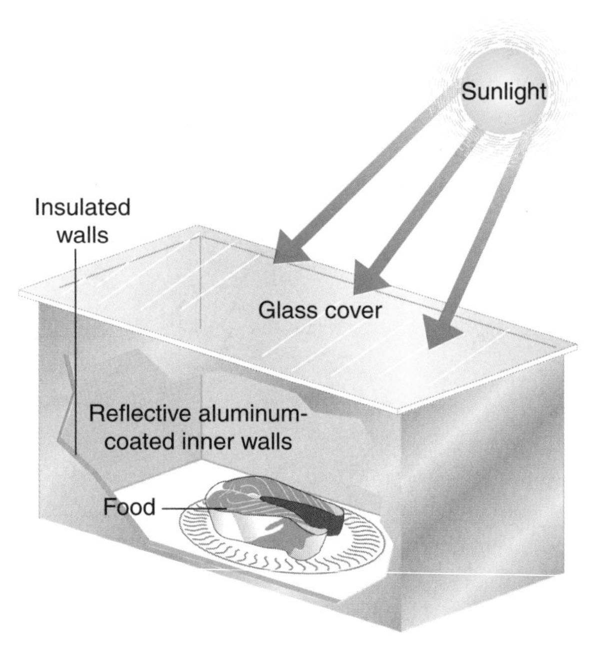

Figure 33-4 A solar cooker uses reflected energy from the sun.

with food are placed inside the box, where temperatures reach 120°C. Cooking time is longer than with conventional methods, but time is saved by not having to search for fuelwood. These solar cookers are most efficient in regions that have little cloud cover throughout the year.

33.1 Section Review

1. Explain the differences between direct solar energy and indirect solar energy.
2. Why are heat collectors usually made of dark-colored materials?
3. Design a solar-heated home that uses collectors made of natural materials found in your area.

33.2 ACTIVE SOLAR HEATING

Active solar heating systems also use solar collectors, but pumps move heat energy through the system. As shown in Figure 33-5, the collectors consist of dark, insulated tanks that contain pipes filled

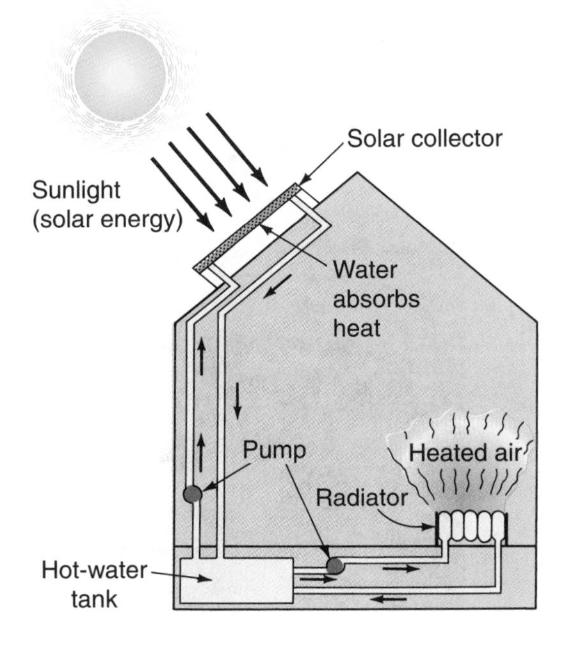

Figure 33-5 **Active solar heating is a nonpolluting way to heat a building.**

with water or another heat-absorbing liquid. Collectors are mounted on the roof, where they can receive maximum exposure to sunlight. The pipes in the collector are connected to a heat storage system inside the building. To increase efficiency, collectors may be motorized so that they always are directed toward the sun. During daylight hours, the collectors absorb sunlight and convert it to heat. The heat is transferred to the fluid in the pipes and then circulated, by means of a pump, to an interior storage tank. A second system circulates the heated water from the storage tank throughout the building by means of pumps. In some systems, the heated liquid is used to warm air, which is circulated by fans. By reversing the process, it is possible to use this system to cool buildings during summer months.

Generating Electricity

Solar generators rely on mirrors to concentrate the sun's energy. They are similar in function to solar stoves. Solar One is a direct solar energy electric-generating plant. This plant operated experimentally for several years in Barstow, California. It consisted of a central tower surrounded by a large array of mirrors that focused solar energy on water tanks on top of the tower. The high temperatures changed the water in the tanks to steam, and the steam drove turbines that generated electricity. (See Figure 33-6.) Since diffuse

Figure 33-6 **This facility in Barstow, California, is a direct solar energy electric-generating plant.**

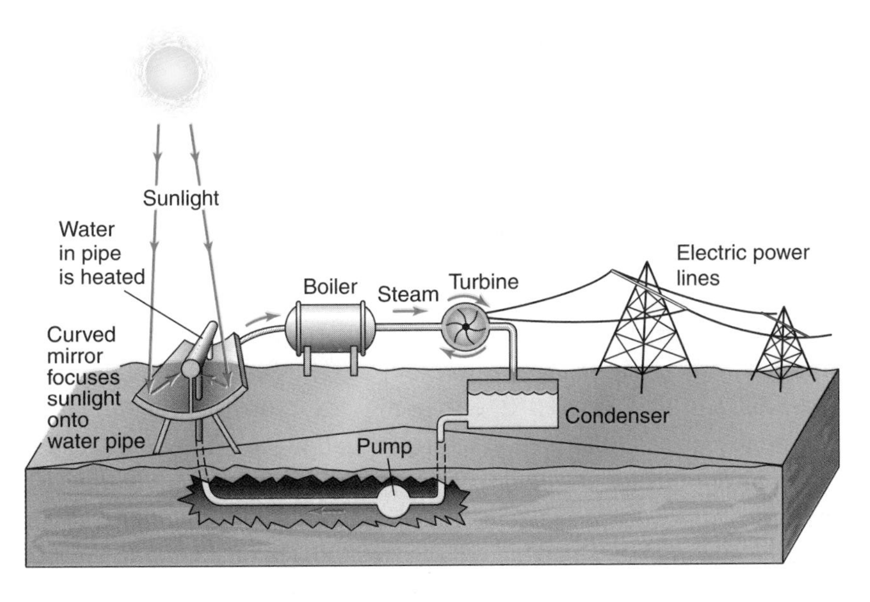

Figure 33-7 **Sunlight can be used to generate electricity.**

light reduces the efficiency of operation, this type of system functions best in areas with long periods of direct sunlight. A system of this type is completely nonpolluting but still has disadvantages.

- The cost of construction is high.
- Large numbers of generators are needed, thus requiring huge tracts of land.
- The mirrors must be cleaned frequently.
- Solar-powered electric motors must be used to keep the mirrors properly aligned with the sun as Earth turns.
- Since solar plants function best in a cloudless environment, they may have to be constructed far from where the energy is needed. This can require long electrical transmission lines and thus increase costs.

Another type of solar power–generating plant uses parabolic reflecting troughs. The troughs focus solar energy on fluid-filled pipes within them. The heated fluid in the pipes is used to drive an engine that generates electrical energy. (See Figure 33-7.)

33.2 Section Review

1. What are some of the disadvantages of solar generators?
2. Why are motorized collectors more energy efficient than stationary collectors?

3. Survey your home to locate areas that would be most suitable for the placement of a solar hot-water-heating system.

33.3 PHOTOVOLTAIC ENERGY

Solar cells, or **photovoltaic cells**, are electronic devices that change sunlight directly into electricity. (See Figure 33-8.) They are composed of thin wafers of different semiconductor materials. A **semiconductor** material contains electrons that are emitted easily. When light energy strikes the cell, electrons are emitted. This process is known as the photoelectric effect. Electrons flow between the two layers, creating an electric current. Solar cells are used to power satellites in space. When completed, the International Space Station will have almost 0.4 hectare (1 acre) of solar panels to provide it with electricity.

Figure 33-8 Photovoltaic cells use light energy to generate electricity.

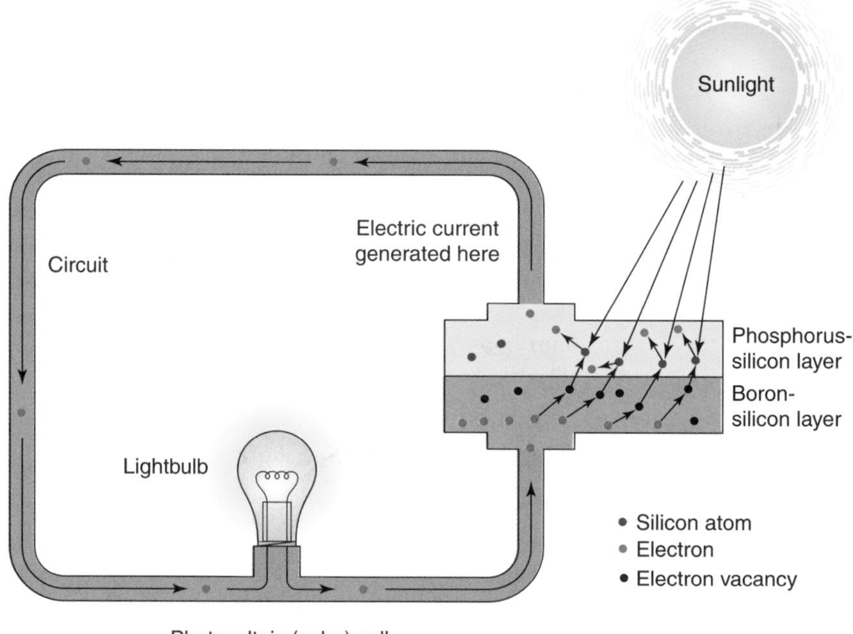

Using Solar Cells

Solar cells are simple devices that operate without any moving parts. They require little maintenance and are nonpolluting; however, production of solar cells uses resources and generates pollution. Solar cells have many applications. They power small handheld calculators and watches, control focusing in cameras, regulate traffic signals, power billboard lighting, and supply energy to operate remote communication devices such as emergency telephones along highways. The main obstacle that prevents solar cells from being used as the source of all our electricity is that they are not economically practical. Even though solar energy is free, the cost of producing solar cells is too high to make photovoltaic energy competitive with more conventional forms of energy production.

The efficiency with which solar cells convert sunlight to electricity poses another obstacle. Solar cells convert energy at a relatively low efficiency. Solar cells convert only 12 to 17 percent of the sunlight that reaches them into usable energy. As a result, large expanses of solar cells are needed to supply even modest energy requirements.

It is projected that sometime in the future engineers will be able to design and build satellites capable of collecting solar energy and transmitting it to Earth. These satellites could supply all of the energy needs to large cities. This form of energy production would require 31 square kilometers of solar panels. The satellite would be placed in orbit at an altitude of 35,400 kilometers, which would allow its revolution to be synchronized with Earth's rotation. Thus it would remain directly over the city that it was supplying with energy. Electrical energy produced by solar cells would be transmitted from the satellite to Earth in the form of microwaves.

33.3 Section Review

1. Why are solar cells considered to be simple devices?
2. Describe some of the solar-powered devices you have used.
3. Why is solar energy an inefficient power source for many urban regions?

33.4 STORING ELECTRICAL ENERGY

The storage of large amounts of electrical energy poses technological and economic problems. This limits the development of photovoltaic technology as well as other forms of electric power production. A storage battery is a device used to store electrical energy. Dry cells, which power flashlights, portable radios, and handheld electronic games, are examples of storage batteries. Traditional lead-acid storage batteries, used in cars and trucks, are heavy, and for their weight and size (mass and volume) they can store only moderate amounts of energy. (See Figure 33-9.) In addition, the lead from discarded batteries can pose a serious environmental health threat.

The standards for energy use in emerging technologies require batteries that meet the following requirements.

- Production costs must be reasonable.
- They must be efficient for an extended lifetime.
- Materials used in construction must be readily available.
- Use and disposal should pose no environmental hazards.
- They must have efficient ratios of energy production to mass and volume.

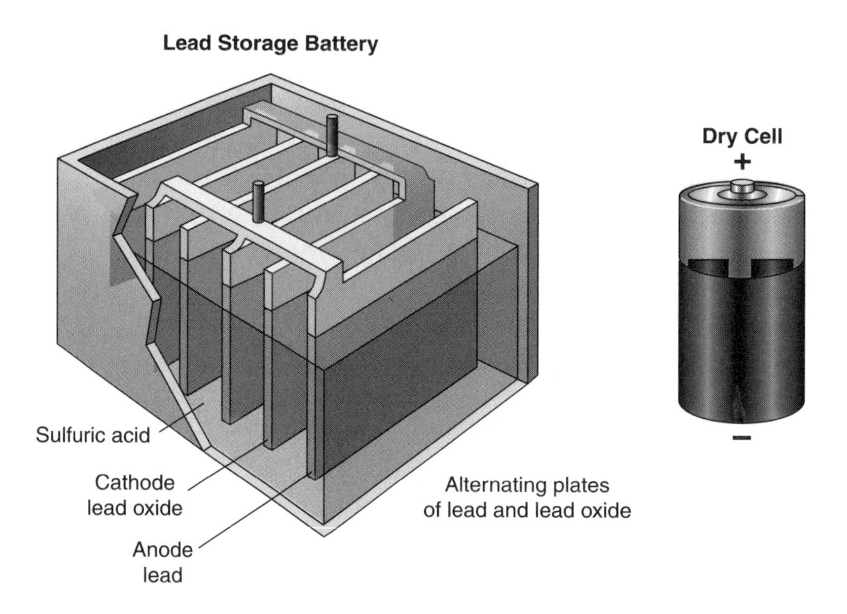

Figure 33-9 Storage batteries change energy in chemical bonds to electric energy.

A method that has been proposed for the storage of photovoltaic energy is to use the electricity for the hydrolysis of water. **Hydrolysis** is chemically breaking down water into hydrogen and oxygen gases. These gases can be liquefied and stored, like natural gas, to make them easier to distribute. Hydrogen gas is highly explosive, and oxygen supports combustion; therefore, they must be handled carefully. Hydrogen can be used as a fuel for internal combustion engines to run vehicles or to drive turbines that produce electricity. When hydrogen is burned it produces heat and gives off water as its only by-product; therefore, it is nonpolluting.

Fuel Cells

A fuel cell is an energy-producing device that acts like a battery. Unlike batteries, fuel cells neither run down nor require recharging; they produce electrical or heat energy as long as there is fuel. Fuel-cell systems can utilize the hydrogen from any hydrocarbon source, such as methane, alcohol, or gasoline. The energy in these hydrocarbon sources originated as sunlight, which was stored as chemical energy during photosynthesis. Fuel-cell systems are clean and nonpolluting, since they do not rely on combustion. In the fuel cell, the overall reaction is combining hydrogen with oxygen from the air to form water. (See Figure 33-10.) Since fuel cells operate silently, they also reduce noise pollution. The waste heat from fuel cells can provide hot water and space heating for homes.

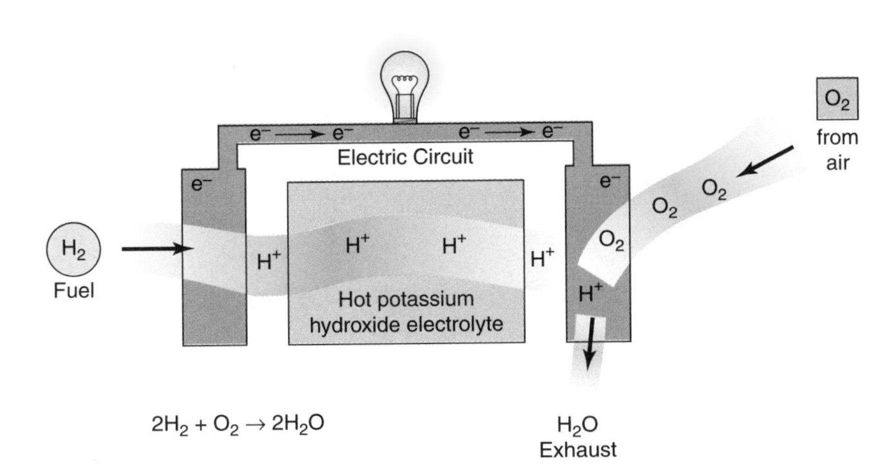

Figure 33-10 A fuel cell provides energy without causing air, water, or noise pollution.

There are a variety of applications for fuel-cell technology. In wastewater treatment plants and landfills, fuel cells use methane gas, produced by decomposition, to generate electricity. Many businesses, schools, and hospitals throughout the world use fuel cells to generate power or to supply backup power in emergencies. Fuel cells can reduce energy costs by as much as 40 percent. In 2003, President Bush set a goal to make fuel-cell vehicles practical and affordable by 2020 and reduce the need for imported oil by making hydrogen fuel available at corner gas stations. All major automobile manufacturers are developing vehicles that use fuel-cell technology. Electronics manufacturers are developing miniature fuel cells to power cell phones, laptops, cameras, recorders, and other devices.

33.4 Section Review

1. Why are automotive storage batteries inefficient for use in storing photovoltaic energy?
2. What are some requirements that batteries should meet in emerging technologies?

LABORATORY INVESTIGATION 33

Air Pollution and Solar Energy

PROBLEM: *What effect does air pollution have on the amount of solar energy that reaches Earth's surface?*

SKILLS: *Measuring, observing*

MATERIALS: *Microscope substage lamp, 500-mL beaker, radiometer, 1% methylene blue, meterstick, graduated cylinder, watch, water*

PROCEDURE

1. Copy Table 33-1 on page 614 into your notebook. Record all data in the table.
2. Arrange a microscope lamp, a 500-mL beaker, and a radiometer, as shown in Figure 33-11 on page 614. These items represent the sun, Earth's atmosphere, and Earth, respectively. The distance between the lamp and the radiometer should be about 250 centimeters. Do not perform this experiment in bright sunlight, as it will interfere with your observations.
3. Fill the 500-mL beaker with water.
4. Make sure the vanes on the radiometer are not moving, then turn on the microscope lamp. Count and record the number of turns, or revolutions, the radiometer makes in 1 minute (rpm). Repeat and record this measurement two more times. Average and record your results.
5. Using the graduated cylinder, measure 1 mL of methylene blue. Add it to the water in the beaker and stir the solution. Repeat and record the measurements as in step 4. Calculate and record the average rpm as before.
6. Add another 1 mL of methylene blue to the water in the beaker. Repeat your measurements and record your data.
7. Add an additional 3 mL of methylene blue. Repeat your measurements and record your data.

TABLE 33-1 AFFECT OF COLOR INTENSITY ON RPMS

Trial	Rpm Clear Water	Rpm 1 mL Dye	Rpm 2 mL Dye	Rpm 5 mL Dye
1				
2				
3				
Total rpm				
Average rpm				

OBSERVATIONS AND ANALYSES

1. At which concentration of methylene blue does the most energy reach the radiometer? The least?
2. Construct a bar graph to display the data you collected.
3. If the methylene blue represents pollutants in the atmosphere, how do these pollutants affect the amount of solar energy reaching Earth?

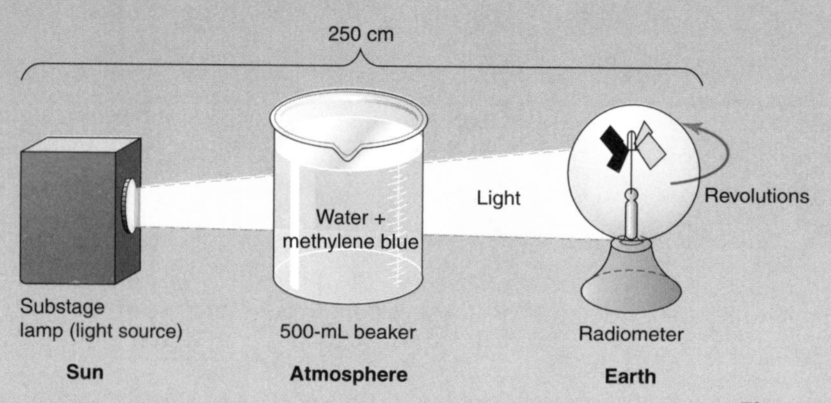

Figure 33-11

Chapter 33 Review

Answer these questions on a separate sheet of paper.

Vocabulary

The following list contains all the boldfaced terms in this chapter.

active solar heating, collectors, hydrolysis, parabolic mirror, passive solar heating, photovoltaic cells, radiometer, semiconductor, solar cells, solar stove, solar generators

Fill In

Use one of the vocabulary terms listed above to complete each of the following sentences.

1. ________ is the most direct and inexpensive method of using solar energy in a home.
2. Devices that convert sunlight into stored heat are ________.
3. Electronic devices that change sunlight directly into electricity are ________, or ________.
4. ________ rely on mirrors to concentrate the sun's energy.
5. A(an) ________ is a material that contains electrons that are easily released.

Multiple Choice

Choose the response that best completes the sentence or answers the question.

6. Which of the following is *not* a disadvantage of solar generators? *a.* Cost of construction is high. *b.* Mirrors must be kept clean. *c.* They work best in a cloudless environment. *d.* Solar energy is renewable.
7. An electronic device that works by converting sunlight into electrical energy is a *a.* radiometer. *b.* light meter. *c.* solar collector. *d.* solar cooker.
8. A semiconductor material must be able to *a.* hold electrons. *b.* emit electrons. *c.* use protons. *d.* emit protons.
9. The waste product from the combustion of hydrogen is *a.* oxygen. *b.* carbon dioxide. *c.* particulates. *d.* water.
10. An example of the indirect use of sunlight to supply energy is *a.* biomass energy. *b.* tidal energy. *c.* hydroelectric energy. *d.* all of the above.
11. Solar energy can be used to produce *a.* light. *b.* heat. *c.* electricity. *d.* all of the above.
12. The glass of a greenhouse *a.* traps light. *b.* emits light. *c.* traps heat. *d.* reduces heat.

Short Answer (Constructed Response)

Use the information you learned in this chapter to respond to the following items.

13. How does a photovoltaic cell work?
14. Why are solar collectors often motorized?
15. Name a passive-cooking device that eliminates the need for wood.
16. What is a radiometer?

Essay (Extended Response)

Use the information in the chapter to respond to these items.

17. What are some inexpensive methods of converting an existing home into one that relies on solar energy?
18. How is a passive solar heating different from active solar heating?
19. How would the switch from fuelwood to solar stoves benefit the whole biosphere?
20. What are the advantages and disadvantages of using hydrogen as a fuel?
21. In the library or on the Internet, research and report on Solar Two, the next phase of the experimental direct solar energy-generating plant in Barstow, California.

Research Projects

- Research and report on the use of solar cells in spacecraft.
- Explore and report on the recent advances in semiconductor technology. What is the energy efficiency of some of the new semiconductors?

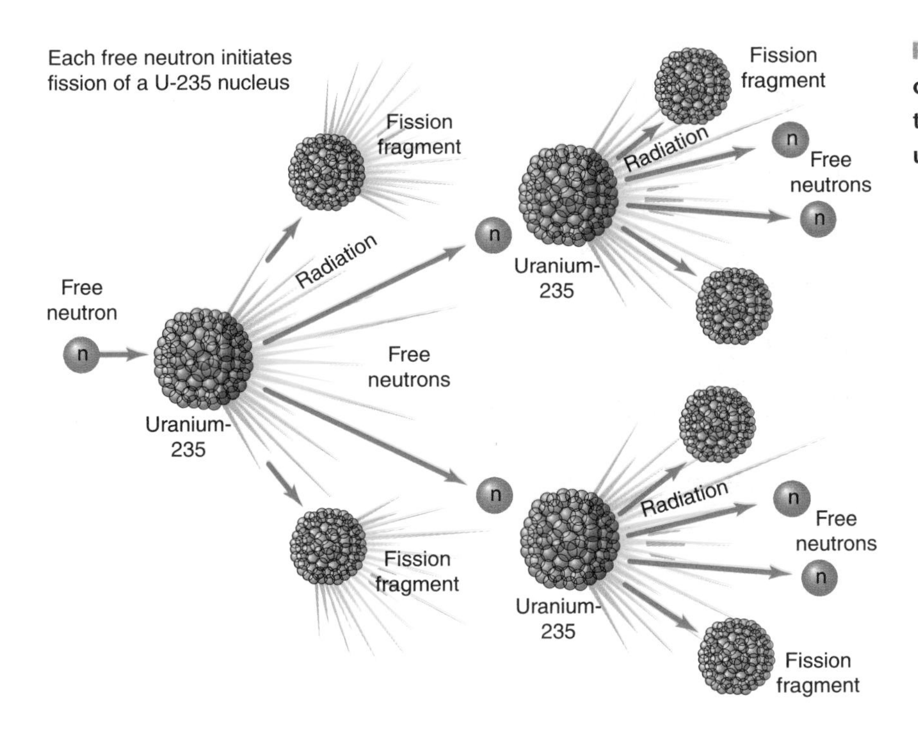

Figure 34-1 This chain reaction shows the breakdown of uranium-235.

walls to absorb the radiation produced. A cooling system bathes the core in a liquid that absorbs heat and regulates the flow of energy. Heat generated by the reactor is used to make steam to drive the turbines that produce electricity. (See Figure 34-2 on page 620.)

Other types of reactors produce weapons-grade materials for the military or radioisotopes used in medicine and research. **Breeder reactors** produce more nuclear fuel than they consume. They make radioactive plutonium and thorium from abundant uranium. The starting material used in these reactors is spent plutonium from conventional reactors. There are many safety problems in the use of breeder reactors. The most serious problem is that they produce excess plutonium, one of the most toxic substances known and a component of nuclear weapons.

A primary environmental concern arising out of nuclear energy production is the accumulation of radioactive wastes, called **spent fuel**. Composed of used fuel rods from reactors, these wastes remain highly radioactive and are dangerous for thousands of years. Since spent fuel and decommissioned reactors are sources of long-term radiation, the methods used for their disposal lead to controversy with regard to the impact they have on the environment.

Figure 34-2 A fission reactor uses a controlled chain reaction as a source of energy.

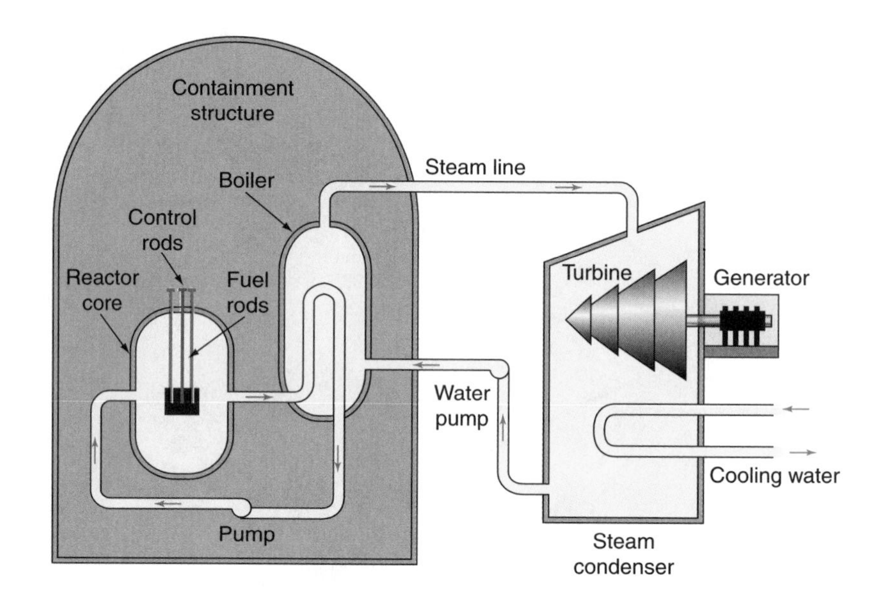

The Decline of Fission Power in the United States

At first, electricity produced by nuclear energy was much cheaper than energy derived from fossil fuels. But operating costs have risen rapidly, and electricity from nuclear power plants is now more than twice as expensive as that produced by other fuels. From 1970 to 1975, in the United States, plans were filed to build an average of 12 nuclear power plants per year; from 1975 to 1978, the number dropped to 9 plants per year. Since 1978, no new plants have been planned and several that had been planned were canceled.

Several factors have contributed to the decline of the nuclear industry in the United States. The main drawbacks to nuclear power are increased construction costs, public opposition to nuclear power, and concerns about safety and the disposal of spent fuel. There are 104 nuclear plants in operation in the United States, most of them located in the eastern part of the country. Approximately 20 percent of our energy production comes from nuclear power.

By 2000, 440 nuclear plants generated almost 20 percent of the electrical power in the world. In some industrialized nations, the use of nuclear energy is very high. France, for example, generates more than 75 percent of its electrical energy with its 59 nuclear reactors. However, developing nations cannot afford this expensive power source. Fewer nuclear plants are now being constructed, and attention is focused on other means of electrical energy generation.

34.1 Section Review

1. Explain how control rods are used to regulate nuclear reactors.
2. What are some of the factors that have caused a decline in the nuclear power industry?
3. What are some of the problems associated with the use of breeder reactors?

34.2 OBSTACLES TO NUCLEAR POWER PRODUCTION

The world is naturally radioactive. In the United States, 87 percent of all radiation exposure comes naturally from rocks and soil and from outer space. Air, soil, and water contain small amounts of naturally radioactive elements. Crops may absorb these elements. When these crops are consumed as food, the radioactive materials are absorbed and remain in our body.

Radioactive elements, or **radioisotopes**, emit fast-moving particles and rays called radiation. Radiation can zip through tissues and cells, causing damage to cellular structures and changes in genes. This can interfere with normal life processes and lead to disease and death. Everyone slowly accumulates the effects of natural radiation over time. (See Table 34-1 on page 622.)

About 11 percent of the radiation we are exposed to comes from medical sources, such as X rays. Less than 2 percent is from nuclear power plant wastes. Additional hazards from nuclear energy production only add to our exposure.

The Waste Problem

Nuclear wastes often contain radioactive isotopes of common elements, such as iodine and cesium. Radioactive isotopes that have a

TABLE 34-1 PROPERTIES OF NUCLEAR RADIATION

Radiation	alpha (α)	beta (β)	gamma (γ)
Identity	helium nucleus	electron	high-energy radiation
Approximate velocity (c*)	≤10% c	≤90% c	100% c
Shielding required	paper, cloth	30 cm wood, aluminum foil	10 cm lead, 30 cm concrete
Penetrating power	low, stopped by skin	medium, ~1 cm of flesh	high, passes through body

*c is the symbol for the velocity of light, 3×10^8 m/s^2.

long half-life can persist in the environment for many years. The half-life of an element is the amount of time it takes for half of the atoms of a radioactive substance to decay into other elements. The half-life of iodine-131 is 8 days, while cesium-137 has a half-life of 30 years. After several weeks, most iodine-131 has been changed into harmless substances, but cesium-137 persists in the environment and remains dangerous for more than 100 years. The longer a radioisotope persists, the more danger it poses to all living things. (See Table 34-2.)

The technical problems of handling nuclear wastes can be solved. However, the problem of where to put the wastes is huge. Attempts have been made to dispose of this waste in the safest possible manner, usually by burying it in chambers deep within Earth's crust. But radiation danger exists in the areas that surround these sites. In addition, there is a potential for accidents caused by natural occurrences such as earthquakes. Any damage to waste containers can have serious consequences. Containers have been known to corrode or break and leak radioactive materials into the environment. These materials then enter the ecosystem through natural processes, such as circulation of groundwater. There are also problems, such as traffic or rail accidents, associated with transporting wastes to a storage facility.

Radioactive wastes are classified by the level of radiation they emit. High-level wastes remain dangerous for tens of thousands of years. A good portion of high-level wastes consists of spent fuel rods. These wastes are usually encased in glass blocks, a process

TABLE 34-2 HALF-LIFE OF SOME RADIOACTIVE ISOTOPES

Isotope	Half-Life
Uranium-238	4.5 billion years
Plutonium-239	25,000 years
Carbon-14	5700 years
Radium-226	1620 years
Cobalt-60	5.25 years
Polonium-210	138 days
Phosphorus-32	14.3 days
Iodine-131	8.07 days
Sodium-24	14.9 hours
Iron-53	8.9 minutes

called vitrification, and stored until a safe method of disposal can be devised. Intermediate-level wastes are dangerous for several hundred years. They usually are buried in chambers deep within Earth's crust. Low-level wastes are those that remain dangerous for several decades and whose total radiation hazard is small. However, when they are stored or disposed of in one location, their cumulative effects can be great. At one time, low-level wastes were stored in drums and buried or dumped at sea. But it was discovered that environmental conditions could cause drums to leak and contaminate the environment. As part of an international treaty protecting the marine environment, the dumping of radioactive materials at sea was banned in 1983.

Since the beginning of the Nuclear Age, millions of liters of radioactive wastes have accumulated in temporary storage facilities around the country. In addition, spent fuel rods pile up in reactor buildings. Some of this material is stored in disposal sites, where there is a question of safety. Inappropriate disposal methods and leaking storage tanks are ever-growing problems. There is a possibility of wastes exploding and spreading radioactive materials over a wide area. A waste dump in Kyshtym, Siberia, exploded in 1957, contaminating hundreds of square kilometers with radiation and poisoning thousands of people. There is a high-level radioactive waste facility in Hanford, Washington, near several large metropolitan areas.

As nuclear plants continue to generate waste and spent fuel rods at reactor sites, the need for high-level radioactive waste disposal sites grows. After 25 to 50 years of operation, nuclear reactors become radioactive and must be decommissioned, or shut down. No one has yet devised a technology to make reactors safe after they have been decommissioned. Waste stockpiles continue to grow without a solution for their disposal.

Radioactive waste disposal problems are not unique to the United States. It is feared that the former Soviet Union has permanently contaminated the oceans surrounding it. Millions of liters of radioactive wastes and at least 18 decommissioned nuclear reactors have been dumped into these waters. These nations still have numerous storage facilities that contain aging tanks overflowing with radioactive wastes. However, there are insufficient funds to monitor or maintain their condition. The entire world is threatened by this crisis, since few disposal options are available.

Nuclear Waste Policy Act

The Nuclear Waste Policy Act (NWPA) is a federal law enacted in 1982. This act set up a schedule to identify and explore potential sites for the construction of an underground radioactive waste storage facility. It was proposed that wastes could be buried deep within Earth's crust where they would remain unexposed to the effects of groundwater for tens of thousands of years until they became safe.

In February 2002, after 20 years of debate and a $4 billion study, President George W. Bush designated Yucca Mountain in Nevada as the repository for the nation's nuclear waste. Congress approved the facility, designed to store up to 77,000 tons of waste, in July 2002. Yucca Mountain was chosen because it is an isolated location on government-owned land. (See Figure 34-3.) Now there will be a place to store the nuclear waste that has piled up at 131 sites in 39 states. It is hoped that the Yucca Mountain site can open by 2010, because by that time the United States will have produced enough nuclear waste to fill the facility.

The state of Nevada and concerned citizen groups have voiced many safety concerns.

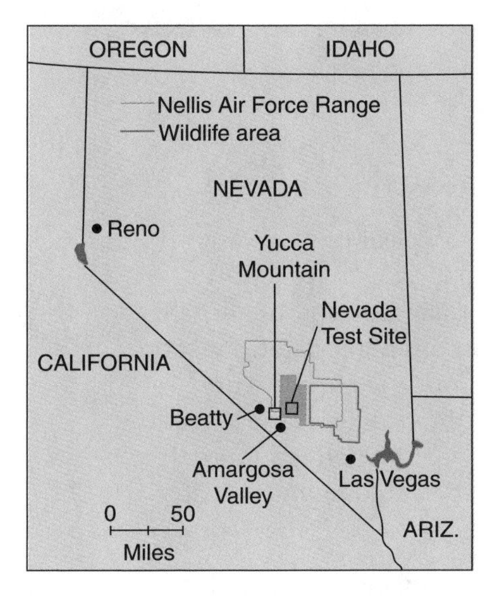

Figure 34-3 The Yucca Mountain nuclear waste facility located on the Nellis Air Force Range.

- Accidents during the transport of nuclear waste
- Possible damage to the site due to natural disasters (There are seven volcanoes and numerous fault lines in the area.)
- Danger to future inhabitants of the area, since it will remain radioactive for many tens of thousands of years
- The potential for the containers to leak
- The effects of future climate change on the site

The Incident at Three Mile Island

On March 28, 1979, one of the reactors at the Three Mile Island Nuclear Plant in Harrisburg, Pennsylvania, suffered a partial meltdown. (See Figure 34-4 on page 627.) Five days passed before the accident was safely under control. Human error and the failure of safety devices built into the cooling systems caused the incident. Overheating caused the failure of pressure-release systems, which led to a gas leak within the containment building. With systems out of operation, core temperatures rose rapidly, which caused the partial meltdown. A **meltdown** is the melting of the reactor core and its fuel, caused by a loss of coolant. This is the worst of all possible reactor accidents. If the molten material escapes from the containment

DISCOVERY

The Chernobyl Disaster

On April 26, 1986, core number 4 of the Chernobyl nuclear power plant in Ukraine experienced two huge explosions. Dangerous radioactive debris spread over the immediate region and was carried to Europe and the rest of the world by atmospheric circulation. Radioactive isotopes infiltrated the food supply and seeped into groundwater. Within days, much of Europe was recording high levels of radiation. The first news of the disaster came several days after the actual event. However, it was from Swedish scientists more than 2000 kilometers away from the explosion who had detected high levels of radioactive fallout coming from Ukraine.

The final death and injury toll from the effects of the radiation could only be guessed at. Even after more than 15 years, the toll remains uncertain, since new cases of radiation-caused leukemia and cancers continue to appear. In some regions, the rate of thyroid cancer in children is 20 times higher than before the accident. Radioactive isotopes in the soil and water still pose constant threats to human health.

When the Iron Curtain lifted, the world finally learned the extent of the damage the explosions had caused. A meltdown, the worst possible nuclear plant accident, had occurred. The core of the reactor had melted, and the radiation containment building had been blown apart. More than 135,000 people had to be evacuated from their homes in the surrounding villages and 20,000 had become ill from radiation exposure.

After the Chernobyl disaster, many governments began to rethink the benefits and dangers of nuclear power. The problems associated with the disposal of waste and the chance of accidents became all too real. The Swedish government, which had decided to gradually phase out nuclear power before the accident, increased the rate of their planned phase-out.

Figure 34-4 At Three Mile Island, Pennsylvania, the core of reactor 2 overheated, leading to the release of radioactive gases.

building, it can have severe environmental consequences. A gas explosion that released a small amount of radioactive material into the environment followed the partial meltdown. Though environmental contamination was slight, a reactor was destroyed. This nearly bankrupted the utility that owned the plant.

The accident led the Nuclear Regulatory Commission to issue new safety regulations and mandate changes for all existing plants. This incident stimulated a great deal of antinuclear sentiment. It was an important factor in checking the growth of nuclear power in the United States.

34.2 Section Review

1. Why should we protect the environment from exposure to radioactive isotopes with a long half-life?
2. What environmental threats are posed by the continued production of nuclear energy?
3. Why are some people opposed to the construction of a nuclear waste storage facility in Nevada?

34.3 NUCLEAR FUSION

Nuclear fusion is the process that combines the nuclei of small atoms to form larger atoms. (See Figure 34-5.) Hydrogen is the fuel for fusion reactions within the sun and other stars. Fusion in the sun produces the energy that supplies Earth with heat and light. The fusing of atomic nuclei requires extremely high temperatures and pressures. Fusion reactions release a far greater amount of heat energy than fission reactions.

The energy potential from commercial fusion power plants could satisfy the world's needs far into the future. Nuclear fusion power promises to be a safer, cleaner, nonpolluting source of energy than present energy sources. Fusion reactors use atoms of deuterium and/or tritium, heavy isotopes of the element hydrogen. The most common form of hydrogen, hydrogen-1, has a single proton in its nucleus; while **deuterium**, hydrogen-2, has one proton and one neutron; and **tritium**, hydrogen-3, has one proton and two neutrons in its nucleus. Deuterium and tritium are present in water in very small quantities. Because these substances can be extracted from seawater, fusion power offers a virtually endless source of inex-

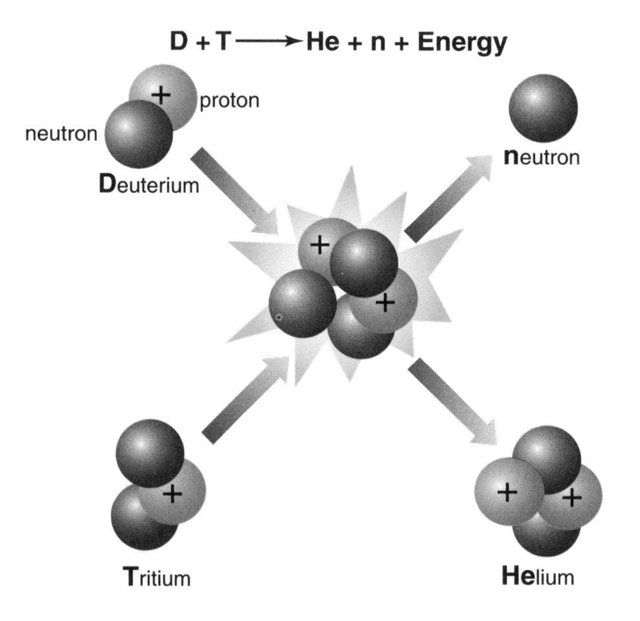

Figure 34-5 **In nuclear fusion, a deuterium (1*p* + 1*n*) nucleus fuses with a tritium nucleus (1*p* + 2*n*) to form helium (2*p* +2*n*) and a neutron (*n*).**

pensive energy. One gram of deuterium has the potential to yield as much energy as 10 tons of coal.

Reactor walls made of matter could not survive exposure to the high temperatures necessary for fusion. Therefore, other means are needed to confine a fusion reaction. There are two main types of fusion reactors: magnetic confinement reactors and inertial confinement reactors.

Tokamak fusion reactors utilize magnetic confinement. A tokamak consists of a torus, or doughnut-shaped chamber in which a plasma is confined by a very strong magnetic field. The plasma is an extremely hot ionized gas that can reach temperatures necessary for fusion. The magnetic field keeps the plasma in a constantly looping path, which does not allow it to touch the walls of the chamber. The magnetic field is produced by an electric current passing through coils of wire wound around the reaction chamber. Several versions of these devices have been produced, but none has been able to sustain a reaction or produce more energy than it consumed. (See Figure 34-6.)

Inertial confinement uses a high-energy laser or ion beam to fuse atoms in a deuterium-tritium pellet. The beams evaporate the outer

Figure 34-6 **Tokamak fusion reactors, such as this one in Lisbon, Portugal, use a magnetic field to contain the plasma.**

layer of the pellet, producing energetic collisions that drive part of the pellet inward. The density and temperature of the interior of the pellet increase, reaching the ignition point for fusion. Lawrence Livermore Laboratory at the University of California in Berkeley has two experimental laser-fusion devices called Shiva and Nova.

Pluses and Minuses

Although fusion yields more energy than fission, it is much more difficult to control. Fusion reactions require extremely high temperatures in the millions of degrees Celsius and pressures of billions of atmospheres. Due to these factors, containment of fuel within the walls of reactors is a problem that has yet to be solved completely. The present cost and energy efficiency of fusion devices are both technologically and economically impractical. As yet, researchers are unable to pass the break-even point, where more energy is produced by the system than it takes to start it.

The production of fusion energy would have little negative impact on the environment. Fusion reactions produce few radioactive wastes, and those produced are low-level wastes. In addition, the generation of energy from fusion reactions eliminates the production of fissionable materials that could be made into bombs. Importantly, fusion reduces the potential for serious nuclear accidents, because runaway chain reactions cannot occur. Elimination of the problems associated with maintaining a hazardous fuel supply (uranium or plutonium) is another positive aspect of fusion power.

34.3 Section Review

1. In what ways is fusion power production preferable to that of fission power?
2. What are some factors holding back the development of fusion power reactors?

LABORATORY INVESTIGATION 34

Model a Chain Reaction

PROBLEM: *How can you use dominoes to model a chain reaction?*

SKILLS: *Observing, modeling*

MATERIALS: *Box of dominoes*

PROCEDURE

1. Line up the dominoes in five or six rows. Place one domino in the first row, two dominoes in the second row, three in the third, etc., so they form an equilateral triangle with the first domino at the apex. Each row should be about one-half a domino length behind the row preceding it. (See Figure 34-7.)
2. Knock over the first domino so that it strikes the two in the second row. Observe what happens.
3. Repeat steps 1 and 2 with the domino rows a full domino length apart.
4. Remove one of the center dominoes from each row, except the first and second. Repeat steps 1 and 2. This represents a moderated chain reaction.

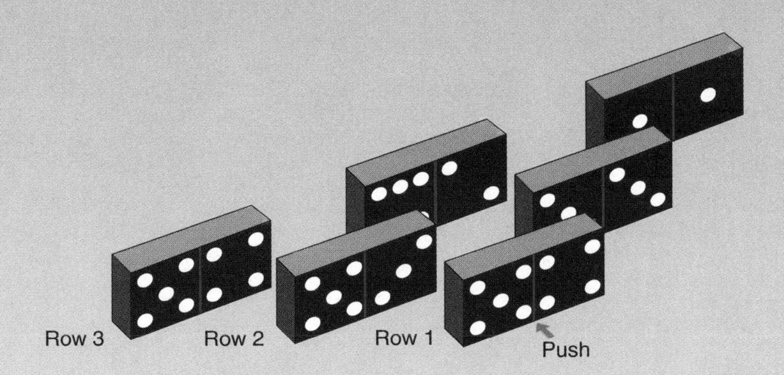

Figure 34-7

OBSERVATIONS AND ANALYSES

1. Describe your observations of the chain reaction produced in step 2.
2. Describe the chain reaction produced in step 3.
3. Describe the moderated chain reaction produced in step 4.
4. Explain what was done to moderate the chain reaction in step 4.
5. Do the dominoes represent atoms or neutrons participating in a reaction? Explain your answer.

Vocabulary

The following list contains all the boldfaced terms in this chapter.

breeder reactors, chain reaction, control rods, core, deuterium, isotopes, meltdown, neutrons, nuclear fission, nuclear fusion, nuclear reactor, protons, radioactive, radioisotopes, spent fuel, tritium

Fill In

Use one of the vocabulary terms listed above to complete each sentence.

1. The process of "splitting atoms" is also known as ______.
2. Reactors that produce more nuclear fuel than they consume are called ______.
3. ______ are atoms of the same element that have a different number of neutrons in their nucleus.
4. A(an) ______ is the worst possible nuclear reactor accident.
5. Used fuel rods removed from nuclear reactors are called ______.

Multiple Choice

Choose the response that best completes the sentence or answers the question.

6. The fissionable radioisotope used as fuel in fission reactors is *a.* uranium-235. *b.* plutonium-235. *c.* uranium-238. *d.* plutonium-238.
7. The fuel for a fusion reactor could be *a.* hydrogen and helium. *b.* uranium and tritium. *c.* plutonium and deuterium. *d.* deuterium and tritium.
8. Radioactive iodine has a half-life that is a few *a.* minutes. *b.* hours. *c.* days. *d.* years.
9. How many years is the useful life of a nuclear fission reactor? *a.* 5 to 10 *b.* 10 to 20 *c.* 25 to 50 *d.* 50 to 100
10. The accident at Three Mile Island nuclear plant resulted in *a.* closing all nuclear power plants. *b.* new safety regulations and changes in all plants. *c.* a complete meltdown. *d.* demand for more nuclear power plants.
11. The deuterium nucleus contains *a.* one proton only. *b.* one neutron only. *c.* one proton and one neutron. *d.* one proton and two neutrons.

Chapter 34 Review

Answer these questions on a separate sheet of paper.

12. How much of the average person's total radiation exposure comes from medical sources, such as X-ray machines?
    a. 2 percent *b.* 11 percent *c.* 25 percent *d.* 87 percent
13. Which of the following is a radioisotope that poses a long-term environmental hazard? *a.* iodine-131 *b.* cesium-137 *c.* uranium-238 *d.* carbon-12

Short Answer (Constructed Response)

Use the information you learned in this chapter to respond to the following items.

14. What is a radioisotope?
15. How is fission different from fusion?
16. What are the three isotopes of hydrogen?
17. Where will the nation's nuclear waste facility be constructed?
18. List three safety concerns expressed by opponents of the nuclear waste facility.
19. List two technological obstacles to building nuclear fusion reactors.

Essay (Extended Response)

Use the information in the chapter to respond to these items.

20. Describe the role of neutrons in a chain reaction.
21. How is the chain reaction regulated in a fission reactor?
22. In what ways are breeder reactors different from nuclear power reactors?
23. How does nuclear radiation harm cells?
24. Why was the ocean dumping of low-level radioactive wastes discontinued?
25. What are some of the arguments against fission power?

Research Project

Contact your state's atomic energy commission to find the answers to the following questions. How does your state plan to transport radioactive wastes to the nation's storage facility? What other states will be involved in the transport process? Are there any geologic or atmospheric hazards along the transport route? How will ecosystems be affected by the transport of radioactive materials. On a state map, plot the location of your state's nuclear power plants. Trace the transport route on a map of the United States.

UNIT NINE

PRESERVING EARTH FOR THE FUTURE

The human species has existed for but a brief period in Earth's long history, yet we have played a major role in shaping our planet's environment. Humans may face the loss of the very resources that make Earth unique in the solar system. Today, we pave over, dig up, drain, and pollute the natural areas that provide habitats for Earth's creatures.

Scientists have learned that, with a little help from its human inhabitants, Earth has the ability to heal itself. However, if we are unable to slow our headlong rush toward environmental disaster, we may, in time, destroy our own life-support system and become, like the dinosaurs, yet another entry in a long list of extinct species recorded in Earth's history book.

CHAPTER 35
Destroying Habitats

When you have completed this chapter, you should be able to:

Explain why forests are important to all life.

Identify the causes of desertification.

Discuss the problems caused by the destruction of wetlands.

In the United States, the quality of life for most people is good. However, this is not true for many people in other parts of the world and even for some people in this country.

Advances in modern technology often have increased the rate of environmental destruction. The commercial development of wetlands, areas that are often the nurseries for marine life, has harmed marine communities. The destruction of tropical rain forests may adversely affect the entire biosphere by causing global climate changes. The impact of human activities often upsets Earth's complex ecosystems. In this chapter, you will learn that there are often unpleasant prices to pay for the good life we enjoy.

35.1 SHRINKING FORESTS

Forests play a vital role in regulating climate, controlling the runoff of water into rivers and streams, purifying the air, and providing food and shelter for wildlife. They are the source of valuable resources such as timber, fuelwood, pulpwood, foods, and pharmaceuticals. They are also of scenic, cultural, recreational, and historic value.

Before large-scale human intervention, forests covered more than 40 percent of Earth's land area. Now, more than one-third of these forests have been converted to farmlands, pastures, cities, and deserts. Timber harvesting has fragmented many forests and impacted on the natural biota. Fragmenting breaks large forest tracts into small areas surrounded by barren land or meadows. This increases forest edge habitat but at the same time reduces forest habitat. (See Figure 35-1.)

Throughout history, humans have had a much greater impact on temperate deciduous forests than on any other type of forest. These forests are concentrated in areas where the soil and climate are good for agriculture. By about the year 1000, the fragmenting of forest lands in England led to the extinction of many species that depended on large tracts of forest habitat for survival, including bears, lynx, and several species of birds.

Smog and acid rain from industrialized nations are decimating

Figure 35-1 Timber harvesting.

forests throughout the world. Atmospheric pollution has damaged up to 20 percent of the trees in some European forests. Acid rain has killed almost 25 percent of the spruce forests in the eastern United States. The destruction of forestlands causes economic and aesthetic losses for human society.

As forests are destroyed, native species die off. With the loss of native plant species goes their potential for improving agricultural diversity or as a source of new products. The bark of the Pacific yew, once considered a trash tree by loggers, has been found to contain a chemical useful in treating some cancers. Can we foresee the potential locked up in a forest species on the road to extinction?

Old-Growth Forests

An **old-growth forest** is a forest that has not been cut or disturbed by humans for hundreds of years. The composition of old-growth forests is very different from that of young, or secondary-growth, forests. As shown in Figure 35-2, old-growth forests are composed of a broad range of trees of varying size and height. Many of the trees are very old. Because of the variety of habitats they contain, these forests have a distinctive community associated with them.

Figure 35-2 **The trees in this old-growth forest in the Pacific Northwest are covered with moss.**

A mature forest can take more than 200 years to develop before it takes on old-growth characteristics. The older trees can grow to be very tall. The old-growth forests of the Pacific Northwest contain Douglas firs that are more than 1000 years old and 91 meters tall. Old-growth forests contain many trees with broken, decaying tops and large dead trees that remain rooted and standing upright. Large fallen trees may remain on the forest floor for many years while they slowly decay.

The varieties of trees in old-growth forests are important habitat resources for forest fauna. The pileated and the ivory-billed woodpeckers need decaying trees in which to hollow out nesting holes and as a source of insect larvae to feed on. The northern spotted owl nests in cavities in old trees and feeds on small animals that live in the canopy and on the forest floor. The endangered marbled murrelet requires old-growth forests for nesting and raising its chicks.

By 1940, almost all of the old-growth temperate deciduous forest in the eastern United States had been cleared for agriculture or by logging. These forests are particularly attractive to loggers because one large tree contains as many board feet of lumber as many smaller trees do. The practice of clear-cutting has also been very destructive to old-growth forests. **Clear-cutting** is a logging technique that cuts down a wide area of forest, leaving a bare scar, rather than cutting only selected trees. (See Figure 35-3.) Without tree roots to hold the soil, rain causes landslides, washes the soil into nearby streams, clogging them with silt and killing the fish. Once the lumber is harvested, the remaining vegetation and felled trees are often burned, which adds to air pollution.

Figure 35-3 A clear-cut forest shows bare areas where the trees were cut.

There are just a few remnants of old-growth forests in the Northeast and in the Pacific Northwest. The future of the northwest forests is the center of a major political controversy. In 2002, the United States Department of Agriculture led the fight to remove federal protections enacted by the Clinton administration. New legislation seems to favor the interests of the timber industry. Timber cutting on public lands increases while protection of wildlife decreases.

The logging industry replanted many of the forests it cut down. However, uniform stands of economically preferred species were planted, instead of replacing the wide variety of trees a forest normally contains. This is known as **monoculture**, or growing a single

DISCOVERY

The Marbled Murrelet

Old-growth forests are characterized by a large percentage of mature trees. Some species of animals can live only in this type of forest. The marbled murrelet is one. For many years, naturalists were unable to find the nesting sites of the marbled murrelet, a small seabird. They finally were found nesting inland, in the old-growth forests of the Pacific Northwest. The murrelet is unique among seabirds in its habit of nesting away from the sea.

The marbled murrelets' nesting ground in the forested Santa Cruz Mountains was not discovered until 1974. The first nest the researchers found was located more than 40 meters above the ground in the crown of a Douglas fir. Since the discovery of that first nest, several others have been found. All the marbled murrelet nests are located 8.5 to 16 kilometers inland, in old-growth forests.

Little is known about marbled murrelets. Their brown and white coloration allows them to blend into the forest canopy. They usually fly to and from their nests in the canopy of old-growth trees at dawn and dusk. This makes it difficult to spot them. These small birds spend most of their lives as sea, feeding mainly on fish. They return to the old-growth forest during nesting season to raise their young. Solitary pairs of murrelets lay a single egg on a moss-covered tree branch. The parents take turns sitting on the egg and flying out to sea to feed. When the chick hatches, the parents continue this alternating feeding routine until the chick is old enough to fly to sea and feed with them.

Marbled murrelets need old-growth forests to raise future generations. The old-growth forests in the Pacific Northwest contain the large, moss-covered trees necessary for the marbled murrelet to nest in. In the United States, only 5 percent of these old-growth forests still exist. Logging activities and replanting programs have converted these forests into younger, same-age trees that are not suitable for the murrelets to nest in.

A coalition of environmental organizations has emerged to protect the habitat of the marbled murrelets. They believe that protecting wildlife is a public trust, and if the marbled murrelet is to survive, the remaining old-growth forests must remain intact.

plant species over a wide area. Monoculture is a common agricultural practice used in growing wheat, corn, cotton, and other crops. These artificial ecosystems are an efficient method of producing desired raw materials. However, they are highly susceptible to disease. Insects that feed on a particular tree can have a field day, since their food source is so abundant. Monoculture, as well as the overplanting of preferred species, has led to forest destruction caused by diseases spread by insect pests. These diseases have nearly eliminated the American chestnut, the American elm, and the butternut from the North American deciduous forest biome.

New seedlings that are planted to replace the harvested trees all mature at the same rate. This leads to stands of trees that are essentially the same age, called **even-aged stands**. Forests planted in the early part of the twentieth century are composed of trees that are from 60 to about 100 years old. Under natural conditions, stands of trees vary greatly in age. They are **uneven-aged stands**. Ecosystems composed of even-aged stands do not have the habitat diversity found in natural ecosystems.

The Disappearing Chaparral

Los Angeles and its suburbs are located in lowlands and stream channels at the base of brush-covered hills that form a chaparral biome. Fires and floods associated with the chaparral ecosystem are constant threats to this area. The population of the Los Angeles region has faced many natural disasters due to its location in the chaparral.

In populated areas of the chaparral, where fires have been prevented for many years, combustible litter piles up. Sooner or later, a careless act by an individual or a bolt of lightning will touch off a raging fire. Huge areas burn, and their protective covering of vegetation is lost. If heavy rains come before the growth of new grasses, which anchor the soil, major flooding and mudslides occur. This causes widespread destruction of property.

The chaparral ecosystem is considered undesirable by many people and is often replaced with grasslands and forests. However, since the area is too dry for forest trees and too wet for most grasses to compete successfully with native shrubs, it often reverts back to chaparral brush. Most of the chaparral landscape consists of steep hills and canyons. Therefore, the chaparral is highly resistant to the encroachments of urban populations.

Tropical Rain Forest Destruction

Today's exploding human population is having a disastrous effect on the tropical rain forests. In the rain forest, most of the nutrients are in the living plants, and the soil is of poor quality. Each day large areas of tropical rain forest are cut down to provide land for agriculture. The poor quality of rain forest soils limits crop production to three or four years at most. When the soil loses its fertility, more of the rain forest must be cut down for new farmland. The abandoned patch of nutrient-poor soil cannot support the natural rain forest flora; it remains bare and subject to erosion.

Tropical rain forests play an important role in limiting erosion, reducing seasonal flooding, protecting the quality of the fragile soil, and preventing silt buildup in rivers and streams. The rain forests also play an important role in maintaining the composition of the atmosphere, and it is feared that their disappearance will harm Earth's water cycle and oxygen–carbon dioxide cycle. Because of rain forests' role in regulating Earth's climate, their destruction may hasten global warming.

The biological diversity within the tropical rain forests is high. (See Figure 35-4.) Within a 10-hectare (25-acre) area, there can be as many as 300 species of trees and thousands of animal species, including mammals, birds, reptiles, amphibians, and insects. Compare this with a temperate forest, which may contain 20 species of trees and a correspondingly reduced number of animal species. The rain forest ecosystem is stressed by deforestation, which may be caused by agricultural growth, livestock grazing, timber needed for construction, and fuelwood used to supply energy.

The most significant threat to the biological diversity of forest ecosystems is the destruction of tropical rain forests. These forests are being cut down at an alarming rate, and many species of animals are being affected by their disappearance. Almost 1 percent of the world's rain forests is cut down each year. It is estimated that rain forest destruction has led, by the beginning of the twenty-first

Figure 35-4 Many different types of plants grow on the floor of the tropical rain forest.

century, to the loss of 20 percent of the world's animal species. The diversity of tropical plants and animals could hold untold chemical treasures within their gene pools. These chemicals may hold future benefits for the human population in the shape of new medicines, agricultural products, or industrial materials.

At the **Earth Summit Conference** held in Rio de Janeiro in June 1992, attendees acknowledged that deforestation is a grave problem affecting the whole world. Nonbinding strategies were developed to manage, conserve, and sustain development of tropical, temperate, and boreal forests. At the follow-up Johannesburg Summit, held in South Africa in 2002, energy and sanitation issues and their relationship to sustainable development were among the critical elements discussed. Global action was initiated to fight poverty and protect the environment.

35.1 Section Review

1. Describe the characteristics of old-growth forests.
2. Why did early human civilizations have an enormous impact on the temperate deciduous forests?
3. Why are fires common in populated areas of chaparral biome?

35.2 GROWING DESERTS

Even though deserts are the least productive ecosystems, they are often used to provide forage for grazing animals or are planted with crops. These human activities often lead to the spread of desert regions. Poor land management can cause changes in the environment by destroying fragile grasslands at the desert's edge. Productive land can lose its ability to support vegetation. This leads to desertification, the formation of a desert caused by human activities.

Overuse of rangeland for grazing can degrade and compact soil, destroy plant roots, and kill natural ground cover. Elimination of grasses, which protect soil through an extensive system of shallow roots, subjects soil to erosion by wind and rain. This erosion can lead to desertification of the environment.

Around the Sahara in Northern Africa, the growing human population and the shortage of grazing land have led to the overuse of existing resources. This added pressure has destroyed the ecosystem and caused desert conditions to spread. On average, more than 100 square kilometers (40 square miles) of Earth's once-productive land become desert each day.

Overuse of groundwater and the cutting of trees and shrubs for firewood also contribute to desertification. In developing countries, trees and shrubs have economic value as sources of fuel or a means of generating income. Removing vegetation destroys the soil's protection against erosion by wind and rain. (See Figure 35-5.) Improv-

Figure 35-5 During the 1930s, a long drought caused the soil to blow away in a large area of the United States.

ing farming practices, limiting the number of grazing animals, and using alternate forms of energy can protect the fragile soil, prevent soil erosion, and slow or reverse the process of desertification.

35.2 Section Review

1. Explain some of the causes of desertification.
2. How do economic activities lead to desertification?

35.3 LOSS OF WETLANDS

Wetlands are lowlands that are permanently or temporarily covered with shallow water. They include marshes, swamps, bogs, potholes, sloughs, salt marshes, and floodplains. Wetlands are natural sponges that retain large quantities of water. Because they absorb water, they are critical in preventing flooding during spring runoff and periods of heavy rain. Many areas have experienced loss of wetland habitats. (See Figure 35-6.)

Coastal wetlands act as buffers against beach erosion and shoreline damage by absorbing the destructive effects of waves during severe storms. The flow of groundwater through coastal marshes prevents the intrusion of salt water into the water table and protects our drinking water against contamination.

Mangrove swamps protect coasts from erosion by absorbing the

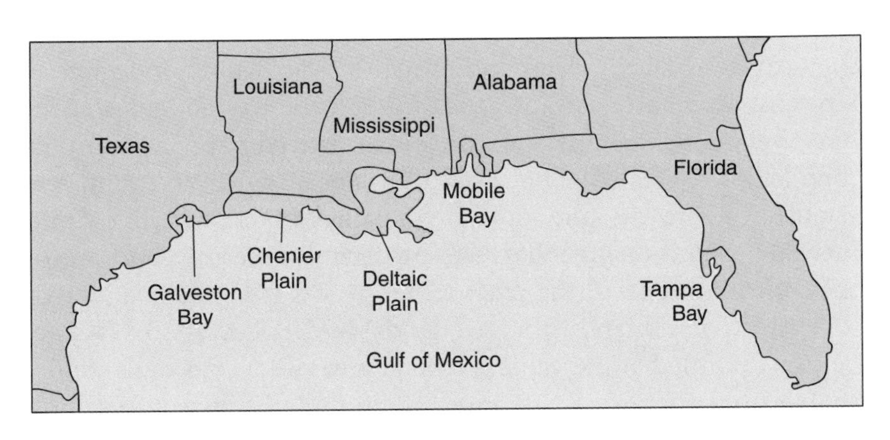

Figure 35-6 **The area around the Gulf of Mexico has experienced loss of wetland habitats.**

Figure 35-7 Shoreline of a mangrove swamp.

shock from hurricanes and the pounding from heavy waves. (See Figure 35-7.) When mangrove swamps are destroyed, coasts and inland areas become vulnerable to damage from tropical storms. Since coastal development often destroys wetlands and barrier islands, their natural value to the ecosystem must be weighed against their value for real estate or commercial development.

Wetlands are the cradle of the world's seafood industry. Fish and shellfish need salt marshes and estuaries as spawning and nursery grounds. These areas are of great value to the United States economy, yet more than half of the wetlands that support many of America's fisheries have been used for landfill operations and real estate developments, and more are in danger of disappearing. Commercial fish and shellfish harvests are declining steadily, and one of the contributing factors appears to be the loss of coastal wetlands.

For most of human history, wetlands were looked upon as wastelands with little natural value. Wetland areas were considered undesirable since they were the breeding grounds for mosquitoes and other insect pests. In the United States, the customary practice was to fill or drain these areas so they could be developed for agriculture or real estate. More than half of the original wetland areas that existed in the United States before 1600 have been destroyed. Their disappearance has led to the extinction of many wetland species, and many others are now in danger.

Early governmental policies only accelerated the destruction of wetlands. The **United States Swampland Act of 1850** allowed individuals to buy swamps and marshes for as little as 10 cents an acre. During the 1930s and 1940s, the government subsidized the drainage of wetlands for conversion to agriculture. Prior to 1977, local governments dumped landfill into wetlands as a convenient way of disposing of wastes, while at the same time creating space for new highways and housing.

Wetlands located close to cities are subjected to environmental pressures caused by the discharge of treated and untreated sewage. Each day, billions of gallons of sewage may be discharged into coastal waters. The sewage contains large quantities of harmless inorganic materials. These materials are filtered out of the water by the wetland vegetation, increasing their productivity. When the plants use up the available nutrients, they die and decay. Their decay removes most of the oxygen from the water. The low concentration of oxygen then contributes to the death of the wetland fauna.

Sewer discharge also may contain dangerous substances, for example, bacteria, viruses, PCBs, and heavy metals, such as mercury, zinc, and lead. These harmful materials may be taken in by wetland plants, passed on to marine organisms, and then passed along the food chain to humans.

The first governmental measure that protected wetlands was a law passed in 1899 preventing the dumping of rubbish into navigable waterways. But it was not until the early 1970s that federal and state governments began enacting laws that recognized the enormous value of wetlands. Our remaining wetlands are now protected under the **Clean Water Act of 1972**. To protect the quality of surface water and our supply of drinking water, this law prohibits the pollution and filling in of wetlands. Through these laws, the government is trying to save existing wetlands and reestablish them in areas where they have been destroyed. Today, wetlands provide people with recreational opportunities such as hunting, fishing, bird-watching, and camping.

Coastal Wetlands are ecologically important for several reasons:

- Control erosion and form a wave barrier to protect shorelines
- Control sediments by binding particles to form mud
- Provide fish and wildlife with habitats for feeding and spawning
- Aid in maintaining composition of the atmosphere
- Protect freshwater supplies from intrusion by salt water
- Provide pollution control by filtering pollutants from water

35.3 Section Review

1. Describe the ecological value of wetlands.
2. How have our attitudes toward the importance of wetlands changed?
3. Explain how the discharge of untreated sewage can destroy wetland ecosystems.

LABORATORY INVESTIGATION 35

Planting Trees

PROBLEM: *What is the growth rate of a tree?*

SKILLS: *Observing, manipulating, measuring*

MATERIALS: *Young tree or sapling about 15 to 20 centimeters tall (or if tree is to be grown from seed: flowerpot or plastic container, compost or soil, and a selection of seeds), shovel or trowel, water, heavy wooden stake, wire, masking tape, mulch or compost, chicken wire (50 centimeters square)*

PROCEDURE

1. Choose a sapling, tree seedling, or the seeds of a tree that is native to your region and able to tolerate the environmental conditions of the location in which it will be planted.

2. If you are starting with a seed:

   a. Plant the seed in a pot filled with moist compost or soil.

   b. Water the seed regularly.

   c. When the seed germinates, move it to an area where it will be exposed to regular periods of sunlight.

   d. When the seedling is 15 to 20 centimeters tall, it is ready for planting.

3. Choose the area where the sapling or seedling will be planted.

4. Dig a hole slightly larger than the pot holding the sapling or seedling and deep enough for the soil to reach to the tree's collar, where the stem and the root meet. Place a layer of mulch in the bottom of the hole. Add enough water to moisten the mulch.

5. Remove the sapling from the pot and loosen the soil around the roots. Place the sapling upright in the hole and fill the hole with soil, tamping it down gently around the roots.

6. For protection and support, place the stake in the ground next to the tree. If the tree is large enough, attach it to the stake with wire. The wire should be loosely attached to the tree. Masking tape can be wound around the trunk to protect it from the wire.
7. Roll the chicken wire into a cylinder and attach the ends together. To protect the sapling from animals, slip the wire cylinder around the sapling and bury the end of the wire cylinder in the ground. The cylinder can be attached to the stake.
8. Thoroughly water the area around the tree.
9. Make sure the tree is watered regularly and that the soil around the tree is free of weeds or competing plants.

OBSERVATIONS AND ANALYSES

1. What species of tree did you choose? Why did you choose that species?
2. If you started with a seed, how long did it take to germinate?
3. How long did it take for the germinated seedling to grow 15 to 20 centimeters in height?
4. How long did it take the sapling tree to grow 2 centimeters?
5. Describe the characteristics of some of your local trees. What kind of seeds do they have?
6. Obtain the seeds of local trees and test methods of germinating them.

Vocabulary

The following list contains all the boldfaced terms in this chapter.

Clean Water Act of 1972, clear-cutting, Earth Summit Conference, even-aged stands, monoculture, old-growth forest, uneven-aged stands, United States Swampland Act of 1850

Fill In

Use one of the vocabulary terms listed above to complete each sentence.

1. A(an) ________ is a forest that has not been cut or disturbed for hundreds of years.
2. ________ is a logging technique that cuts down wide forest areas.
3. ________ are forests in which the trees vary in age.
4. Growing a single plant species over a wide area is called ________.
5. When trees that are planted at the same time replace logged forests, it results in ________.

Multiple Choice

Choose the response that best completes the sentence or answers the question.

6. A species that requires old-growth forests for its survival is the *a.* marmoset. *b.* horned owl. *c.* pileated woodpecker. *d.* mangrove.
7. Which of the following is an example of a wetland habitat? *a.* rain forest *b.* swamp *c.* lake *d.* river
8. Which of the following is an ecological function of wetlands? *a.* protect freshwater from saltwater intrusion *b.* expose wildlife to danger *c.* destroy sediments *d.* provide safe areas for the disposal of toxic wastes
9. Which of the following is a cause of desertification? *a.* poor land management for agricultural purposes *b.* floods *c.* natural forest fires *d.* planting trees for firewood
10. Why are old-growth forests attractive to loggers? *a.* They are easier to find than other forests. *b.* They contain many large trees. *c.* They contain only one species of tree. *d.* They are even-aged stands.

Chapter 35 Review

Answer these questions on a separate sheet of paper.

Short Answer (Constructed Response)

Use the information you learned in this chapter to respond to the following items.

11. Name two resources obtained from forests.
12. List two environmental problems caused by monoculture.
13. Briefly describe three environmental problems caused by deforestation.
14. Name two environmental benefits of restoring wetlands.

Essay (Constructed Response)

Use the information in the chapter to respond to these items.

15. How is an old-growth forest different from an even-age stand?
16. Explain how clear-cutting degrades the environment.
17. How does the marbled murrelet, a seabird, depend on the old-growth forest?
18. What threats face urban areas located within the boundaries of the chaparral biome?
19. Why is the chaparral biome not readily convertible to forests or grasslands?
20. Why were wetlands once considered to have little economic value?
21. Why is it important that we maintain the biological diversity present in tropical rain forest ecosystems?

Research Projects

- List three important environmental benefits your parents' generation passed on to your generation. Describe how these benefits were achieved. Prepare a second list of three ways your parents' generation degraded the environment. Describe one method your generation might use to counteract the effects of the degradation.
- Develop a plan to help preserve the forest areas and wetlands in your state. Compare your plan with those of your classmates; choose the best ideas and send them to your state legislator.
- Explore the effects of the loss of all the remaining old-growth and tropical forests in North America. How would the loss impact the atmosphere, the geosphere, and the biosphere?

CHAPTER 36
Protecting Wildlife and the Environment

When you have completed this chapter, you should be able to:

Explain the importance of biodiversity to the biosphere.

Discuss the development of the conservation movement.

Even if you live in a big city, such as New York, Chicago, or Los Angeles, you can see grass, trees, and birds. In many places, you can keep animals, such as dogs and cats, as pets. If you live in the suburbs or in the country, you are surrounded by nature.

Imagine what life would be like if there were no grass or trees, you couldn't keep pets or even breathe the air without wearing a mask for protection from pollution, and you could see animals only in a zoo. Scientists have made us aware of the problems we face preserving biodiversity, conserving wildlife, and maintaining the biosphere as we know it. In this chapter, you will learn what is being done to preserve our world and what you can do to help.

36.1 PRESERVING BIODIVERSITY

Biodiversity is a measure of the number of different species in an area, or of the genetic variations found within a species. The stability of ecosystems is directly related to the biodiversity of the biotic community. The presence of large numbers of organisms and species, which can adapt to a wide range of environmental conditions, allows an ecosystem to adjust to changes that occur. The extinction, or loss forever, of any species impacts on the biodiversity of the entire ecosystem.

The Earth Summit Conference held in Rio de Janeiro in 1992 recognized the importance of biodiversity to the biosphere. One outcome of the conference was a resolution that resources required by Earth's human population depend on the continued presence within the biosphere of a varied, ever-changing combination of genes, species, populations, and natural ecosystems. To protect Earth's biodiversity, the conference stressed the importance of achieving a long-term, sustainable balance between population growth and available resources. The United Nations Framework Convention Treaty on Climate Change was signed by 153 nations. It calls for an all-out international effort to reduce greenhouse gas emissions. This marked the first formal, international recognition of the threats that human activities pose to world climate and thus to the biosphere.

Figure 36-1 If thistle escapes from a garden into a wild area, it may compete with native plants and reduce the biodiversity of the region.

The biodiversity of an ecosystem cannot be maintained simply by the presence of large numbers of plants and animals. There is little biodiversity in a community composed mainly of large numbers of opportunistic native species such as weeds, rats, crows, or coyotes, or in one overrun by alien species, such as kudzu, English sparrows, water hyacinth, or zebra mussels. In those ecosystems where the natural populations have been displaced, biodiversity is greatly reduced. (See Figure 36-1.) The goal of biodiversity should be to protect a balanced mix of native species in their natural habitats. Protection must be a priority in those habitats that contain species whose populations are dwindling. One way to protect plant species is through seed banks, where seeds are stored for the future. The protection must begin before their numbers reach the point of no return and species are lost forever.

The Endangered Species Act of 1973 (ESA) has become the most powerful tool in the United States for protecting biodiversity. The act attempts to prevent the extinction of any species by recognizing **endangered species**, those that are in danger of becoming extinct, or **threatened species**, those at risk of becoming endangered. The act provides for the protection of these species and of their natural habitats. ESA further allows citizens to file lawsuits against federal projects that do not provide for the protection of endangered or threatened species. There are more than 1000 native plants and animals in the United States that are listed officially as endangered or threatened. In addition, more than 500 foreign species have such protection.

Exotic Species and Biodiversity

One obstacle to the preservation of biodiversity is the introduction of new, nonnative species into an ecosystem. These species are referred to as **exotic species**, or alien species. They are a form of biological pollution. Exotic species are often opportunistic invaders that are aggressive and prolific. Thus they are able to take hold in environments that are disrupted by human intervention. (See Figure 36-2.) They displace native species either through competition or predation. The lack of natural predators to check population growth

Figure 36-2 Exotic species, such as the mongoose and kudzu, are serious competition to our native species.

allows exotic species' numbers to soar. In island ecosystems, such as the Hawaiian Islands and Guam, exotic species have led to a large number of extinctions. These species are also a growing problem throughout many areas of the continental United States.

Human activities often have led to the destruction of natural ecosystems. At times, seemingly harmless acts have had disastrous consequences. The importation of exotic species to improve hunting or fishing, as ornamental plants, as pets, or to prey on agricultural pests has led to the displacement of native species. Many European trout species that were introduced into United States waters have replaced the natural fish populations. The mongoose, introduced into Haiti to prey on rats in the sugarcane fields, has become a threat to native birds and domestic poultry. Honeysuckle, introduced into the United States as an ornamental plant on estates, now infests many areas and is a problem for gardeners. Hydrilla and water hyacinths, aquatic plants that were introduced into Florida's waters, now form dense growths that clog waterways.

The brown tree snake arrived in Guam as an uninvited guest by stowing away in the cargo holds of planes from New Guinea. With no predators on the island to hold its numbers in check, the snake's population quickly grew. Agile climbers, they prey on nesting birds and have quickly decimated the native bird populations.

The zebra mussel, which is native to the area around the Caspian Sea, was accidentally brought to the Great Lakes in ballast water in the holds of ships. Since 1988, the mussel has spread into the rivers that empty into the lakes and become an environmental hazard. Zebra mussels clog pipelines and have replaced many native species. In recent years, environmental organizations have called for a concerted federal effort to combat the threat to biodiversity due to exotic species.

36.1 Section Review

1. Explain the outcomes of the Earth Summit Conference of 1992.
2. How does an endangered species differ from a threatened species?
3. In a particular ecosystem, what are some of the factors that may lead to an exotic species replacing a native species?

36.2 WILDLIFE CONSERVATION

Over the course of recorded time, humans have imposed great changes on the environment. We have become the most widely distributed and most numerous of all large mammal species. In its need for food, Earth's ever-increasing human population has reshaped the planet. Forests have been cut and replaced by farmland, the characteristics of species have been modified to better supply food and clothing needs, animal species have been lost forever through intensive hunting and fishing, and changes in the environment have led to changes in global climate.

Extinctions: Going, Going, Gone Forever

In most food chains of which humans are a link, we are the top predator. At times, we consume food at a rate that depletes the environment of other animal and plant species. Here is an example. The dodo was a large flightless member of the pigeon family that inhabited the island of Mauritius in the Indian Ocean. During the 1500s, the dodo was used as a source of fresh meat for sailors on passing ships. Since this bird was not able to fly, it could not escape pursuit and easily was clubbed to death. When the last dodo died, the species was gone forever. The dodo was the first animal in modern times driven to extinction by humans. (See Figure 36-3.)

Figure 36-3 **The dodo (left) and the passenger pigeon (right) are examples of organisms driven to extinction by people.**

Since then, many other animal species have become extinct, including the passenger pigeon, moa, great auk, Tasmanian wolf, and Steller's sea cow. Other species now face extinction; still others are threatened with becoming endangered. Preservation of any species benefits all species, including our own. We are all part of the web of life that makes up the biosphere.

Scientists theorize that the hunting practices of Paleo-Indians, who arrived in North America 12,000 years ago, led to the extinction of many large mammal species. Extensive hunting is thought to have killed off mammoths, cave bears, giant bison, ground sloths, saber-toothed cats, and dire wolves. In New Zealand, 27 species of flightless birds became extinct after settlement by the Maori, who arrived from their Polynesian homeland about 1000 years ago.

Some endangered or threatened species have become flagship species, symbols of conservation programs organized for their protection. These include the Florida manatee, giant panda, African elephant, Indian (Bengal) tiger, and mountain gorilla, species on which extensive conservation efforts are being expended.

Some species are called **indicator species**, since the population size and general vigor of the members of these species indicate the overall health of their ecosystem. For example, the spotted owl is an indicator of the health of the old-growth forests of the Pacific Northwest, since it depends on the forest to feed and shelter it.

Mussels are types of bivalves; freshwater species inhabit many river ecosystems. These filter feeders are indicators of a river's water quality. It is estimated that two-thirds of the freshwater mussel species in the United States are endangered, and some have already become extinct. Their decline has been attributed to a variety of causes, including agricultural runoff, improper erosion control, dredging, and invasion by Eurasian zebra mussels.

Umbrella species are indicator species that require extensive areas of habitat for their survival. By preserving large areas for these species, other species in the same ecosystem are indirectly protected. Large carnivores such as the African lion and grizzly bear, as well as large herbivores such as the African elephant and rhino are examples of umbrella species. (See Figure 36-4.)

Keystone species are indicator species that are of pivotal importance to the biodiversity of an ecosystem. The loss of these species can affect the entire ecosystem and may lead to the extinction of

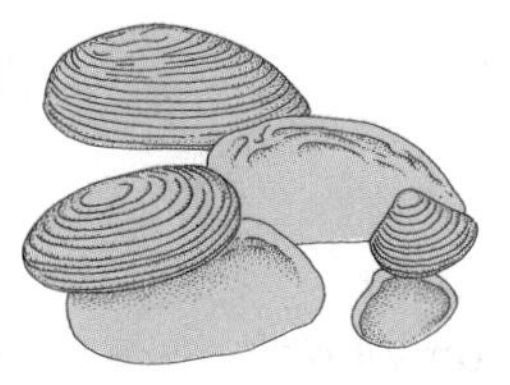

Figure 36-4 Indicator species, umbrella species, and keystone species are shown.

other species. The beaver is a keystone species in some wetland ecosystems; moose fill this role in boreal forest ecosystems; and African elephants maintain the biodiversity of their savanna biome.

Sentinel species are organisms whose health and vitality reflect the environmental conditions in an ecosystem. These organisms are used to monitor toxic substances in the ecosystem. Exposure to environmental contaminants in air, water, or soil reduces their populations or causes physiological damage to these species. The factors that affect these organisms usually affect the human population as well. The effects of bioaccumulation of pollutants in sentinel species will often suggest beforehand the effects on the human population. By studying sentinel species, scientists can monitor pollution buildup before it reaches levels that are harmful to humans.

Honeybees are excellent monitors of environmental pollutants. They have been used to determine the levels of heavy metals, pesticides, and industrial solvents that may accumulate in the environment. By equipping hives with sensors, scientists can determine the composition of pollutants picked up as the honeybees forage for food.

The populations of as many as 200 amphibian species are declining, and many individuals are found with physical abnormalities. Scientists are trying to discover the environmental cause for these changes. Investigators have not yet been able to determine whether pollutants or climatic factors are causing the changes. Possible culprits are pesticides or exposure to ultraviolet radiation due to thinning of the ozone layer. Herbicides are suspected of causing abnormalities in leopard frogs.

The Conservation Movement

The growing human population places increasing pressure on natural habitat areas within the biosphere. The destruction of any

CONSERVATION

Silent Spring

Rachel Carson (1907–1964) was a marine biologist and skillful writer who alerted the public to the dangers of DDT and other pesticides. Her book *Silent Spring,* published in 1962, detailed the damage pesticides had caused to the environment. It signaled the beginning of the environmental movement in the United States.

Carson's writing inspired scientists, nature lovers, and the news media to broadcast the message of the dangers to human health and the environment posed by pesticides. The book raised a storm of controversy regarding DDT and led to a search for alternative methods of pest control.

Silent Spring presented evidence that pesticides killed more than pests. Other organisms in the community were eating pesticide-contaminated insects and plants, thus concentrating the pesticides in their own bodies. This posed problems for all life, including people. Pesticides in agricultural runoff were reaching the waterways. It was found that the concentration of DDT in fish was so high that they were unfit for human consumption. High concentrations of DDT were also found in the tissues of fish-eating birds, such as ospreys and brown pelicans. The high levels of DDT caused them to produce thin-shelled eggs, which broke when they were incubated by the adult birds. Thus fewer baby birds lived, reducing the reproductive success of the species. The DDT residues found in milk were a concern because of potential effects on infants who drank the milk.

By 1971, the use of DDT was restricted by law in the United States. Safer, biodegradable pesticides were developed, and biological controls were exploited to eliminate insect pests. Rachel Carson's legacy to all of us is a lasting public awareness of the dangers of pesticides and a deep interest in environmental conservation. However, DDT is still used in many developing countries.

habitat leads to the elimination of all species that depend on that habitat. The loss of all animal species is known as **faunal collapse**. Since habitat loss can lead to extinction, it must be avoided. The loss of any genetic resource within the biosphere is an unacceptable price to pay for human population growth. Humans must develop conservation policies that allow for sharing Earth's resources with all other organisms. Earth is their home as well as ours. To protect endangered species, habitat areas must be conserved and set aside for their survival. Conservation is the careful, organized management and use of our natural resources. In developed countries, unspoiled habitat areas are becoming increasingly rare, and conservation practices must be put in place now while there is still time.

For many years, wildlife conservation was carried out on a species-by-species basis. But recent environmental investigations have revealed that, by singling out individual species for protection, we were missing the forest for the trees. The approach in conservation has now shifted to securing the long-term preservation of entire habitats and ecosystems.

Wildlife Refuges

Wildlife refuges, or preserves, are habitat areas set up to preserve whole ecosystems. In the United States, a complex system of national wildlife refuges shelter a wide array of species within a variety of biomes. Caribou, musk oxen, and snowy owls thrive on protected areas of Arctic tundra. Taiga refuges shelter moose and wolves. Whales, seabirds, and sea turtles find protection in marine refuges. Bison, pronghorn antelopes, and prairie dogs inhabit protected areas of the grasslands.

America's national wildlife system was among the world's first, and it stands today as the largest land management system for the conservation of wildlife in the world. It was instituted in the late 1800s as a result of concern over the disappearance of colonies of birds, such as egrets, herons, terns, and pelicans. These birds were being killed for their feathers, which were used in the manufacture of hats and quill pens. By 1903, reserves were being set aside to protect wildlife. In 1916, the Migratory Bird Treaty with Canada recognized the need to protect the waterfowl that migrated across

Figure 36-5 The proceeds from the sale of the Migratory Bird Hunting Stamp are used to purchase and protect wetland habitats.

international boundaries. To protect these species, reserves were established on waterfowl routes, or flyways, in both the United States and Canada.

In the 1920s and 1930s, periods of drought and draining of wetlands caused severe declines in waterfowl populations. The Migratory Bird Hunting Stamp Act was passed in 1934. It created a special fund to buy wetland acreage to increase the size of refuges at key points along major waterfowl migration routes. Continued loss of wetlands led Congress to enact a new program in 1961 to accelerate the purchase and protection of wetland habitats. (See Figure 36-5.)

Game Laws

Humans have always been hunters. Ecologically, hunters assume the role of predator in an ecosystem. Where natural predators are missing, this is an important role. Hunting deer or elk reduces their population, reduces competition for food, and prevents starvation of individuals within a herd. When properly managed, hunting and fishing do not adversely affect natural ecosystems. Since healthy populations produce a surplus of individuals, removing this surplus does not affect the species as a whole.

States regulate hunting and fishing by issuing licenses and setting game laws. Hunting and fishing regulations determine which species can be taken; the age, sex, and number of individuals that can be taken; and the time and duration of the hunting and fishing seasons. Limited seasons insure that a species will not be disturbed during breeding time or when young are being cared for. Limits insure that a species is not overhunted. Many problems arise when hunting is unregulated, or when **poaching**, or unlawful hunting, trapping, or fishing, depletes species' populations. State license fees help cover the costs of wildlife management and allow for the purchase of land for preserves. The 1980 Fish and Wildlife Conservation Act provided states with matching funds for the protection of nongame species.

The international trade in wildlife takes in billions of dollars each year. Some endangered tropical species are caught and smuggled to industrialized nations where they are sold as pets. Many of these animals die in the filthy, crowded conditions they are subjected to

Figure 36-6 The African elephant (left) and black rhino (right) are endangered species.

while in transit. Other species are killed by poachers and their skin, horns, bones, and organs turned into useless commercial products. Some endangered plants are smuggled abroad to be sold as houseplants. The United States Fish and Wildlife Service and the United States Department of Agriculture have joint authority over the import and export of plants protected by international agreements. But only if everyone agrees to stop buying products obtained from endangered or threatened species will this trade stop.

The African black rhino is dangerously close to extinction due to the trade in its horn. (See Figure 36-6.) Rhino horn is exported to the Far East, where it is ground to make traditional medicines, and to the Middle East, where it is carved to make dagger handles. East Africa's wild elephant population has declined in the past 25 years due to habitat loss and poaching for its ivory. Africa's wild animal parks usually lack sufficient staff to stop poaching.

In 1973, the Convention on International Trade in Endangered Species of Wild Fauna and Flora (CITES) was established through an international agreement. CITES prohibits the trade in endangered species, restricts the trade in threatened species, and regulates the trade in protected species. More than 1200 animal and plant species are on the CITES protected list.

Figure 36-7 Breeding programs have brought back the red wolf and black-footed ferret from near extinction. These species are being reintroduced to their original ecosystems.

In recent years, there have been intensive efforts to save endangered species, such as the whooping crane, the California condor, the black-footed ferret, and the red wolf. (See Figure 36-7.) Loss of habitat, overhunting, pesticide poisoning, and egg collecting have lowered populations of these species to dangerous levels. Captive-breeding programs have enabled biologists to increase populations. Breeding pairs may be released into suitable habitat areas, where they can be protected and monitored until a stable population builds up. Setting aside undisturbed habitat is essential for the survival of these species. The return of the whooping crane shows the effect concentrated conservation efforts can have in saving an endangered species from extinction. The population of whooping cranes in the wild has grown from 15 individuals to more than 200. (See Figure 36-8.)

Zoos are often the focus of conservation efforts to protect species whose numbers have been severely reduced in their natural habitat. Many zoos have captive-breeding programs that increase populations while at the same time allowing for genetic diversity to be maintained. Others are living museums that maintain rare breeds of domestic animals. Botanical gardens serve as gene banks for plant species by maintaining plants and stores of viable seeds.

36.2 Section Review

1. How has the approach to wildlife conservation changed in recent years?

Figure 36-8 Captive-breeding programs have increased the number of California condors (left) and whooping cranes (right) in the wild.

2. What are some of the factors that led to the institution of a national wildlife system?
3. What effect does managed hunting have on species' populations?

36.3 MAINTAINING THE BIOSPHERE

The World Conservation Strategy, launched in 1980, presented a revolutionary approach to global problems. This document was based on three important propositions. First, species' populations must be helped to retain their capacity for self-renewal. Second, measures must be taken to conserve Earth's basic life support systems—climate, the water cycle, and soil. Third, the genetic diversity within the biosphere must be preserved. These concepts serve as the basis for the conservation policies of more than 50 nations.

In 1991, a new strategy was put forth in the book *Caring for Earth.* It calls for a commitment from governments to conserve Earth's vitality and diversity. The book proposed that by the end of the twentieth century, all countries adopt comprehensive strategies to safeguard their biodiversity and establish a system of biosphere reserves that cover at least 10 percent of each of their ecological regions.

In many ecosystems, one species is the key to the stability of the entire system. These may not be the most abundant organisms, but they are the most influential. They hold the ecosystem together, and their removal creates a ripple effect that impacts every species in the community. Keystone species often are not recognizable as such, and thus their importance to the ecosystem may not be seen until they have been removed. This may be too late to protect the entire ecosystem from collapse.

National Parks

The concept of national parks was begun in 1872 with the establishment of Yellowstone National Park, in Wyoming. To preserve wildlife and habitat areas, many nations have since established parklands and reserves using Yellowstone as a model. The United

Figure 36-9 Devils Tower in Wyoming and Montezuma Castle in Arizona are national monuments.

States national parks retain a record of our nation's history, preserve scenic wonders, and are sanctuaries for wildlife.

In the 1800s, the westward expansion of the United States awakened the nation to the awesome beauty and natural wonders of our landscape. Books by Henry Thoreau and John Muir raised the public consciousness regarding conservation of the natural environment. President Theodore Roosevelt made conservation a public issue. He established forest reserves and wildlife sanctuaries, as well as designated historic and scientific areas as national monuments, including Devils Tower, Montezuma Castle, and El Morro. (See Figure 36-9.)

By an act of Congress, the National Park Service was established as part of the Department of the Interior in 1916. The service was set up to create the national parks that would conserve the nation's scenery, natural and historic locations, and wildlife. This would leave them unspoiled so they could be enjoyed by future generations. In recent years, additional parklands were opened in urban areas to provide opportunities for people to retreat into natural settings. (See Figure 36-10.) More than 300 million people a year visit our national parks.

Today, many of our parks are islands surrounded by barren lands that have been exposed to the ravages of development. Some environmentalists have suggested creating "green corridors" between the national parks to allow species to migrate from park to park. This would help maintain the predator-prey balance. Commercial enterprises often cluster at park entrances, detracting from the park's natural beauty. Agriculture, grazing, and mining activities that surround parklands pollute streams and groundwater within the park. Heavy traffic on park roads and trails can lead to the deterioration of the landscape. Air pollution from large cities, carried to park areas through prevailing weather patterns, reduces the visibility in scenic areas. In recent years, because of a shrinking budget, the National Park Service has found it difficult to maintain the parks.

The United States Forest Service is an agency within the Department of Agriculture. The Forest Service manages almost 200 million acres of national forest lands. To sell timber rights, the Forest Service builds and maintains roads on these lands. The forests also allow recreational use, such as fishing, boating, and camping.

Wilderness refers to those areas where the ecosystem is undisturbed by human activities and where people visit but do not

Figure 36-10
National parks of the United States are shown on this map.

remain. In 1964, Congress passed the Wilderness Act, which established the National Wilderness Preservation System. Certain areas were set aside as wilderness areas. To maintain their wild state, these large tracts of land have restricted use. Activities banned on wilderness lands include the use of all motorized vehicles or motorized equipment and timber cutting, except when needed to control fires, destructive insects, and wildlife diseases. Some activities that were allowed to continue on these lands are livestock grazing and mining for fuel and metal resources.

The designation of wilderness areas often creates controversy. Timber and mining interests oppose these designations because valuable resources are locked up. Environmentalists press for more land to be set aside. Visits to our national parks, forests, and wilderness areas are all valuable experiences, catering to the recreational needs of many citizens. However, visitors often find these areas crowded, and frequently encounter restrictions set by the Forest Service to protect the wilderness environment. The very presence of crowds, which are attracted to these areas, often ends up destroying that experience.

Biosphere Reserves

The program Man and the Biosphere created by United Nations Educational, Scientific, and Cultural Organization in 1971 deals with the interactions of human society and the environment. Among the concepts it introduced was that of **biosphere reserves,** regions established to protect and develop natural ecosystems. They allow human activities but only through the rational use and development of the natural resources of the ecosystem.

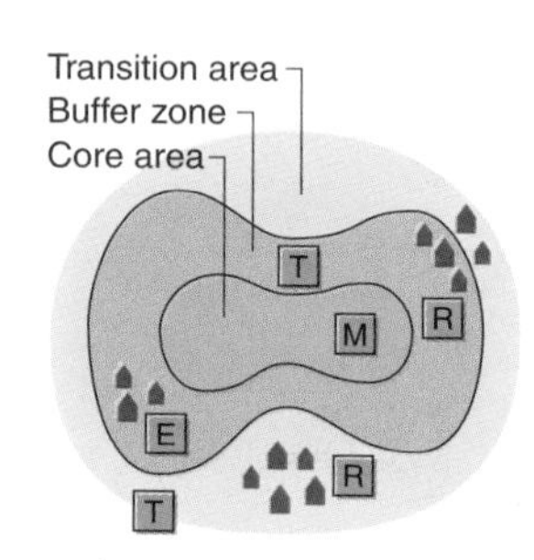

A biosphere reserve has a nucleus, or central area, set aside for conservation and limited scientific investigations. Surrounding this nucleus is a buffer zone in which people live and use the resources on an ecologically sound basis. (See Figure 36-11.) These concepts set biosphere reserves apart from national parks, in which people are only observers. Biosphere reserves are especially important in developing countries, where conservation and economic development are of equal importance to the population.

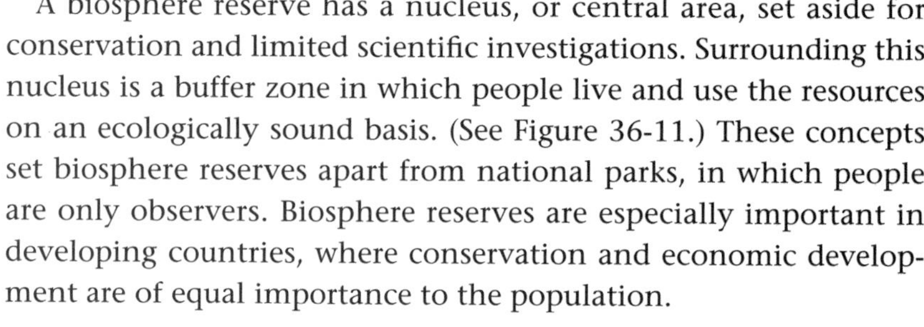

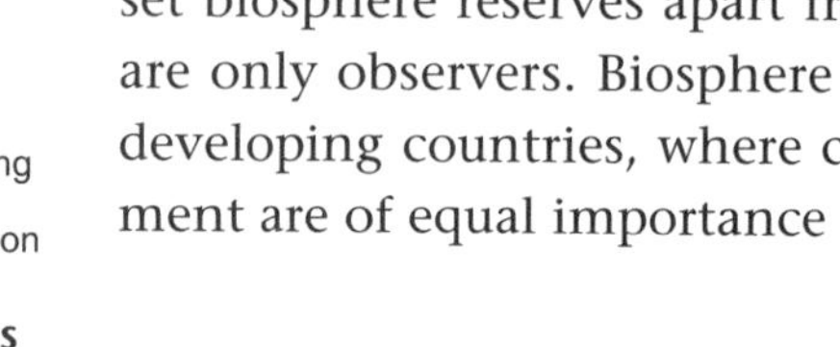

Human settlements
R Research station or experimental research site
M Monitoring
E Education and training
T Tourism and recreation

Figure 36-11 Zones in a biosphere reserve.

Antarctica

Antarctica is the driest, coldest, and windiest continent. Precipitation averages only about 3 centimeters per year. The continent is an ice-covered landmass situated in the South Polar region. The Antarctic ice cap holds 70 percent of Earth's freshwater reserves. Only 2 percent of the continent's surface is free of ice. Temperatures range from cold to very cold. However, the region supports a wide variety of plants, animals, and insects. The nutrient-rich waters that surround Antarctica contain a highly specialized ecosystem. Phytoplankton and krill fill the waters and serve as a source of food for penguins, seals, whales, and many species of fish. (See Figure 36-12.)

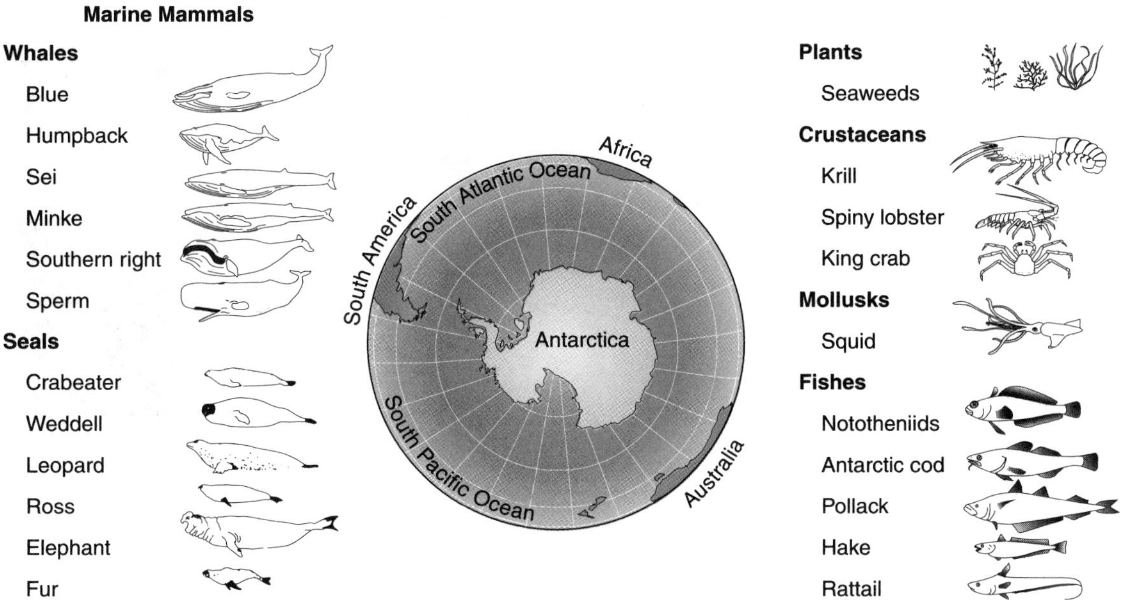

Figure 36-12 Many organisms are adapted to life in the Antarctic ecosystem.

International cooperation for the preservation of Antarctica was begun during the International Geophysical Year of 1957–1958. This ultimately led to a 1991 treaty to preserve Antarctica for peaceful and scientific projects. Nations have agreed to make no territorial claims in Antarctica. No mining or oil exploration will be allowed on the continent for 50 years. To monitor Earth's climate and the human impact on the Antarctic ecosystem, the treaty countries maintain 69 research stations on the Antarctica. Many of Earth's major environmental problems, such as global warming, the ozone hole, and the atmospheric buildup of carbon dioxide, were first recognized in Antarctica.

The Antarctic environment is very fragile. A footprint can remain for more than 200 years. The continent has been subjected to tourism in recent years, and it is feared that tourists will harm the environment and disrupt scientific research.

In 2001, the United Nations Panel on Climate Change stated that during the past century worldwide temperatures had climbed more than 2°C. This warming has caused many glaciers to retreat or disappear from mountain ranges around the world. Between 1900 and 2000, three of Antarctica's largest glaciers have thinned by 50 meters.

The effects of warming on Antarctica have been documented

Figure 36-13 **The breakup of the Larsen Ice Shelf in February 2002.**

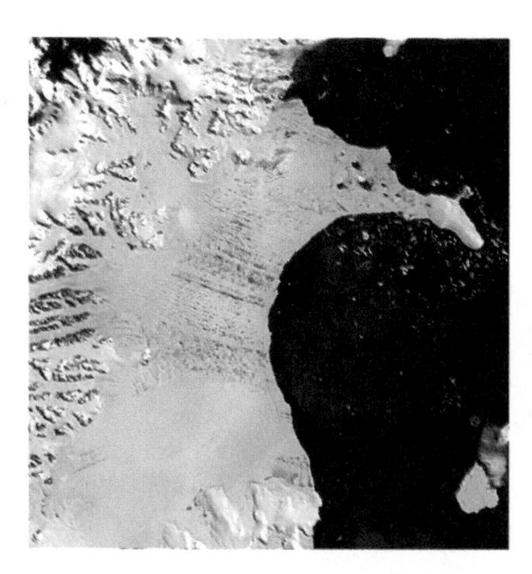

through satellite imagery. In 2000, three huge icebergs broke away from the continent. In 2001, another huge piece of ice was lost. These giant icebergs have reduced the plankton population in the Ross Sea, impacting the region's marine ecosystem. Thousands of penguins have died due to a shortage of krill, which feed on phytoplankton.

In March of 2001 along the Weddell Sea, a large part of the 250-meter-thick Larsen B Ice Shelf collapsed, reducing the sea ice habitat. (See Figure 36-13.) The disintegration of the shelf scattered thousands of icebergs throughout that sea. This caused the Adélie penguin population to drop by 50 percent and the emperor penguin colonies to have reduced breeding success in 2002. Birds had to travel farther to find food and icebergs obstructed their movement.

Over the past 50 years, the Antarctic Peninsula has warmed by 2.5°C. There is evidence of significant warming of the ocean waters flowing beneath the floating ice. In the last 20 years, the annual melt season has increased by 2 to 3 weeks. However, other measurements indicate that contrary to expectations, Antarctica's harsh desert valleys have been growing cooler since 1980, and in parts of the continent ice is thickening rather than melting.

36.3 Section Review

1. Explain the three propositions of the World Conservation Strategy.
2. How is a biosphere reserve different from a national park?

LABORATORY INVESTIGATION 36

Observing Birds

PROBLEM: *What are the food preferences of the species of birds that live in your area?*

SKILLS: *Observing, recording data*

MATERIALS: *Bird feeder, notebook, pen or pencil, binoculars, a guide to the birds of your area*

PROCEDURE

1. Set up a feeding station where native birds can be observed daily.
2. Supply the feeding station with a selection of foods. Include a variety of large and small seeds, such as sunflower, millet, and cracked corn, as well as small pieces of fruit and berries.
3. Observe birds as they feed. Note their food preferences.
4. Determine the techniques the birds use to pick up food and whether the food is eaten on the spot or carried away.

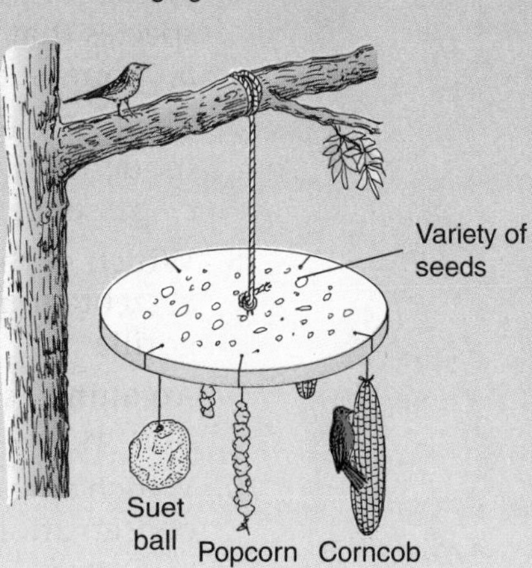

OBSERVATIONS AND ANALYSES

1. Determine the species of birds that are native to your area.
2. How are beak size and food preferences related?
3. What niche does each observed species occupy?
4. Which birds do you think may be nesting? Why?

Chapter 36
Review

Answer these questions on a separate sheet of paper.

Vocabulary

The following list contains all the boldfaced terms in this chapter.

biodiversity, biosphere reserves, endangered species, exotic species, faunal collapse, indicator species, poaching, sentinel species, threatened species, umbrella species, wilderness, wildlife refuges

Fill In

Use one of the vocabulary terms listed above to complete each sentence.

1. Nonnative species are called ________.
2. Species that are in danger of becoming extinct are said to be ________.
3. The measure of the genetic variety within an ecosystem is known as its ________.
4. ________ is unlawful hunting, trapping, or fishing.
5. Species that are used to describe the condition of the environment are ________.

Multiple Choice

Choose the response that best completes the sentence or answers the question.

6. Which of the following is an example of an opportunistic species that is native to the United States? *a.* zebra mussel *b.* honeysuckle *c.* kudzu *d.* coyote
7. The first species to be driven to extinction by humans in modern times is the *a.* mammoth. *b.* dodo. *c.* passenger pigeon. *d.* dinosaur.
8. Which of the following is an example of a keystone species? *a.* zebra mussel *b.* beaver *c.* giant panda *d.* mountain gorilla
9. An animal species that is endangered due to illegal trade in its horn is the *a.* lion. *b.* rhino. *c.* moose. *d.* blue whale.
10. Which species was saved from extinction due to intensive conservation efforts? *a.* blue heron *b.* whooping crane *c.* Canada goose *d.* imperial woodpecker

Short Answer (Constructed Response)

Use the information you learned in this chapter to respond to the following items.

11. List five exotic species that have been introduced to North America.
12. List five species that have become extinct in the past two centuries.
13. What is an umbrella species?
14. Name an indicator species.
15. What is a keystone species?
16. Name two ways the breaking off of giant icebergs has impacted penguin colonies.
17. To what degree has Antarctica warmed over the past 50 years?

Essay (Extended Response)

Use the information in the chapter to respond to these items.

18. How does poaching impact the biosphere?
19. Why do scientists use sentinel species to monitor environmental conditions?
20. How is habitat destruction related to faunal collapse?
21. Why are exotic species considered to be forms of biological pollution?
22. Describe three problems faced by our national parks.
23. Explain how international efforts have served to protect the Antarctic ecosystem.

Research Projects

- Investigate the ecosystem in a national park, wilderness area, wildlife refuge, or local park. List those species that may be endangered. Identify any exotic invaders and the impact they have on the ecosystem. Propose solutions to the problems you encounter.
- Prepare a time line to show the effects of global warming on Antarctica over the past 50 years.

CHAPTER 37
Preserving Our Food Supply

When you have completed this chapter, you should be able to:

Explain what is meant by the green revolution.

Describe how the use of pesticides and fertilizers is both beneficial and harmful.

Define the green movement.

Corn is food for people and animals. Farmers must be aware of the benefits and problems associated with growing crops. Chemicals used to increase production may pose health threats, especially for children. Runoff from farms pollutes many of our waterways and freshwater supplies. Mismanagement can deplete the soil and cause erosion. To increase crop production, wetlands and wildlife habitats are converted to farmland. In this chapter, you will learn how productivity is increased and the food supply is protected.

37.1 WORLD FOOD RESOURCES

The rapid growth of world population is due in part to the success of modern agriculture. The yield of food per hectare has grown steadily over the years. The dramatic increase in harvests is due to the use of modern machinery, fertilizers and pesticides, and the introduction of new crop varieties that produce better yields. The revolution in agricultural practices has been essential for the growth of developing nations that have large and expanding populations.

The Green Revolution

The introduction of new varieties of cereal grains and the use of fertilizers to increase their yield have become common agricultural practices throughout the world. The **green revolution** describes the widespread use of high-yield crop varieties, pesticides, and better management techniques. Planting new high-yield, rapidly maturing varieties of grain allows farmers to produce multiple harvests during the year. With this practice, some nations have more than doubled farm production. However, the high productivity has its costs. These new varieties usually require large amounts of fertilizers, pesticides, and irrigation water. The need for more fertilizers increases costs and slowly decreases farm income. In addition, the practice of growing just these new varieties has reduced the genetic diversity of crops. The loss of genetic diversity leads to the potential for widespread destruction by insect pests or unfavorable climatic conditions.

By transplanting cereal grasses to new regions, people have changed whole ecosystems. Wheat is a grass that originated in Asia and was introduced to the midwestern United States as an agricultural crop. There it replaced the original prairie grasses. Instead of the mixture of plant species the native grassland originally contained, these new cultivated wheat fields are composed of just one species of grass.

Monoculture of wheat consists of vast areas planted with identical crops. This can accelerate the spread of diseases such as wheat blight

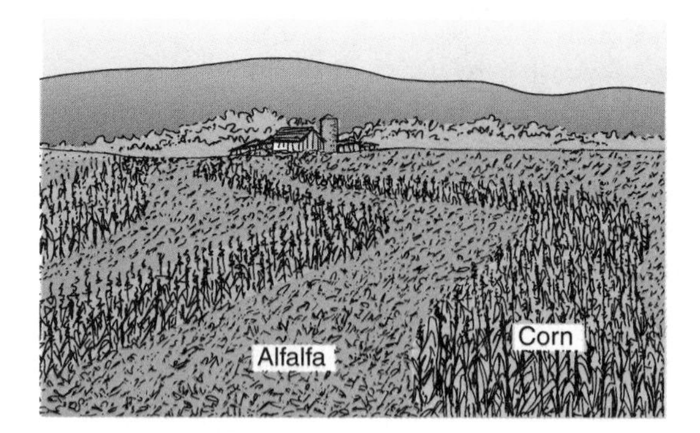

Figure 37-1 **A mixed planting of several crops helps (left) insures that whole fields will not be lost to a disease or pest. Fields planted in only wheat (right) are at risk from wheat blight.**

and wheat rust and kill an entire crop. (See Figure 37-1.) The threat of crop loss has sparked the development of hardier, disease-resistant varieties of wheat. Research into new seed varieties and agricultural techniques is an ongoing process. Even though plant geneticists may be able to breed new disease-resistant varieties, they take many years to develop. It is a race against nature. Mutant forms of disease organisms always seem to be one step ahead. This requires the development of new disease-resistant varieties of wheat. A new disease-resistant variety has only about five years before the disease organism adapts to it.

In the United States and Canada, highly successful grain harvests have resulted in surpluses. Therefore, wheat from North America is exported to many countries that are unable to produce enough grain to support their populations. Thus the prairies of central North America have come to be known as the "breadbasket of the world."

At one time, grasslands covered half of Earth's land area. Today, most of Earth's grasslands are used for grazing cattle and sheep and growing wheat and corn. They have been lost as natural ecosystems. In some places, their overuse has created barren deserts.

In the 1930s and again in the 1950s, widespread droughts affected the midwestern United States for several years. With cultivated crops destroyed, winds blew away the region's rich topsoil. This led to dust bowl conditions. When the rains returned, neither the native grasses nor cultivated crops could be reestablished in the poor soil that remained.

Growing the same crop on the same land for many years drains the land of nutrients essential for plant growth. Fertilizers become necessary to aid the growth of crops. When fertilizers are washed out of the soil, they can contaminate water supplies with nitrates and phosphates.

Major Food Crops

Wheat is the world's most important cereal grain crop. Cereal grains are the seeds of plants that belong to the grass family. (Look at Figure 23-4 on page 409 again.) Wheat has a high nutrient content, which makes it a **staple food**, or basic food, for more than one-third of the world's population. It is grown throughout most of the temperate regions of the world.

Rice, another cereal grain, is the principal food crop in Asia. It is usually cultivated year-round in water-filled fields, called paddies, and on terraced areas of steep mountainsides.

Corn, or **maize**, is a cereal grain that is the staple crop throughout South America and Africa. The United States is the world's largest producer of corn, with the major portion of the crop used as livestock feed. Barley, oats, rye, and millet are other important cereal grains.

Potatoes grow best in the cool, moist portions of the temperate regions. The edible potato tuber is an underground storage stem. The potato's high starch and low protein content make it attractive mainly as a secondary food source. Sweet potatoes are underground storage roots that are rich in starch and protein. They grow best in the wetter regions of the world. (See Figure 37-2.)

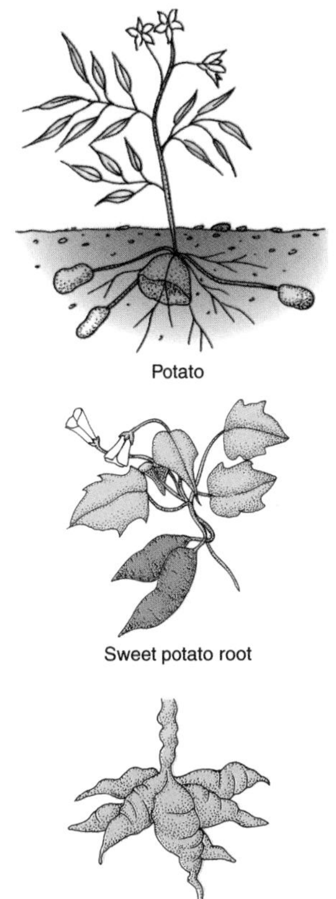

Figure 37-2 Potato, sweet potato, and cassava are food crops.

Cassava is an important food crop in Africa. The plant is highly resistant to drought. The edible root has a high starch and low protein content.

Soybeans are the seeds of a legume. Their high protein content makes them an attractive food source in developing nations. Since legumes enrich the soil with nitrogen, they are often rotated with other crops.

Livestock Resources

The domestication of animals began at about the same time as the development of agriculture. Humans discovered that it was easier to trap and fence in the wild animals they used as food than to hunt

Figure 37-3 **The domestication of cattle has produced breeds that are used for beef and others that are used for the production of milk.**

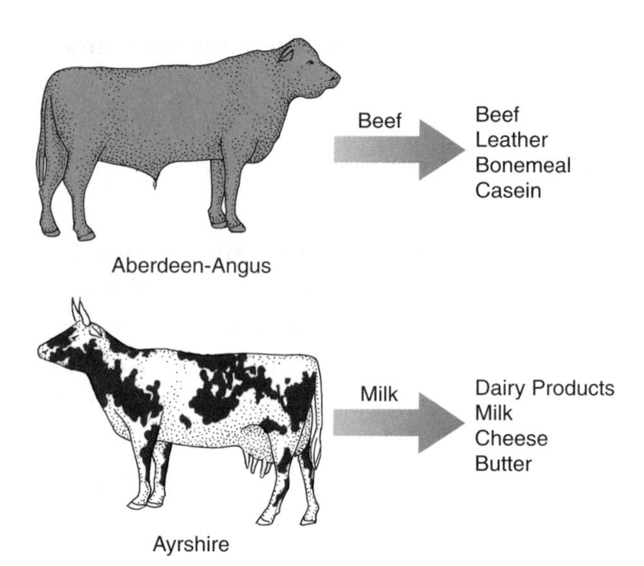

for them. The first animals to be domesticated were probably caught foraging in cultivated fields. Over time, a wide variety of animals have been domesticated as sources of meat, milk, or eggs. (See Figure 37-3.) They include cattle, sheep, goats, camels, pigs, chickens, turkeys, ducks, and geese. These animals also supply humans with hides, skins, wool, feathers, and bristles.

Most domestic animals forage on plants that are not used as food for people. Thus they do not compete with us for food. Except for goats, which can strip the land bare of all vegetation and cause great environmental damage, animals can usually graze on lands without adversely affecting the natural environment. Domestic animals that graze on pastureland efficiently convert unusable plant materials into usable protein and energy. However, grain-fed animals take in 10 calories of usable food energy for every 1 calorie they produce. This is viewed by some as an inefficient and costly squandering of food resources.

37.1 Section Review

1. How have modern agricultural practices damaged the environment?
2. What is the green revolution?
3. Why are cereal grains used as staple foods?

37.2 INCREASING PRODUCTIVITY

The green revolution while providing more food has also harmed the environment. Runoff that contains fertilizers and pesticides damages waterways, lakes, and underground aquifers. It may also pollute municipal water supplies and reduce the number of fish in commercial fisheries. Agriculture makes land vulnerable to erosion and depletes the soil of minerals. In particular, pesticides have been responsible for the death of many mammals, fish, and birds, as well as the loss of reproductive ability in many species.

Chemicals That Aid Production

The use of agricultural chemicals, such as fertilizers, herbicides, and pesticides, has increased dramatically in recent years. Since 1960, fertilizer use is up 300 percent, while herbicide and pesticide use is up more than 150 percent. Fertilizers are compounds that are added to the soil to provide plants with adequate nutrition for growth. They provide nitrogen, in the form of nitrates; phosphorus, in the form of phosphates; potassium; sulfur; and trace minerals, such as magnesium, copper, zinc, and boron. (See Table 37-1.) Fertilizers may contain just one nutrient or a combination of nutrients. Agricultural runoff that contains fertilizers now pollutes many of the nation's lakes, streams, and groundwater supplies.

Pesticides kill, drive away, or change the behavior of a pest so that it is no longer a problem. There are a variety of substances that fall into this category. Some are **insecticides**, compounds that kill insects, such as DDT, chlordane, aldrin, malathion, or sevin. Runoff that contains these compounds pollutes many of our rivers and streams. Another type of pesticide is the microbial insecticide, composed of bacteria and viruses, which are used to kill specific insect

TABLE 37-1 NUTRIENTS ESSENTIAL TO PLANTS

Primary nutrients	Nitrogen, phosphorus, potassium
Secondary nutrients	Calcium, magnesium, sulfur
Micronutrients	Boron, chlorine, copper, iron, manganese, molybdenum, zinc

pests. **Fungicides** are used to treat fungal diseases of plants, for example rust or blights. **Herbicides**, such as 2,4-D and paraquat, rid cultivated areas of weeds and other unwanted plants. Many of these pesticides are dangerous because they accumulate in animal and plant tissues, are highly toxic, or persist in the environment for long periods.

Irrigation Water

For thousands of years, canals, dams, dikes, and storage reservoirs have been used to transport water to irrigate desert areas for the growth of crops. The development of civilization often has been spurred on by the increased agricultural production achieved through irrigation. In recent times, the area of irrigated agricultural land in the world has doubled. Often the process of diverting large amounts of water to these projects creates water shortages in other areas. The low nutrient content of desert soils often requires the addition of fertilizers to sustain yields. Runoff from these regions has a high nutrient content, which can pollute water supplies downstream.

Desert soils often contain high salt levels, since there are few streams available to dissolve minerals and carry them away. Therefore, large quantities of irrigation water are needed to dissolve and dilute these salts before the land is suitable for agriculture. Continued irrigation often forms salt deposits on the surface of the land. Excess salts harm some plants and competes with roots for water.

Agriculture's greatest impact on the nation's freshwater supplies results from extensive land irrigation practices. (See Figure 37-4.) By reimbursing landowners for much of the costs, United States policy encourages the use of water for irrigation. This has allowed previously barren land to be exploited for agriculture. In some regions, groundwater supplies have been pumped dry. The Ogallala Aquifer has been partially depleted due to agricultural water usage.

Soil Conservation Practices

Plowing the land each spring makes it vulnerable to drying. When soil is dry, it may be washed into lakes and streams by heavy rains,

Figure 37-4 **Irrigation in the desert produces green areas that contrast with the desert soil.**

or blown away by strong winds. In many parts of the world, this leads to desertification, or the creation of deserts.

To resolve these problems, the United States Soil Conservation Service recommends that, each year, 25 percent of all farmland be allowed to go unplowed. This practice recycles nutrients and prevents the soil from drying out. It also increases production since soils are given a chance to regain lost nutrients.

Other conservation practices are used, such as contour plowing, or crosswise tilling of hills, and crop rotation. (See Figure 37-5.) Treebreaks are planted to reduce wind erosion. These methods protect both soil and water resources. Diversifying operations to include a variety of crops and livestock can increase productivity of the land while at the same time reducing production costs.

Figure 37-5 **Contour plowing (left) and windbreaks (right) reduce soil erosion.**

Camel

Water buffalo

Poultry

Figure 37-6 Animals have been domesticated to serve a wide range of human needs.

Breeding New Plants and Animals

Domestication is a process by which populations of animals are selectively bred over many generations to adapt them to life in a human-made habitat. Once domesticated, these species rely on human care to satisfy most of their needs. Humans learned early that they could extend their uses for domesticated animals. Through selective breeding, people could manipulate the characteristics of an animal to emphasize those traits that were most desirable. Size, length or thickness of coat, type of meat, amount of fat, intelligence, and strength are some traits that can be controlled through breeding. Cattle varieties have been specifically bred to exhibit the best characteristics for meat or milk production, or as draft animals. (See Figure 37-6.)

Many varieties of horses have been developed through selective breeding. Arabians have been bred for speed and endurance, trotters and thoroughbreds for racing, the quarter horse for handling cattle, Clydesdales for pulling heavy wagons, and Shetland ponies originally to work in mines and now as pets.

By domesticating plants and animals, humans are able to select those varieties that best meet their needs. Plants can be cultivated that produce greater yields, are resistant to certain diseases or pests, or are able to tolerate adverse climate conditions. Widespread planting of these varieties can lead to the creation of new species over time.

Vegetables that appear to be very different from each other may actually be varieties of the same species. As shown in Figure 37-7, cauliflower, broccoli, Brussels sprouts, and kale all were developed from the wild cabbage. Each stores starch in a different part of the plant. As a result of human influences, the species has been modified to produce several different varieties.

Agricultural Biotechnology

One of the modern marvels of technology is the increase in agricultural productivity through the practice of genetic modification of organisms. These modifications have led to improvements in production, crop varieties, animal breeds, and food quality.

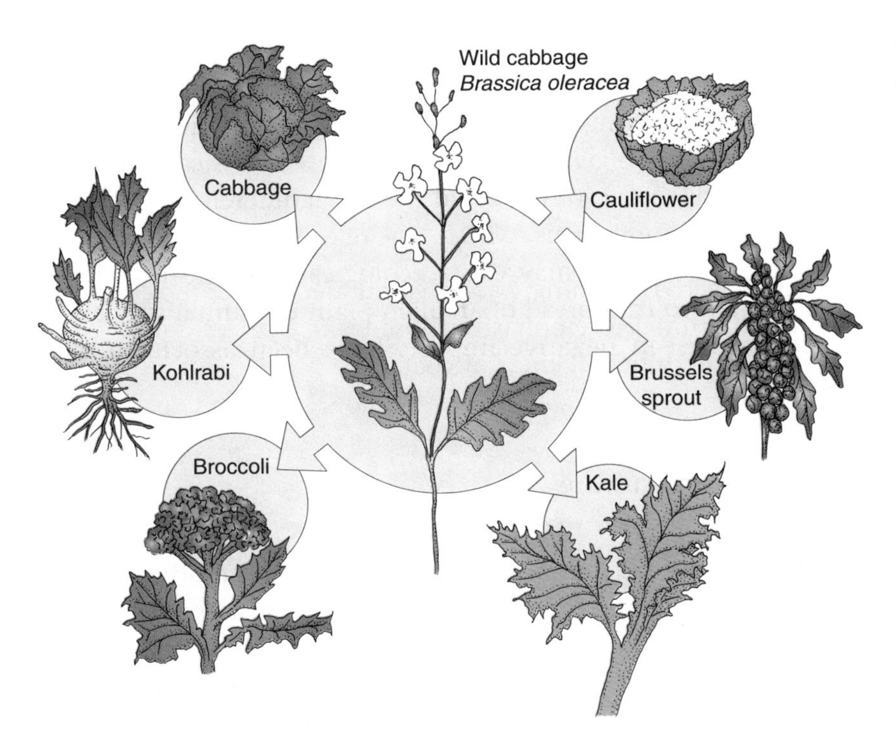

Figure 37-7 Varieties of wild brassica yield many different food crops.

Genetic engineering is the manipulation of DNA to develop gene combinations that have never existed before. Scientists use parts of the gene structure of living organisms to modify life-forms, giving them desirable traits. By changing the information stored in the genes of an organism, scientists also alter the traits of future offspring. Since species barriers are overcome, the transfer of genetic information can be made between organisms that would not normally exchange genes.

One outcome of genetic engineering is the production of hormones and vaccines that can increase food production. Pigs and fish have been supplied with genes that increase their rate of growth and produce leaner meat. A microorganism has been created that kills caterpillar pests. A bacterial gene has been transferred to corn to provide resistance to attacks by the European corn borer, an insect pest. Tomatoes have been bred that are resistant to attack from hornworms.

A major global problem is the control of genetic engineering. These experiments change the information stored in genes of an organism, and the characteristics of the new generation are thus altered. This can pose a danger to the environment if a potentially deadly organism is produced and accidentally released.

The federal government, as well as various state governments, regulates all research in genetic engineering. Research is carried out in self-contained laboratories to avoid the accidental release of genetically engineered organisms (GEOs). The release and spread of GEOs into the environment may cause ecological complications with results similar to the spread of an alien plant or animal species. So far, no unforeseen negative impacts have been associated with GEOs.

Alternative Agriculture

Modern farming practices include a variety of alternative agricultural methods ranging from organic farming to sustainable agriculture. **Organic farming** replaces the use of agricultural chemicals with natural fertilizers such as manure and crop residues and uses biological pest controls rather than chemical pesticides. Despite long-held beliefs among farmers that agricultural chemicals are essential for boosting crop yields, some farmers have found they can make a profit when adopting alternative methods.

In the practice of crop rotation, the type of plants grown in a particular area is changed on a regular cycle. This practice limits weed growth, deters insect infestations, and reduces the effects of soil erosion. At the same time, crop rotation that utilizes nitrogen-fixing legumes increases soil nitrogen. Production costs are cut through the reduction in purchases of fertilizers, pesticides, and other farm chemicals.

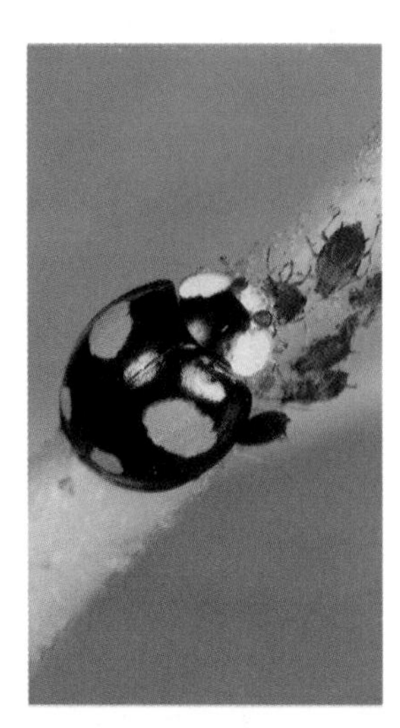

Figure 37-8 Many gardeners use ladybugs to eat aphids rather than using chemical pesticides to kill them.

Integrated pest management, through the use of biological pest controls rather than pesticides, decreases the presence of chemical contaminants in food crops. **Biological pest control** is the use of a naturally occurring predator or parasite (bacteria or virus) to control an agricultural pest. An example is the use of a predatory insect such as ladybugs to consume a crop pest such as aphids. (See Figure 37-8.) These practices are essential to organic farming, where crops are grown under natural conditions without the use of chemical additives. Other methods of pest management include mixed planting and clearing of pest-breeding areas. The intentional release of sterile male pests and the spraying of insect hormones both control pests by disrupting their breeding cycle.

Government Aid to Agriculture

The United States Department of Agriculture was established in 1862. The original role of the department was to distribute plants and seeds to farmers. Today its functions also include subsidizing the nation's farmers to protect crop prices as well as overseeing the quality and safety of meats and agricultural products.

The payment of money in the form of **farm subsidies** has encouraged farmers to intensify or diminish their growth of certain crops. Subsidies often result in huge yields of produce and the payment of large sums to farmers. Some farmers may derive as much as half their income from subsidies. Farm subsidies are often based on the average number of acres a farmer plants in a single crop over a five-year period. Thus farmers benefit more over this time period when they plant one crop. Subsidies often create huge surpluses and lead to waste. Production incentives due to subsidies can lead to soil depletion and soil erosion.

37.2 Section Review

1. Describe the dangers associated with the continued use of fertilizers to increase the productivity of a field.
2. Explain why selective breeding is used to produce new varieties of a domesticated species.
3. What are some important soil conservation practices?

37.3 BIOTECHNOLOGY AND THE ENVIRONMENT

Biotechnology allows farmers to employ more environmentally friendly farming techniques to manage agricultural pests and diseases, making the soil more productive and at the same time conserving our land and water resources. By genetically altering crops to make them more hardy, farmers are able to reduce soil erosion, decrease the impact of pesticides on the environment, increase yields, use less energy, produce more nutritious crops, and make better use of low-yield land.

Through biotechnology, it is possible to produce agricultural products that contain edible vaccines. These vaccines can be efficiently administered to large populations at a fraction of the present costs. Products can be engineered to contain lower amounts of saturated fats and higher amounts of antioxidants, to fight heart disease and cancer. Using biotechnology, scientists can inactivate allergens in foods, protecting those who may suffer from food allergies.

A genetically altered tomato plant has been developed that can grow in salty soil (high sodium content) and still produce tasty tomatoes. A DNA sequence from another plant was inserted into the tomato plant, so that the tomato produces more of a certain protein that can store sodium in the leaves of the plant.

37.3 Section Review

1. List three improvements farmers will see when they employ environmentally friendly farming techniques based on biotechnology.
2. How have tomato plants been genetically altered?

37.4 THE GREENING OF AMERICA

Many people feel that if we are going to save our planet, we must make major changes in our way of life. Many organizations and political groups that campaign on environmental issues put forth this view. These groups are part of the **green movement**. They share a perception that the world is in the midst of a major ecological crisis that requires political and social change. The wave of ecological concern has boosted the appeal of membership in many environmental groups. With large memberships and funds to provide resources, these groups have a huge impact on national politics and policies.

Companies around the nation are cleaning up their act and getting into the business of making their products **environment-friendly**, that is, doing the least damage to the environment in production, use, and disposal of their products. These businesses are trying to prevent future pollution problems rather than simply correcting past problems. Environmental spending by these companies includes research into methods to make their manufacturing processes more efficient, discovering ways to conserve energy and fuel, and changing over to the production of nonpolluting or biodegradable "green" products and packaging.

Some environmental groups are lobbying for environmental taxes to attack the problems created by pollution. Environmental taxes would be imposed on products that pollute or degrade the environment or deplete natural resources. In addition to taxing pollution, tax incentives may be offered to promote conservation or the efficient use of natural resources. These tax policies could be a powerful tool for protecting the environment as well as encouraging wise use of natural resources. Indirectly, the taxes would lead to improved public health, increased employment, and benefits to the nation's economy.

Ecolabeling is advertising of a product's environmental benefits on the product or packaging. One of the most common ecolabels is the "chasing arrows" symbol, used to denote that a product is recyclable. (See Figure 37-9.) Opinion polls show that most consumers prefer products that do not harm the environment. Another example of ecolabeling is packages that are marked "biodegradable." These products are preferred for disposal in landfills since they readily decompose.

Recycling symbol

Figure 37-9 Ecolabels indicate environmentally friendly products.

Cities across the United States have instituted recycling programs that collect newspapers, corrugated cardboard, aluminum cans, steel food cans, plastic milk containers and soda bottles, yard wastes, tires, lead-acid auto batteries, and used motor oil. (See Figure 37-10 on page 690.) But recycling involves a complex chain of activities that only begins with the collection process.

Once collected, recyclable materials must be sorted. Then they are prepared for sale to interested buyers. This may require washing, shredding, compacting, or bleaching. The next step is transporting the material to manufacturers, where it can be made into new products. Unless there is a consumer demand for these secondary products, the whole system breaks down.

Figure 37-10 **This pie chart illustrates the make up of garbage.**

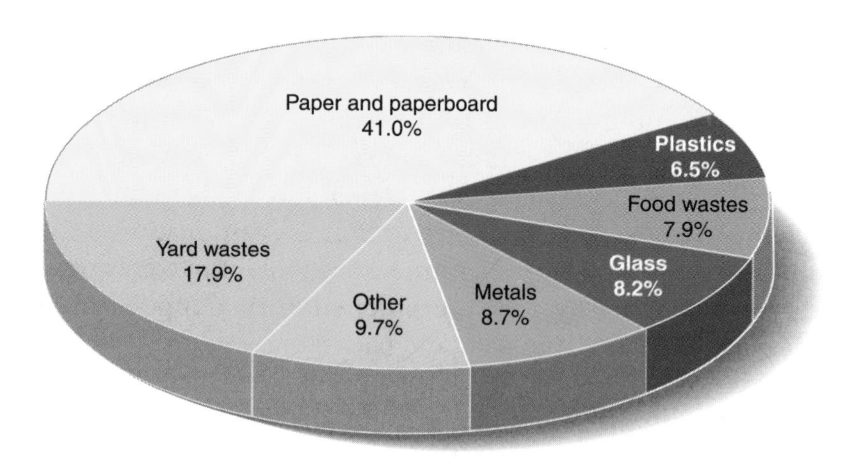

One of the main reasons for our destructive impact on the environment is our consumerism. People who live in industrialized nations tend to use resources at a rate that is out of proportion to their percentage of the world population. We must reassess our level of consumption and find options that reduce the impact we have on the environment. We should try to select products that are environment-friendly, since the use of these products can have a huge impact on the protection of the environment. The following is a list of the characteristics of environment-friendly products.

- Wrapped in minimal packaging
- Packaged in materials that can be recycled or reused
- Able to be reused over and over, for example, cloth napkins, washable dishes and utensils, pens with replaceable cartridges, razors with replaceable blades, cameras with replaceable film, and flashlights with rechargeable batteries
- Made to last; do not have built-in obsolescence
- Composed of biodegradable substances
- Manufactured using a minimum of energy resources

Americans have become more energy conscious. Home insulation and solar water heaters have reduced the use of fuels. The use of energy-efficient appliances is becoming more widespread as states have adopted efficiency standards. Manufacturers are producing appliances to meet the public demands for reduced electric use. (See Figure 37-11.) Automobile engines are designed to consume less fuel. Many people are willing to pay more for products or packaging that does not

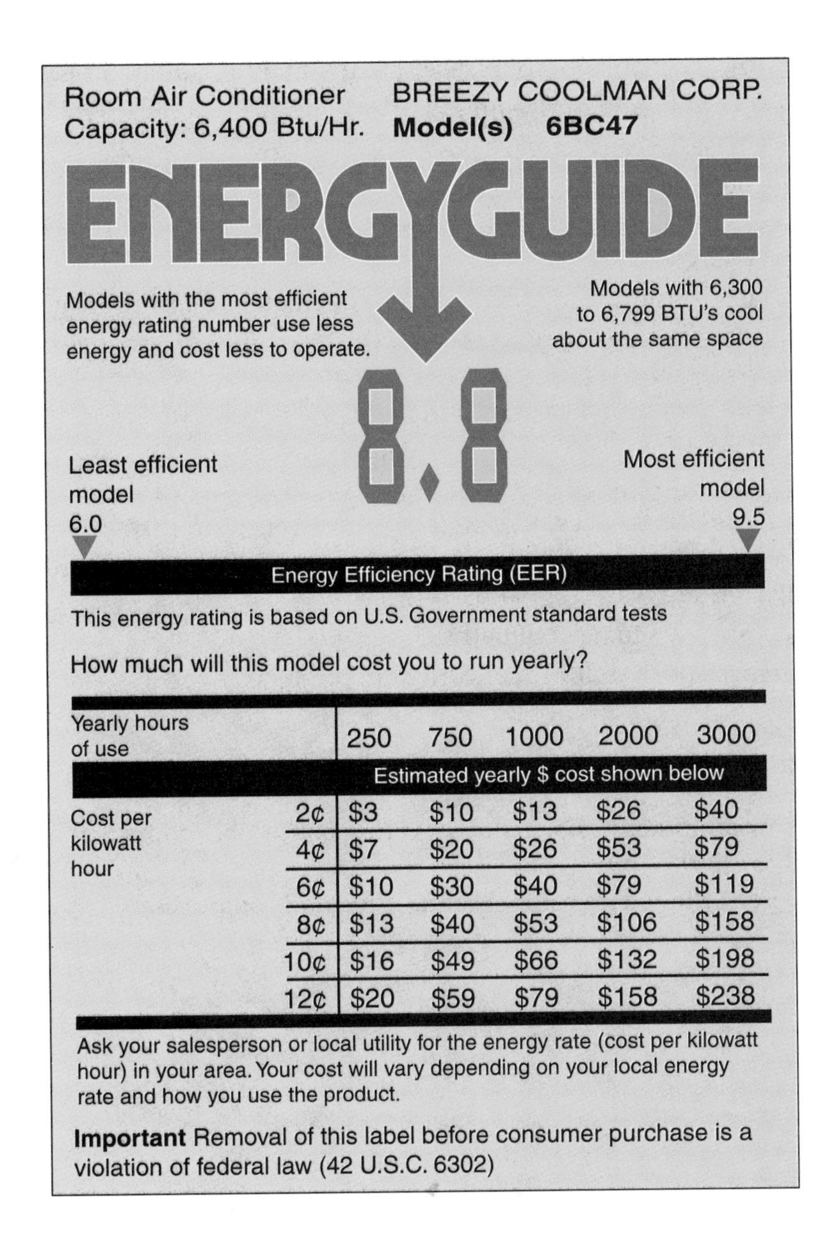

Room Air Conditioner
Capacity: 6,400 Btu/Hr.

BREEZY COOLMAN CORP.
Model(s) 6BC47

ENERGYGUIDE

Models with the most efficient energy rating number use less energy and cost less to operate.

Models with 6,300 to 6,799 BTU's cool about the same space

8.8

Least efficient model 6.0

Most efficient model 9.5

Energy Efficiency Rating (EER)

This energy rating is based on U.S. Government standard tests

How much will this model cost you to run yearly?

Yearly hours of use		250	750	1000	2000	3000
		Estimated yearly $ cost shown below				
Cost per kilowatt hour	2¢	$3	$10	$13	$26	$40
	4¢	$7	$20	$26	$53	$79
	6¢	$10	$30	$40	$79	$119
	8¢	$13	$40	$53	$106	$158
	10¢	$16	$49	$66	$132	$198
	12¢	$20	$59	$79	$158	$238

Ask your salesperson or local utility for the energy rate (cost per kilowatt hour) in your area. Your cost will vary depending on your local energy rate and how you use the product.

Important Removal of this label before consumer purchase is a violation of federal law (42 U.S.C. 6302)

Figure 37-11 The Energy Guide label helps consumers identify energy-efficient appliances.

harm the environment. By altering our lifestyles and thinking "green" we may be able to reverse some of the conditions that have degraded the environment. Here is a list of environmental conservation practices you can follow. (See Figure 37-12 on page 692.)

- Buy fewer consumer products.
- Extend the life of products we buy and use.

Figure 37-12 **People of all ages can practice environmental conservation.**

- Borrow from the library instead of buying books, tapes, or CDs.
- Give hand-made cards and gifts.
- Use reusable bags to carry purchases.
- Buy necessities in bulk.
- Walk, ride a bicycle, or use public transportation to conserve fuel and energy.
- Reduce consumption of electricity, water, and fuel in the home.
- Dry clothes on a drying rack or line instead of using gas or electric clothes dryers.
- Carpool or use public transportation to school or work.
- Reduce use of water for gardens and lawns.
- Replace plumbing that consumes large amounts of water with water-saving devices. Repair dripping faucets promptly.
- Do not let water run continuously when washing dishes, brushing teeth, or washing hands.
- Take showers instead of baths.
- Purchase only environment-friendly products.

37.4 Section Review

1. How do industries benefit by making consumer products that are environment-friendly?
2. Describe how environmental taxes could promote conservation.

LABORATORY INVESTIGATION 37

Chemical Fertilizers

PROBLEM: *What are the effects of fertilizer on plants?*

SKILLS: *Measuring, observing, interpreting data*

MATERIALS: *5 plant pots (7 cm × 7 cm), commercial fertilizer, graduated cylinder, 200-mL beaker, 4 stoppered 100-mL bottles, marking pencil, balance, seeds (radish, pepper, or squash), sand, 5 saucers or halves of petri dishes, water, centimeter ruler*

PROCEDURE

1. Use the balance to measure 10 grams of solid fertilizer. If using liquid fertilizer, use the graduated cylinder to measure 10 mL.
2. Place the fertilizer in a 200-mL beaker and add 100 mL of water. Stir until the fertilizer dissolves. Pour the mixture into one of the 100-mL bottles. Mark this bottle "10-to-1 fertilizer mixture." Clean the beaker.
3. Measure 10 mL of the 10-to-1 mixture. Pour it into the clean 200-mL beaker. Add 100 mL of water. Stir the mixture and pour it into an empty 100-mL bottle. Mark this bottle "100-to-1 fertilizer mixture." Clean the beaker.
4. Repeat step 3, using the 100-to-1 fertilizer mixture. Mark the third bottle "1000-to-1 fertilizer mixture."
5. Repeat step 3, using the 1000-to-1 fertilizer mixture. Mark the fourth bottle "10,000-to-1 fertilizer mixture."
6. Fill the five pots with sand to within 2 centimeters of the rim. (Paper cups can be used instead of pots, but a hole must be made in the bottom to allow for drainage.) To catch water that drains, place each pot on a saucer or petri dish half. Label the pots A through E.
7. Plant 10 seeds in each pot. Add 10 mL of water to each pot. Water regularly until the seeds sprout and begin to show above the sand.

8. Begin watering with 10-mL amounts of the fertilizer mixtures. Water pot A with the 10-to-1 mixture, pot B with the 100-to-1 mixture, pot C with the 1000-to-1 mixture, pot D with the 10,000-to-1 mixture, and pot E with 10 mL of plain water.

9. Water the pots every three or four days. Copy Table 37-2 into your notebook and use it to record the watering dates and measure and record the growth of the plants in each pot. Look for residue that may form in the saucers or on the surface of the sand.

TABLE 37-2 EFFECT OF FERTILIZER ON THE GROWTH OF PLANTS

Date Pots Watered	Height of Plants, Pot A	Height of Plants, Pot B	Height of Plants, Pot C	Height of Plants, Pot D	Height of Plants, Pot E

OBSERVATIONS AND ANALYSES

1. What was the chemical composition of the fertilizer you used? It should be listed on the package.

2. Describe any residue you observed.

3. Under normal use in a garden or farm, what effects would the residue have on the environment?

4. What concentration of fertilizer was best at promoting plant growth?

Chapter 37 Review

Answer these questions on a separate sheet of paper.

Vocabulary

The following list contains all the boldfaced terms in this chapter.

biological pest control, ecolabeling, environment-friendly, farm subsidies, fungicides, genetic engineering, green movement, green revolution, herbicides, insecticides, maize, organic farming, pesticides, staple food

Fill In

Use one of the vocabulary terms listed above to complete each sentence.

1. ______ are used to rid cultivated areas of weeds and other unwanted plants.
2. The government program that encourages farmers to plant or cut back on certain crops is ______.
3. The manipulation of genes to create new life-forms is called ______.
4. The agricultural practice that does away with the reliance on agricultural chemicals is called ______.
5. The advertising of a product's environmental benefits on a package is called ______.

Multiple Choice

Choose the response that best completes the sentence or answers the question.

6. Which of the following is a cereal grain? *a.* spinach *b.* rice *c.* kale *d.* beans
7. Which of the following is a staple food? *a.* beef *b.* cauliflower *c.* cabbage *d.* corn
8. Which of the following is an important compound supplied to plants in fertilizers? *a.* aldrin *b.* phosphates *c.* paraquat *d.* DNA
9. Which of the following is a domestic breed of horse? *a.* donkey *b.* zebra *c.* Clydesdale *d.* eohippus
10. Which of the following was developed from wild cabbage? *a.* maize *b.* spinach *c.* broccoli *d.* millet
11. Genetic engineering can create new life-forms with desirable traits through the manipulation of *a.* DDT. *b.* DNA. *c.* 2,4-D. *d.* GEOs.
12. Which is an example of a biological pest control method? *a.* parathion *b.* fungicide *c.* ladybugs *d.* all of the above

Short Answer (Constructed Response)

Use the information in the chapter to respond to these items.

13. What is the green revolution?
14. Name three staple food crops and the part of the world in which they are produced.
15. Why are goats not environment-friendly farm animals?
16. What is the purpose of a pesticide?
17. What country is the world's largest producer of corn?
18. List two problems caused by growing the same crop on the same land for many years.
19. Name three soil conservation practices.

Free Response

Use the information in the chapter to respond to these items.

20. Describe the effects that the introduction of wheat had on the North American prairies.
21. How is organic farming different from ordinary farming?
22. Explain how the payment of farm subsidies can sometimes lead to soil depletion.
23. How has the American lifestyle contributed to the misuse of resources and damaged the environment?
24. What are two possible dangers of using ladybugs to control aphids?
25. What are two possible advantages of using ladybugs to control aphids?

Environmental Issue

Consumers have come to expect a readily available, abundant, safe, and healthy food supply. At present, there is no evidence suggesting that genetically altered foods are unsafe. In some states, legislation has been passed that bans or regulates genetically modified crops. Hold a class discussion to answer the following questions. Does your state regulate or ban genetically modified crops? Do you think people should be kept from eating genetically engineered food? Should people be allowed to choose whether or not to eat genetically engineered food?

CHAPTER 38
The Exploding Population

When you have completed this chapter, you should be able to:

Explain how birth rate, death rate, emigration, and immigration affect population growth.

Identify the means available to control population growth.

Distinguish between the health care crisis in developed and developing nations.

When Christopher Columbus arrived in the New World in 1492, Earth's total population was slightly more than 400 million people. In the five centuries since then, the population has grown to more than 6 billion people. The present size of the global population stresses the limited resources of the planet and creates in many areas crowded conditions that contribute to food shortages, famine, and the spread of disease.

38.1 STANDING ROOM ONLY

Throughout much of its history, the human species coexisted in harmony with the other species in the biosphere. This relationship was based on the dependency of humans on the resources the biosphere offered. However, within the past 500 years, human society has changed this relationship. Developing cultures advanced a new philosophy in which humans were placed at the center of the universe, different and apart from all other living things. People exploited the biosphere and modified it for their own uses. As the human population grew, it intruded on and converted more and more of Earth's natural ecosystems to meet its increasing needs.

Suddenly, we have discovered what primitive humans knew all along: we are dependent on the other forms of life in the biosphere. What affects them also affects us, since Earth is home to all creatures. We must think of ourselves not as lords of the biosphere but as guardians of the biosphere. In effect, the human species is in the process of consuming and polluting the limited resources of the biosphere and possibly bringing about the destruction of all life.

The population explosion describes the unusually rapid growth the human population has experienced during the past 100 years. (See Figure 38-1.) This growth is the result of improved living conditions and disease control in many of the nations of Latin America, Africa, and Asia, along with increased food production and crop yields. It has been a decrease in the rate of deaths, rather than an increase in the rate of births, that has created this explosion.

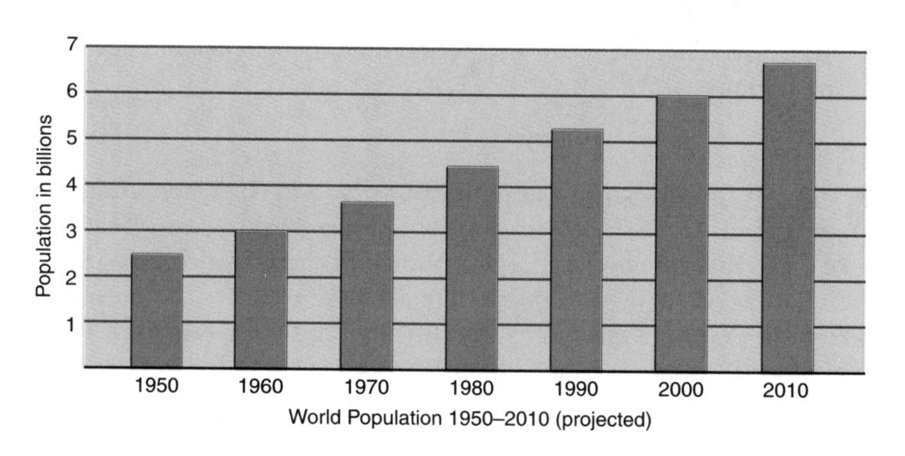

Figure 38-1 What happened to the world population from 1950 to 2000?

Population Statistics

The **birth rate** is a population statistic that is determined by the number of births per thousand people in a population, while the number of deaths per thousand people in a population determines the **death rate**.

The population growth rate of a nation is the yearly growth rate for the population expressed as a percentage. This includes births, deaths, immigration, and emigration. **Immigration** is the movement of people into a country, while **emigration** is the movement of people out of a country.

Both social and economic forces impact on populations. A **stable population** occurs when the age and sex distributions are constant because birth rates and death rates are constant. However, as shown in Figure 38-2, stable populations still experience population growth. Nations with a large population of young people experience growth even when the birth rate is low, because the mortality rate is also correspondingly low. A **stabilized population** is one that has achieved a zero population growth rate or a negative growth rate. **Zero population growth** (ZPG) is the replacement of existing members of a population, that is, two children per family. Ecologists have advocated ZPG as a solution to the problems of overcrowding and environmental degradation.

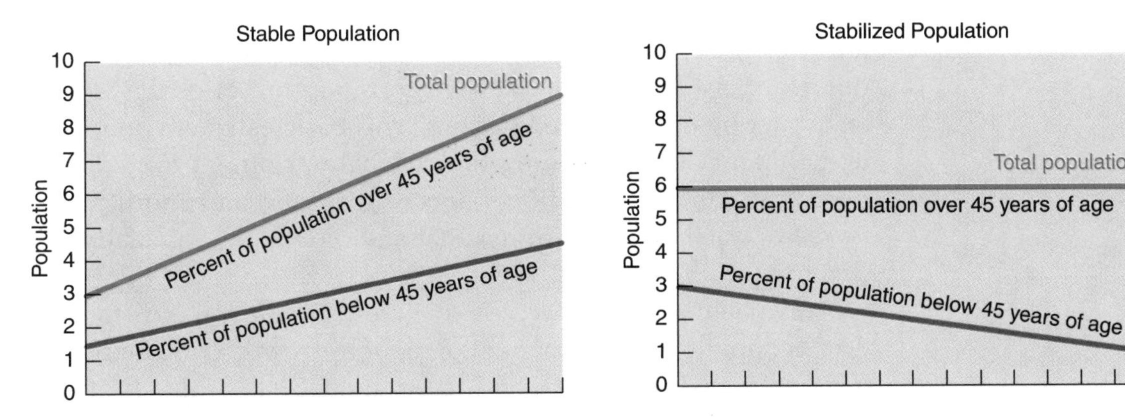

Figure 38-2 **Stable population—the percent of individuals above and below 45 years of age remains constant, but the total population increases. Stabilized population—the total population does not change, but the age distribution within the population changes due to a declining birth rate.**

Only a handful of countries, all in Europe, have reached zero population growth or a negative population growth. Even though China has advocated a one-child policy for every family, the nation has not yet reached ZPG. This is due to reduced mortality rates, which in turn have led to an aging population.

The United States Census Bureau-International Database estimated that as of December 2004, the world population was 6.48 billion. They project the population to rise to 8.6 billion by 2025 and 9.2 billion by 2050. These projections show that the population will eventually stabilize at about 11.5 billion, but only if the birth rates continue to decrease throughout Latin America, Africa, and Asia. If the present downward trend in the birth rate should reverse, the population could reach 20 billion by the end of this century. And should the current world population growth rate continue to grow for the next 200 years, the world's population could exceed 100 billion. The question arises, "How many are too many?" What is the optimum population of people that Earth can support? This is difficult to estimate.

Overpopulation and the Biosphere

The most pressing global problem today is the threat that overpopulation presents to the biosphere. The world's growing population is evidence of the conquest of many diseases. However, as mortality rates decreased, famine and overcrowding increased. Unsound technologies have led to the excessive consumption of energy, food, and water by industrialized nations. This has created environmental pollutants at an alarming rate. Global warming, ozone depletion, acid rain, deforestation, vanishing species, and shortages of food and water are all warnings that environmental degradation is occurring.

Many scientists are concerned that rapid population growth hinders economic development. Population growth contributes to poverty, forced migration, and internal conflicts in many of the world's developing nations. If population growth is not curtailed, famine, disease, and infant mortality figures will surely increase. There must be global cooperation toward population stabilization during this century.

38.1 Section Review

1. What has been the main cause for the "population explosion" during the past 100 years?
2. How is the population growth rate of a nation determined?
3. Explain how the human species has changed the carrying capacity of its environment.

38.2 MANAGING THE POPULATION

More than 90 percent of the world's population growth takes place in developing nations. The factors that maintain these high rates are deeply rooted in religious, cultural, social, and economic biases that have existed in these regions for many generations. For some poor parents, children represent potential workers who will someday contribute toward the household income. Others see children as caregivers who will support them in old age.

In those nations with a high population growth rate, the **fertility rate**, or the average number of childbirths per woman, is four, while in industrialized nations with stabilized populations the fertility rate is two. Population stabilization can be achieved only through methods that reduce fertility rates and thus the birth rate. In poor societies, high fertility rates and the accompanying population growth are detrimental to both women and their children. As the population increases, the resources, especially food, available to each person decrease. **Birth control**, or methods that prevent or terminate pregnancies, can help stabilize populations. These methods include the avoidance of sex during fertile periods, using mechanical or physical barriers that prevent conception, surgical methods, chemicals that interfere with fertilization, and abortion. **Contraception** is the use of a device or chemical substance that prevents **conception**, or the fertilization of an ovum.

Family Planning

Family planning is a strategy that allows couples to determine the number and spacing of their children. It does not necessarily mean fewer children, but it does imply that couples will control their reproductive lives and make conscious decisions on the number of children they will have and when these children will be born. Birth control is an essential aspect of family planning since it eliminates chance from the process of conception.

Family-planning services are available to 95 percent of the world's population. However, the United Nations Population Fund finds that more than 300 million women in developing nations still do not have access to safe or effective means of contraception. Due to the economic pressure to increase the number of available workers, many nations ignore contraception. This often increases the number of individuals that are forced into poverty and raises the rate of child mortality.

In many nations, professionally run family-planning services have improved the health of both mothers and children. Spacing births has reduced infant mortality, since mothers can breast-feed their infants for longer periods. Healthier children create a more vigorous population, thus improving the economy of a nation. Family planning increases the personal choices available to women, while improving their health and reducing fertility rates.

Some say that the right to reproduce should be curtailed when it interferes with the welfare of society in general. But is it ethical to control population? Shouldn't individuals have the right to reproduce freely? This is probably the most hotly debated issue that faces the biosphere. Population control can be seen as a violation of deep religious beliefs, an intrusion on individual privacy, or a means of genocide toward a minority population. Or it can be viewed as a means of protecting the biosphere and thus all living things.

There have been some well-documented cases of abusive practices carried out as part of a nation's population program. This is typified by the program of forced sterilization instituted in India in 1975 and later overturned, as well as coerced abortions practiced in other nations. But on the whole, family-planning programs have been well received by citizens. The distribution of contraceptive information and services has proved to be a cost-effective strategy for limiting population growth. Today, many millions worldwide use contraception as their primary family-planning strategy.

At one time, China's population was among the fastest growing in the world. Thanks to an effective national family-planning program,

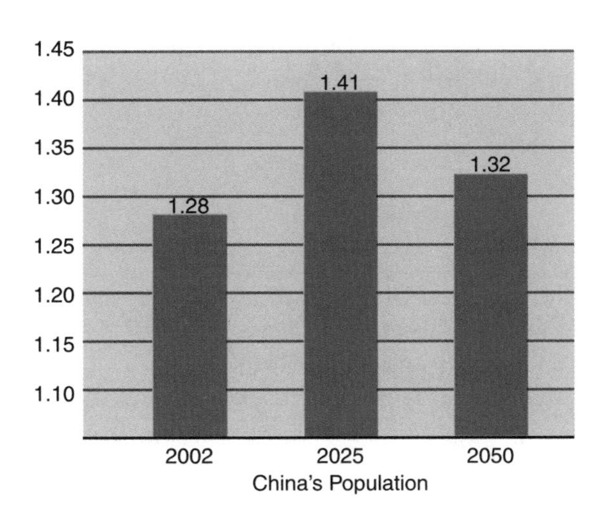

Figure 38-3 **Projected growth from 1997 to 2050 of China's population (in billions).**

coupled with programs to improve the health and education of its people, China's population growth rate has been greatly reduced. China limits couples to two children. The one-child family program offers couples incentives to pledge to have only one child. (See Figure 38-3.)

38.2 Section Review

1. What are some of the cultural biases that contribute to high fertility rates?
2. Explain how family planning can reduce infant mortality rates.
3. Describe two birth control methods that are in use today.

38.3 THE WORLD HEALTH CRISIS

In prehistoric times, human health and survival were primarily limited by the availability of shelter, food, water, and habitat that offered protection from predators. The development of cities, accompanied by crowded living conditions, led to infectious and parasitic diseases becoming the leading causes of sickness and death. During the twentieth century, many of these diseases were brought under control, doubling human life expectancy. Today we find that throughout the world's industrialized nations the major causes of human death are degenerative disorders, such as cancers, and heart and circulatory diseases.

Environmental Health Hazards

A **disease** is a harmful change in the condition of an organism due to an infectious agent, a toxic substance, a physical factor, or psychological stress. (See Table 38-1.) In the world's developing countries, pathogens pose the greatest threats to human health. Pathogens are disease-causing organisms, such as viruses, bacteria, fungi, protozoa, and worms. These organisms cause infectious diseases, which are spread in several ways.

- Direct contact: contact with open sores of an infected person. Sexually transmitted diseases are spread by direct contact during sex.

TABLE 38-1 MAJOR PATHOGENIC DISEASES

Disease	Type of Pathogen	Deaths per Year Worldwide
Respiratory diseases:		6,000,000
tuberculosis	Bacterium	
influenza	Virus	
pneumonia	Virus, bacterium	
whooping cough	Bacterium	
SARS	Virus	
Diarrhea	Protozoan, bacterium	5,000,000
Measles	Virus	2,000,000
Malaria	Protozoan	2,000,000
AIDS	Virus	1,000,000
Tetanus	Bacterium	600,000
Polio	Virus	200,000
Worm infections:	Worm	200,000
schistosomiasis,		
river blindness,		
filariasis, tapeworm		

- Indirect contact: touching doorknobs, food, hypodermic needles, or clothing used by an infected person.
- Aerosols: inhaling pathogens that have been discharged into the air during a cough or sneeze.
- Biological vectors: insects may transmit disease pathogens between an infected individual and healthy individuals through a bite.

Malnutrition, diarrhea, and infectious diseases are the primary causes of infant mortality. Preventive medications to guard against these child killers are inexpensive and easy to administer. However, governments often expend little funding toward solving these problems.

Overcrowding has serious social and environmental effects. Crime rates usually increase, while the quality of life decreases. The effects of overcrowding include stress, unemployment, poverty, pollution, disease, and deteriorating social systems.

In many developing nations, the high mortality rates, especially infant mortality, drain the economy. Illnesses and disabilities caused by malnutrition, poor sanitation, and poverty can affect up to one-third of the population. Insufficient expenditures for health care, a shortage of hospital facilities, and a lack of trained health care professionals add to the problems faced by these nations.

The bulk of expenditures for health care goes toward curing illnesses rather than finding and eliminating the cause of disease. Primary health care and preventive medicine are usually underfunded, while high-technology medicine consumes huge budgets.

The most serious health problems in industrialized nations are a variety of cancers and cardiovascular diseases, many of which arise from factors in the environment and modern lifestyles. In industrialized nations, cardiovascular disease accounts for half of the deaths from disease, while cancers are the cause of about 20 percent. Illnesses related to smoking and tobacco are responsible for 80 percent of all premature deaths. To maintain or improve their health, many people have increased their level of exercise by walking or running.

Increased longevity has caused a crisis in health care because many elderly people require nursing home or hospital care. There is concern that in a nation with an aging population there will not be enough workers to support the economy.

Pandemics

During an epidemic, disease spreads through a large portion of a population in a limited area. An epidemic that spreads over a wide area and affects a large portion of the population is called a **pandemic**. The influenza pandemic of 1918 killed 21 million people worldwide, including 700 thousand people in the United States. Recently, Ebola epidemics in Africa have been kept under control by the combined health resources of the world's nations. However, Ebola is a deadly disease. Should we fail to control the spread of its next epidemic, it could be the world's next pandemic.

AIDS, or acquired immunodeficiency syndrome, is a viral disease that affects the immune system. It is caused by the human immunodeficiency virus (HIV). The spread of AIDS has reached pandemic proportions. As of 2003, worldwide 37.8 million people were living with AIDS/HIV and there were 2.9 million deaths from AIDS. Many new cases of AIDS develop from sexual relations with infected partners. In some central African communities, one person in five is suspected of carrying the virus. At present, we are unable to predict what effect AIDS may have on population growth rates, but there is a fear that it will impact the economic progress of developing nations.

Emerging Infectious Diseases

During the twentieth century, advances in medicine eliminated or greatly reduced many **infectious diseases**, those diseases caused by microbes. It was thought that infectious diseases were a thing of the past. But now, new infectious diseases are emerging and old ones are returning.

New diseases, such as AIDS, Ebola virus, hantavirus, and Lyme disease resist efforts to keep them under control. In the spring of 2003, severe acute respiratory syndrome (SARS), which is caused by a virus, appeared in southern China. It was quickly carried around the world by air travelers. Tuberculosis has made a comeback, as the bacterium has become resistant to those drugs that used to control it. A "bird flu" virus began affecting humans, killing many people in Hong Kong. A "flesh-eating" form of streptococcus bacteria resists treatment by most antibiotics. Once confined to Africa, West Nile virus, carried by mosquitoes, has spread throughout much of the eastern United States. Mad cow disease, transmitted by eating beef from infected cattle, appeared in the British Isles, infecting many people and causing the government to destroy millions of head of cattle. Foodborne disease outbreaks caused by bacteria, viruses, and parasites have sickened many people throughout the nation and led to product recalls.

These unseen enemies are gaining in strength and numbers, posing a challenge for our health care system and threatening our population. Public health professionals will need to increase their understanding of infectious diseases and develop new and better means of dealing with them.

38.3 Section Review

1. How have the factors that affect human health changed since prehistoric times?
2. Explain how infectious diseases may be spread.
3. What is a pandemic?

38.4 SPACESHIP EARTH: FILLING BASIC NEEDS

Figure 38-4 The International Space Station is a step forward in the colonization of space.

Earth can be thought of as a giant spaceship that follows a path around the sun and through the Milky Way galaxy. As with any spaceship, for example, the International Space Station, it holds finite resources within a finite space. (See Figure 38-4.) The biosphere can be thought of as the contents of the spaceship. As the population grows, the space available for each individual becomes smaller and the resources become scarcer. Each year, the world has to find enough food to feed 100 million more people. To maintain the functioning of the whole system, population size, consumption of resources, and accumulation of wastes must be managed. Population control and the recycling of resources are methods of preserving the integrity of this ecosystem.

In spite of its environmentally destructive tendencies, the human race is actually one of Earth's greatest natural resources. The skills, energy, and creativity of people are inherently renewable far into the future. To protect this resource, the full potential of every individual must be developed, nurtured, and realized. However, this potential cannot be realized when many millions of people lack the basic life necessities of food, water, fuel, clothing, and shelter. While some countries have made progress in meeting the basic needs of their citizens, others have not come close to eliminating poverty and hunger. In general, modern societies must be faulted for their failure to capitalize on the abilities of women and minorities, their neglect of children and the impoverished, and their indifference toward the unemployed and homeless. To harness the full potential of every human, social, economic, and technological commitments must be made on a global scale.

An examination of the economic and social conditions that exist throughout the world indicates that hundreds of millions of people are not supplied with the basic elements that we in the United States take for granted. United Nations statistics show that:

- There are more than 800 million people in the world who are starving or malnourished.
- Industrialized countries consume more resources per person than the populations of developing nations.
- Throughout the world, almost 75 percent of public housing is substandard, due mainly to the lack of effective sanitation facilities.
- In the Northern Hemisphere, of every 1000 children born, 12 die before reaching one year of age. In the Southern Hemisphere, of every 1000 children born, 71 children die within the first year of life.
- Though many of the world's developing nations have made progress in promoting literacy, more than one of every four adults worldwide are illiterate, the majority of these being women. More than 90 percent of the world's children start school, but 40 percent of these drop out before completing their primary education.
- In many nations, citizens lack many of the civil liberties and political freedoms we take for granted.
- At least 33 percent of the world's potential workforce is unemployed. Many nations lack social programs to supply the basic needs of the unemployed and impoverished.

Economic Trends

Nations are becoming aware of the importance of the biosphere to all living things. Cooperation among nations is necessary to solve problems of environmental degradation and climate change. The end of the Cold War toppled political barriers between many countries and increased international cooperation on common environmental problems. Nations are working to establish a balance between economic development and the preservation of natural resources and habitats. However, the ultimate solution to many of these problems will come about only through global cooperation. The industrialized nations will have to drastically modify their consumption patterns while the developing nations will have to eliminate poverty and slow population growth.

38.4 Section Review

1. In what ways can the human race be considered one of Earth's natural resources?
2. What percentage of the world's population is starving or undernourished?
3. What might be the reasons that infant mortality rates are higher in the Southern Hemisphere than in the Northern Hemisphere?

LABORATORY INVESTIGATION 38

Graphing Population Trends

PROBLEM: *How is a graph of population trends drawn?*

SKILLS: *Reading, graphing data*

MATERIALS: *Graph paper, pencil, ruler, population statistics 1950–2010 (Figure 38-1)*

PROCEDURE

1. Use the bar graph of population statistics to prepare a line graph that describes the growth of the world's population from 1950 to 2010.
2. Calculate the percentage increase in population for each 10-year period:

$$\text{population increase} = \frac{\text{population B} - \text{population A}}{\text{population A}} \times 100$$

3. Project the population figures for 2020, 2030, 2050, and 2100.

OBSERVATIONS AND ANALYSES

1. Which decade had the greatest percentage increase in population?
2. Which decade had the smallest percentage increase in population?
3. What is your projected world population for 2020? Have you accounted for trends in population growth?
4. What is your projected world population for 2030? For 2050? For 2100?

Chapter 38 Review

Answer these questions on a separate sheet of paper.

Vocabulary

The following list contains all the boldfaced terms in this chapter.

AIDS, birth control, birth rate, conception, contraception, death rate, disease, emigration, family planning, fertility rate, immigration, infectious diseases, pandemic, stabilized population, stable population, zero population growth (ZPG)

Fill In

Use one of the vocabulary terms listed above to complete each sentence.

1. The average number of childbirths per woman in a population is the ______.
2. The use of a device or chemical substance to prevent conception is known as ______.
3. ______ is a strategy that allows couples to determine the number and spacing of their children.
4. ______ is a sexually transmitted viral disease that affects the immune system.
5. A(an) ______ is an epidemic that has spread over a wide area and affects a large portion of the population.

Multiple Choice

Choose the response that best completes the sentence or answers the question.

6. Worldwide, the major cause of substandard housing is *a.* old buildings. *b.* rats. *c.* ineffective sanitation facilities. *d.* roaches.
7. The percent of the world's potential workforce that is unemployed is *a.* 10 percent. *b.* 25 percent. *c.* 33 percent. *d.* 50 percent.
8. What type of pathogen causes polio? *a.* virus *b.* bacterium *c.* protozoan *d.* worm
9. What pathogen is responsible for malaria? *a.* mosquito *b.* protozoan *c.* virus *d.* bacterium
10. The outbreak of influenza in 1918 was *a.* an epidemic. *b.* a pandemic. *c.* spread by mice. *d.* spread by bats.
11. When Christopher Columbus arrived in the New World, the Earth's human population was approximately *a.* 4,000,000. *b.* 40,000,000. *c.* 400,000,000. *d.* 4,000,000,000.

12. Which of the following has contributed most to the population explosion? *a.* increased rate of birth *b.* increased rate of deaths *c.* decreased rate of death *d.* none of the above

Short Answer (Constructed Response)

Use the information in the chapter to respond to these items.

13. What does ZPG mean?
14. How is immigration different from emigration?
15. What is a disease?
16. Describe one factor that has caused tuberculosis to make a comeback.
17. How is mad cow disease transmitted to people?

Essay (Extended Response)

Use the information in the chapter to respond to these items.

18. What is the major factor that will lead to the stabilization of the world's population?
19. What are the main factors that contribute to the high world-wide rate of illiteracy?
20. What are some of the social problems that arise from over-crowding?
21. Is it ethical for a country to legislate population control?
22. How is a pandemic different from an epidemic?

GLOSSARY

abiotic environment nonliving factors in the environment, e.g., water and temperature

absolute time measured time; identifies the actual date an event occurred and determines its absolute age

acid rain precipitation that contains sulfuric and/or nitric acids, which form when sulfur and nitrogen compounds combine with moisture in the air

acquired immune deficiency syndrome a viral disease that affects the immune system; AIDS

active solar heating a heating system that uses solar energy collectors and pumps to move the heat energy through the system

adaptation a trait, or characteristic, that an organism inherits

aeration a disinfecting process in which water is sprayed into the air; allows oxygen in the air to dissolve in the water, which kills anaerobes and improves water quality

aerobe an organism that requires oxygen for respiration

aerobic a process that requires oxygen

agrarian a farming society

air filter a mesh of interwoven fibers of cotton, spun glass, or cellulose that traps particulates and aerosols in air

air pressure pressure caused by the weight of the atmosphere pressing on Earth

Air Quality Index a modification of the Pollution Standards Index that measures the concentration of a number of dangerous air pollutants, including ozone, sulfur dioxide, carbon monoxide, nitrogen dioxide, and particulates

alien organism exotic species; an organism introduced into a new environment from another area

alpine meadow the area above the timberline where grasses, wildflowers, and miniature, slow-growing shrubs are found

alpine tundra the area above the alpine meadow; looks much like the Arctic tundra, but lacks permafrost and short days

amphibian a vertebrate that begins life in the water and as an adult can live on land

anaerobe an organism that uses chemicals other than oxygen to release energy

anaerobic a process that does not require oxygen

animal waste digester a device that uses decomposer organisms to produce methane from animal wastes

annual a plant that completes its life cycle in one growing season

anthropologist a scientist who studies the development of the human species and its cultures

aquaculture the cultivation of fish or shellfish in ponds or pools

aquatic describes organisms that live and feed in the water; water-based ecosystems

aquifer an underground area of permeable rock, gravel, or sand that is saturated with water

arboreal describes organisms that live in trees

archaeologist a scientist who studies the remains and artifacts of early civilizations

arroyo a gully formed by runoff from rain, usually found in desert areas

artifact an object made by humans or influenced by human activities

atmosphere the blanket of gases, usually called "air," that covers the lithosphere and hydrosphere

autotroph a self-nourishing organism; usually a green plant

autumnal equinox the day the sun is directly overhead at the equator; in the Northern Hemisphere, it is on or about September 21

axis an imaginary line that runs through the center of Earth from North Pole to South Pole

beach wrack piles of natural debris washed in by the tide that collect on the beach at the strand line

benthic zone the division of the marine ecosystem that consists of the ocean floor

benthos organisms that live in or on the ocean bottom

bioaccumulation in a food chain, the process by which the concentration of a pollutant increases at each trophic level

biodegradable a substance that can be broken down by decomposers within the environment

biodiversity a measure of the number of different species in an area, or of the genetic variations found within a species

biological index small aquatic invertebrates sensitive to pollutants that can be used to monitor water quality

biological pest control the use of a naturally occurring predator or parasite to control an agricultural pest

biomass energy the solar energy stored by plants during photosynthesis

biome a large region that has its own typical flora and fauna; a major habitat type

biosphere the part of Earth where there are living organisms; the overlapping mosaic of ecosystems

biosphere reserve region established to protect and develop a natural ecosystem

biota the organisms in an area

biotic environment everything in an organism's surroundings that is or was alive

bipedal gait walking on two feet

birth control methods that are used to prevent or terminate pregnancies

birth rate a population statistic determined by the number of births per thousand people in a population

bloom a dramatic increase in algal growth, usually caused by an increase in nutrients in the water

boreal forest another name for the taiga region

brackish water water with a lower salinity than seawater, and a higher salinity than freshwater

breeder reactor a nuclear reactor that produces more fuel than it consumes

brood parasitism the practice in which birds, such as the cuckoo and cowbird, lay their eggs in the nests of other birds; the host parents unknowingly raise the foreign chicks as their own

buoyancy the upward force water exerts on floating objects

buttress a wide, spreading prop root that grows from the base of the trunk and into the soil to keep a tree from falling over

calm a period when there is no wind

camouflage coloration that enables an animal to blend into its surroundings

canopy layer the upper layer of branches and leaves in a forest

capillary action the physical property that pulls water up through the vascular tissues of plants

carbohydrates compounds composed of carbon, hydrogen, and oxygen

carbon sinks large reservoirs of materials that contain carbon or carbon dioxide

carburation the process by which fuel and air (oxygen) are mixed before the fuel is burned

carcinogen a compound that causes cancer

carnivore an animal that eats another animal; a secondary or higher order consumer

carrion the remains of dead animals

carrying capacity the number of organisms that can be supported by the resources within an ecosystem

cast a pile of material that an earthworm pushes up from below ground

catalytic converter a device that lowers air pollutants released from an automobile's exhaust system

cell the basic unit of all living things

cellular respiration the process by which living things use oxygen from the atmosphere to release energy in food and, in the process, give off carbon dioxide as a waste gas

Cenozoic Era the most recent era, it is divided into two periods

chain reaction a self-sustaining nuclear reaction

chaparral a specialized, woodland ecosystem characterized by cool, wet winters and long, dry, hot summers

chemosynthesis the process that uses energy stored in chemical bonds to make food

chlorination a disinfectant process for drinking water that uses chlorine to kill bacteria and remove objectionable tastes and odors

chlorophyll the green pigment in plants that traps the energy in sunlight; plants use the energy to make carbohydrates

chloroplast organelles in plant cells that contain chlorophyll

city a community that has abundant resources and a population large enough to allow its people to specialize in a wide variety of occupations

Clean Water Act of 1972 protects the quality of surface water and our supply of drinking water by prohibiting the pollution and filling in of wetlands

clear-cutting a logging technique that cuts down a wide area of forest, rather than cutting only selected trees

climate the long-range or yearly weather patterns over a particular area

climax community the end-product of ecological succession, a long-term, stable ecosystem

cloud seeding a process that uses high-flying aircraft to sprinkle dry ice or potassium iodide crystals on saturated clouds to make rain or to lessen the intensity of storms

coagulation a chemical process that uses alum and filtration to remove dirt and debris particles from water

coal gasification a process that produces combustible gases from coal

coal liquefaction a process that produces an oil from coal

coevolution as one species evolves, the changes in that species affect the selection pressures on another species, causing it to evolve, too

cogeneration the simultaneous production of steam for heating and electricity for lighting and powering appliances

cohesion the tendency of molecules to attract one another

coliform bacteria bacteria that live in the intestines of humans and other animals; these bacteria are used to indicate recent contamination by untreated human wastes

collector a device constructed of dark-colored materials used to convert light into heat

commensalism a form of symbiosis in which one species benefits from the association without harming or benefiting the other

community all the interacting populations in an ecosystem

competition a relationship in which organisms struggle with one another and with the environment to obtain the essentials for life

competitive exclusion the principle that in the competition for the same food, energy, or shelter, the species that is better-adapted to use the resources available in that habitat will displace the less well-adapted species

conception the fertilization of an ovum

condensation the process by which water vapor changes to drops of liquid water

condensation nuclei small particles upon which water vapor collects

conifer a tree that has needles instead of broad leaves; most conifer species produce seeds in cones

coniferous forest an ecosystem dominated by conifers

conservation the careful, organized management and use of our natural resources to prevent their exploitation, destruction, or neglect

continental climate describes the climate in the interior of continents; these areas experience a wide range of temperatures during the year

continental drift the theory that the continents were at one time joined in a single supercontinent that later broke apart, eventually drifting to their present positions

contraception the use of a device or chemical substance that prevents conception

control rod a rod composed of boron, graphite, or cadmium, that absorbs neutrons and is used to control the rate of a nuclear fission reaction

convergent evolution in similar ecosystems, different species have adapted in similar ways to the same lifestyle to make use of available resources

core in a nuclear reactor, the chamber in which fission takes place

correlation the process by which geologists match rock strata in different locations to determine if they formed at the same time

cover places that offer protection from predators, shelter from the weather, and sites for nests, dens, or homes

crepuscular describes organisms that search for food only during the cooler periods of dusk and dawn

crop rotation to replace lost nitrates, legumes are planted every two to three years in place of other crops

culture human activities passed from generation to generation by teaching and learning

current a moving river of ocean water

cyanobacteria blue-green bacteria that make their own food

day the time it takes Earth to make one rotation on its axis; 24 hours

death rate a population statistic determined by the number of deaths per thousand people in a population

decay to break down the remains of dead tissues and wastes into simpler compounds that are returned to the environment

decibel the unit used to measure noise levels

deciduous tree the type of tree that drops its leaves seasonally

decomposer an organism that breaks down dead animals and plants, recycling the organic and inorganic compounds in them

denitrifying bacteria bacteria that act on soil nitrates to produce nitrogen gas, which returns to the atmosphere

desalinization the process of converting salt water to freshwater

desert a region where the annual rate of evaporation is greater than the annual rate of precipitation

desertification the process that changes productive land to desert; caused by overgrazing, deforestation, poor irrigation techniques, soil depletion, and global warming

detritivore an organism that feeds directly on detritus, e.g., earthworms, wood lice, millipedes, snails, and slugs

detritus the organic remains of plants and animals, animal droppings, and partially decomposed materials

detritus food chain an alternate method of energy transfer in which the primary consumers are decomposer microorganisms

deuterium the isotope of hydrogen that has one proton and one neutron in its nucleus; hydrogen-2

disease a harmful change in the condition of an organism due to an infectious agent, a toxic substance, a physical factor, or psychological stress

disinfection the process that destroys pathogens, i.e., harmful microorganisms, in water

diurnal describes organisms that feed and are active in the daytime

domestication the process of adapting populations of plants and animals to live in association with, and to the advantage of, people

drainage basin the area drained by a river and its network of tributaries; the watershed

dune a mound, or hill, often formed as sand is blown inland, away from the beach

Earth Summit Conference 1992 conference held in Rio de Janeiro at which a statement of principles acknowledged that deforestation was a grave problem affecting the whole world

earthquake shaking of the ground caused by crustal movements that occur along faults

echolocation high-pitched sounds emitted by animals that are used as an aid in navigation

ecolabel advertising a product's environmental benefits on a product or packaging, e.g., "biodegradable"

ecological niche an organism's activities as well as its general relationships within the community

ecological succession the gradual replacement of one group of organisms by another group

ecologist a scientist who studies the interactions that occur within the biosphere

ecology the study of the relationships that exist between the living and nonliving things in environments

ecosystem all the living and nonliving things that interact within a certain area; a web of life

ectoparasite a parasite that lives on the outside of its host

effluent the liquid that remains after large solids have been removed from raw sewage

electrostatic precipitator a device used to control particulate emissions from power plants

emergents plants that grow half in and half out of water; the tallest trees in the rain forest, having crowns that grow above the canopy

emigration the movement of people out of a country

emission standard limits the level of pollutants that may be discharged by a source, e.g., standards for automobile exhaust

endangered species a species in danger of becoming extinct

endoparasite a parasite that lives within the body of its host

energy efficiency the ratio of useful energy produced to energy used for production

environment an organism's surroundings

environment-friendly doing the least damage to the environment in production, use, and disposal of the products

environmental sink a feature within the environment that traps a particular substance

environmental tobacco smoke cigarette smoke released from burning cigarettes or exhaled by smokers, and inhaled by persons nearby; also called secondhand smoke

enzyme a compound used to break down food

epiphyte a plant that grows on another plant and whose roots are able to absorb moisture from humid air

equator an imaginary line that divides Earth into Northern and Southern Hemispheres

era the largest division of geologic time; boundaries between eras mark extreme changes that occurred in Earth's surface, climate, or biota

erosion the geological process that wears away Earth's crust

estivation a state of suspended animation during which animals achieve minimum water loss underground; used to escape drought conditions

estuary the area where freshwater of a river joins and mixes with salt water from the ocean

eukaryotes organisms whose DNA is located within a nucleus enclosed by a nuclear membrane

eutrophic lake an older lake that has a very high nutrient content and productivity

eutrophication a natural process that enriches waters with dissolved minerals needed to support plant and animal life

evaporation the process by which liquid water changes into water vapor

even-aged stand a forest in which all trees are approximately the same age

evolution the process of change that occurs in living things over time

exotic species a species that is not native to an ecosystem; an alien species

extinct no longer existing

fall overturn in the fall, the surface water layer of lakes cools, becomes more dense than the water below it, and sinks; less dense layers below are displaced upward

family planning a strategy that allows couples to determine the number and spacing of their children

farm subsidy the payment of money to encourage farmers to alter their production of certain crops

fault cracks in Earth's crust along which movement has occurred

fauna animal life

faunal collapse the loss of all animal species in an area

fecal material solid wastes produced by animals

fermentation the anaerobic release of energy

fertility a measure of the soil's ability to support a plant community

fertility rate the average number of childbirths per woman

fertilizer a chemical compound spread over the land to replace lost nutrients

fieldwork the study of animal and plant species in their natural setting

filtration a process that removes solid particles from water

flora plant life

flyway a well-defined migration route followed by birds

flywheel a massive wheel that spins like a top; used to store kinetic energy

foliage the leaves of plants

food chain the flow of matter and energy through the community; each food chain begins with a producer and ends with a decomposer

food web a collection of food chains that indicates all the interactions occurring among the producers and consumers within an ecosystem

forage to look for food

fossil fuel the hydrocarbon remains of plants or animals that have been changed by natural processes

fuel efficiency a measure of the consumption of fuel per unit of distance (kilometer or mile) traveled

fuelwood wood used for fuel

fungi organisms with cell walls, a nucleus, but lacking chloroplasts; fungi must get their nutrients from other organisms

fungicide a compound used to treat fungal diseases of plants, for example, rusts or blights

galaxy a star cluster

garbage food waste from domestic and commercial sources

gasohol an oxygen-rich fuel composed of 90 percent gasoline and 10 percent ethanol (ethyl alcohol)

gene the part of the genetic molecule that codes for an inherited trait

gene pool the total of all the genetic variations within a population or species

genetic engineering the manipulation of DNA in an organism

geographic isolation the process by which new species evolve when a barrier, such as a mountain range or river, separates a population into two distinct groups

geothermal energy heat derived from radioactive processes and from the Earth's core

glacial period recurrent cycles of worldwide cooling

glacial striation scratches made by glaciers on the surface of rocks

glacier a large, permanent mass of ice or snow

global warming the increase in Earth's average temperature caused by the greenhouse effect

gradualism the slow and steady process of evolution

grazing food chain a food chain in which energy stored by green plants is first transferred to a grazing herbivore

green movement organizations and political groups that campaign on environmental issues

green revolution describes the rapid, widespread use of high-yield crop varieties, pesticides, and better management techniques

greenhouse effect the trapping of heat in the atmosphere; caused by carbon dioxide and methane

greenhouse gases carbon dioxide and methane, the gases responsible for the greenhouse effect

ground cover the layer of wildflowers, ferns, and mosses that grow on the forest floor

groundwater the underground supply of freshwater

guano deposits of bird and bat feces that are mined to be used as fertilizer

habitat a specific environment that contains an interacting community

half-life the amount of time it takes for half of the atoms of a radioactive substance to decay into other elements

hard water water that contains dissolved salts of magnesium, calcium, and iron

hazardous substance a substance that poses a threat to human health or to the environment

headwaters the source of a river

heavy metal metallic elements, such as cadmium, lead, mercury, nickel, arsenic, and selenium, and their compounds

herbaceous plant a small, nonwoody plant; e.g., weeds

herbicide a compound used to kill weeds and other unwanted plants

herbivore an animal that eats plants; a primary consumer

heterotroph an organism that relies on an outside source of food; a consumer

hibernation a state of suspended animation during which an animal's respiration, metabolism, and heartbeat are slowed; used to escape the cold

high tide the bulging of the ocean caused by the moon's gravity; the highest point seawater moves up a beach

home range the area over which an individual animal or a family group travels in search of food and cover

hominids early humans and humanlike species as well as modern humans

host an organism on which other organisms live

humidity a measure of the moisture content of the air

humus dead and decaying organic matter; part of soil

hunter-gatherers humans who make their living by hunting animals, gathering plants, or scavenging the remains left by predators

hydroelectric power electricity generated by using the kinetic energy of moving water

hydrogen bond the attraction of the hydrogen atoms of one water molecule for the oxygen atom of another water molecule; a weak electrostatic force of attraction between adjacent molecules

hydrologic cycle the water cycle; i.e., the circulation of water through the biosphere

hydrolysis the decomposition of water into hydrogen and oxygen

hydrosphere the layer made up of surface water, and the water that exists below ground

hypothesis a proposed solution to a problem or answer to a question

ice age a period of worldwide glaciation

immature river a young river, in the early stages of its development; a young river follows a straight course and has rapids and waterfalls

immigrant a person who moves into a country

immigration the movement of people into a country

incineration the burning of solid waste

index fossil the fossil remains of organisms that existed for a brief geologic time and had a wide geographic distribution

indicator species a species that is an indicator of the health of the ecosystem it inhabits

Industrial Revolution the change that occurred in manufacturing, from small-scale production by hand to large-scale production by machine

industrialization the process of transforming the economy of a nation or region through the development and application of technology

infectious disease a disease caused by a microbe

infrared radiation heat energy

insecticide a compound that kills insects, such as DDT, chlordane, aldrin, malathion, or sevin

instinct an inborn response

interspecies interaction an interaction that occurs among members of different species

interspecific competition competition among organisms that belong to different species

intertidal zone the division of the benthic zone that includes the beach region between the highest and lowest tide lines

intraspecies interaction an interaction among members of the same species

intraspecific competition competition among members of the same species

invertebrates animals without backbones

irradiation a method of disinfection that treats water with ultraviolet light

isotopes atoms of an element that have the same number of protons but a different number of neutrons in their nucleus

juvenile an organism at an immature stage of development

keystone species in a community, the one species linked to most other species, either through the food web or by some other interaction

kinetic energy the energy possessed by moving objects

kingdom one of the five major categories of the classification system

landfill an area used to bury trash and garbage

lava magma that flows out onto Earth's surface during a volcanic eruption

leeward the side of a mountain that faces away from prevailing winds

legume plants, such as beans, peas, alfalfa, and clover, that have nodules on their roots that contain nitrogen-fixing bacteria

life zone on a mountain, a belt of vegetation that has its own distinct flora and associated fauna

light pollution light from cities that interferes with observations of the night sky

limestone a mineral made of calcium carbonate

limiting factor an ecological condition that controls population size

lithosphere the dense, solid outer layer of Earth

litter layer the bottom-most layer of decaying leaves, branches and other vegetation that covers the soil of a forest floor

living fossil a species that has survived unchanged for millions of years

loam naturally fertile soil that contains particles that range in increasing size from clay to silt to sand

low tide the lowest point seawater usually reaches on a beach; occurs between two high tides

magma hot, molten rock in Earth's interior

maize corn, a cereal grain that is the staple crop throughout South America and Africa

marine ecosystem an oceanic ecosystem

maritime climate climate modified by being near the ocean; this climate experiences relatively small variations in temperature throughout the year

marshland land areas that are partially or periodically covered with water during most of the year

mature river an old river that has a broad, flat valley; it has eroded away its waterfalls and rapids, and follows a curving path

maximum concentration level (MCL) standard for the maximum concentration of dangerous pollutants allowed in the air

meander a loop or curve in the path of a mature river

megacity a city with a population in excess of 10 million people

megafauna very large animals that lived about 25,000 years ago

meltdown the melting of the reactor core and its fuel, caused by a loss of coolant; the worst of all possible nuclear-reactor accidents

mesotrophic lake a middle-aged lake, which contains a larger amount of nutrients than an oligotrophic lake

Mesozoic Era the third era of geologic time; it is broken down into three periods

metamorphosis a change in form accomplished through a series of stages

mid-ocean ridge a large system of underwater, volcanic mountains that extends throughout the oceans

midden the trash heap of an ancient household

migrate to travel from one area to another to find food, water, or safe areas to raise young

mimicry the adaptation of one species in which it resembles another species

mitochondria the sites of energy production in a cell

monoculture the growing of a single plant species over a wide area

montane a specialized mountain ecosystem

month division of a year that is loosely based on the time it takes the moon to revolve around Earth

mortality rate the annual number of deaths per thousand in a population

motile able to move easily from place to place by walking, crawling, flying, or swimming

multicellular an organism made of many cells

musk a strongly scented secretion produced by a special gland in some animals

mutagen a substance that causes mutations

mutation a change in a gene that controls a specific trait

mutualism a symbiotic relationship that benefits the two species involved

mycorrhizal fungi fungi that live in a symbiotic relationship with the roots of many types of plants

National Ambient Air Quality Standards (NAAQS) the specific pollution levels set by the EPA

native organism an original inhabitant of an ecosystem

natural resource a useful material that living things get from the environment

natural selection the process by which those organisms that are best adapted to their environment live to reproduce

neap tide the lower-than-normal tides that occur during the first- and third-quarter phases of the moon

nekton animals that can swim and control their motions

neritic zone the division of the pelagic zone that includes the coastal waters above the continental shelf

neutron electrically neutral particle in the nucleus of atoms; neutrons are assigned a weight of 1 atomic mass unit

niche an organism's activities as well as its general relationships within the community

nicotine a poisonous compound produced by the tobacco plant that is used in some insecticides; it is an addictive drug found in tobacco products

nitrate a compound that contains nitrogen, oxygen, and at least one other element, e.g., sodium nitrate, $NaNO_3$

nitrifying bacteria bacteria that take in ammonia and convert it to nitrites

nitrogen cycle a series of natural processes that cycle nitrogen through the biosphere

nitrogen-fixing bacteria bacteria that convert free nitrogen in the air into nitrates

nitrogenous wastes waste products that are composed mainly of urea and uric acid

nocturnal describes organisms that feed and are active at night

noise an annoying or undesirable sound

nomads people who wander through a large home territory in bands, clans, or tribes

nonbiodegradable pollutants that are not broken down by organisms

nonmotile not able to move on its own

nonpoint source pollution source that cannot be traced to a single point of origin

nonrenewable resource a resource that is depleted with use

nuclear fission the process in which an unstable nucleus splits into two or more stable nuclei, releasing energy

nuclear fusion the process that combines the nuclei of small atoms to form larger atoms while releasing energy

nuclear reactor a device that uses controlled nuclear fission to produce heat, which is used to make steam to generate electricity

nucleus the cell's control center; contains the cell's genes

oasis a fertile area in a desert that supports dense vegetation due to the presence of water

observation information gathered directly by the senses, by measurements, or by experiments

oceanic zone the division of the pelagic zone that includes the deep waters beyond the continental shelf

old-growth forest a forest that has not been cut or disturbed by humans for hundreds of years

oligotrophic lake a young lake that has a low nutrient content

omnivore an organism that eats plants and animals

orbit the elliptical path of a planet or satellite

organic farming replaces the use of agricultural chemicals with natural fertilizers, such as manure and crop residues, and uses biological pest controls rather than chemical pesticides

organism a living thing

overburden the surface layer of dirt and rock that covers a coal bed

oxbow lake a lake that forms when a meander is cut off from a river

oxidation a chemical reaction in which oxygen combines with another element

oxygen–carbon dioxide cycle all the processes that occur within ecosystems that make use of and renew the biosphere's supply of oxygen and carbon dioxide; the sum of respiration and photosynthesis

ozone the triatomic form of oxygen; in the stratosphere, acts as a shield to protect Earth from ultraviolet radiation; also a photochemical oxidant

ozone hole a large area over the polar region that shows a seasonal decrease in ozone concentration

paleontologist a scientist who studies the fossil record to reconstruct the life-forms that existed in the past

Paleozoic Era the second era of geologic time; it is divided into six periods

pandemic an epidemic that spreads over a wide area and affects a large portion of the population

parabolic mirror a mirror curved to reflect light so that it can be focused at a specific point

parasite an organism that is dependent on its host as a source of nutrients

parasitism an extreme form of symbiosis in which the symbiont is dependent on the host as a source of nutrients

particulates solid materials or tiny droplets of liquid that are suspended in air

passive solar heating a heating system that uses sunlight to heat air, floors, or the walls of a house or to heat water

pathogen an organism that causes disease in the host organism

pathogenic disease-causing

peat moss thick mats of decayed plant material, mostly sphagnum moss, that form in the tundra

pelagic zone the division of the marine ecosystem composed of the ocean waters

perennial a plant that lives for many years

period block of geologic time characterized by the appearance, disappearance, or dominance of various life-forms in the fossil record

permafrost the permanently frozen layer of soil found in the tundra

persistent chemicals chemicals that break down slowly and remain in the environment for a long time; these chemicals are insoluble in water but highly soluble in fat—once ingested, they cannot be excreted easily from the body

pesticide a substance that kills, drives away, or changes the behavior of a pest so that it is no longer a problem

phosphate a compound that contains phosphorus

phosphorus cycle the movement of phosphorus through the biosphere

photochemical smog smog most often caused by the action of sunlight on motor vehicle exhaust emissions that forms ozone

photosynthesis the process by which green plants use chlorophyll and energy from the sun to change inorganic compounds into organic compounds (food)

photovoltaic cell an electronic device that changes sunlight directly into electricity; a solar cell

phylum the largest division within a kingdom

phytoplankton microorganisms that produce food by photosynthesis

pioneer organism the first plant species to colonize a disturbed or barren area

plankton small organisms that live near the surface of the water

plate a rigid slab of crustal rock that floats on Earth's mantle

plate tectonics the theory that describes the movements of the plates that make up Earth's crust

plow a tool used to cut, lift, and turn over the soil for planting

poaching unlawful hunting, trapping, or fishing

point source a pollution source that can be traced to its point of origin

polar compound a compound whose molecules have an unequal charge distribution

pollination the transfer of pollen from the male cone or stamen of a flower to the female cone or pistil; an important step in the reproductive process of gymnosperms and angiosperms

pollutant a chemical or physical agent (such as light or heat) that when added to the environment threatens people, wildlife, plants, or the normal functioning of an ecosystem

pollution unwanted environmental change, usually caused by human activity

pollution indicator organism a species of animal, plant, or microorganism that is not normally present in an aquatic environment unless the water is polluted

population all organisms of the same species that live in a particular ecosystem

population explosion a great increase in the population of a species

population growth rate the annual surplus of births over deaths

positive crankcase ventilation (PCV) the system that channels oil that escapes from the pistons and unburned gasoline back to the engine to be burned, thus reducing air pollution

potable water water that is fit for drinking

pottery articles made from baked clay

prairie a temperate grassland biome; a relatively flat area, usually in the interior of a continent, in which grasses are the dominant form of vegetation

Precambrian Era The earliest division of geologic time; it represents 87 percent of geologic time, or 4 billion years

precipitation water that falls from the sky to Earth's surface as rain, sleet, or snow

predator an animal that kills another animal for food

prey an animal that is killed by another animal for food

primary consumer an organism that uses green plants as food

primary pollutant a substance released directly into the environment

primary succession ecological succession that occurs in areas that have never been occupied by living things

producer usually a green plant or an alga, because they use photosynthesis to make their own food from inorganic compounds; some producers use chemicals to make food

productivity a measure of the amount of biological materials that producers are able to make using incoming solar radiation

prokaryote an organism whose DNA is distributed in the cytoplasm of its cells, not in a nucleus

protein a compound formed from amino acids

proton positively charged particle in the nucleus of atoms

punctuated equilibrium a theory of evolution which states that species may undergo little or no change for long periods of time then undergo a period of sudden and rapid changes

pyramid of mass a diagram that shows the biomass available at each trophic level

pyramid of numbers a diagram that shows the number of organisms at each trophic level in an ecosystem

qualitative data observations that describe qualities such as color, odor, sound

quantitative data observations obtained by direct measurement

radioactive atoms that have an unstable nucleus that releases particles and radiation

radioactive dating the process that uses the rate of decay of naturally occurring, radioactive isotopes of elements to determine the ages of rocks and fossils

radiocarbon dating the process that uses carbon-14 (half-life 5700 years) to date materials that are less than 50,000 years old

radioisotope an isotope that releases fast-moving particles and radiation; a radioactive element

radon an inert gas that is colorless, odorless, and radioactive

rain shadow a region of reduced rainfall on the leeward side of a mountain

rapids an area of fast-moving water in a river

reclamation after mining is completed, the land is restored to its original vegetation and appearance

recycle to use again

recycling a program of collecting and processing wastes for reuse

reformulated gasoline produced by altering the refining process, which changes the chemical composition of the fuel

relative time a dating method that puts events in sequence, but does not give their actual age

renewable energy resource an energy source that is regenerated by natural processes at a rate that exceeds its rate of use

renewable resource a resource that can be replaced through regeneration and growth of individual organisms or by methods that increase crop populations; usually a biotic resource

reproductive isolation the process by which new species evolve when a population is isolated from the main population, breeds among its own members, and changes over time

reservoir a large area used to store water

rhizome underground runner used by some plant species, especially certain grasses, to propagate asexually

ribosome a site of protein synthesis within the cell

rural area an area in which most residents depend on agriculture for their livelihood

salinity a measure of the amount of salt in water; the concentration of salt in seawater

satellite a smaller object that orbits a larger object

savanna a tropical grassland that also contains scattered trees

scavenger an animal that feeds on dead organisms it did not kill

scent marking a chemical message in the form of a spray of urine, a pile of dung, or a deposit of musk used by some animals to mark their territory

scientific method a systematic approach to solving a problem or answering a question

season the changing weather pattern caused by the tilt of Earth's axis

secondhand smoke cigarette smoke exhaled by smokers and inhaled by persons nearby; also called environmental tobacco smoke

secondary consumer an organism that feeds on primary consumers

secondary pollutant a substance that enters the air and is made hazardous by reaction with natural atmospheric components

secondary succession ecological succession that occurs in an area where an ecosystem existed previously, but was destroyed

sedimentation the accumulation of eroded rock in a body of water; a process in which water is allowed to stand undisturbed, so that particles such as sand and dirt can settle to the bottom and be removed

seed dispersal the movement of seeds away from where they were produced

semiconductor a material that contains electrons that are released easily

sewage wastewater from homes, businesses, or industry that contains cooking, cleaning, or bathroom wastes

shale oil a type of petroleum recovered by processing certain kinds of shale

shrub layer the layer of shrubs that grows beneath the understory

silt tiny particles of organic and inorganic material

slash-and-burn agriculture clearing land for planting by cutting and burning all vegetation, then mixing the ashes with the soil

sludge treated solid sewage or organic matter produced by sewage treatment plants, paper mills, and refineries

smog an atmospheric condition in which visibility is reduced due to air pollution that contains high levels of particulates or photochemical oxidants

soil a complex mixture of organic materials and inorganic minerals

solar cell an electronic device that changes sunlight directly into electricity; photovoltaic cell

solar cooker a passive solar heating device that reduces reliance on fuel wood and reduces air pollution; also called solar stove

solar energy electromagnetic energy from the sun

solar generator an electric generator that uses light from the sun, focused by mirrors, to boil water, produce steam, and turn a generator

soluble a substance is soluble when it dissolves in another substance

species a group of similar organisms that can interbreed and produce fertile offspring

specific heat the amount of heat energy needed to raise the temperature of one gram of a substance by one degree Celsius

spent fuel used fuel rods from nuclear reactors

sphagnum moss a plant that grows in bogs in the tundra

spring equinox the day the sun is directly overhead at the equator; in the Northern Hemisphere, it marks the first day of spring and occurs on or about March 21

spring overturn rising temperatures melt the layer of ice on a lake; water density is greatest when its temperature is 4°C and it sinks; this water—rich in oxygen—displaces the less dense water below it, causing that water to move to the surface

spring tide the very high tides that occur during the new and full moon phases

stabilized population a population that has achieved a zero population growth rate or a negative growth rate

stable population a population in which age and sex distributions are constant because birth rates and death rates are constant

standing crop the amount of organic material at each trophic level

staple food a basic food, one with a high nutrient content

stomata pores through which leaves absorb carbon dioxide and water vapor, and release water vapor and oxygen

Stone Age a period in human history marked by the use of stone tools

strategic mineral a mineral that a country needs but does not produce itself

stratified having layers; often refers to the layers found in a forest

subduction the process in which the seafloor plunges through the crust into Earth's interior, forming a trench

subsidence a sinking of Earth's surface in the area around natural gas wells

subterranean underground; used to describe organisms that live and feed under Earth's surface

subtidal zone the division of the benthic zone that begins below the lowest tide line and extends out along the continental shelf

subtropical desert deserts that lie along the Tropics of Cancer and Capricorn

succulent plants with thick, fleshy stems and leaves that store water

summer solstice the day that has the longest period of daylight; in the Northern Hemisphere, it is on or about June 21

superposition the principle that says sediments are deposited underwater in horizontal layers, and since the bottom layers are deposited first they are the oldest layers, and succeeding layers are progressively younger

supratidal zone the beach region above the highest tide line

symbiont an organism that lives close to, on, or in another organism

symbiosis living together; a close prolonged, physical relationship between two or more species

synfuel a synthetic fuel

taiga the northern coniferous forest biome that extends across the northern reaches of the temperate zone

tapetum an opaque layer behind the retina in the eye that reflects light back through the retina for greater efficiency in seeing

tar sands sandy surface deposits that contain bitumen, a petroleumlike substance

taxonomy the method of classifying organisms according to common characteristics and evolutionary relationships

technology the ways in which society solves problems posed by the environment

temperate deciduous forest biome the biome dominated by broad-leafed deciduous trees

temperate grassland biome see **prairie**

temperate rain forest rain forest found where there are mild, wet winters and warm summers; average yearly precipitation is 200 to 400 centimeters

terracing cutting terraces into steep hillsides that are susceptible to erosion to diminish the effects of runoff and protect against loss of topsoil

terrestrial used to describe organisms that live and feed on the ground

terrestrial ecosystem a land ecosystem

territory the area an organism defends against intruders to protect its food sources or living space

tertiary consumer an animal that preys on secondary consumers

theory a logical explanation of an event that is based on facts that were gathered

thermal energy heat

thermal pollution waste heat released into the environment

threatened species a species at risk of becoming endangered

tidal power electricity generated by using the kinetic energy in tides

tide the daily rise and fall of ocean water caused by the moon's gravitational attraction and modified by the gravitational attraction of the sun

timberline on a mountain, the upper limit at which the climate is suitable for the growth of trees

tool a hand-held device used to perform a specific function

top predator the large organism at the end of the food chain

topography the physical features of Earth's surface, such as mountain ranges, valleys, plains, and plateaus

topsoil the upper layer of the soil that contains particles of weathered rock and organic matter

toxic substance a substance that can damage living tissues through contact or absorption

transpiration the process by which water evaporates from pores in the leaves of plants

trash nonfood waste, such as glass, aluminum, some plastics, and tin cans

tritium the isotope of hydrogen that has one proton and two neutrons in its nucleus; hydrogen-3

trophic levels the feeding stages in a food chain

tropical rain forest biome biome located along or near the equator; temperature and humidity are high

tsunami an ocean wave or a series of waves usually associated with an earthquake, often quite large and dangerous

tundra biome vast treeless polar desert that has low year-round temperatures, long months of darkness, and low precipitation because cold air holds little water vapor; it stretches along the Arctic reaches of North America, Europe, and Asia

turbidity a measure of the cloudiness of water

umbrella species an indicator species that requires extensive areas of habitat for its survival

understory layer the small trees that grow beneath the canopy

uneven-aged stand a forest in which the trees vary in age

unicellular a one-celled organism

United States Swampland Act of 1850 allowed individuals to buy swamps and marshes for as little as 10 cents an acre

urban area an area in which the population is not dependent on an agricultural or natural resource-based economy to make their living

urban biome the artificially created environments in cities that have become unique habitats for many plants and animals

variation the differences that occur in individual organisms

vertebrates animals with backbones

village a collection of rural households linked by culture, family ties, or occupational relationships

volcano an opening in Earth's crust through which magma can reach the surface

wash a desert streambed that is dry most of the time and only contains water after heavy rains

waste stream describes the flow of waste materials into the environment

water budget the water cycle that occurs over a particular region

water cycle the circulation of water through the biosphere

water pollution any physical, chemical, or biological change in water quality that adversely affects living things or the environment

Water Quality Criteria standards that define the maximum concentration level (MCL) of contaminants permissible in drinking water

water table the boundary between the saturated and unsaturated soil

water treatment the processing of well water or surface water before distribution to the public

watershed the area drained by a river and its network of tributaries; the drainage basin

waves rhythmic motions of water caused by winds

weather the short-term (hours or days) conditions that affect the troposphere

wetlands areas where the water table is at or near the surface of the land; lowlands that are permanently or temporarily covered with shallow water

wilderness those areas where the ecosystem is undisturbed by human activities and where people visit but do not remain

wildlife refuge a habitat area set up to preserve an ecosystem; preserve

wind farm a group of windmills used to generate electricity

windward the side of a mountain that faces the prevailing winds

winter solstice the day that has the shortest period of daylight; in the Northern Hemisphere, it is on or about December 21

xerophyte a type of plant that is adapted to collect and conserve water and is therefore able to survive dry conditions

year the time it takes Earth to make one complete revolution around the sun; approximately 365 days

zero growth rate no change in the size of a population

zooplankton microorganisms that feed on phytoplankton

INDEX

N

S

ART AND PHOTO CREDITS

American Museum of Natural History, Special Collections Library: 188, Frank Puza; 338 center, Bruce Hunter; 338 right, C. Sibley; 353, A.J. Emmerich, 665 left

Ken Bates, Chief Environmental Engineer, Washington State Department of Fish & Wildlife: 581

Center for Nuclear Fusion, Superior Technical Institute, Lisbon, Portugal: 629

Corbis: ii–iii, Craig Tuttle/Stock Market; 212 bottom, Tom Brakefield; 494, Japack Company; 560, Roger Ressmeyer; 668 bottom, Tom Bean; 692, Mug Shots

John Deere & Company: 419

Thomas F. Green: 224, 299, 383

Hadel Studio: 6, 11, 127, 131, 167, 197, 239, 278, 280, 344

Carol Davidson Hagarty: 250

Steven Hamara: 299 left, 457, 464

Eric Heiber: All art not otherwise credited

Kent Sea Farms Corporation, San Diego: 512

Jonathan Knowles: 642

Image Works: 636–637, Larry Kolvoord

Library of Congress: 646

Ronald J. Mabile: 132, 174 right, 189, 207 left, 250 right, 323 bottom, 428, 442, 656

NASA: 2–3, 19, 21, 25, 47 bottom, 53, 55, 60 bottom, 99, 105, 138–139, 157, 182, 461, 505, 552, 558–559, 672 left, 672 right, 707

Olympic National Park Visitor's Center: 640

Don Pearce: 96, 110, 122, 140, 204, 211, 227, 237, 256, 332, 415, 427, 435 bottom, 551

Margaret Pearce: 181, 343

Photo Researchers: 4, Simon Fraser/SPL; 37, Soames Summerhays; 76, Mehau Kulyk/SPL; 89 top, Stephen J. Krasemann; 89 bottom, Tom McHugh; 94–95, Alan & Sandy Carey; 98, Chris Butler; 117, Tim Davis; 173, James Amos; 187, M. P. Kahl; 202–203, Dan Suzio; 212 top, Kjell B. Sandved; 213 top, Michael Tweedie; 213 bottom, Michael Tweedie: 219, Alan Carey; 267, Scott Camazine; 275, Kenneth Thomas; 294–295, Dennis Flaherty; 296, Stephen J. Krasemann; 311, S. E. Cornelius; 316, Leonard Lessin; 323 top, Francois Gohier; 358, Dennis Flaherty; 371, Paul Stepan; 400–401, Rafael Macia; 402, John Reader/SPL; 433, Andy Levin; 445 center, Tom and Pat Leeson; 445 bottom, James H. Robinson; 450–451, Spot Image; 452, Tom McHugh; 469, Alan L. Detrick; 475 Grapes/Michaud; 490, David Frazier; 499, Calvin Larson; 506, John Cancalosi; 517, Simon Fraser/SPL; 528, David Parker/SPL; 540, Garry McMichael; 564b right A, Murton/Southampton Oceanography Centre/Science Photo Library; 566, Wayne G. Lawler; 571, David Frazier; 576, Herve Donnezan; 591, Jim Zipp; 600, SPL; 606, Georg Gerster; 617, Mark Burnett; 638, Simon Fraser/SPL; 639, Earl Roberge; 641, Calvin Larson; 655, Mark Newman; 662, Tom and Pat Leeson; 666 top, Tom and Pat Leeson; 666 bottom, Tom and Pat Leeson; 668 top, James Steinberg; 683, Francois Gohier; 686, Nigel Cattlin; 697, Mark Burnett

Martin Schachter: 645, 648, 676

Storms Laboratory: 60 top

U.S. Department of Agriculture/APHIS: 445 top

U.S. Department of Energy: 580, 583, 627

United States Department of the Interior: 664

Visuals Unlimited: 300, Richard Thom; 564 left, A. Boonyaratanakornkit & D.S. Clark, G. Vrdoljak, EM Lab, U of C Berkeley

Estate of Lawrence Weisburg: 435 top, 458

The Wildlife Conservation Society at the Bronx Zoo: 269, 338 left, 665 right